Marine Molluscs of Victoria

HANDBOOK No. 2
NATIONAL MUSEUM OF VICTORIA

Marine Molluscs of Victoria

J. HOPE MACPHERSON
Curator of Molluscs, National Museum of Victoria

and

C. J. GABRIEL
Honorary Associate in Conchology
National Museum of Victoria

Drawings by G. J. Browning

MELBOURNE UNIVERSITY PRESS
in association with
THE NATIONAL MUSEUM OF VICTORIA

FIRST PUBLISHED 1962

PRINTED AND BOUND IN AUSTRALIA
BY BROWN, PRIOR, ANDERSON PTY LTD,
430 LITTLE BOURKE STREET, MELBOURNE C.1, FOR
MELBOURNE UNIVERSITY PRESS, PARKVILLE N.2, VICTORIA

REGISTERED IN AUSTRALIA FOR TRANSMISSION
BY POST AS A BOOK

London and New York : Cambridge University Press

FOREWORD

OF ALL the people who find interest in the varied branches of natural history, it seems to me that by far the greater number are attracted by conchology. Collectors are legion, and for many the subject forms a serious and full-time hobby. Apart from these, however, there are great numbers of people who are so fascinated by shells that they gather them and carry them home—perhaps without quite knowing why.

The reason for this is, of course, that most shells have beauty—in their flowing contours, in the delicate modelling of their sculpture, in the glow and pattern of their coloration, or in a combination of these qualities. Man, consciously or unconsciously, always desires to acquire beautiful things.

If this be a commencement for the dawning conchologist, the next logical step is to find out something about his prizes, their names and how and where they live. This is not always easy, for on the subject of Australian shells very little popular literature, which is the foundation for all such enquiry, has been published.

I am sure it is with these thoughts in mind that the present authors have written this volume. It is a book for use by the beginner. Almost every species is illustrated, thus simplifying the recognition of any shell, and the difference between each species is set out in non-technical terms. For the more advanced collector a detailed scientific description of each species is added, and, although this is of necessity in technical language, a very full glossary explains all the terms. The text discusses where the shells may be found—both the locality and habitat—and tells of the relationship between the different kinds. Finally, it is a book which is useful to the specialist in that it contains essential references to problems of taxonomy and advanced study.

The authors are well qualified to execute such a multifarious undertaking. Miss J. Hope Macpherson is Curator of Molluscs at the National Museum of Victoria, and Mr C. J. Gabriel, Honorary Associate in Conchology at this institution.

CHARLES W. BRAZENOR
Director
National Museum of Victoria

PREFACE

FOR MANY YEARS there has been a need—even a demand—for a book on Victorian molluscs which will be of use to the interested layman as well as to the scientist. We hope that in this volume each kind of reader will find some satisfaction. As is always the case, there will be specialists who will feel we have been superficial, and some amateurs who will consider there is a preponderance of unnecessary detail. Our answer is that this is a handbook, generally simple and non-technical enough to be used by the beginner, but containing such information as will give the advanced collector or specialist some explanation of our reasons for controversial matters such as taxonomic changes. We hope and believe that we have succeeded in steering this middle course.

Victorian molluscs were originally listed in *The Catalogue of the Marine Shells of Victoria* produced by G. B. Pritchard and J. H. Gatliff and issued through nine parts of the *Proceedings of the Royal Society of Victoria*, from 1898 to 1906. By means of intermittent papers it was continually revised and augmented through the years, first by the original authors, later by Gatliff and Gabriel, and after Gatliff's death by Gabriel alone. This is still a most useful work for the specialist, but because of its piecemeal publication, is difficult for the amateur to discover and use.

In 1936 one of us (C.J.G.) issued through the Field Naturalists Club of Victoria a small book of Victorian shells. This was necessarily limited in size, gave a popular description of only the commoner sea shells, and in any case it has long been out of print. This present volume is an elaboration and expansion of this book. Detailed descriptions and illustrations cover all the common species, together with their range and distribution in Victoria. Their points of difference are stated simply so that identification is made easy, and any point of special interest is discussed. There is a definition of each family and of the principal genera within it, including the type species and author of the genus. A complete systematic list of all species recorded from Victoria is added.

Nearly 500 illustrations of shells have been drawn by Mr G. J. Browning and are accurate to the last degree. We are grateful for his patience and attention to detail. Our thanks are also extended to Miss Patricia Hoggart, Assistant in the Department of Mollusca at the National Museum, whose help has been generous and very valuable.

J.H.M.
C.J.G.

National Museum of Victoria
Melbourne, 1961

CONTENTS

INTRODUCTION

THE ANIMAL KINGDOM is divided into a number of major groups (phyla) and one of the largest and most widely distributed is the phylum Mollusca. In it are grouped the mussels, scallops, pipis, clams, sea and land snails, slugs, chitons or coat-of-mail shells and squids, cuttle-fish and octopuses. A probable estimate of the living forms is a hundred thousand, and there are many thousands of fossils as well, for molluscan remains are found in the most ancient rocks (Cambrian).

Until 1957 the primitive molluscan type was only known from limpet-like fossils in some of the ancient fossil-bearing rocks (Cambrian to Devonian). In that year the Danish *Galathea* Expedition brought up from a depth of 5,000 metres (approximately 3 miles) off the Pacific coast of Mexico living examples of this very primitive animal. Dr H. Lemche named it *Neopilina galathea* and placed it in a separate class Monoplacophora. It has a flat, saucer-shaped shell, two inches long, with eight muscle scars on each side. The animal has a ventral foot, a mouth in front surrounded by jelly-like appendages and a posterior anus. The five pairs of gills lie in a shallow groove between the mantle and the foot.

The living forms of the phylum Mollusca are divided into five classes which, in external appearance at least, differ greatly. However, they are distinguished from all the other phyla of the animal kingdom in having, typically, an external shell composed of three layers. Some present-day forms have partially or completely lost their shells but their general anatomy shows their molluscan affinity. The molluscan animal is elongate and unsegmented. It has a head region, and the underside of the body forms the foot upon which it moves (Figure 1). The internal organs are within a body-hump which is usually covered by the shell.That portion of the skin covering the visceral hump is called the mantle and secretes the largely calcareous shell. It is produced into free flaps which extend beyond the body-hump and partially enclose a cavity into which the respiratory, excretary and reproductive organs discharge.

Except in the Pelecypoda the mouth is furnished with a rasping organ or radula. It is a ribbon-like membrane to which are attached many small hard teeth; the mollusc uses it like a rasp, driving it backwards and forwards to break up its food. The length of the radula and the number and shape of the teeth attached vary in the different families, genera and even species.

All molluscs lay eggs which usually are deposited and hatch later although in a few instances they are retained in the mantle cavity till hatched. The molluscan shell is secreted by the mantle and the method is the same for all groups. The original embryonic shell is formed by the mantle of the juvenile animal and is extended, as the animal grows, by additions to its edge. The outer horny layer (periostracum) is produced by the cells within a groove at the mantle's edge and then the usually thick calcareous layer (ostracum), which consists of either calcite or aragonite, is formed beneath it by neighbouring cells. The innermost pearly or nacreous layer (hypostracum) is produced by the cells of the whole mantle surface and therefore

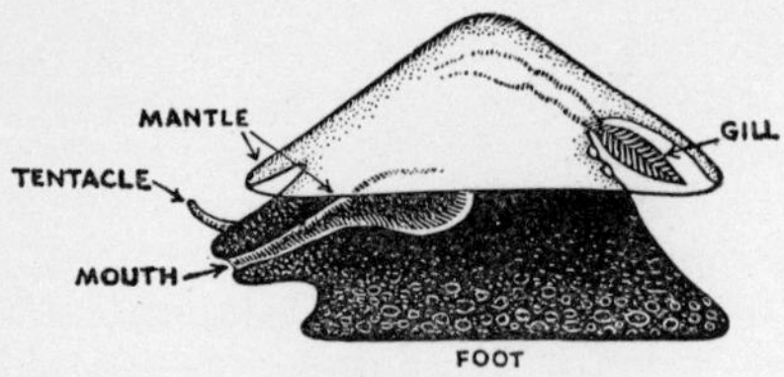

Figure 1

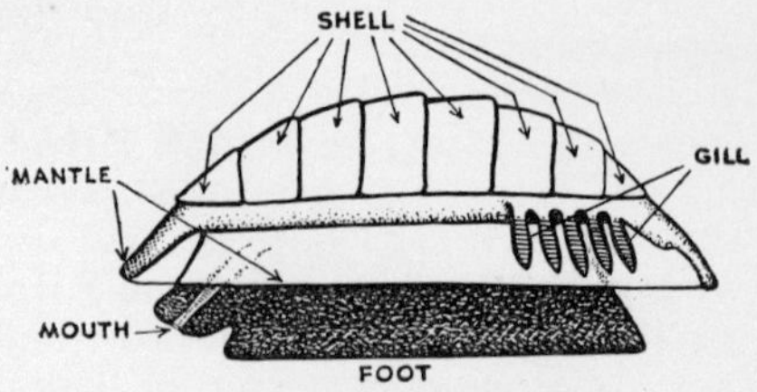

Figure 2

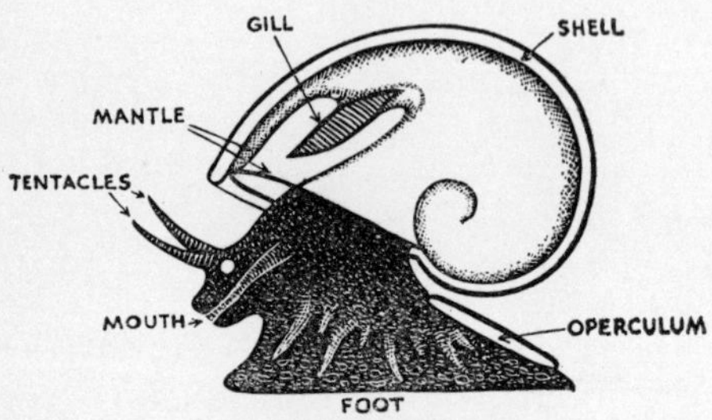

Figure 3

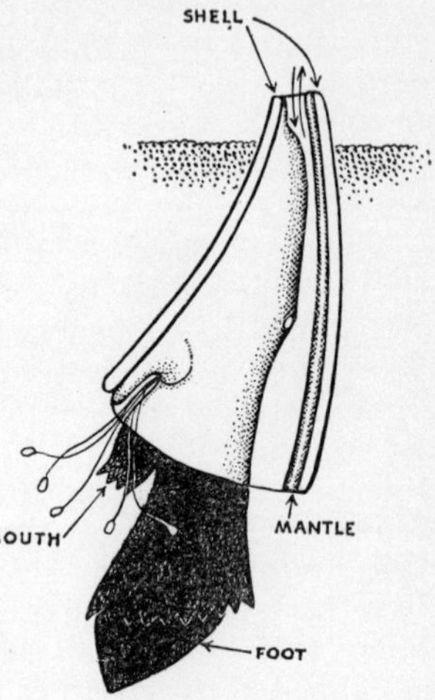

Figure 4

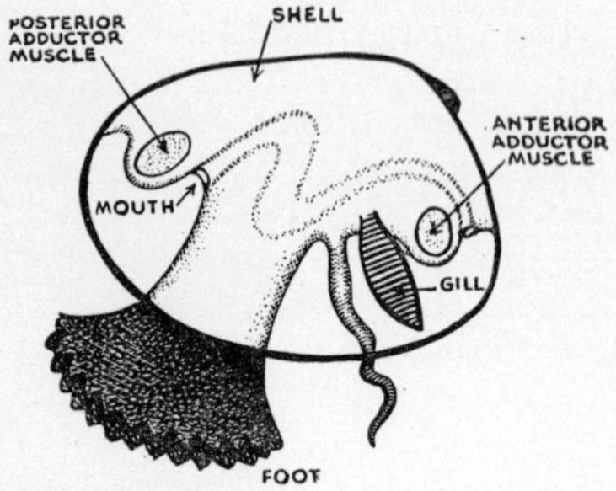

Figure 5

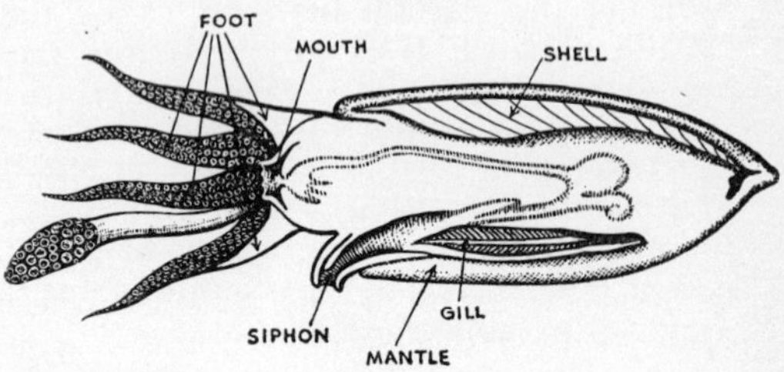

Figure 6

1. Primitive mollusc
2. Amphineura or Crepipoda = base foot
3. Gastropoda = stomach foot
4. Scaphopoda = boot foot
5. Pelecypoda = hatchet foot
6. Cephalopoda = head foot, the foot being modified into tentacles and siphon

can be added to indefinitely, thus in some older shells this layer may be greatly thickened.

The molluscs are divided into five major groups or classes for which different authors have used different descriptive names. Here for the sake of uniformity we have used names based on the form of the foot (Figures 2-6).

It should be noted that the Crepipoda are sometimes known as Loricata, and in much of the literature the Pelecypoda are called Lamellibranchiata, a name based on the form of the gills.

Collecting

Shell collecting is usually commenced when one is very young and the gathering is quite indiscriminate. Any piece of broken coloured shell—or even stone—is treasure. Unfortunately, there are a few people who continue to collect in this manner.

The forming of a good collection necessitates in the first place some knowledge of the animals that construct shells, and also where they can be found. The richest collecting ground of easiest access is that area of the shore between high and low tide, called the littoral zone. Here the inhabitants are exposed to the air twice daily, the period of exposure depending on whether they occur near high or low tide mark, and varying from several hours to but a few minutes. In Victoria there are about twenty different species of molluscs living in the littoral zone of even the most inhospitable rocky shore. Where shelter and food are abundant this number may be greatly increased. Sandy beaches usually are more sparsely inhabited than rock platforms, and many ocean beaches are so pounded by the surf that nothing can survive except a few creatures, such as worms, that live by burrowing deeper than the normal scouring force of the waves.

Space does not permit describing each type of habitat in detail, and really it is unnecessary as collectors must learn these things for themselves. However, a few essentials apply to every collecting place. Only a few species can live in extreme exposure, the remainder must protect themselves by hiding under boulders, weeds, or pieces of driftwood; others, again, will burrow into the substratum or hide among clusters of other animals, e.g. *Kellia australis*, the small bivalve, hides in colonies of the beaked mussel (*Brachidontes rostratus*) and also amongst the tubes of the serpulid worm *Galeolaria caespitosa*. When collecting, all stones, etc., rolled over in the hunt should be returned to their original position so that the general environment is disturbed as little as possible. In this way the habitat is preserved, and future collecting is assured.

The hidden presences of Gastropods is indicated on sand beaches and muddy flats by faint trails in the sand, often with a hump at the end revealing the position of the animal. The presence of bivalves hidden beneath the sand is indicated by small holes made by the siphon.

Many marine creatures are noctural, and night collecting with the aid of a powerful torch is often very profitable because such creatures are out feeding at night.

Preservation

Most collections of molluscs consist of hard parts (shells) only, but today there is a growing realization of the necessity to study the animal that is responsible for constructing the shell. Unless the collector is very familiar with the species a note should be made of just where and how the animal is living. Some species always live under rocks; others may be in the open but in shade, others are found only on certain types of algae, etc. Such information should be written on the label with the name and locality of the specimen.

If the animal as well as the shell is to be preserved, those species without an operculum can be killed with the body expanded by narcotizing in one of several substances. A cheap and easily handled method is to sprinkle a small quantity of Epsom salts ($MgSO4$) on the surface of not more than half a pint of water. The animal should be left undisturbed until it no longer flinches on touching. Nicotine sulphate or menthol may be used instead of Epsom salts.

Molluscs with an operculum tend to withdraw and close the shell opening, so are most satisfactorily handled in water or under moist conditions until flaccid. If handled in this way the animal will not contract as much when put into preservative, and is more easily dissected. The animal should be fixed by placing in a solution of 100 parts 70 per cent alcohol and 6 parts 40 per cent formalin for forty-eight hours, or, if this mixture is not available, in methylated spirits. After forty-eight hours the specimen should be removed and stored in a fresh 80 per cent solution of ethyl alcohol or methylated spirits in which it can be stored indefinitely. Molluscs should never be stored in formalin unless it has been neutralized, as it is slightly acid and will damage the shell.

An Introduction to Taxonomy

From the time man first began to study the natural world he realized the necessity for the division of the objects of his study into allied groups. First came the major sorting into mineral-plant-animal, and then a further division into smaller sections and individual kinds. For many ages the attempted arrangements were uncoordinated and haphazard, but in the early part of the eighteenth century a Swedish scientist, Carl Linnaeus, devised the system upon which all modern nomenclature is based. He divided the animal kingdom first into primary groups which he called phyla, and these he subdivided into progressively smaller groups which he called classes, orders, families, genera and species.

Animals are known by the names of the last two of the groups and the system is called 'binomial nomenclature' because of this. Linnaeus, as was the practice of his time, used Latin for his names, and that usage has continued. The first or generic name is usually a Latinized noun in the nominative case and is written with a capital initial; the second or species name, usually descriptive, is an adjective agreeing grammatically with the first and written with a lower case initial. If the author of the species and the year of authorship be added, it is written in ordinary roman type. Thus the name of the Red Ear Shell is written *Haliotis ruber* Leach, 1814. However, if the generic name is, for any reason, subsequently changed, the author's name should be set in brackets as *Notohaliotis ruber* (Leach, 1814).

In nomenclature the law of priority is rigidly applied, and the accepted name must be the first published name that can be definitely attached to the species. This applies for all names in, and later than, Linnaeus' tenth edition of his *Systema Naturae* published in 1758; no name prior to this is acceptable.

A generic name cannot be used more than once in the animal kingdom, nor a species name more than once in each genus. Should such a happening occur in error, the last usage must be replaced by a new name.

When an author describes a species he must select an individual specimen which becomes the *type* and this should, if possible, be lodged in an institution where other workers have access to it. In making a new genus a typical species must be selected and this is then known as the *type species.*

In 1895, the Third International Zoological Congress held at Leyden appointed an International Commission on Zoological Nomenclature to legislate in controversial cases. This Commission continues to operate and is the final arbiter on all problems of taxonomy.

In the present volume orders, families, genera and species are listed and where the families are large their division into subfamilies is also shown.

The diagnostic characters of most families and important genera are discussed and common and striking species are described in full. Related species are listed and described briefly, while the very small and uncommon forms are recorded under the appropriate genera with details of locality.

The type species of each genus may be shown in one of two ways. Where there is a full generic description it is placed immediately after, following the words 'type species'. If the genus is merely listed, the type species (in brackets) follows immediately after the genus, author and date.

The order Nudibranchiata has been treated in slightly less detail. These shell-less molluscs present problems in both preservation and classification and have been largely neglected except by the specialist. So, like the families of smaller shells, only the commonest have been discussed in detail.

Class AMPHINEURA

Chitons or Coat-of-mail Shells

THE CLASS Amphineura is divided into two subclasses, the Crepipoda or Polyplacophora (chitons) and Aplacophora. The latter is a very specialized group which has not as yet been recorded for southern Australia.

Chitons have a flattened oval-shaped body covered dorsally by a shell consisting of eight transverse valves surrounded by a leathery or fleshy girdle (Figure 7).

The valves are of three types—the head or anterior, six central or median, and the tail or posterior valve, and they are numbered one to eight. The valve sculpture and colour pattern varies from species to species. The median valves are divided into two lateral and two central areas connected by a dorsal area which may project posteriorly into a beak. The anterior valve is undifferentiated and is usually similar in pattern to the lateral areas of the median valves. The posterior valve is divided into two areas—the ante-mucronal and post-mucronal—separated by the central projection or mucro.

Each valve consists of two layers, the inner projecting beyond the outer to form insertion plates to which the girdle is attached and over which the immediately anterior valve overlaps. The shape of the insertion plates varies for each species, and they are dissected by slits which are also approximately constant for a species. The projections of the insertion plate between the slits are called teeth and may be variously shaped.

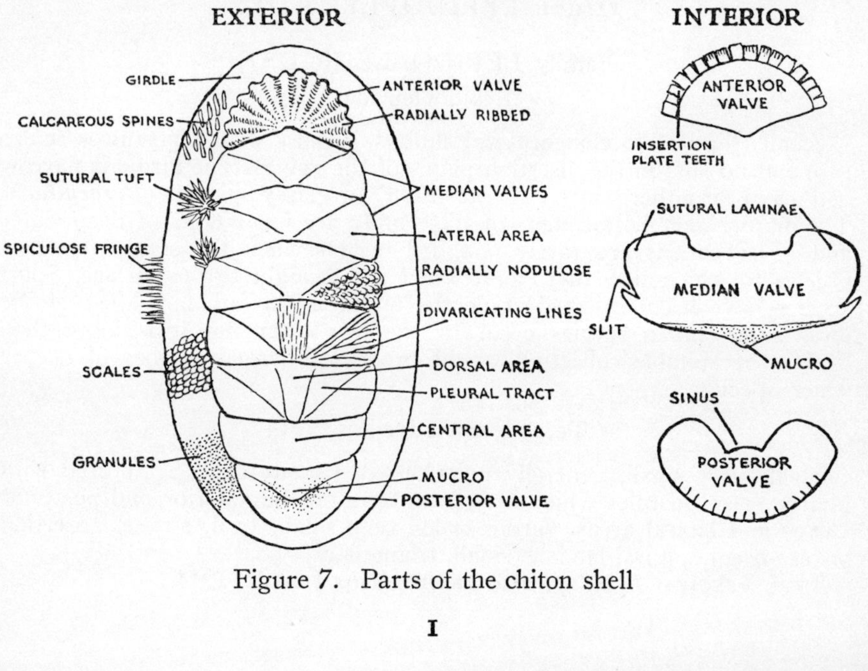

Figure 7. Parts of the chiton shell

The girdle consists of muscular tissue in which the edges of the valves or insertion plates are embedded. The upper surface of the girdle is tough, leathery, and may be covered with scales, calcareous spines, spicules, hairy processes, or a combination of two such coverings.

Collection and Preservation

The chitons, because of their habit of rolling into a ball like a pill-bug when removed from their substratum, need special handling immediately they are collected. They should be removed from the rocks carefully by slipping a knife beneath the foot, care being taken not to damage the girdle. Immediately they are removed from the rock and while still flat they should be placed on a slip of smooth wood or glass and gently but firmly bound on to it with tape, care being taken to spread the mantle edge. In the case of large species the animal can be removed before tying the shells down. This can be done by running a sharp knife round the body, beginning at the tail. As each slip is filled with specimens it can be dropped into a jar of methylated spirit. Later the specimens can be removed and allowed to dry out; they can then be removed from the slips and stored either in boxes or mounted on cards for cabinet display.

The vermiform species such as *Cryptoplax* must have the animal removed and the flexible girdle packed with a boat-shaped piece of wood cut to the right length. Tape is then bound tightly round the whole.

Many papers have been written on the Australian chitons. Tom Iredale and A. F. Bassett Hull published 'A Monograph of the Australian Loricates' which originally came out in parts in the *Australian Zoologist* and was published in book form in 1927.

Order LEPIDOPLEURIDA

Family LEPIDOPLEURIDAE

Lepidopleurids

Small elliptical to elongate-oval chitons, usually with a pustulose sculpture and no slits in the insertion plates of the valves. The girdle is narrow and may bear either scales (*Terenochiton*) or glassy spicules (*Parachiton*). The majority of the members of this family are inhabitants of deep water and so specimens are rarely obtained except when the stones or other objects to which they may be attached are brought to the surface. Some species have been collected at depths up to 2,000 fathoms. Most of the known Australian species occur in shallow water but it is more than likely that suitable collecting would produce a larger number of deeper water species.

Terenochiton Iredale, 1914

Shells very small, generally unicoloured, usually oval. Sculptured with minute erect pustules which may form rays on the anterior and posterior valves and lateral areas. Girdle scales very small, oval, striate. Insertion plates absent, sutural laminae small, triangular.

TYPE SPECIES *Lepidopleurus subtropicalis* Iredale, 1914.

Brown Terenochiton

Terenochiton badius (Hedley & Hull, 1909)

(Figure 8)

Shell small, ovate, subcarinate; colour shades of light brown. Sculptured with small pustules which are arranged in rows on the central areas, lateral areas scarcely differentiated. Girdle covered with very minute, elongate, oval, striated scales, which give it a furry edge. Size: ⅓-inch.

Localities: Torquay; Portland.

This species hides under stones embedded in sand below low-water mark and is also recorded from New South Wales and South Australia.

liratus (H. Adams & Angas, 1864); a slightly larger species (⅜-inch) with the pustules forming close radial rays on the anterior valve and concentric growth-lines very strongly marked on the outer edges of all valves. This species occurs in Port Phillip; Western Port; Torquay; and Portland.

matthewsianus (Bednall, 1906). Exact locality not recorded.

Parachiton Thiele, 1909 (*Parachiton acuminatus* Thiele, 1909).

columnarius (Hedley & May, 1908); a small (⅓-inch) pinkish species sculptured with very fine pustulose rays and a spiculose girdle. A deep-water species, it has been taken off the Ninety Mile Beach, Victoria, and has also been collected from Tasmania and South Australia.

collusor Iredale & Hull, 1925; from Victoria and South Australia, is a slightly larger species (¾-inch) with narrower valves than the preceding species.

profundis (May, 1923); 8 fathoms off Point Cook.

Order CHITONIDA

Family LEPIDOCHITONIDAE

Lepidochitons

Shell uniformly granulose, rarely with weak radial or longitudinal ribs or a special arrangement of larger granules. Anterior and posterior valves with a varying number of slits; edge of insertion plate scalloped. Girdle has small elongate scales or spicules, with single or groups of longer spicules interspersed.

Subterenochiton Iredale & Hull, 1924

Small shells, elevated elongate-oval. Sculptured with small pustules. Girdle scales small, smooth, imbricating. Insertion plates small, slit in anterior valve only.

TYPE SPECIES *Ischnochiton gabrieli* Hull, 1912.

Gabriel's Chiton

Subterenochiton gabrieli (Hull, 1912)

(Figure 9)

Shell small, elongate-oval, elevated, subcarinate; pale cream or pink mottled with deeper colour. Whole shell sculptured with evenly scattered granules, lateral areas elevated and with the grains slightly more regular

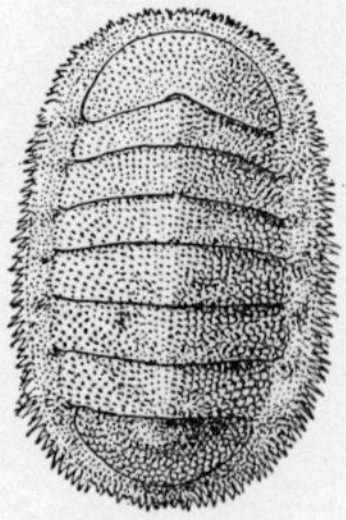

Figure 8

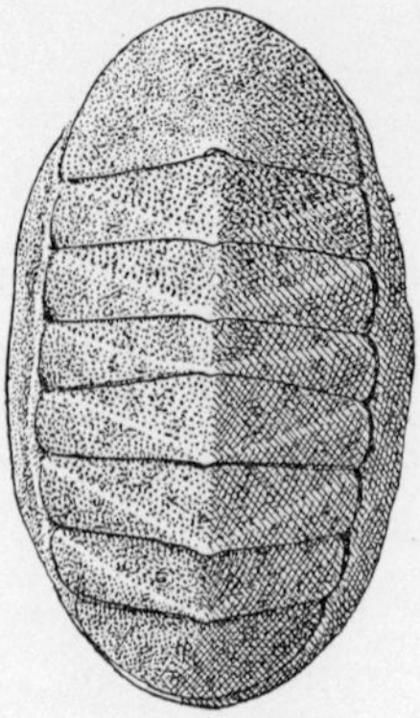

Figure 9

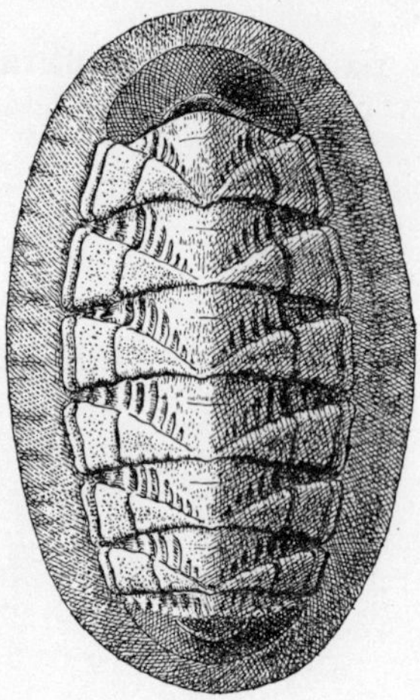

Figure 10

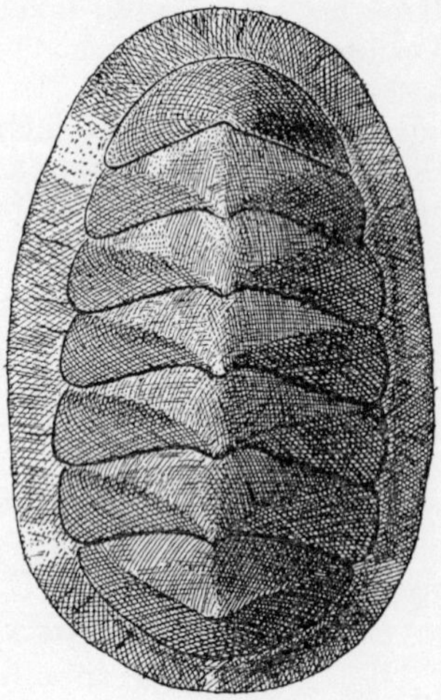

Figure 11

8. *Terenochiton badius* × 8
9. *Subterenochiton gabrieli* × 8
10. *Acutoplax mayi* × 3
11. *Paricoplax crocinus* × 3

towards the edge. Girdle wide with very small scales. Size: ¼ inch.

Locality: Western Port (5 fathoms).

This small species is also recorded from New South Wales and is fairly common in Sydney Harbour.

Acutoplax Cotton & Weeding, 1939

Medium-sized shells, valves large. Sculptured on central areas with deep, longitudinal grooves. Sutural laminae and insertion plates well developed, insertion plate of posterior valve with ten to sixteen slits.

TYPE SPECIES *Callochiton mayi* Torr, 1912.

May's Chiton

Acutoplax mayi (Torr, 1912)

(Figure 10)

Shell medium-sized, ovate, keeled; colour greenish with few brown splashes; end valves brown. Sculpture minutely granulose; lateral areas strongly elevated and granulose; central areas and ante-mucronal area with deep, longitudinal grooves which decrease in length towards the dorsum. Girdle covered with long, slender, curved spicules that lie flat. Size: ¾ inch.

Locality: Portland.

This species is rare but has been taken in three states, South Australia and Tasmania as well as Victoria. It is found under stones below low tide and the difficulty of access to the habitat may account for its rarity.

rufa Ashby, 1900, a second species, was dredged off Port Phillip Heads; it is rounder and has fewer, weaker and shorter grooves than *mayi.*

Paricoplax Iredale & Hull, 1929

Small to medium-sized chitons. Sculptured with small granules. Lateral areas only slightly raised, covered with a glutinous periostracum. Girdle wide, covered with long, slender, curved spicules.

TYPE SPECIES *Chiton crocinus* Reeve, 1847.

Yellow Chiton

Paricoplax crocinus (Reeve, 1847)

(Figure 11)

Shell varies considerably in size, elevation and colour according to the environment (the colour is lost on drying). Entire shell sculptured with fine granules forming a coarse matt surface. Girdle scales very small, elongate. Girdle wide. Size: 1½ inches.

Localities: Portland; Mallacoota.

This species lives on stones below low tide. In life it is protectively coloured in browns, greens, reds or pinks according to the environment but the colour seems to be at least partially contained in the periostracum and is lost with death. It occurs from southern Queensland to south-western Australia and Tasmania.

In the past it has been confused with *Chiton platessa* Gould, 1846 and *Chiton versicolor* A. Adams 1852.

Eudoxoplax Iredale & May, 1916 (*Chiton inornatus* Tenison Woods, 1881).
inornata (Tenison Woods, 1881); 5 fathoms off Hastings (collector Mrs J. Plant).

Family MOPALIIDAE

Mopalid Chitons

Shell with grooved or netted sculpture and as a rule radial ribs fanning from the scalloped shell-edge. Anterior valve with at least eight slits. Sutural laminae large to very large, sometimes united. Girdle with small spicules or scales interspersed with different longer hair-like structures.

Poneroplax Iredale, 1914

Large chitons, oval. Sculpture of fine wrinkles. Anterior valve with eight radial ribs; insertion plates and sutural laminae large. Anterior valve with eight slits and no insertion plate on posterior valve which is heavily calloused. Girdle broad, leathery with long corneous spicules.

TYPE SPECIES *Chiton costatus* Blainville, 1825.

White Poneroplax
Poneroplax albida (Blainville, 1825)
(Figure 12)

Shell large, ovate, round-backed. Whole shell sculptured with fine wrinkles which on the lateral areas are visible to the naked eye. Anterior valve with edge slightly elevated, radial ridges overlying the basic wrinkles. Colour dark green or brown, flamed with yellow and white; beaks of valves often tipped with white. Girdle broad, leathery, with long corneous processes. Size: 3½ inches.

Localities: Cape Bridgewater; Cape Otway; Lorne; Port Phillip; Western Port; Walkerville; Mallacoota.

This species is very common on the upper surface of exposed rock platforms at and just above low tide. It occurs also in Western Australia, South Australia, Tasmania, New South Wales and south Queensland; in the latter two states the wrinkling is very well marked and persistent even on the central areas. This form has been distinguished as *P. paeteliana* Thiele, 1909, but having examined a large series of Victorian and New South Wales specimens we consider that the differences are not constant and occur to a greater or lesser extent in any population.

costata (Blainville, 1825); the wrinkling is reduced so that it is only discernible at ten magnifications and the shell appears smooth to the naked eye. It occurs in similar situations to the above species but perhaps prefers a slightly higher level. In Victoria this species is not recorded east of Walkerville but its southern and western range is Tasmania, South and Western Australia.

Kopionella Ashby, 1919

Medium-sized shells, subcarinate, oval. Sculptured with wavy lines. Girdle with a posterior notch, covered with small erect scales and complex, club-tipped corneous processes.

TYPE SPECIES *Plaxiphora matthewsi* Iredale, 1910.

Matthews' Kopionella

Kopionella matthewsi (Iredale, 1910)

(Figure 13)

Medium-sized chiton, moderately elevated, subcarinate, oval; colour deep green marked with darker colour. Sculptured with wavy lines that form ribs on the anterior and posterior valves, lateral areas with indistinct ribs, central areas indistinctly waved. Insertion plates reduced to a callus in the posterior valve; ten slitted in anterior valve and with a single slit in the median valves. Size: ¾ inch.

Localities: Mornington; Portland. Also recorded from Tasmania and South Australia.

Family CRYPTOPLACIDAE

Hidden Shelled Chitons

Shell sculptured with granules which may form longitudinal lines in the dorsal area, lateral and central areas show no differentiation. Insertion plate of anterior valve often very wide with five or three slits that correspond to the radial ribbing of upper surface. Girdle well developed, leathery, covered with shorter or longer spicules or scales, and with sutural tufts at the junction of the valves.

Craspedoplax Iredale & Hull, 1925

Small chitons with nodular sculpture. Anterior insertion plate short, five-slitted median valves small for the family; insertion plate of posterior valve small, with six to ten slits in juveniles, disappearing with age and a callus ridge developing. Girdle covered with scale-like spicules and with sutural tufts.

TYPE SPECIES *Hanleya variabilis* H. Adams & Angas, 1864.

Variegated Craspedoplax

Craspedoplax variabilis (H. Adams & Angas, 1864)

(Figure 14)

Shell small, depressed oval; colour green with splashes of yellow or black, all black or black and white. Anterior valve broad and shallow, sculptured with small round pustules and five indistinct rays. Median valves of most adults uniformly pustulose but juveniles have the pustules of the central and lateral areas arranged in definite patterns, those on the dorsal area being in longitudinal lines, while those on the lateral areas are very small and only slightly raised. Posterior valve, pustulose with the mucro post-central. Girdle with fine scales and glossy sutural tufts. Size: ⅜ inch.

Localities: Kilcunda; San Remo; Shoreham; Cowes; Mornington; Barwon Heads.

This species occurs below half-tide and usually in the calcareous algae (*Lithothamnion*) zone. It is recorded throughout southern Australia from Queensland to south Western Australia and Tasmania.

cornuta Torr & Ashby, 1898; a smaller species with a distinctive sculpture, the dorsal area is longitudinally striate and separated by a series of

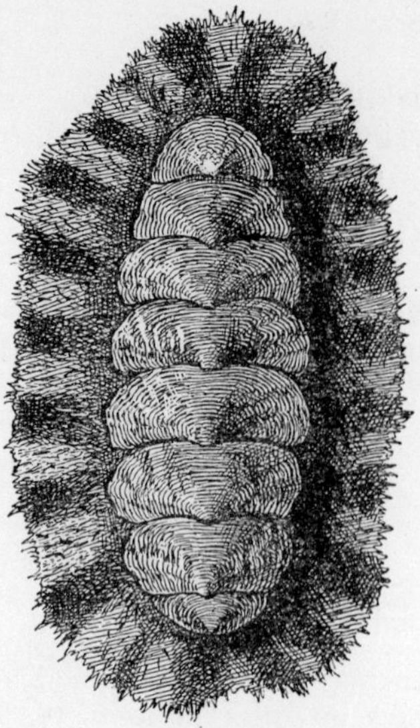

Figure 12

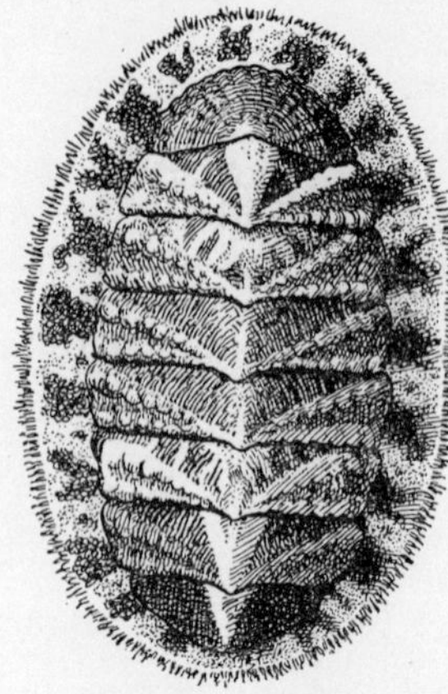

Figure 13

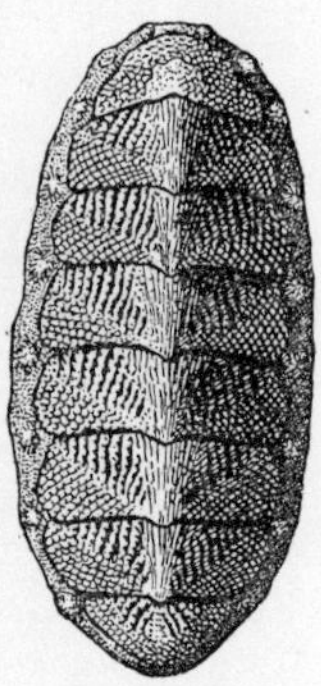

Figure 14

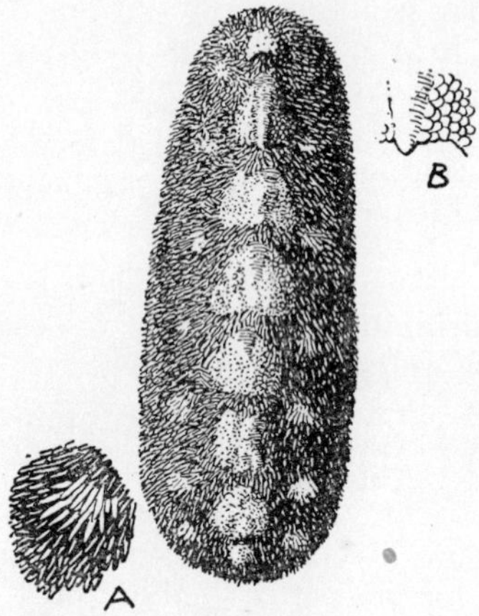

Figure 15

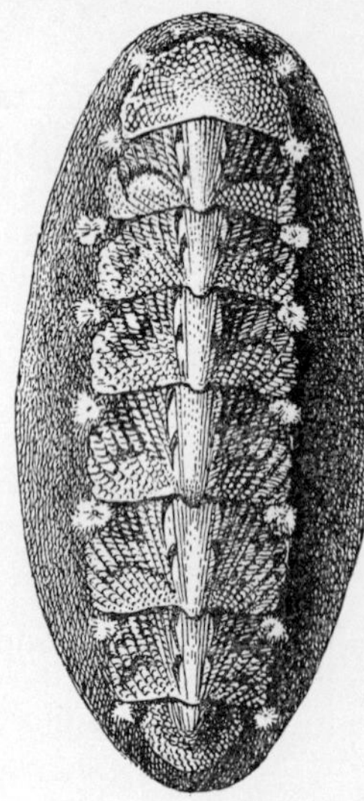

Figure 16

12. *Poneroplax albida* × 1⅓
13. *Kopionella matthewsi* × 4
14. *Craspedoplax variabilis* × 4
15. *Notoplax speciosa* × 2
    A Girdle spicules (enlarged)
    B Sculpture of valve (enlarged)
16. *Bassethullia matthewsi* × 6

five pits from the rest of the central area which is covered by ten to twelve rows of larger pustules arranged longitudinally; lateral areas minutely pustulose. Girdle minutely scaled, sutural tufts consist of long spicules. Size: ½ inch. Originally discovered from South Australia and dredged off Sorrento by C.J.G.

Notoplax H. Adams, 1861

Elongate oval chitons with small shell valves and a large girdle. Sculptured with more or less distinct radials. Anterior valve with five ribs. Insertion plate of posterior valve has many slits. Girdle wide, leathery, covered with dense spicules and obscure sutural tufts, or spicules present only as sutural tufts.

TYPE SPECIES *Chiton violaceus* Quoy & Gaimard, 1835.

Beautiful Notoplax
Notoplax speciosa H. Adams, 1861
(Figure 15)

Shell large, valves very narrow, elongate; colour red brown to yellow or dark brown. Sculptured with pustules which on the anterior valve are round and flat-topped, the radial ribbing being represented by larger elongate pustules; the small lateral areas are limited by a row of oval pustules, the narrow dorsal areas are smooth. Girdle very broad and covered with long, thin, delicate spicules; the sutural tufts are not very prominent. Size: 3 inches.

Localities: Port Phillip; Portland; Western Port.

This species which is dredged in shallow water, generally living in sponges, is recorded also from New South Wales, Tasmania and South Australia.

costata (H. Adams & Angas, 1864); recorded from Portland, has the same range as the above species but it is distinguished by smaller size (approximately 1 inch), less elongate valves, and the minutely spiculose girdle with prominent sutural tufts.

rubrostratus (Torr, 1913), Sorrento, Port Phillip (5 fathoms).

Bassethullia Pilsbry, 1928

Elongate, keeled shells. Sculptured with longitudinal rays. Insertion plates and sutural laminae large. Anterior valve with five slits, median valves with one, and posterior multi-slit. Girdle wide, leathery with fine spicules and small tufts.

TYPE SPECIES *Acanthochites matthewsi* Pilsbry, 1894.

Iredale and Hull, 1925, erected *Glyptelasma* for this species but Pilsbry pointed out that he had previously used it in 1907 and replaced it with the above name.

Matthews' Bassethullia
Bassethullia matthewsi (Pilsbry, 1894)
(Figure 16)

Shell medium-sized, elongate, keeled; colour greenish white marked with darker green. Sculptured with pustules which on the anterior valve form irregular wavy lines and on the median valves parallel lines which may lose their pustulose character except on the lateral areas; dorsal areas almost smooth. Girdle large, leathery, with fine spicules and sutural tufts of fine silky spines. Size: 1 inch.

Localities: Port Phillip Heads; Palm Beach; Western Port.

This species, originally described from South Australia, is found at and below low water to approximately twenty feet.

glypta (Sykes, 1896); dredged by J. Bracebridge Wilson at Port Phillip Heads; it is distinguished by a complete lack of nodular sculpture and is smooth except for a few slanting cuts on the central areas. The girdle is leathery with inconspicuous tufts.

Acanthochiton Gray, 1821

Shell small, elongate. Sculpture pustulose, no radial ribbing on anterior and posterior valves. Insertion plate and sutural laminae large. Usually five slits in anterior valve and one in median and posterior valve, occasionally absent. Girdle leathery, covered with spicules which are usually bunched at the sutures.

TYPE SPECIES *Chiton fascicularis* Linné, 1758.

Granular-striated Acanthochiton

Acanthochiton granostriatus Pilsbry, 1894

(Figure 17)

Shell of medium size, elongate, subcarinate; colour brown to grey with yellow to red markings which become subdued in dried specimens. Sculptured with oval, flat-topped pustules. Anterior valve with five, low, rounded undulations, dorsal area trigonal, smooth. Girdle densely covered with minute greenish spicules, large tufts at the sutures. Size: ¾ inch.

Locality: Port Phillip.

This species is found under stones from half-tide to below low water mark and has a range from southern Queensland to South Australia and northern Tasmania.

bednalli Pilsbry, 1894; dredged at Port Phillip Heads, is distinguished from the above species by its usually larger size (1 inch), coarser sculpture, thick grey green sutural tufts, lack of waving on the anterior valve, and the fine longitudinal sulci on the dorsal areas. It also is recorded from Tasmania and South Australia.

gatliffi Ashby, 1919; a small species (⅜ inch), coarsely sculptured with pustules and with the dorsal area lined and pitted. It, like the preceding species, is recorded from South Australia and Tasmania as well as Port Phillip, Victoria.

Cox's Acanthochiton

Acanthochiton coxi Pilsbry, 1894

Shell large, elongate semi-carinate; colour dark green, flamed with white or yellow. Sculptured with crowded, elongate, pear-shaped pustules. Girdle covered with dense coarse tapering spicules, large tufts of long spicules at sutures. Size: 1½ inches.

Locality: Portland.

This species is recorded from New South Wales, Tasmania and Victoria but the southern examples have been separated as *coxi lachrymosus* May & Torr, 1912 (Figure 18), because of slight differences in the shape of the pustules and the finer girdle scales. It is found hiding under the edges of stones in sandy or muddy waters of sheltered bays.

Pilsbry's Acanthochiton

Acanthochiton pilsbryi Sykes, 1896

(Figure 19)

Shell medium, elongate, depressed; colour red brown or black. Pustulose sculpture, pustules elongate, convex. Anterior valve without ribbing, junction of latero-central areas indicated by a slight elevation, dorsal areas narrow and smooth. Girdle broad in life, covered with long spicules, sutural tufts small. Size: ⅝ inch.

Localities: Shoreham; Cowes; Port Phillip; Portsea pier.

This species like other members of the genus is often found attached to the underside of rocks in muddy situations. It is recorded also from New South Wales, Tasmania and South Australia.

sueurii (Blainville, 1825); recorded from Benison Island, Corner Inlet, and Port Phillip, is sculptured with small oval pustules even on the dorsal areas; the girdle is leathery with long asbestoid golden or silvery tufts. Size: 1 inch. This species is usually found adhering to stones embedded in sand and is recorded also from Western Australia, South Australia and Tasmania.

kimberi Torr, 1913; differs from the above species, having large elongate pustules, small longitudinal grooves on the dorsal area and in its smaller size ⅜ inch. It is recorded from Western Australia to New South Wales and Tasmania.

wilsoni Sykes, 1896; a medium-sized shell (1 inch), sculptured with oval, flattened pustules; the dorsal area is marked with both longitudinal and transverse striae. The girdle is large and densely spiculose. It is recorded from Victoria, west of Wilson's Promontory, and South Australia.

Meturoplax Pilsbry, 1894

Very small chitons with pustulose sculpture and finely spiculose girdle with prominent sutural tufts. Anterior valves one slit, posterior valve two slits.

Type Species *Acanthochites retrojectus* Pilsbry, 1894.

Hidden Meturoplax

Meturoplax retrojecta (Pilsbry, 1894)

(Figure 20)

Shell small, flat, rounded; colour dark green or grey with blotches of lighter colour. Sculptured with small, round, convex pustules which are smaller on the dorsal areas. Girdle furry with long, silky, sutural tufts. Size: ½ inch.

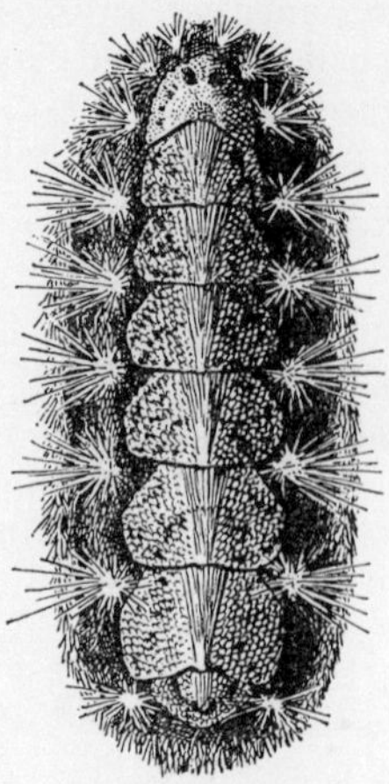

Figure 17

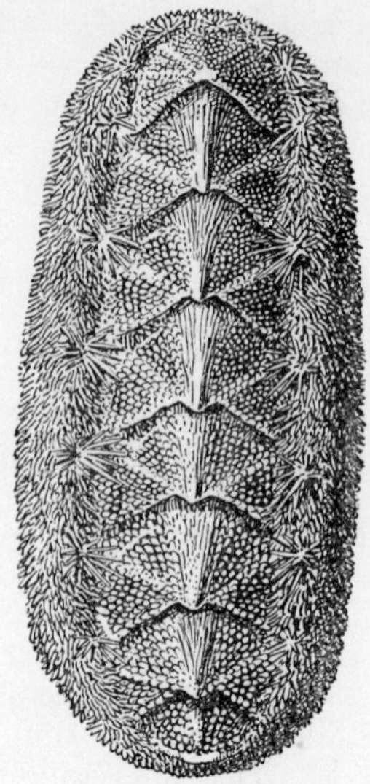

Figure 18

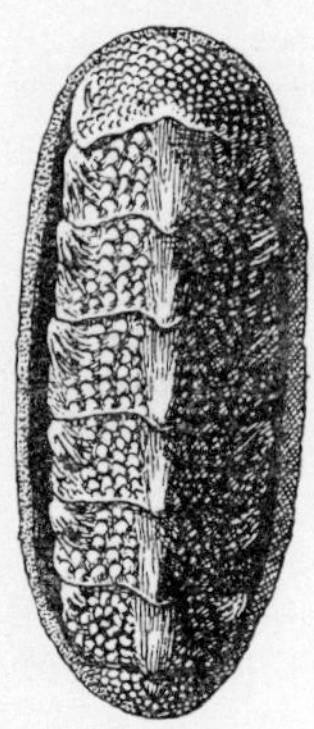

Figure 19

Figure 20

17. *Acanthochiton granostriatus* × 4
18. *Acanthochiton coxi lachrymosus* × 3
19. *Acanthochiton pilsbryi* × 8
20. *Meturoplax retrojecta* × 6
21. *Cryptoplax striata* × 1
    A Single valve (enlarged) B Girdle spicules

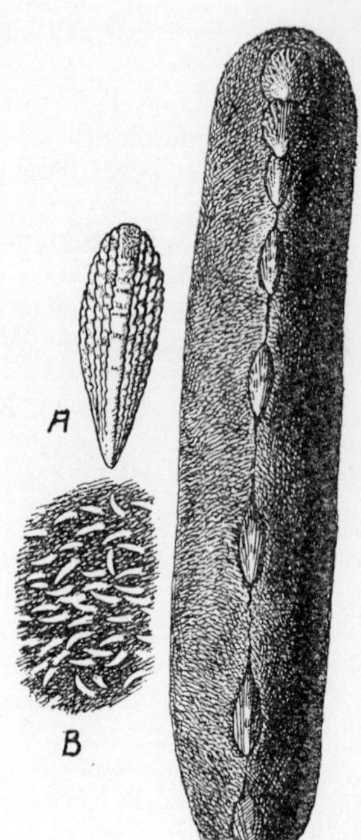

Figure 21

Localities: Mallacoota; Wilson's Promontory; Phillip Island; Port Phillip; Cape Otway.

This species is common on exposed rock platforms below mid-tide level. It cannot stand extreme exposure but seeks shelter in cracks and among other animals such as the tubes of the Serpulid worm *Galeolaria caespitosa* Lamarck and at the base of the Ascidian *Pyura stolonifera* (Heller).

Cryptoplax Blainville, 1818

Elongate worm-like chitons with a narrow shell embedded in a large fleshy girdle which is densely, finely spiculose and has small indistinct sutural tufts. The valves are in contact in juveniles but tend to become separated and embedded in the fleshy girdle as the animal grows. Sculpture is granulose, granules sometimes forming chains of lozenges. Slits are absent except in the anterior valve in which there may be three.

TYPE SPECIES *Chiton larvaeformis* Burrow, 1818.

Striated Cryptoplax

Cryptoplax striata Lamarck, 1819

(Figure 21)

Shell very large, elongate; colour dull fawn or brown, girdle sometimes banded with light and dark colour. Sculptured with longitudinal ribbing. Valves long and narrow and posterior four disconnected in adult specimens. Girdle covered with dense coarse spicules. Size: 4-5 inches.

Localities: Port Phillip; Western Port.

This species is fairly common under stones in rock-pools and below low-tide mark. It is recorded also from Tasmania, South Australia and Western Australia.

iredalei Ashby, 1923; has the same range; it is easily distinguished by the narrower valves and very fine spicules of the girdle which give it a felt-like texture.

Family ISCHNOCHITONIDAE

Ischnochitons

Chitons with the girdle covered with scales only and with the valves finely sculptured. There are insertion plates on all valves and more than eight slits on the anterior valve; the teeth are smooth and not scalloped. The members of this family live from below mid-tide level to approximately twenty-five fathoms but few occur below one fathom, and most of them prefer rock surfaces.

Ischnochiton Gray, 1847

Shells of medium size, variously coloured, elongate, oval. Sculpture of pustules or irregular lines. Girdle scales varying in size, oval, usually striate. Insertion plates sharp, more than eight slits on end valves, one slit on each side of each median valve; sutural laminae large, separated.

TYPE SPECIES *Chiton textilis* Gray, 1847.

Elongate Ischnochiton

Ischnochiton elongatus (Blainville, 1825)

(Figure 22)

Shell medium, rather elongately ovate. Sculpture weak in juveniles, well expressed in adult examples. Colour and colour pattern extremely variable: olive green, black, grey, red, blue, brown or cream and variously marked with dots, V-shaped markings, etc.

The sculpture of the anterior valve consists of radiating rows of elongated pustules; the lateral areas of the median valves are similarly marked and the central areas covered with fine curving lirae. The girdle scales are medium-sized, uniform, rounded oval; horizontally deeply grooved, about six to a scale. Size: 1 inch.

Localities: Walkerville; Lorne; Port Phillip; Western Port.

Common under stones in sheltered pools on all Victorian rock platforms. Also New South Wales, South Australia and Tasmania.

crispus Reeve, 1847; considered by Iredale and Hull to be the eastern variety of this species and differs in having smaller scales and weaker sculpture. It is recorded from Mallacoota, Victoria, and New South Wales.

Lined Ischnochiton

Ischnochiton lineolatus (Blainville, 1825)

Large, elongate-oval, coloured white or yellowish with brown or black longitudinal splashes down the middle and sometimes on the outer margin of the lateral areas. Median valves sculptured with zigzag ridges which are cut on the lateral area by irregular, elongate, radiating nodules; anterior valve rayed with numerous fine semi-nodulose ribs; posterior valve with mucro ante-median; ante-mucronal area same as median central areas but finer; post mucronal area similar to anterior valve but more nodulose (Figure 23). Girdle scales regular, rounded, fairly large with six to eight grooves. Size: 2 inches.

Localities: Point Addis; Lorne.

Common under stones below mid-tide level from King George Sound, Western Australia; South Australia; Victoria; Tasmania.

contractus Reeve, 1847; smaller (1½ inches) and differs in having the anterior valve, lateral areas and post-mucronal area sculptured with separated, rounded pustules. Found below mid-tide level under stones, from Fremantle, Western Australia, to central Victoria and islands of Bass Strait.

Variegated Ischnochiton

Ischnochiton versicolor (Sowerby, 1840)

(Figure 24)

Shell large, elongate-oval; colour variable usually dull green with a light dorsal stripe or may be unicoloured white, rose, blue or brown. Central areas, and ante-mucronal area sculptured with fine wavy linear ridges; lateral areas with three or four divaricating ribs; interstices finely punctate. Girdle scales large, rounded, oval with six to eight grooves. Size: 2½ inches.

Localities: Point Nepean; Shoreham; San Remo.

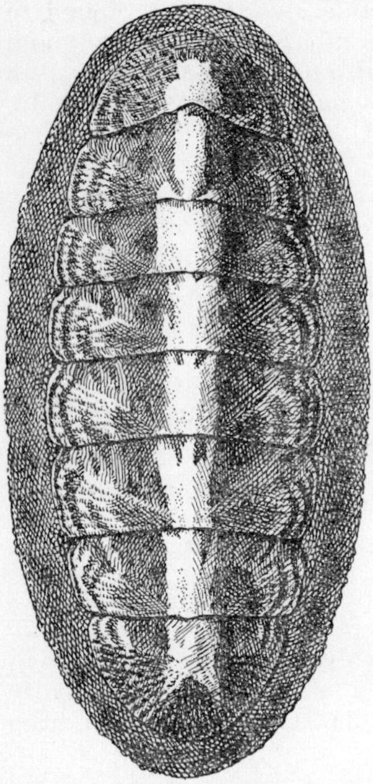

Figure 22

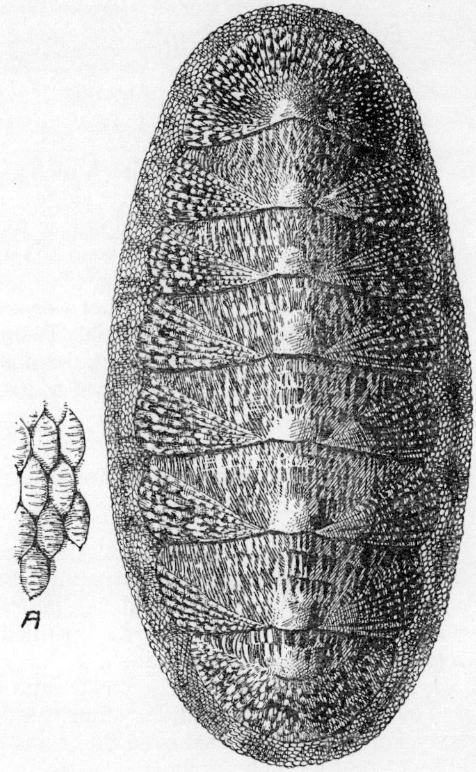

Figure 24

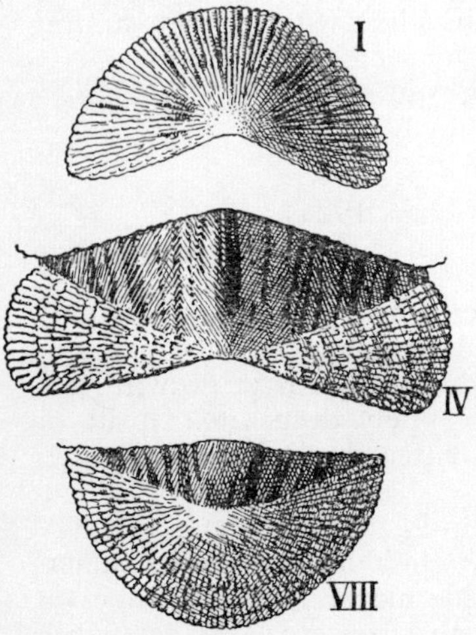

Figure 23

22. *Ischnochiton elongatus* × 2
23. Single valves of *Ischnochiton lineolatus* × 3
24. *Ischnochiton versicolor* × 2
    A Girdle scales (enlarged)

These chitons live under stones below mid-tide level and are found in New South Wales, Victoria, Tasmania, and South Australia. Iredale and Hull split this species into two subspecies: *versicolor s.s.*, the eastern form from Mallacoota, Victoria, and *versicolor milligani* Iredale & May, 1916, the larger coarser sculptured western form that occurs in South Australia, western Victoria and Tasmania.

falcatus Hull, 1912; a small (¾ inch) yellow brown species. The anterior valve is rayed with twenty-five ridges, the lateral area elevated with two prominent ribs, and the posterior saw-toothed. Girdle scales are small with approximately ten shallow grooves. It is dredged in shallow water in Victoria and Tasmania.

tateanus Bednall, 1897; another deeper water species occurring in southern New South Wales, Victoria, Tasmania and South Australia and is distinguished by strong sculpture, the posterior edge of the lateral areas and the lateral edge of the anterior valve strongly serrated.

Atkinson's Ischnochiton

Ischnochiton atkinsoni Iredale & May, 1916

(Figure 25)

Shell small, round-backed; colour uniform buff. Anterior valve and antemucronal area quincuncially punctate; median valves with the central areas coarsely quincuncially granulose, lateral areas elevated and irregularly pustulose; post mucronal area pustulose. Girdle scales very small oval, regularly grooved. Size: ½ inch.

Localities: Lorne; Apollo Bay; Cape Otway.

The usual habitat is under stones mid-tide level and the recorded range is western Victoria, Tasmania and King Island.

Autochiton Iredale & Hull, 1924

Elongate-oval chitons with variable sculpture and small closely imbricating striated girdle scales. Insertion plates of median valves with one slit, sutural laminae widely separated, projecting forward.

TYPE SPECIES *Ischnochiton torri* Iredale & May, 1924.

Torr's Autochiton

Autochiton torri (Iredale & May, 1924)

(Figure 26)

Shell elongate-oval, round-backed; colour cream or light brown flamed with darker colour. Lateral areas and anterior valve sculptured with radiating ribs (about fifty on anterior valve) traversed by concentric growth-lines forming elongate pustules; central areas very finely decussate. Girdle of moderate width, scales minute, elongate oval, flattened, closely imbricating, striate. Size: 1¾ inches.

Localities: Port Phillip; Ocean Beach, Phillip Island; Portland.

This species is very active and moves quickly when disturbed in its habitat beneath stones in tidal pools and below low-tide mark. It is also recorded from South Australia. The two other Victorian species of this genus are easily distinguished.

virgatus (Reeve, 1847); small (⅓ inch). olive, spotted with blue green and the whole shell is minutely quincuncially punctate.

wilsoni (Sykes, 1896); medium-sized (¾ inch), smooth, glossy, light coloured, longitudinally marked with darker lines.

Both these latter species are recorded from South Australia but are not recorded east of Wilson's Promontory.

Heterozona Dall, 1878

Shells large. Sculptured with longitudinal lines and rays. Girdle scales striate, larger near valves and irregularly elongate grading to small regular scales at the edge.

TYPE SPECIES *Heterozona cariosa* Dall, 1878.

Corroded Heterozona

Heterozona cariosa Dall, 1878

(Figure 27)

Shell large, elliptical, round-backed; colour varies from pale yellow to buff and brown. Sculptured with irregular rays on the anterior valve, the lateral areas and the post-mucronal area, but rays of the latter are pustulose except in extreme old age; central areas and ante-mucronal area strongly longitudinal, ribbed particularly on the sides. Girdle with scales in two series, large against the shell, and small, sometimes tall and striate on the outer margin. Size: 2 inches.

Localities: Mallacoota; Phillip Island; Port Phillip Heads; Port Fairy.

This species is fairly common under stones in intertidal pools of the open coast. It is recorded from northern Tasmania, South Australia and south Western Australia as well as all along Victorian coastline.

Greenish Heterozona

Heterozona subviridis Iredale & May, 1916

(Figure 28)

Shell medium size; colour usually blue green with white on the dorsal areas. Sculpture of anterior valve and lateral areas flat, closely set rays; post-mucronal area with more nodulose rays; central areas and ante-mucronal area finely wrinkled with zigzag lines. Girdle scales in two series but much smaller than preceding species. Size: 1¼ inches.

Localities: Walkerville; Phillip Island; Apollo Bay.

This species is also found under stones between tide marks but is restricted to this state and Tasmania. Iredale and Hull state that it is very plentiful at King Island.

fruticosa (Gould, 1846); has a range from southern Queensland to South Australia and Tasmania. It is distinguished by the very strong ribbing, divaricating towards the margin of the anterior valve, and on the lateral and post-mucronal areas; also by the very small girdle scales which become microscopic at the outer edge. This seems to favour muddier situations than the previous species.

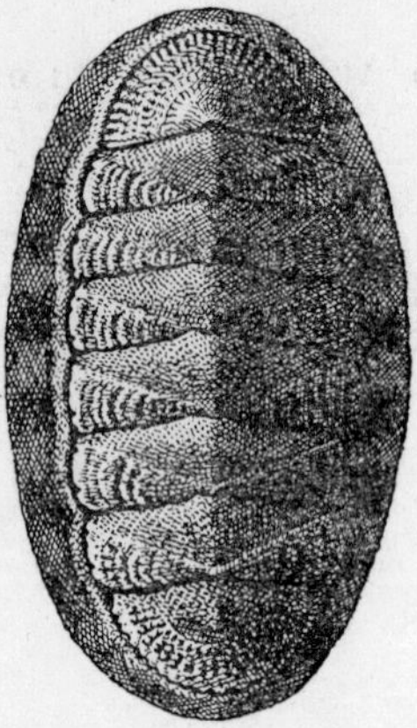

Figure 25

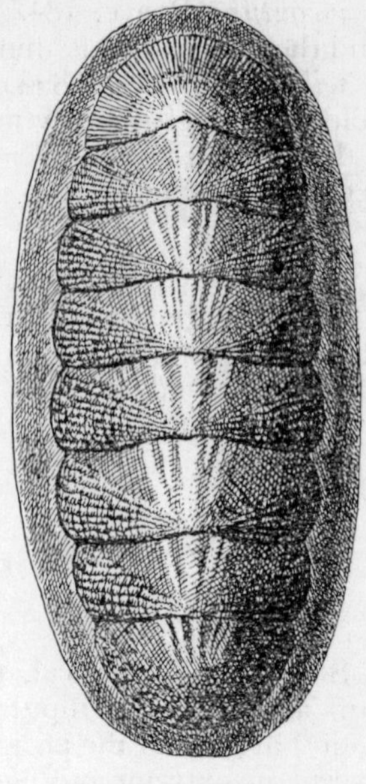

Figure 26

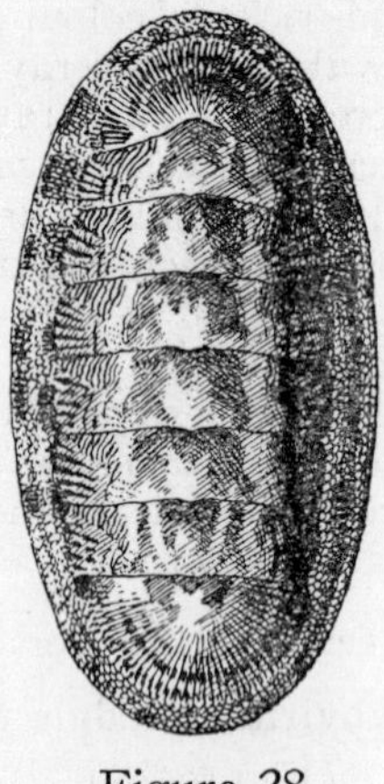

Figure 28

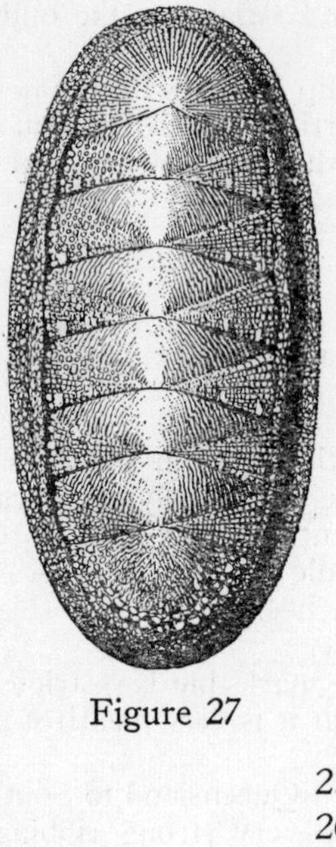

Figure 27

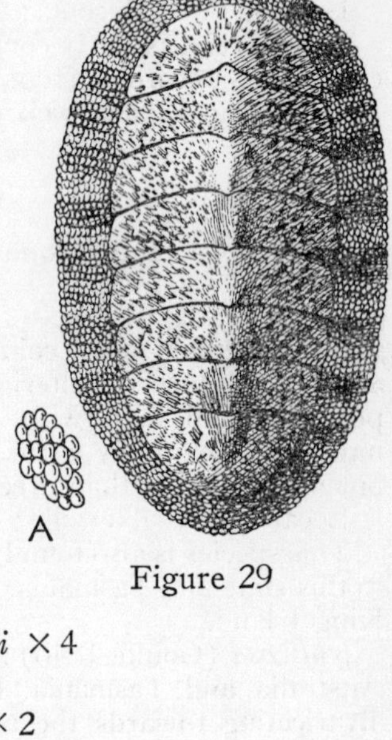

A

Figure 29

25. *Ischnochiton atkinsoni* ×4
26. *Autochiton torri* ×2
27. *Heterozona cariosa* ×2
28. *Heterozona subviridis* ×2
29. *Haploplax thomasi* ×4
    A Girdle scales (enlarged)

Haploplax Pilsbry, 1894

Shells of medium size for the family, often with brilliant colouring, smooth, shining. Girdle scales rounded, polished, smooth or faintly striate.

TYPE SPECIES *Lophyrus smaragdinus* Angas, 1857.

The members of this genus are all strikingly coloured but because of its microscopic character the full beauty of the pattern is only seen with the aid of a lens.

Thomas' Haploplax

Haploplax thomasi Bednall, 1897

(Figure 29)

Shell elongate-oval, subcarinate, smooth, glossy; colour pale green peppered with wedge-shaped white markings. Sculpture is microscopically quincuncially punctate, lateral areas slightly raised. Girdle scales small, glossy, slightly elongate. Size: ¾ inch.

Localities: Inverloch; Torquay; Portland.

This distinctive species lives under stones between tide marks and is also recorded from South Australia.

pura (Sykes, 1896); has a small, pure white, glossy shell and is recorded from eastern Victoria and southern New South Wales.

resplendens (Bednall & Matthews, 1906); bluish green, sculptured with coarse reticulations that form wavy lines. It is recorded from Victoria, South Australia and south Western Australia.

lentiginosa (Sowerby, 1840); Mallacoota, Lakes Entrance.

smaragdina (Angas, 1867); eastern Victoria.

Ischnoradsia Shuttleworth, 1853

Large chitons, dark coloured. Entire surface finely punctate; end valves and lateral areas with elevated ridges. Girdle scales large, round, with peaked edge, faintly horizontally striate.

TYPE SPECIES *Chiton australis* Sowerby, 1840.

Southern Ischnoradsia

Ischnoradsia australis (Sowerby, 1840)

(Figure 30)

Shell large, flat, oval; colour dark green or greenish blue. Sculpture of anterior valve twenty to thirty irregular radiating ribs; lateral areas with four to six ribs; central areas longitudinally ribbed, ribs becoming coarse and farther apart towards the edge. Girdle medium width. Size: 3 inches.

Localities: Apollo Bay; Lorne; Phillip Island; Walkerville.

This species is common under stones below mid-tide level and occurs on most rock platforms with suitable shelter. It is also recorded from New South Wales and southern Queensland.

evanida (Sowerby, 1840); the western species, has been separated because of the usually completely smooth character of the central areas. However, having examined a large series we are inclined to think this character is not constant but may be partly a result of environment. This latter form according to Iredale and Hull occurs in Tasmania, western Victoria and South Australia.

Stenochiton H. Adams & Angas, 1864

Shells very narrow, elongate; glossy, brightly coloured. Girdle scales small, flat, oval, smooth and shining. Slits of insertion plates numerous in terminal valves, two to four in median valves.

TYPE SPECIES *Stenochiton juloides* H. Adams & Angas, 1864 = *Chiton longicymba* Blainville, 1825.

The various members of this genus show affinities with members of *Autochiton,* and Iredale and Hull suggest that the Stenochitons may have arisen from two or three species of the same genus through adaptation to their new station. This station is a very specialized environment attached to the blades or stems of sea-grasses.

Clasping Stenochiton

Stenochiton longicymba (Blainville, 1825)

(Figure 31)

Shell large, very narrow, elongate, round-backed, elevated glossy, colour purplish brown flamed with white. Sculpture of whole shell minutely reticulate. Girdle narrow, scales very small, lozenge-shaped, glossy. Size: 1½ inches.

Locality: Port Phillip Heads.

These shells are found on the blades and root sheathes of sea-grasses (Zostera, etc.) below low-tide mark. They are also recorded from South Australia and southern Western Australia, the range of all the species of this genus.

cymodocealis Ashby, 1918; a small species (⅝ inch) with a very narrow girdle with which it clasps the stalks of its host plant *Cymodocea.*

pilsbryanus (Bednall, 1897); another small species (¾ inch) with coarser sculpture and larger girdle scales than *longicymba.* It is usually attached to the leaves of sea-grasses.

pallens Ashby, 1900; broader than other species of the genus from which it is also distinguished by slightly larger girdle scales and lack of sculpture except for faint growth-lines. This species lives below low water and has only been taken by dredging. Its less specialized shape suggests a less confined habitat.

Callistelasma Iredale & Hull, 1925

Shells small to medium, elongate-oval. Strongly sculptured with nodulose ribs on end valves and lateral areas, central areas ridged. Girdle wide, scales minute, oval, dull, striate; slits of end valve insertion plates agree in number and position with external ribs; sutural laminae separated.

TYPE SPECIES *Chiton antiquus* Reeve, 1847.

Antique Callistelasma

Callistelasma antiqua (Reeve, 1847)

(Figure 32)

Medium-sized shell, oval; colour cream to red brown, sometimes black or blue green. Sculpture of anterior valve and post-mucronal area elevated

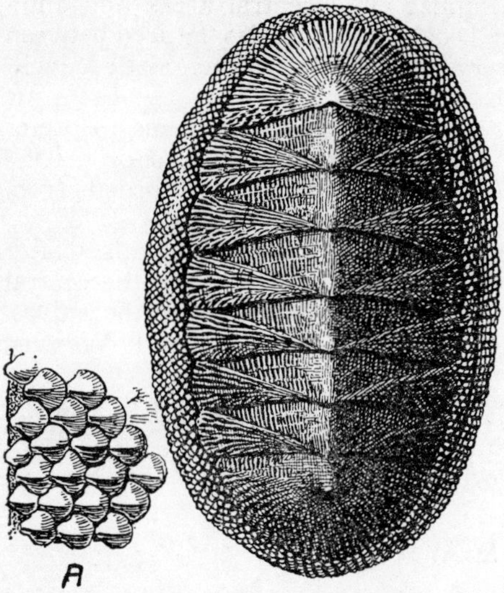

Figure 30

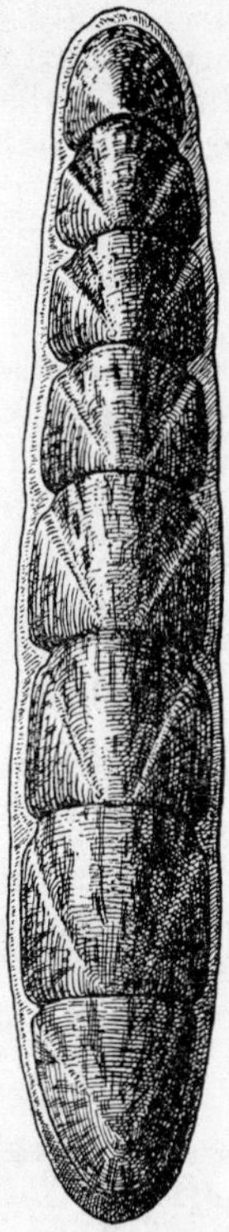

Figure 31

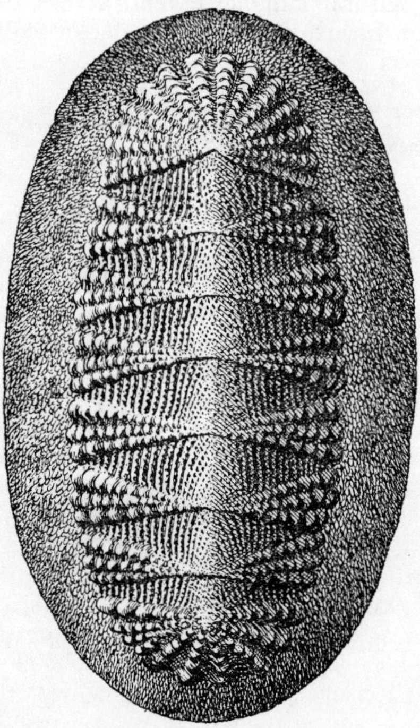

Figure 32

30. *Ischnoradsia australis*
× 1⅓
A Girdle scales (enlarged)
31. *Stenochiton longicymba*
× 2
32. *Callistelasma antiqua*
× 4

nodular ribs; lateral area with two similar ribs; central areas and antemucronal area ridged with rows of small coalescing pustules, latticed between with fine lines. Girdle wide, scales very small, oval, striate. Size: ¾ inch.
Locality: Mallacoota.

This species lives under stones between tide marks and seems to prefer areas where the stones are partially embedded in muddy sand or marine debris. Its colouring usually matches its habitat. It is recorded from southern Queensland, New South Wales and eastern Victoria.

meridionalis (Ashby, 1919); occurs in western Victoria. It is distinguished from the above species by the lack of linear ridges on the central areas, though there are a few slanting lines near the sides of the valves. Its range extends to Tasmania, South Australia and south Western Australia.

Callistassecla Iredale & Hull, 1925 (*Callistochiton mawlei* Iredale & May, 1916).

mawlei (Iredale & May, 1916); Portland.

Aulacochiton H. & A. Adams, 1858

Shell large, elongate-oval, keeled; posterior valve small with recurved terminal mucro, girdle with posterior slit; sculpture of small, separate erect pustules, sometimes forming rays or lines. Girdle covered with unequal ovate, striate scales, and numerous, scattered, spiculose tufts. Anterior valve with eight slits, median valve one slit; sutural laminae extend across the valve except for a small, deep, central gap. Insertion plates of posterior valve reduced to striated callus.

TYPE SPECIES *Chiton volvox* Reeve, 1847.

This genus is better known as *Lorica* H. & A. Adams, 1852, but the name is preoccupied and therefore it is necessary to replace it.

Mud Chiton

Aulacochiton cimolia (Reeve, 1847)

(Figure 33)

Shell large, elongate-oval, strongly keeled; colour brown. Anterior valve and lateral areas sculptured with granular rays; central area with thin, longitudinal ridges. Size: 3½ inches.
Localities: Williamstown; Port Fairy.

This species is found below low tide under stones in muddy situations. It is a sluggish creature and the shells usually carry algae and other growth. It is found also in Tasmania, South Australia and south Western Australia.

Loricella Pilsbry, 1893

Shell large, roundly oval, depressed; colour and texture dull; interior valve very large, posterior valve small; sculptured with rays on anterior and posterior valves and lateral areas. Ridges on central area. Girdle cleft posteriorly and covered with very small round scales among which are tufts of complex, corneous processes.

TYPE SPECIES *Lorica angasi* H. Adams, 1864.

Angas' Loricella

Loricella angasi (H. Adams, 1864)
(Figure 34)

Shell large, roundly oval, depressed; colour dull brown or grey, sometimes banded with darker colour. Anterior valve broad, nearly twice as wide as long. Sculptured with many radiating ribs which may show primaries and subsidiaries or may be all of equal strength; lateral and central areas show similar sculpture but in some examples the ribbing may be obsolete. Girdle as described in the generic description. Size: 2½ inches.

Locality: Rhyll, Western Port (8 fathoms).

This species, which is rare in Victoria, prefers muddy situations in bays and inlets where it is found attached to stones below low-water mark. It is recorded also from New South Wales, Tasmania, and South Australia.

Family CHITONIDAE
Chitons

This family contains the most highly specialized members of the class and they show considerable diversity of form. The diagnostic features of the family are (1) the pectination of the insertion plates, i.e. the edges of the teeth are finely cut into combs; (2) in most genera there are eight or more slits in the anterior valve (*Schizochiton* six or seven).

Rhyssoplax Thiele, 1893

Shells of varying size, elongate-oval, keeled. Sculpture varies from almost smooth to rugose but longitudinal sculpture is present on the central areas. Anterior valve eight-slit, median valves one-slit and posterior valve ten- to twelve-slit, all teeth pectinate. Girdle scales striate and closely imbricating.

TYPE SPECIES *Chiton affinis* Issel, 1869.

There has been some confusion in the placing of the Australian species but reference to the original descriptions of the two genera in question, *Anthochiton* Thiele, 1893, and *Rhyssoplax* Thiele, 1893, and comparison with the respective type species indicates that southern Australian shells belong in *Rhyssoplax s.s.*

Elevated Chiton

Rhyssoplax jugosa (Gould, 1846)
(Figure 35)

Shell large, oval, keeled; colour variable and brilliant. The anterior valve, lateral areas and post-mucronal area are ornamented with coloured concentric lines, ribs of central area reddish brown and interstices blue; valves II and III may be wholly or partly red, while in other shells valves II to VII may be white. Anterior valve, lateral areas and posterior valve smooth; central area with straight square-cut ribs, ten to fourteen on each side of the smooth dorsum. Girdle scales large and obsoletely striate. Size: 1½ inches.

Locality: Mallacoota.

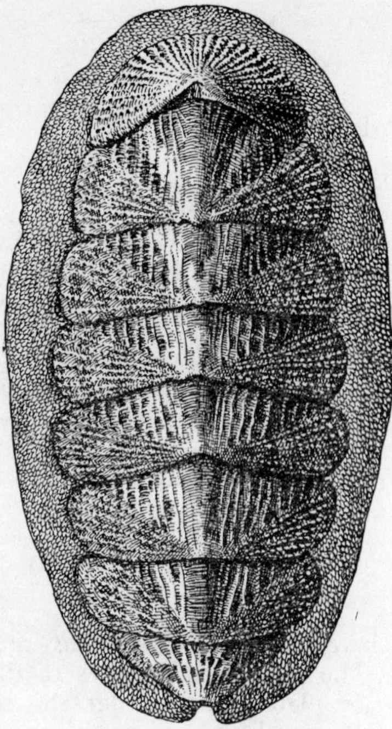
Figure 33

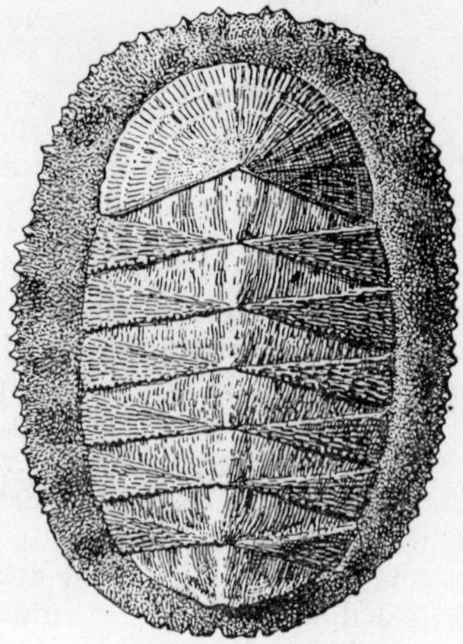
Figure 34

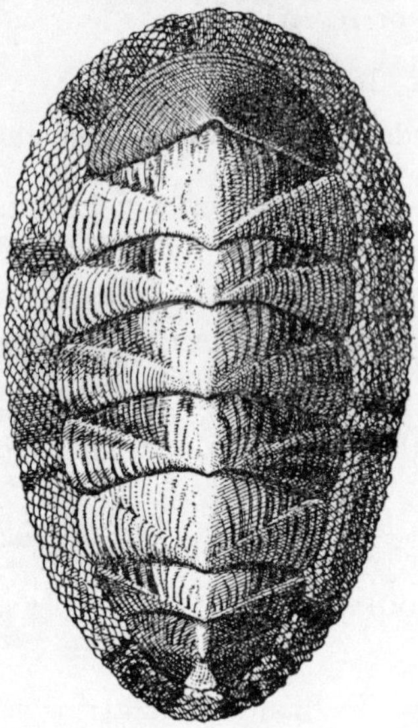
Figure 35

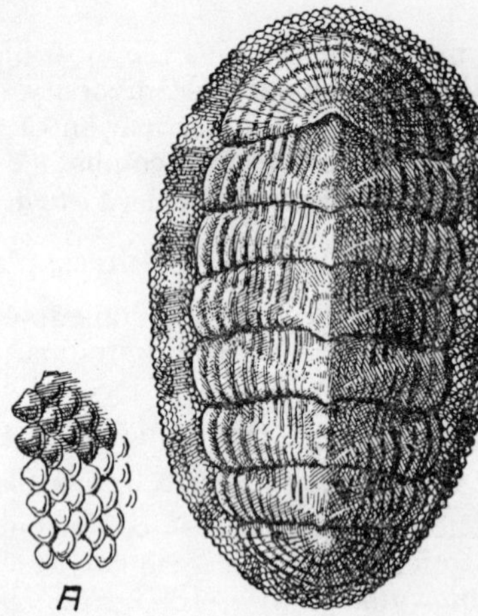
Figure 36

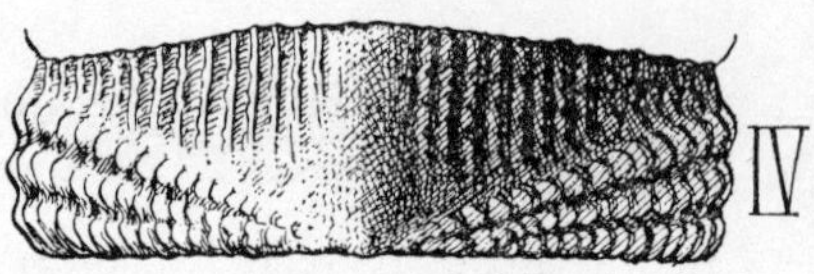

Figure 37

33. *Aulacochiton cimolia* × 1⅓
34. *Loricella angasi* × 1¼
35. *Rhyssoplax jugosa* × 2⅔
36. *Rhyssoplax exoptanda* × 3
A Girdle scales (enlarged)
37. Single valve of *Rhyssoplax tricostalis* × 5⅓

This species is found under stones below low-water mark and is recorded from eastern Victoria and New South Wales.

diaphora Iredale & May, 1916, has been confused with it but is easily separated by the weaker ribbing of the central areas, more than half the ribs not extending across the valve. This species is recorded from Torquay, Port Fairy and Portland in Victoria as well as from Tasmania and South Australia.

Desirable Chiton

Rhyssoplax exoptanda (Bednall, 1897)

(Figure 36)

Shell large, elongate-oval, keeled; colour reddish with concentric lines of darker colour on end valves and lateral areas. Sculpture of anterior valve, lateral areas and post-mucronal area flattened smooth ribs, interstices narrow and pitted; central areas crossed by eight, bold, well-separated, longitudinal ribs followed by six decreasing in length and more closely spaced, leaving a wedge-shaped smooth dorsum. Girdle scales large, faintly apically striate. Size: 2 inches.

Locality: Western Port.

This species is dredged in shallow water and has been taken on the upper surface of stones in deeper sandy bottomed pools. It is also recorded from South Australia and Western Australia.

bednalli (Pilsbry, 1895); with the same range and habitat has been taken at Port Phillip Heads. It is distinguished by the more elongate shape, stronger radial sculpture (twelve ribs on the central area) and paler colouring.

Tricostate Chiton

Rhyssoplax tricostalis (Pilsbry, 1894)

Shell medium size, elongate-oval, keeled; colour variable, either unicoloured red, brown or green or with valves II and III white; girdle generally striped. Sculpture of anterior valve, lateral areas and post-mucronal area strongly nodulose radial ribs; central area with approximately fifteen longitudinal ribs all but one or two extending across valve leaving a narrow dorsum (Figure 37). Girdle scales small, imbricating, striate. Size: 1 inch.

Localities: Port Phillip Heads; Port Fairy; Torquay; Portland.

This species is readily distinguished by its strong, nodular sculpture; it is found under stones below low-water mark and is recorded also from northern Tasmania, South Australia and Western Australia.

calliozona (Pilsbry, 1894); Portland; Ocean Beach, Phillip Island.

orukta (Maughan, 1900); Flinders.

Mucrosquama Iredale & Hull, 1926 (*Chiton carnosus* Angas, 1867).

verconis (Torr & Ashby, 1898); Port Fairy.

Class GASTROPODA

Univalves

THE LARGEST and most varied class of molluscs, typically with a snail-like shell in both marine and land forms. Exceptions are the flat-shelled limpets and land and marine slugs which may be shell-less or have an internal shell beneath the mantle.

The typical animal is snail-like, with a foot running the entire length of the ventral surface of the body. The organs—digestive, reproductive, etc.—are coiled in a spiral which is enclosed in, and is protected by, the shell. There is a head with a mouth, tentacles, and eyes which may be sessile or stalked. The entire body may be withdrawn into the last (largest or body) whorl of the shell and in many marine and some land snails the opening may be sealed with a horny or calcareous plug (operculum) which is attached to the posterior dorsal surface of the body.

The typical shell is a cone coiled round a central axis as a spiral. This central axis is called the columella and at the anterior end of the shell forms the inner lip of the mouth opening. It may be drawn out to form with the anterior edge of the mouth a canal (anterior canal) sheathing the proboscis. If, as in some species, there is a notch between the outer lip and the body-whorl, this is called the posterior canal. The shape of the spiral of shells varies considerably as also does the form of the mouth; the latter may be small and round, or if the body-whorl is large, it may extend for the whole length of the shell. The lip may be thickened or sharp and may or may not be indented with teeth.

At the apex of the spiral is the embryonic shell (protoconch) and it often has a different form and ornament from the adult shell (Figure 38).

Order ARCHAEOGASTROPODA

Family HALIOTIDAE

Ear Shells or Abalones

Shell spiral, depressed, saucer-shaped, with a large body-whorl and small spire placed off centre and to the right posterior. A row of pores on the left margin, the earlier ones are closed as the shell enlarges. Interior pearly and often iridescent. Animals live attached to rocks by the muscular foot. Known in Australia as Ear Shells or Mutton Fish, in America as Abalone and in the Channel Islands as Ormer. Animals are used extensively as food, and the shells for the manufacture of ornaments, pearl buttons and mother-of-pearl inlay.

Notohaliotis Cotton & Godfrey, 1938

This genus was erected for Haliotis with rounded, rough, irregularly sculptured shells with large and conical marginal pores and no spiral dorsal rib.

TYPE SPECIES *Haliotis ruber* Leach, 1814.

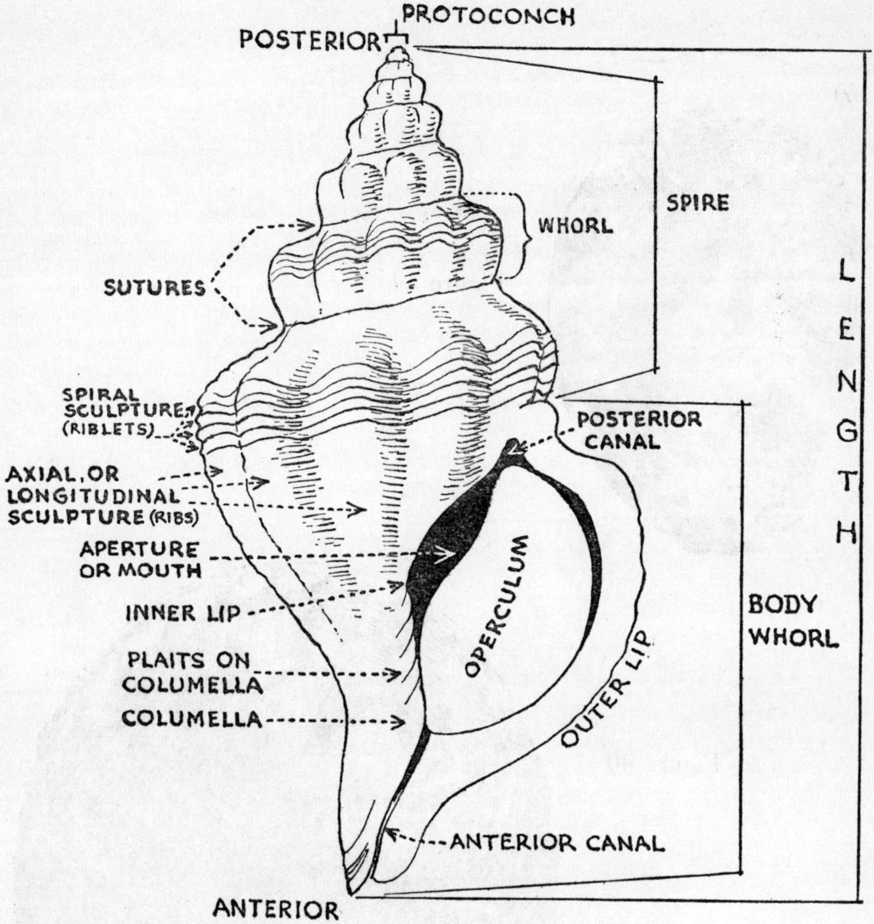

Figure 38. Parts of univalve shell

Scarlet-rayed Ear Shell
Notohaliotis coccoradiata (Reeve, 1846)
(Figure 39)

Shell oval, fragile, depressed, colour varied with blotches of scarlet and olive green. Spire pink and closed perforations scarlet. Sculptured by irregular, concentric ridges which are cancellated by fine longitudinal striae with growth-lines plainly visible. Perforations small, almost rounded, very slightly raised, five to six being open. Interior silvery with pink and green outstanding. Size: 1½ inches.

Locality: Mallacoota.

A very delicate, flat, highly coloured species with short spire. The animal is of greenish colour, and apparently is a common species on the New South Wales coast.

Figure 39

Figure 40

39. *Notohaliotis coccoradiata* × 1⅓
A Lateral elevation

40. *Notohaliotis conicopora* × ⅔

Conical-pored Ear Shell

Notohaliotis conicopora (Peron, 1816)
(Figure 40)

Shell strong, broadly ovate; colour of earlier whorls fawn with blotches of brown and green, which gradually coalesce developing into a uniform dark brown towards the outer lip. Spire very short. Body-whorl flatly convex and radiately rugosely ridged, ridges more or less following the contour of the lip. The whole surface further ornamented with numerous, irregular, concentric closely scaled riblets. Perforations very prominent, tubiform, five to six being open. Interior iridescent, outer lip fairly sharp. Size: 5 inches.

Localities: Shoreham; Flinders.

The internal coloration displays an extraordinarily rich iridescence with green and pink as the predominating colours. The species is a close ally of *ruber* Leach but broader, with the spiral whorls showing greater elevation; body-whorl is flatter, and the perforations, as the author of the species remarked, are 'prominent and formed like so many cones opened and truncated'. *Haliotis granti* Pritchard & Gatliff, 1902, is a synonym. It appears frequently in South Australia and south Western Australia.

Red or Knotted Ear Shell

Notohaliotis ruber (Leach, 1814)
(Figure 41)

Shell large, scarlet red, streaked with light green, ovate, obliquely wrinkled, spirally ridged, some ridges larger than others; this concentric sculpture being crossed by close, feeble growth-lines. At the row of perforations, slightly angular, below this a decided excavation is evident. Internally corrugated, silvery, brilliantly iridescent. Size: 5 inches.

Localities: Frankston; Mordialloc; Sorrento; Pope's Eye, Port Phillip, and coast generally.

This is our commonest Ear Shell and, although in colour and ornament showing considerable variation, is easily distinguished from all other Victorian forms. The species is traced to New South Wales and South Australia.

Iredale, 1924, has named the western form *improbulum* as he considers it a more rounded variety. However, we do not consider this feature constant and therefore are not prepared to separate the eastern and western shells.

Marinauris Iredale, 1927

Shells round, depressed with large, conical, marginal pores, sculptured with more or less developed spiral cords and a medium dorsal rib. A tendency to develop axial folds is obsolete in Queensland members of the genus but becomes more marked in southern and western Australian species. *Neohaliotis* Cotton & Godfrey is a synonym of this genus.

TYPE SPECIES *Marinauris melculus* Iredale, 1927.

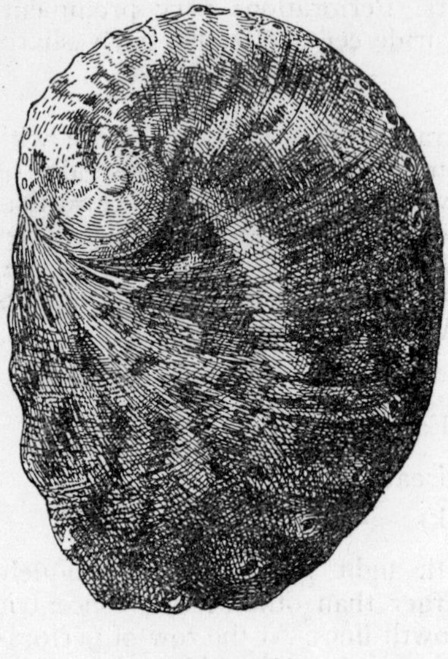

Figure 41

Figure 42

Figure 43

41. *Notohaliotis ruber* × ⅔
42. *Marinauris emmae* × ⅔
43. *Marinauris roei* × ⅔
A Lateral elevation

Emma's Ear Shell

Marinauris emmae (Reeve, 1846)

(Figure 42)

Shell orange red, obliquely rayed with white zones; somewhat rounded; spirally, finely, granosely ridged, with an obtuse keel almost centrally situated; radiately plicated round the rather conspicuous spire. Growth-striae distinct. Perforations tubiferous, of which generally six are open; between this row of holes and the outer keel a well-defined channel is seen. Columella shelf narrow, flat. Internally silvery, iridescent. Size: 4 inches.

Localities: Sorrento; Western Port.

This species is easily separated by the radial plications round the spire, and the presence of the central obtuse keel. A striking, brightly coloured species which is also recorded from Tasmania and South Australia.

Roe's Ear Shell

Marinauris roei (Gray, 1827)

(Figure 43)

Shell small, roundly ovate. Spire somewhat elevated. Colour scarlet red with irregular rays of white and blotches of green. Sculpture consisting of strong, unequal, closely scaled concentric ridges with deep grooves between. Perforations almost circular, small, slightly raised, six to eight being open. Interior silvery, iridescent, with pink and green predominating. Size: 2 inches.

Locality: Portland.

An outstanding species and because of its mother-of-pearl interior and strong concentric sculpture no difficulty should be experienced in recognizing it. Beyond Victoria it occurs in South Australia and Western Australia.

Exohaliotis Cotton & Godfrey, 1933

Medium-sized shells, rounded, elevated with small and only slightly conical marginal pores. Sculpture consists of spiral cords overlaid with axial wrinkles more marked near the spire.

TYPE SPECIES *Haliotis cyclobates* Peron, 1816.

Elevated or Round-back Ear Shell

Exohaliotis cyclobates (Peron, 1816)

(Figure 44)

Shell almost circular in outline, very convex, coloured with radiating patches of cream, brown or green. Spire much elevated composed of three rounded whorls. Body-whorl with a blunt keel at the row of holes. Perforations five to six, oval, slightly raised. Sculpture consisting of somewhat irregular concentric, granose ridges and lines of growth which are plainly seen. Interior irridescent with green and pink reflections. Size: 2 inches.

Locality: Portsea.

A characteristic species with a decidedly elevated spire and well-rounded body-whorl, features that immediately separate it from other Victorian forms. Apparently of rare occurrence in this state but common in South

Figure 44

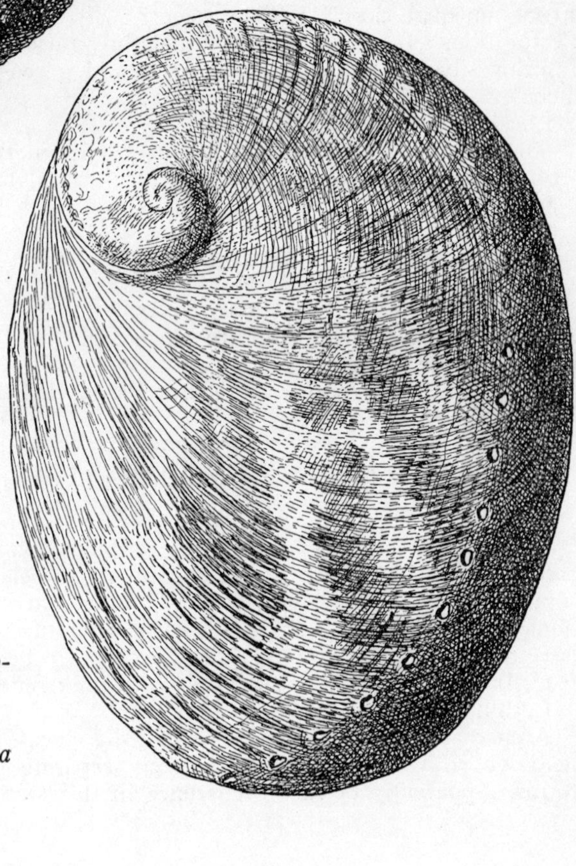

Figure 45

44. *Exohaliotis cyclobates* × 1

A Lateral elevation

45. *Schismotis laevigata* × ⅔

Australia adhering to rocks and other larger shells. *Haliotis excavata* Lamarck, 1822, has been applied to the shell for many years but this falls into synonymy of the above prior name of Peron.

Schismotis Gray, 1856

Shell rounded, large with a smooth surface and small flush marginal pores.

TYPE SPECIES *Haliotis laevigata* Donovan, 1808.

Mutton Fish, Ear Shell or Sea Ear

Schismotis laevigata (Donovan, 1808)

(Figure 45)

Shell large, ear-shaped, broad, spire moderately elevated. Surface sculptured with obsolete spiral lirae. Aperture very large, with a series of small perforations along the left side following the course of the spire. Inside silvery, nacre fairly smooth and minutely wrinkled. Size: 6 inches.

Localities: Sorrento; Portsea; Queenscliff.

This is our largest Mutton Fish and one of the largest Victorian shells. A specimen in the collection of C.J.G. from the first-named locality measures eight inches. The species is also found in South Australia and Western Australia. It is sublittoral in habit, being located in secluded spots among the rocks at below low tide. Externally the coloration varies, but usually consists of continuous oblique stripes of orange scarlet and white. The small perforations and more silvery interior will serve as useful recognition marks. For many years this shell has been known as *Haliotis albicans* Quoy & Gaimard, 1834, but a prior name existed which precludes its use.

Family SCISSURELLIDAE

Slit Shells

Shell minute, unicolored, umbilicated, turbinate or depressed, few-whorled, thin. Aperture oval; upper part of outer lip with a slit which is continued as a differently sculptured band or oval furrow encircling the whorls; operculum corneous, thin, circular with central nucleus.

Schismope Jeffreys, 1856

Scissurella in which the oval slit becomes closed in the adult and transformed into an oblong perforation; furrow short, does not extend beyond one and a half whorls.

TYPE SPECIES *Scissurella striatula* Philippi, 1844.

Atkinson's Slit Shell

Schismope atkinsoni (Tenison Woods, 1876)

(Figure 46)

Shell minute, globose depressed, brown. Spire short, low, blunt. Whorls three, rapidly widening, plane above, the last very deeply descending towards the aperture, with strong keel at shoulder, occupied by the oval furrow, and another keel at the periphery; the space between concave. Below this carina there are a number of spiral lirae extending to the umbilicus; growth lines fine. Size: $\frac{1}{16}$ inch.

Localities: Dredged off Rhyll, Western Port; Flinders; Cape Schanck; Portarlington.

These minute molluscs are dredged alive or are found dead in shell sand when they are usually bleached white.

Several other members of the family Scissurellidae occur in Victoria but they are all minute and most of them are rare.

beddomei Petterd, 1884; Flinders.

pulchra Petterd, 1884; off Rhyll, Western Port.

Scissurella Iredale, 1924 (*Scissurella rosea* Hedley, 1904).

remota Iredale, 1924; Flinders; San Remo; Point Nepean.

Schizotrochus Monterosato, 1877 (*crispata* Fleming, 1832).

australis Hedley, 1903; off Wilson's Promontory.

Family FISSURELLIDAE

Keyhole Limpets

Shell limpet-shaped, porcellaneous, conical, depressed oval at base, the surface usually strongly sculptured, having an anterior perforation, slit, notch or emargination for the passage of the anus.

Scutus Montfort, 1810

Shell oblong depressed, apex directed backwards, no oval groove or slit, but the front margin more or less truncated and sinuous. The shell sits like a saddle on the back of the large dark animal and can be almost concealed by the mantle.

TYPE SPECIES *Scutus antipodes* Montfort, 1810.

Duckbill

Scutus antipodes Montfort, 1810

(Figure 47)

Shell white, oblong, depressed, shield-shaped, apex directed backward. Surface devoid of radiating sculpture. Aperture with anterior edge truncate, arched, and slightly emarginate. Size: 4 inches.

Localities: Sorrento; Portsea; Queenscliff.

This is a frequent shell in Western Port and at various localities along the coast. It is found between tide marks under boulders, the removal of which often yields several specimens. The animal is black and exceedingly tough. Normally the interior of shell is white, but occasionally specimens are obtained tinted with light red. The name of Elephant Fish is frequently applied to this creature. Occurs from New South Wales to Tasmania and southern Western Australia.

The Western Australian members have been divided off as *anatinus* (Donovan, 1820) but examination of a series does not indicate that there are any significant and constant differences.

Figure 46

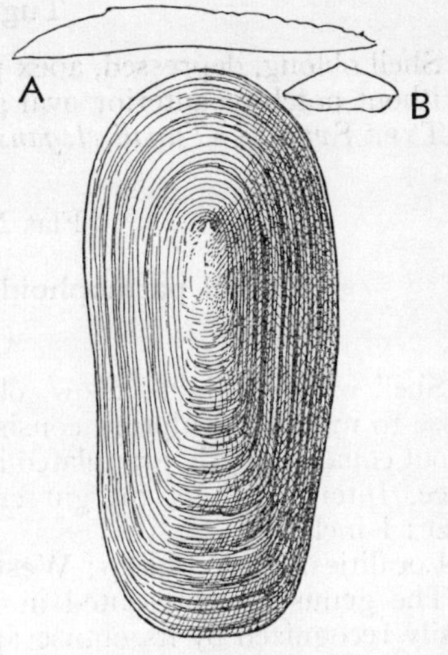

Figure 47

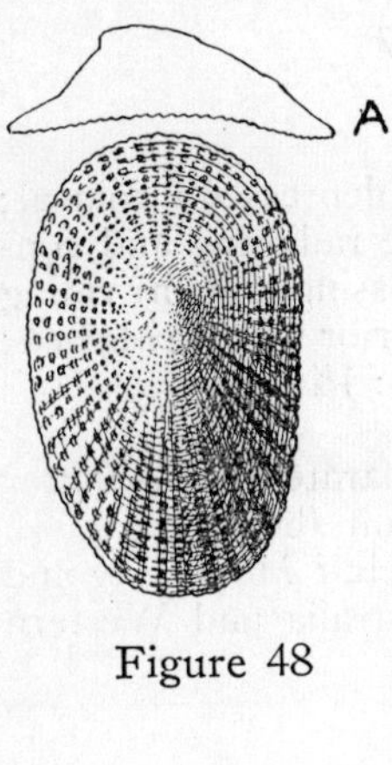

Figure 48

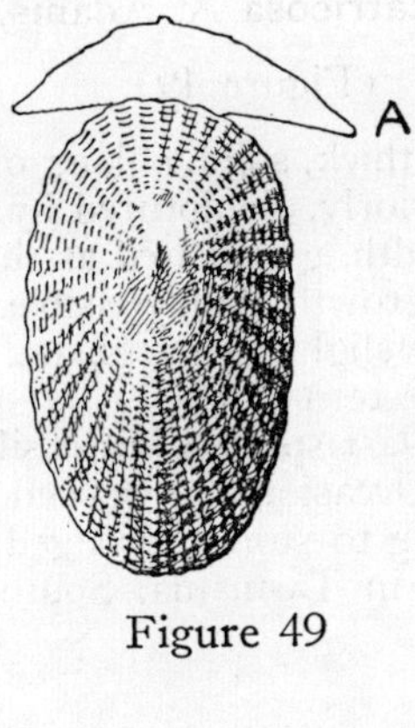

Figure 49

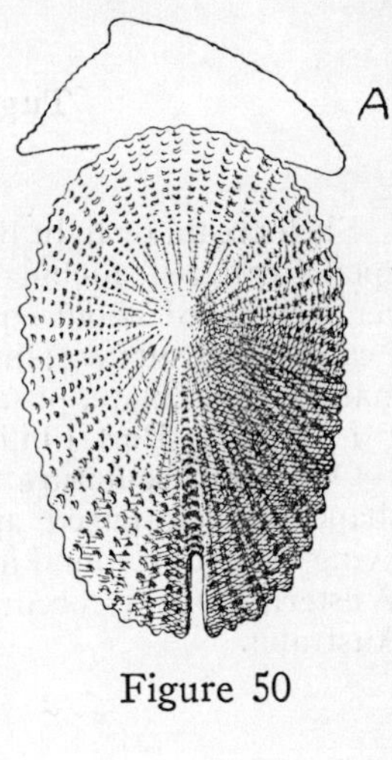

Figure 50

46. *Schismope atkinsoni* × 12
    A View from top of shell
47. *Scutus antipodes* × ⅔
    A Lateral elevation
    B Anterior elevation
48. *Tugali parmophoidea* × 1⅛
    A Lateral elevation
49. *Tugali cicatricosa* × 1⅓
    A Lateral elevation
50. *Notomella dilecta* × 3
    A Lateral elevation

Tugali Gray, 1843

Shell oblong, depressed, apex posterior, front margin rounded or sinuous. Without notch or anterior oval groove.

TYPE SPECIES *Tugali elegans* Gray, 1843.

Flat Notched Limpet

Tugali parmophoidea (Quoy & Gaimard, 1834)

(Figure 48)

Shell white or light yellow, oblong-ovate, arcuate; apex obtuse, situated close to margin. Sculpture consisting of close, radial and concentric ribs of about equal strength, granulated and giving the surface a cancellated appearance. Interior white, margin crenulated, slightly emarginate posteriorly. Size: 1[illegible] inch.

Localities: Port Phillip; Western Port; Anderson's Inlet; Port Albert.

The genus is represented in Victoria by two species, the above being easily recognized by its obtuse apex and cancellated sculpture. Occurs also in New South Wales, Tasmania, through Bass Strait to South Australia.

Scarred Notched Limpet

Tugali cicatricosa A. Adams, 1857

(Figure 49)

Shell cream or light brown, thick, solid, oblong-ovate, depressedly conical; apex cicatrized, inclined anteriorly. Sculptured by close radiating and concentric ribs of about equal width, granulated at the crossing and providing a cancellated surface; lines of growth plainly visible. Interior white, polished; margin denticulate, posteriorly slightly emarginate. Size: 1½ inches.

Localities: Port Phillip, Western Port.

Of similar sculpture to the last species, but easily separated by its flatter shape and cicatrized apex. Occasionally washed up on the beaches. In living condition found adhering to stones, dredged off Half Moon Bay and Western Port. It occurs also in Tasmania, South Australia and Western Australia.

Notomella Cotton, 1957

Shell oval, conical, depressed, apex posterior, turned backwards; radial riblets alternately stronger, latticed by concentric ribs; fissure long and narrow with the filling of its continuation differentiated from the other ribs.

TYPE SPECIES *Emarginula candida* A. Adams, 1851.

The Indo-Pacific Slit-Limpets were originally placed in *Emarginula,* a genus erected for the Mediterranean species *conica* Lamarck, which is taller than its base diameter with a recurved apex. Cotton, 1945, erected the name *Entomella* for the depressed Indo-Pacific groups but as the name was preoccupied by Cossman, 1888, he later altered it to *Notomella.*

Delicate Slit-Limpet

Notomella dilecta (A. Adams, 1851)

(Figure 50)

Shell small, creamy white, oblong-ovate, depressedly conical, arched from apex to anterior end; apex situated about one-quarter the length of shell from posterior margin. Surface of shell cancellated by radial and concentric ridges, the former being slightly stronger. Interior white, margin crenulated. Anterior end with a narrow slit about one-third the distance from anterior edge to apex, the slit continuous with a furrow which extends as far as the apex. Size: ½ inch.

Localities: Port Phillip; Western Port; Lorne; Portland.

This small, ornate species is found in shallow water in New South Wales and south-western Australia as well as Victoria where it is of infrequent occurrence.

amitina (Iredale, 1925); of similar sculpture but a high-arched shell with the apex closer to the posterior margin, is dredged in deep water off Wilson's Promontory.

candida (A. Adams, 1851); perhaps commoner on the Victorian coast than *dilecta* and is distinguished from it by its coarser sculpture and the raised rib covering the oval furrow.

bajula (Hedley, 1913); a smaller, flatter, more delicately sculptured species than any of the above, with a furrow running between the fissure and the apex.

superba (Hedley, 1906); off Wilson's Promontory.

gabensis Gabriel, 1961; off Gabo Island. This is the largest known species of the genus measuring ¾ inch.

Montfortula Iredale, 1915

Shell ovate, conical, elevated; apex slightly posterior, riblets strong and sculptured by scales or nodules giving them a rough texture. Notch very short.

Type Species *Emarginula rugosa* Quoy & Gaimard, 1834.

Rugose Slit-Limpet

Montfortula rugosa (Quoy & Gaimard, 1834)

(Figure 51)

Shell small, ovate, conical; apex almost central, slightly inclined posteriorly; colour greyish brown, worn examples almost white. Sculptured with about twelve to fifteen fairly prominent radiating ribs, the intervals being further ornamented with two to three riblets, crossed by close concentric ridges, cancellating the whole surface, the point of crossing being somewhat granulated. Interior white or greenish; muscle scar outlined with olive green; margin crenulated. Anterior margin with a small slit which is continuous with a distinct groove extending as far as the vertex. Size: ¾ inch.

Localities: Portland; Lorne; Port Phillip; Western Port; Kilcunda.

A very common species on the Victorian coast, located on rocks or amongst *Galeolaria* at low tide. In the interior coloration much variation exists, some examples being white while more often they are more or less green throughout. Easily identified by its cancellated sculpture and small anterior slit. Recorded also from Queensland, New South Wales through Bass Strait to South Australia and Western Australia.

The eastern shells have been separated from the Western Australian as *M. conoidea* (Reeve, 1842) but examination of a series suggests that differences in shape and texture are not constant but differ with environment. Therefore, we prefer to consider *M. conoidea* as a synonym of the above species.

Hemitoma Swainson, 1840

Shell high, cap-shaped, sculpture consists of numerous, knotted, radial ribs crossed by raised irregular encircling growth-lines. Apex subcentral and directed to the posterior, no fissure but anterior margin faintly notched with a strong rib running from the notch to the apex.

TYPE SPECIES *Patella octoradiata* Gmelin, 1891 = *Emarginula tricostalis* Swainson, 1840.

Emarginate Slit-Limpet

Hemitoma subemarginata (Blainville, 1819)

(Figure 52)

Shell creamy, or light brown, solid ovate, conical; apex not quite central, slightly inclined posteriorly. Sculptured with sixteen to eighteen irregularly spaced, fairly prominent, radiating ribs, the intervals further adorned with one, two or three radial riblets, crossed by numerous, close, inequidistant concentric ridges, cancellating the whole surface, the three anterior ribs being much stronger than the remainder. Interior white or pale green, margin sharp, denticulate within. Anterior edge with a small notch or slit continued upwards as an inside groove to the vertex. Size: 1 inch.

Localities: Portland; Kilcunda; Anderson's Inlet; Western Port.

This is the largest of our Slit-Limpets immediately separated by its bolder sculpture, particularly in the three anterior radial ribs. Consistency of shape is not apparent, some individuals being much higher than others and often with the apex much nearer to the posterior margin. Worn examples are occasionally washed up on the beaches, but at the last-named locality, in 5-8 fathoms between Phillip and French Islands, they are dredged alive holding tenaciously on to stones and other objects. The shell is frequently much eroded and often covered with a species of *Lithothamnion*.

Recorded also from Tasmania, South Australia and Western Australia, it was known for many years as *Hemitoma emarginata* (Blainville, 1825).

Vacerrana Iredale, 1958 = *Vacerra* Iredale, 1924 preoccupied (*Puncturella demissa* Hedley, 1904).

demissa (Hedley, 1904); between Phillip and French islands, Western Port.

harrisoni (Beddome, 1882); Western Port.

Amblychilepas Pilsbry, 1890

Shell saddle-shaped, elevated at each end, margin thickened, blunt. Perforation oblong, oval, subcentral.

Figure 51

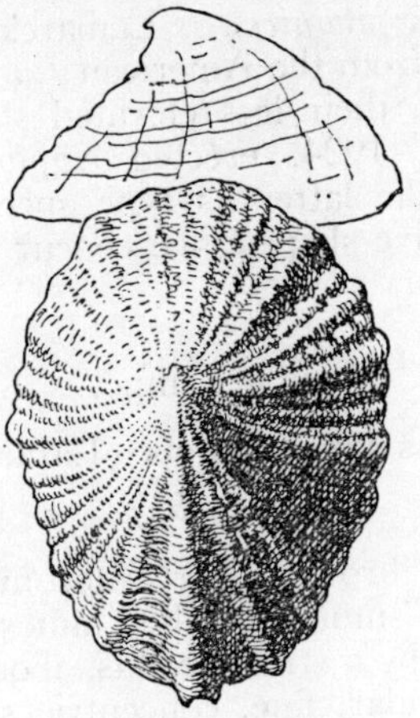

Figure 52

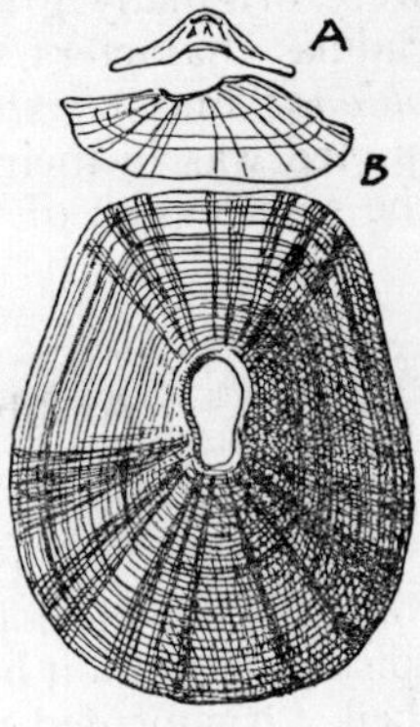

Figure 53

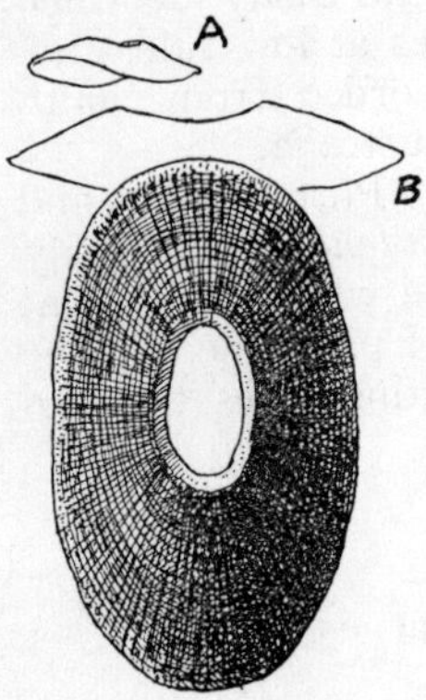

Figure 54

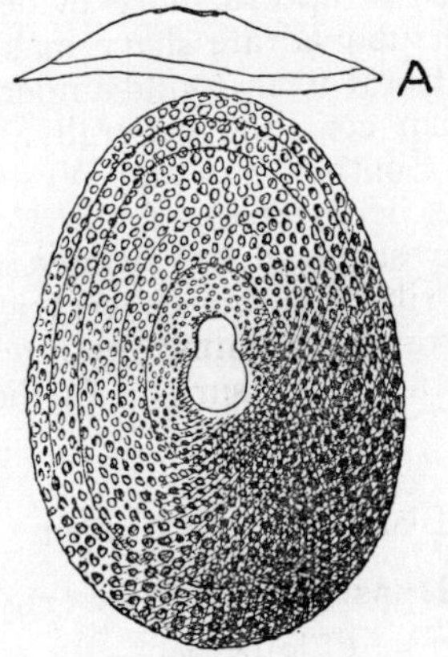

Figure 55

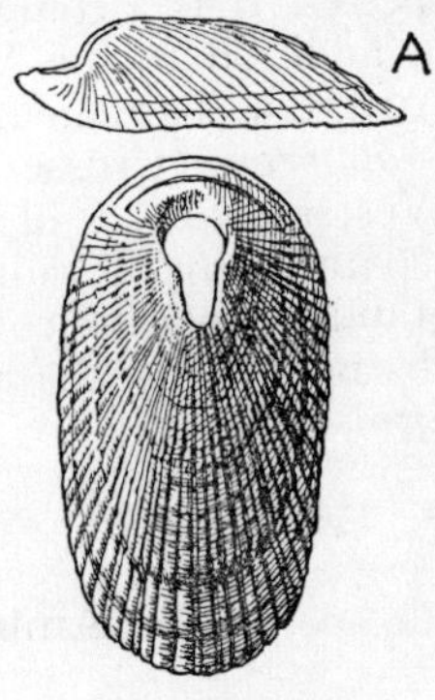

Figure 56

51. *Montfortula rugosa* $\times 2$
A Lateral elevation
52. *Hemitoma subemarginata* $\times 1\frac{1}{3}$
A Lateral elevation
53. *Amblychilepas javanicensis* $\times 1\frac{1}{3}$
A Anterior elevation
B Lateral elevation
54. *Amblychilepas nigrita* $\times 1\frac{1}{3}$
A Anterior elevation
B Lateral elevation
55. *Cosmetalepas concatenatus* $\times 3$
A Lateral elevation
56. *Macroschisma tasmaniae* $\times 1$
A Lateral elevation

TYPE SPECIES *Fissurella javanicensis* Lamarck, 1822.

Australian species differ from the American *Lucapinella,* with which they were originally grouped, in their less elevated shape, finer sculpture and radula characters. Iredale, 1924, erected *Sophismalepas* for *Fissurella oblonga* and *nigrita* with the latter as type species. The only difference quoted was in their respective shapes; we concur with Ludbrook, 1956, in the suppression of this genus.

Keyhole Limpet

Amblychilepas javanicensis (Lamarck, 1822)

(Figure 53)

Shell fawn with rays of brown, short subconical, subquadrate, angles rounded, posterior broader; summit of shell almost central with an oblong apical perforation bordered by a strong callus, about one-fifth the length of shell. Ornamented with regular, fine, concentric striae, a little more conspicuous than the radiating sculpture. Size: 1 inch.

Localities: Sorrento; Portsea; Point Nepean; common in Western Port.

The beautiful decussated sculpture more evident near the summit, the characteristic rays and the subquadrate shape make this an easily identified species. It is a common littoral form found under stones at low tide, and in addition to the Victorian coastline generally, is recorded from South Australia, Tasmania, New South Wales and Western Australia.

omicron (Crosse & Fischer, 1864); from Portsea, Port Phillip and Western Port, is of similar shape to *Cosmetalepas concatenatus* but easily separated by its radiating ribs and fine concentric striae with a somewhat granular appearance towards the summit; the interior of pale blue and the absence of the characteristic pitted surface further distinguishes the two forms.

Black Keyhole Limpet

Amblychilepas nigrita (Sowerby, 1834)

(Figure 54)

Shell brown with irregular rays of darker brown, solid, oblong-ovate, depressed, anterior and posterior edges distinctly raised. Perforation not quite central, situated slightly posteriorly. Sculpture consisting of fairly regular, radiating ribs crossed by close, irregular, concentric striae, a somewhat latticed surface resulting. Interior white or bluish; margin strong; perforation surrounded by a narrow callus rim. Size ¾ inch.

Localities: Anderson's Inlet; Kilcunda; Western Port; Lorne.

A common Keyhole Limpet located on the under surface of stones; the shell, partly covering a rather pretty, pinkish animal, should be easily identified by its oblong-oval shape and strong latticed ornament.

It occurs at Kelso Bay, Tasmania, and in the waters of Queensland, New South Wales, South Australia and Western Australia. The name *nigrita,* which conveys that the shell is black, is perhaps ill-chosen, as examples of this colour are unknown to the writers. Much variation exists in this respect, shades of fawn, light and dark grey predominating.

oblonga (Menke, 1834); its nearest ally. It may be distinguished by its broader contour and proportionately larger and wider perforation.

crucis (Beddome, 1882); a third and much rarer form, appears occasionally at Portsea, Port Phillip; it is a charming and beautifully coloured species of lesser dimensions, with the perforation centrally situated, the original description reading thus 'oval, raised and cancellated; white or yellowish; two red lines on back, forming a cross; aperture oval'.

Cosmetalepas Iredale, 1924

Shell thin, oval depressed, without radiating ribs, ornamented with small rounded pits. Perforation central, large, keyhole shaped.

TYPE SPECIES *Fissurella concatenatus* Crosse & Fischer, 1864.

Pilsbry, 1890, originally grouped all the keyhole limpets in *Megatebennus* but this genus is now reserved for the American species which differ in shape and sculpture.

Pitted Keyhole Limpet

Cosmetalepas concatenatus (Crosse & Fischer, 1864)

(Figure 55)

Shell cream, thin, small, ovate, depressed, possessing a striking ornamentation of rounded and oval pits all over the surface giving it a somewhat honeycombed appearance. Perforation large, oblong, almost centrally situated, barely one-sixth the length of shell. Interior white, yellow, occasionally with narrow brown streaks towards the perforation which is bordered with a narrow, white callus rim. Size: ½ inch.

Localities: Port Phillip; Anderson's Inlet; Kilcunda; Western Port (dredged 5 to 8 fathoms).

This species is readily separated from *Amblychilepas javanicensis* (Lamarck, 1822) by its more oval form, the absence of brown colour-rays and the presence of the honeycombed surface. The species enjoys a rather wide distribution from New South Wales, Tasmania, through Bass Strait to South Australia and Western Australia.

Macroschisma Sowerby, 1835

Shell oblong, with apex, which is near the posterior margin, wholly removed by a large keyhole-shaped fissure, the wider end close to the posterior margin of the shell.

TYPE SPECIES *Macroschisma macroschisma* Sowerby, 1835 = *Macroschisma hiatula* Swainson, 1840.

There has been confusion over the author and genotype of *Macroschisma*. Sowerby erected the genus, in *Conchological Illustrations* 1835 (see Shaw, 1909), for a New South Wales shell, thus predating Swainson, 1840, the usually accepted author. Sowerby referred this shell to *Fissurella macroschisma* Humphrey, a Japanese species which does not occur in south-eastern Australia. The next available name for this misidentification is *M. hiatula* erected by Swainson for Sowerby's shell.

Because Sowerby, 1866, recognized the only two *Macroschisma* known from southern Australia as distinct from his original New South Wales species, the identity of this latter species has been in doubt. However, in 1940 Iredale described a *Macroschisma* from Lord Howe Island, part of the territory of New South Wales, which in shape and colouring is identical with Sowerby's shell. We therefore submit that *Macroschisma enopa* Iredale is a synonym of *M. hiatula* Swainson.

Iredale does not consider the southern Australian species congeneric with *Macroschisma* and he proposed two new genera *Forolepas* and *Dolichoschisma* for *M. tasmaniae* Sowerby and *M. producta* Adams respectively. The present authors do not consider this is warranted and have retained *Macroschisma* for the Victorian species.

Tasmanian Keyhole Limpet

Macroschisma tasmaniae Sowerby, 1866

(Figure 56)

Shell creamy white, fairly strong, oblong, length almost double the breadth, height about one-fourth the length. Perforation wedge-shaped, posteriorly situated, internally surrounded by a thick callus; about one-fifth the length of shell. Sculpture consists of radiate ribbing made irregular by growth-lines; ribs alternately large and small, more obvious at the anterior end. The outer margin curves upward posteriorly and is more blunt and more broadly rounded than the anterior extremity. Size: 1 inch.

Localities: Portsea; Western Port.

Variety of colour is noticeable, shades of yellow and brown being frequently seen. The shell is obtainable on the whole Victorian coastline and extends its range from southern New South Wales to Western Australia. Found alive under rocks at low tide.

Elongated Keyhole Limpet

Macroschisma producta A. Adams, 1850

(Figure 57)

Shell white variegated with light brown; oblong, height about one-quarter the length of shell. Perforation approximately one-quarter the length of shell, narrowly wedge-shaped, situated close to the elevated posterior edge; anterior narrow, produced. Surface of shell cancellated by flatly rounded ribs and numerous, irregular concentric striae. Interior white; margin feebly crenulate; margin of perforation thickened within. Size: 1 inch.

Localities: Sorrento; Lorne; Western Port.

A much rarer species than *tasmaniae* Sowerby from which it may be distinguished by its more elongate shape, finer sculpture and a much narrower wedge-shaped perforation. Also recorded from South Australia, Tasmania and New South Wales.

Elegidion Iredale, 1924

Fissurella with an ovate, conical shell, with bold radiating ribs and encircling rings latticing the ribs. Perforation replaces the apex, subcentral on the anterior slope. The internal perforation callus not truncated posteriorly.

TYPE SPECIES *Elegidion audax* Iredale, 1924.

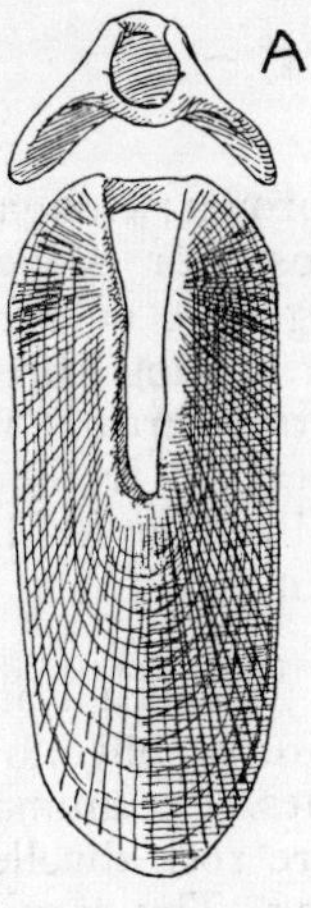

Figure 57

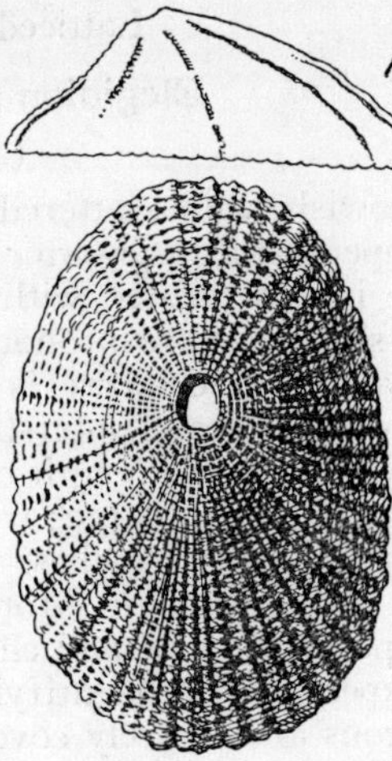

Figure 58

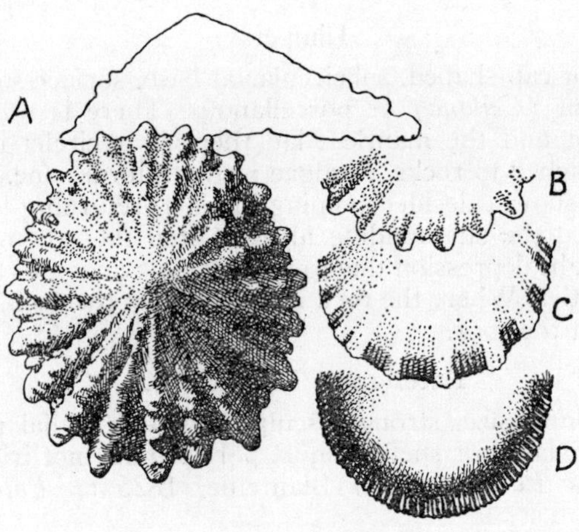

Figure 59

57. *Macroschisma producta* × 2
A Anterior elevation

58. *Elegidion audax* × 1
A Posterior elevation

59. *Patellanax peroni* × 1⅓
A Lateral elevation B,C,D Varieties of marginal ribbing

Latticed Keyhole Limpet

Elegidion audax Iredale, 1924

(Figure 58)

Shell yellowish, with scattered blotches of dark brown; elongately oval, conical, limpet-shaped, anterior slope straight, posterior slope convex. The surface is ornamented with numerous radiating ribs, of which about twenty are slightly more pronounced, the intervals bearing three smaller ribs; this radiating sculpture is crossed by concentric threads which are continuous over ribs and interstices. Perforation oval, in front of centre, about one-twelfth of the length of shell, and internally bounded by a well-defined callus. Aperture large; basal margin level, not elevated at the ends, edge crenulated. Size: 2 inches.

Localities: Sandringham; Mornington; Sorrento; Western Port.

With its peculiar shape and characteristic latticed ornament, no difficulty should be experienced in identifying this ornate species. The animals, which are herbivorous and entirely covered by the shell, are rock dwellers, being obtained under stones at low water, and by dredging. The species is also recorded from New South Wales.

This shell was for many years mistaken for another species *Diodora lineata* (Sowerby, 1841).

Family PATELLIDAE

Limpets

Shell conical or cap-shaped, subcircular at base, surface smooth or ribbed, interior iridescent (*Cellana*) or porcellanous. There is no gill plume but between the foot and the mantle edge there is a circlet of gill lamellae. Animals live attached to rocks, or algae in the littoral zone.

Many species show a decided homing instinct, returning to the same spot after browsing on the surrounding algae. Often these home areas become eroded into a slight depression due to the abrasive action of the shell margin as the animal settles. Where the rock is harder than the shell, the latter takes on the contour of the rock.

Patellanax Iredale, 1924

Shells of medium size, strongly sculptured with radial ribs which may bear scales, inner layer of shell opaque, porcellanous not iridescent.

TYPE SPECIES *Patella peroni* Blainville, 1825 = *Patella squamifera* Reeve, 1855.

Scaly Limpet

Patellanax peroni (Blainville, 1825)

(Figure 59)

Shell large, fairly thick and solid, ovately rounded, tumidly conical; apex situated at about the anterior third; colour dull white tinged with brown or black. Sculpture extremely variable, consisting normally of about twenty-five irregular, conspicuous, radiating ribs which are more or less prickly scaled or beaded and lighter in colour. Interior bluish white, orange tinted near the apex; margin dentate. Size: 1½ inches.

Localities: Port Phillip; Sorrento; Anderson's Inlet; Western Port; Lorne; Apollo Bay.

This very variable shell has been given a number of names. However, the grading of a large series of shells from a single locality shows that all the forms are present, and that intermediates make them impossible to separate into distinct species. At one end of the series is the form known as *Patella squamifera* Reeve with twenty-four large white ribs interspaced with smaller yellow ones. At the other end is the shell Verco called *Patella hepatica* which has jet black uniform radials. The shell Blainville described as *Patella peroni* was between these two extremes and it is fitting that, because it is the oldest, his name is retained for the species.

These shells are very common on the exposed rock platforms of the lower littoral (Pyura) zone and descending to the sublittoral fringe among the holdfasts of the giant kelp *Durvillea potatorum* of New South Wales and Victoria. In Tasmania, South Australia and Western Australia the species occurs in the corresponding algal zone. It is often encrusted with Bryozoa, *Lithothamnion* and seaweed.

Chapman's Limpet

Patellanax champmani (Tenison Woods, 1876)

(Figure 60)

Shell ovate, somewhat broad behind; reddish or scorched and nebulously brown; apex acute, submedian; with eight radiating ribs more or less valid, and depressedly rounded, profusely radiate with very fine lirae, and girdled with irregular sulci; margin angulate, nodulose; white within and clouded pale rose; spatula scarcely visible. Size: ¾ inch.

Locality: Western Port; Port Phillip.

This shell shows considerable variation in form, due to environment. Those on surf-beaten rocks are flat and eight ribbed, while those living in more sheltered situations are taller with the eight ribs far less prominent. The variations have been given a number of names including *Patella perplexa* Pilsbry, 1891, and *Patella alba* Tenison Woods, 1876.

Rare in Victoria, this species is recorded also from Tasmania, New South Wales and South Australia. It originally bore the name *Patella octoradiata* Hutton, 1873, which is preoccupied.

Cellana H. Adams, 1869

Shell solid, apex subcentral, anterior, sculpture of smooth, rounded ribs. Interior translucent with an iridescent nacreous layer.

TYPE SPECIES *Cellana cernica* H. Adams, 1869.

Orange-edged Limpet

Cellana solida (Blainville, 1825)

(Figure 61)

Shell large, grey, orange, but variable in colour with a distinct orange red margin; conoidal, apex subcentral inclining towards the anterior margin. Surface generally much eroded. Sculptured with about twenty-five fairly bold radiating ribs with concentric growth-lines plainly visible. Interior

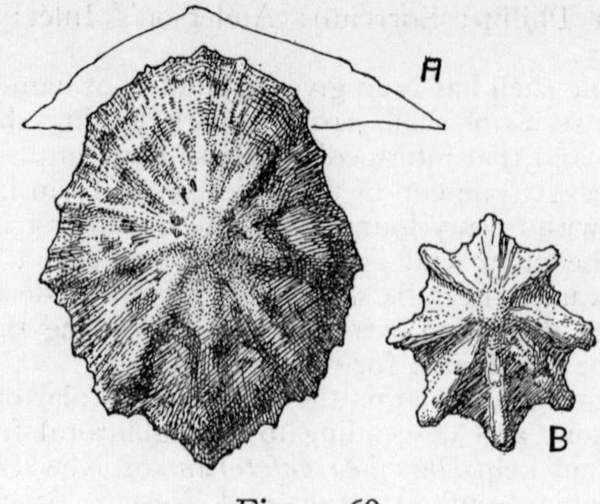

Figure 60

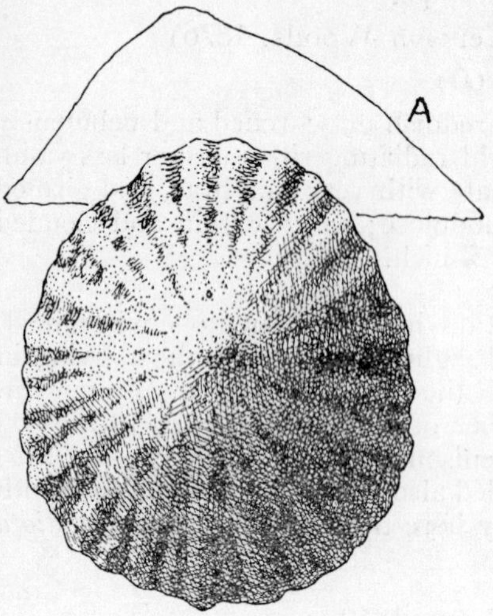

Figure 61

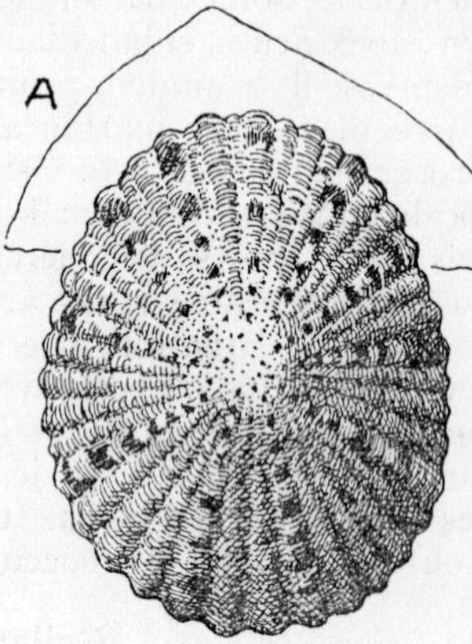

Figure 62

60. *Patellanax chapmani* $\times 2$
    A Lateral elevation
    B Asteroid type $\times 1$
61. *Cellana solida* $\times \frac{2}{3}$
    A Lateral elevation
62. *Cellana tramoserica* $\times 1$
    A Lateral elevation

smooth, muscular impression blue, margin orange and blotched with reddish brown. Aperture ovate. Size: 2 inches.

Localities: Cape Otway; Wilson's Promontory and the off-shore islands.

Its nearest ally in Victoria is *tramoserica* (Sowerby) from which it may be separated by its broader ribs and very definite orange red margin. The species is recorded from South Australia and Tasmania where it grows to a large size. For many years known as *Patella limbata* Philippi, 1849, or *Patella rubroaurantiaca* Blainville, 1825.

Variegated Limpet

Cellana tramoserica (Sowerby, 1825)

(Figure 62)

Shell thick, large, conical, slightly broader posteriorly, apex subcentral. Sculpture approximately thirty-six strong, subcarinate, radiating ribs, and numerous fine, closely set encircling growth-lines; margin sharp, scalloped. The colour is very variable, and may be orange or pink, patterned with brown and white, or brown, patterned with yellow, orange and white; often every third or fourth rib is darker, giving a striped appearance, and in some, the interspaces between the ribs are streaked with dark brown or black. Interior iridescent grey or yellow with the darker exterior colour showing through, particularly near the margin. Spatula white or grey. Size: 2 inches.

Localities: St Kilda; Frankston; Mornington; Point Nepean; Queenscliff; common on entire coast.

This is one of our commonest shells The animal, to a slight degree, is capable of excavating a cavity for itself, more noticeable in some rocks than others. Consistency in shape and colouring is not apparent. Some examples possess a more rounded contour while others are definitely of higher elevation.

Located on exposed rock surfaces in the mid-littoral zone but extending into both the lower and upper littoral zones of eastern South Australia, Victoria, New South Wales and southern Queensland. It is also recorded from the east coast of Tasmania (May, 1923). This shell was for many years given the name *Cellana variegata* (Blainville, 1825) but Sowerby's name is the older and must come into use.

Family ACMAEIDAE

Acmaeid Limpets

Shell conical, usually depressed oval, never iridescent within. The gill plume lies across the head under the mantle. Animals of the shore region living attached to rocks, algae, or other animals.

Patelloida Quoy & Gaimard, 1834

Shell with regular prominent ribs, colour predominantly white, the spatula and interior margin dark.

TYPE SPECIES *Patelloida rugosa* Quoy & Gaimard, 1834.

Tall-ribbed Limpet

Patelloida alticostata (Angas, 1865)

(Figure 63)

Shell conical, solid; approximately eighteen radiating ribs, subcarinate; apex slightly anterior and usually eroded; margin thin, scalloped. Colour white or grey green due to algal growth which seems to attack this species very frequently. Interspaces betwen the ribs marked with fine black cross-lines which are characteristic of this species. Interior porcellanous, white, spatula grey and white, with the strong muscle scar tinged with brown. Margin light brown with the black of the exterior showing through on the indentations. Size: 1½ inches.

Localities: Sandringham; Mornington; Point Nepean; Western Port; common on entire coast.

This species is easily distinguished from other species by the characteristic sharp ribbing and the fine black cross-lines in the interspaces.

Lives on exposed rock surfaces in the lower littoral. Shells are usually eroded and often partially covered with algal growth. Occurs from Geraldton, Western Australia, to southern Queensland and Tasmania.

Lateral-striped Limpet

Patelloida latistrigata (Angas, 1865)

(Figure 64)

Shell small, brown, ovate, conical, but showing variation in this respect, some examples being much depressed and considerably eroded; apex, situated at about the anterior third. Sculpture consisting of twelve to fifteen irregular, rounded ribs some stronger than others. Interior bluish, margin dark brown, extremities of ribs white; spatula very clearly defined, generally light brown spotted or streaked with dark or chocolate brown and bordered white. Size: ½ inch.

Localities: Port Phillip; Western Port; Lorne; Portland.

This is a distinctive little species widely distributed throughout the whole Victorian coastline and easily recognized by its prominent radiating ribs and the characteristic spotted, white bordered spatula. In living condition it is located on exposed flat rock surfaces at mid- and upper-tide level, in association with *Notoacmea granulosa* Macpherson, *Siphonaria diemenensis* Quoy & Gaimard, and *S. funiculata* Reeve. Distribution Tasmania, South Australia.

A variety *submarmorata* (Pilsbry, 1891); common in Port Jackson with a range as far south as Mallacoota and Tasmania possesses rather narrow riblets in place of the few wide prominent ribs of the typical form.

Victorian Liver-coloured Limpet

Patelloida victoriana (Singleton, 1937)

(Figure 65)

Shell dark or greenish brown, moderately elevated; apex obtuse situated at about the anterior third. Sculptured by 100-120 irregular, flatly rounded,

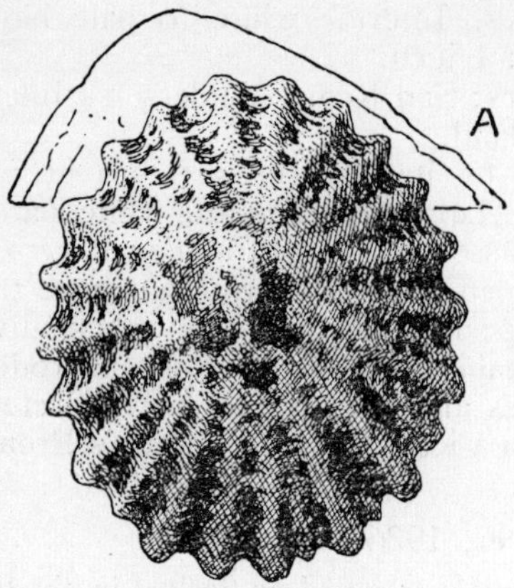

Figure 63

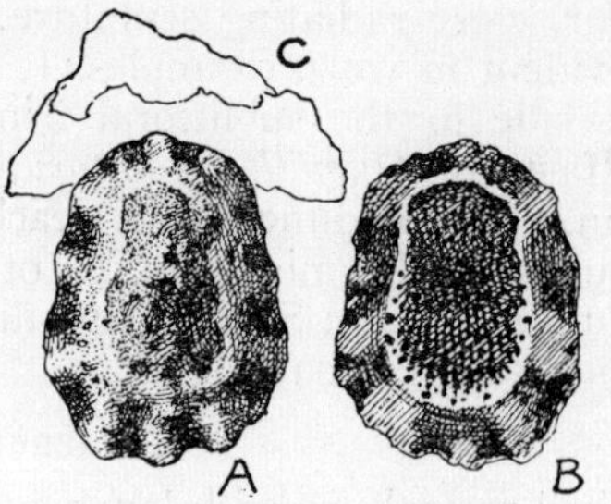

Figure 64

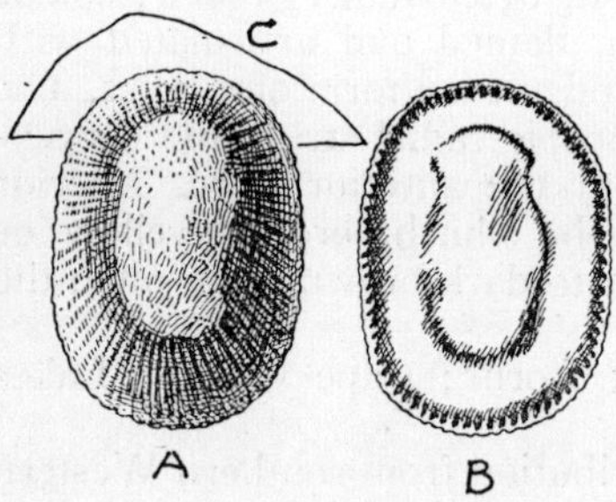

Figure 65

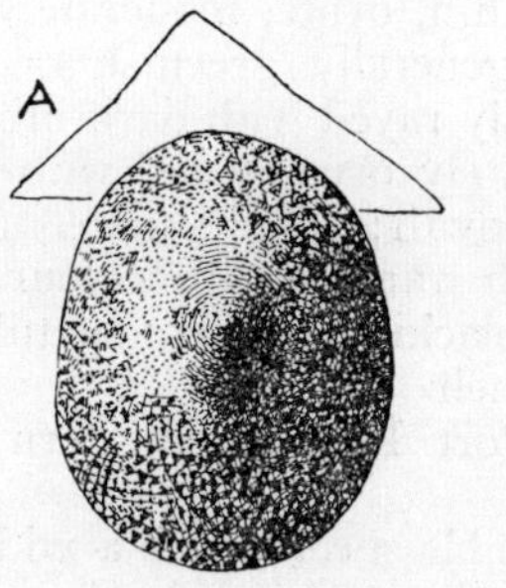

Figure 66

Figure 67

63. *Patelloida alticostata* × 1
A Lateral elevation

64. *Patelloida latistrigata* × 1½
A Dorsal view B Ventral view
C Lateral elevation

65. *Patelloida victoriana* × 1⅓
A Dorsal view B Ventral view
C Lateral elevation

66. *Chiazacmea flammea* × 2
A Lateral elevation

67. *Naccula parva* × 8

radial riblets crossed by fine growth-lines. Interior white to pale blue, margin brown and finely crenulated. Size: 1 inch.

Localities: Sorrento back beach; Flinders; San Remo; Anderson's Inlet; Lorne; Apollo Bay; Lady Julia Percy Island.

This shell should be easily recognized by its dark or greenish brown colour, close radiating sculpture, and the characteristic apical white patch so evident in worn examples. It is found associated with *Patellanax peroni* Blainville in the sublittoral fringe, on open rock platforms amongst the holdfasts of *Durvillea potatorum*. This species is hardly ever taken alive because of the difficulty of reaching its habitat and because when eroded it can easily be mistaken for other species and is therefore not collected. Occurs in eastern South Australia, western Victoria east to Wilson's Promontory, and Tasmania.

Chiazacmea Oliver, 1926

Shell smooth, or with low rounded ribs; colour rayed or flecked in shades of brown, a cross-pattern usually evident; spatula dark.

TYPE SPECIES *Patelloida flammea* Quoy & Gaimard, 1834.

Flamed Limpet

Chiazacmea flammea (Quoy & Gaimard, 1834)

(Figure 66)

Shell small, thin, ovate, moderately conical, often much eroded; colour very variable, generally greenish or cream, flamed and articulated with brown, frequently rayed with dark brown bands in the form of a cross. The surface is delicately ornamented with microscopic radial striae and concentric lines of growth; apex situated at about the anterior third. Interior exhibiting much variation in colour; may be bluish, brown, yellow or creamy with a blackish, border; spatula light to dark brown more often the latter. Size: ½ inch.

Localities: Port Phillip; Western Port; Lorne; Cape Otway; Lakes Entrance.

This is a variable species with a wide distribution from southern Western Australia to southern Queensland and has been divided into a number of subspecies. These latter have not been clearly marked out in terms of modern nomenclature so for the present only one species is recognized here. To add to the confusion another limpet *Notoacmea granulosa* Macpherson, 1954 (p. 51) has been mistaken for this species.

Naccula Iredale, 1924

Shell small, hyaline with anterior apex.

TYPE SPECIES *Patelloida punctata* Quoy & Gaimard, 1834.

Members of this genus were originally included in *Nacella*, a group of Antarctic limpets belonging to the Family Patellidae.

Spotted Limpet

Naccula parva (Angas, 1876)

(Figure 67)

Shell, small, smooth but with fine concentric growth-lines seen only with the aid of a lens; depressedly conical, somewhat cap-shaped; cream with a single row of pale blue spots and crescent shaped opaque markings extending from the apex centrally, more or less along the outer arc of the shell; apex recurved anteriorly, sometimes beyond the margin. Aperture oblong-ovate. Interior white, shining. Size: ¼ inch.

Locality: Portland.

A very striking little species easily identified by the dorsal line of delicate blue spots. In South Australia found living on the seaweed *Cymodocea antarctica* associated with *Asteracmea stowae* Verco, 1906.

Actinoleuca Oliver, 1926

High, conical shells with straight sides and subcentral apex; finely and evenly ribbed, porcellanous throughout.

TYPE SPECIES *Patella campbelli* Filhol, 1880.

Delicate Limpet

Actinoleuca calamus (Crosse & Fischer, 1864)

(Figure 68)

Shell small, ovately rounded; colour dull white or yellow with irregular brown or dark brown markings; slopes almost straight. Interior white, faintly pinkish or purple tinted, with a brown spot at the apex and reddish brown blotches towards the margin. Surface sculptured with numerous, unequal radiating riblets. Size: ½ inch.

Localities: Port Phillip; Western Port.

A very distinctive little species easily recognized by its size, erect, conical form and delicate radial ornament. It is occasionally washed up on the shores, and is dredged alive attached to stones, fragments of oysters, mussels and other bivalves. The shells are often encrusted, sometimes entirely, with a species of *Lithothamnion*. Occurs in New South Wales, Tasmania, through Bass Strait to South Australia and Western Australia.

Notoacmea Iredale, 1915

Shell usually thin; apex usually anterior, the posterior slope convex; usually rayed with yellow or green on a darker ground. Interior light coloured with a dark margin.

TYPE SPECIES *Patelloida pileopsis* Quoy & Gaimard, 1834.

Granulated Limpet

Notoacmea granulosa Macpherson, 1954

(Figure 69)

Shell thick, conical; high apex slightly anterior; anterior slope steep and almost straight; posterior slope convex; surface granulose, with faint radiating ribs. Colour grey, flecked with darker grey and brown; the flecks near the margin may be elongated into short stripes that show through to the inside. Spatula dark blue or brown, region between spatula and margin lighter in colour. Usually superimposed on the interior pattern is a bluish black Maltese cross which may show through to the exterior. Size: ½ inch.

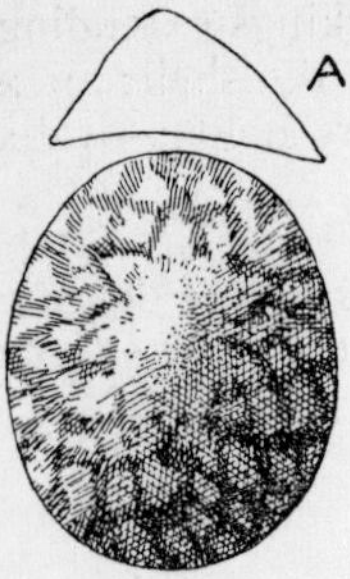

Figure 68

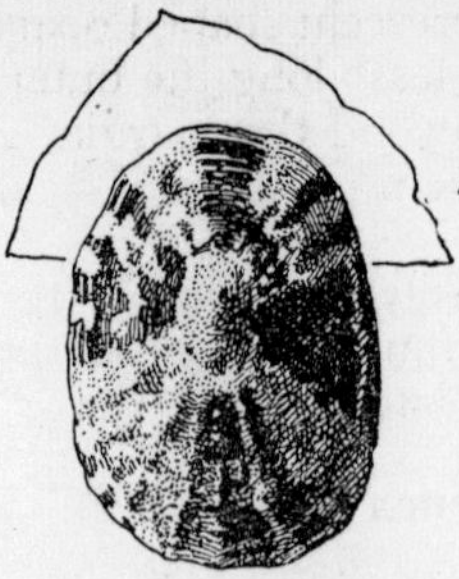

Figure 69

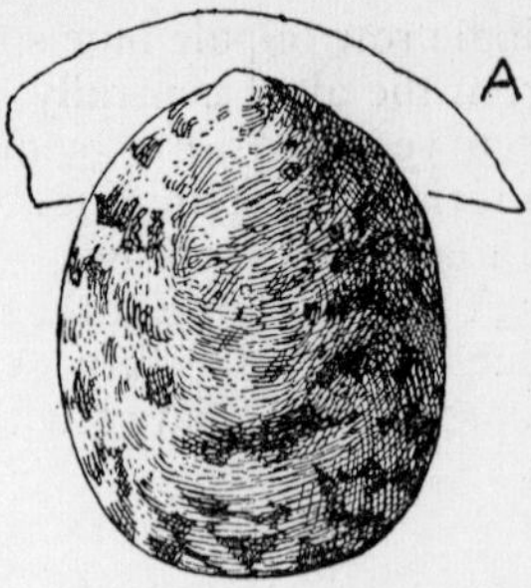

Figure 70

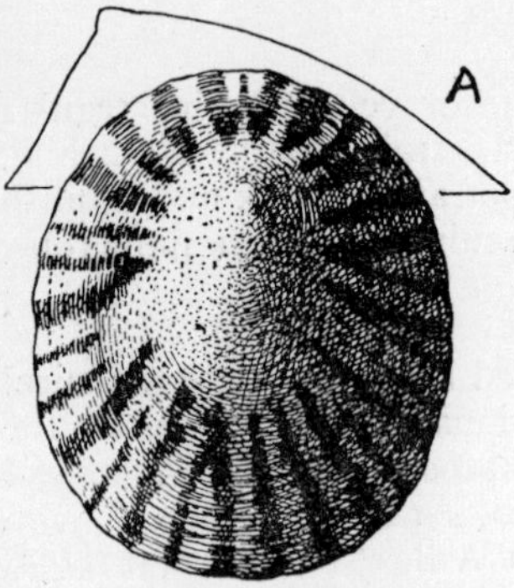

Figure 71

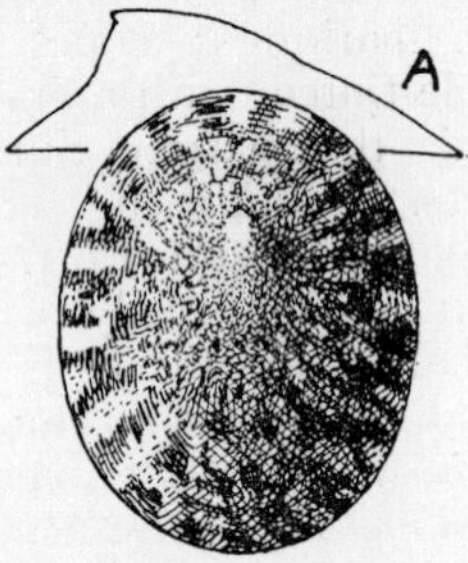

Figure 72

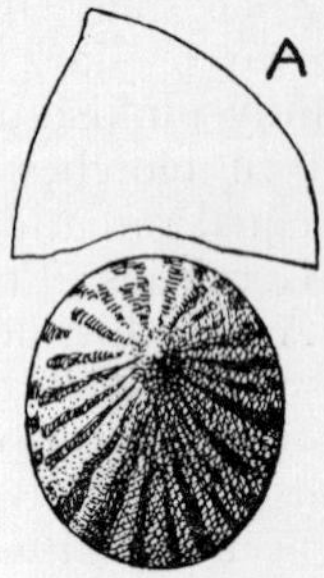

Figure 73

68. *Actinoleuca calamus* × 2
A Lateral elevation
69. *Notoacmea granulosa* × 2
A Lateral elevation
70. *Notoacmea mayi* × 2
A Lateral elevation
71. *Notoacmea petterdi* × 2
A Lateral elevation
72. *Notoacmea scabrilirata* × 2
A Lateral elevation
73. *Notoacmea alta* × 3
A Lateral elevation

Localities: Warrnambool; Cape Otway; Lorne; Phillip Island.

This limpet, which is found in the upper littoral of exposed ocean platforms, has been confused with *Chiazacmea flammea* (or one of its subspecies) in most Victorian collections. The confusion has probably arisen because the present species may carry a Maltese cross which is very similar to that of some forms of *Chiazacmea flammea.* Examination of the radula, which is of the *Notoacmea* type and intermediate in form between *N. mayi* and *N. scabrilirata,* separates it from *Chiazacmea.* Also this is an inhabitant of the upper littoral of ocean coasts, whereas *C. flammea* is only found in the quiet bays and inlets near low-tide mark. Occurs on vertical surfaces of exposed rock faces in the upper littoral zone of ocean platforms, from Cape Bridgewater to Wilson's Promontory, Victoria, and at Kangaroo Island, South Australia.

May's Beetle Limpet

Notoacmea mayi May, 1923

(Figure 70)

Shell small, ovate, rather thin, somewhat depressed; anteriorly obtusely pointed; apex anterior extending beyond the margin and more or less corroded; colour reticulated brown and white showing much variation. From apex to posterior extremity convex. Interior blackish chestnut or dark brown with a lighter colour bordering the spatula and whitish markings at the margin. Size: ¾ inch.

Localities: Portland; Cape Bridgewater; Port Campbell; Sorrento; Phillip Island.

A curiously shaped limpet easily recognized by its flat and obtusely pointed form. The apex extends beyond the anterior margin and this is a consistent and characteristic feature. Common in western Victoria and as far east as Phillip Island, located on vertical rock faces in the upper littoral zone. Previously called *Acmaea cantharus* (Reeve, 1855), a New Zealand shell. Recorded also from Tasmania.

Petterd's Limpet

Notoacmea petterdi (Tenison Woods, 1876)

(Figure 71)

Shell small, broadly ovate, moderately conical; dull white with about thirty-five to forty light brown radiating bands; apex acute, eroded, situated at about the anterior fourth, posterior slope arched; concentric growth-lines plainly visible. Interior bluish white, margin thin, narrowly bordered with dark blue; spatula sharply defined, dark brown. Size: 1 inch.

Localities: Lorne; Pope's Eye, Port Phillip; Mallacoota.

A broadly ovate species, as the author remarks 'somewhat resembling an old and enlarged *septiformis* Quoy & Gaimard, 1834, but its size is larger than that species is ever known to attain'. Found on sloping and vertical rock faces in the upper littoral zone, usually occurring in small thickly populated colonies. Occurs in Tasmania, New South Wales and southern Queensland.

Scaly-lined Limpet

Notoacmea scabrilirata (Angas, 1865)

(Figure 72)

Shell small, oval, depressed; extremely variable in colour, may be yellowish, brown, greenish or in some cases creamy white or nearly black and frequently rayed with dark brown bands; apex situated at about the anterior fourth, and in occasional juvenile examples very forward almost at the anterior margin; posterior slope arched, anterior slope slightly concave. Surface ornamented with fine microscopic radiating striae, very often obsolete. Interior bluish white, margin acute; dark brown, spatula more or less brown. Size: ½ inch.

Localities: Port Phillip; Western Port; coast generally.

The depressed nature of this little shell with its fine radial microscopic striae and characteristic colouring are features which will identify it. In life, found under stones just above low-tide mark. Recorded also in Tasmania and South Australia.

Tall Limpet

Notoacmea alta Oliver, 1926

(Figure 73)

Shell small, fragile, much elevated; apex acute, situated at about the anterior third, anterior slope almost straight; colour dark brown or black with numerous light green bands radiating from the apex, but this is not a constant feature as some examples are almost bandless. Interior brown with radiating lighter coloured bands. Aperture ovate; margin sharp; basal plane arched. Size: ¼ inch.

Localities: Cape Bridgewater; Apollo Bay; Wilson's Promontory; Mallacoota.

Though small and often eroded to a uniform grey brown it is nevertheless easily identified by its extreme conical form. In living condition it is located at mid-tide level adhering to the mussel *Brachidontes rostratus* Dunker. The author described the species from a South Australian specimen.

Asteracmea Oliver, 1926 (*Helcioniscus illibrata* Verco, 1906).
illibrata (Verco, 1906); exact locality not recorded.
stowae (Verco, 1906); exact locality not recorded.

Family TROCHIDAE

Trochus or Top Shells

Shell spiral, conical, or pyramidal. The interior lined with a thick, iridescent, nacreous layer. The opening closed by a thin, horny operculum. Herbivorous sea snails generally found living amongst seaweed in shallow water.

Herpetopoma Pilsbry, 1889

Shell turbinate, conical, solid with a thick lip, spire elevated; operculum multi-spiral.

TYPE SPECIES *Euchelus scabriusculus* Angas, 1867.

Australian shells of this genus were originally included in *Euchelus* which has an operculum with few whorls while that of *Herpetopoma* is multi-spiral.

Pearled Euchelus or Top Shell

Herpetopoma aspersa (Philippi, 1846)

(Figure 74)

Shell cream, sparsely scattered with brownish dots; globose-conic. Spire elevated. Whorls rounded, ornamented with spiral ribs, about eight on the penultimate whorl, the ribs being closely beaded. Aperture rounded, lip crenulated inside, columella with a small tooth at the fore part. Size: ½ inch.

Localities: Point Nepean; Point Lonsdale; Western Port.

This shell, though small, is a beautifully ornate species and should prove easy to recognize. In shape and coarseness of the sculpture it exhibits slight variation, some examples being more elongated. Beyond this state it occurs in Western Australia, South Australia, New South Wales and Tasmania. Known for many years as *Euchelus baccatus* (Menke, 1843) which was preoccupied.

profundior (May, 1915); exact locality not recorded.
pumilio (Tate, 1893); San Remo.
scabriuscula (Angas, 1867); Port Phillip.
tasmanica (Tenison Woods, 1875); Cape Schanck; Western Port.
vixumbilicata (Tate, 1893); 7 fathoms off Phillip Island.

Stomatella Lamarck, 1816

Shell depressed. Aperture excessively oblique. Sculpture consisting of close, equal spiral cords.

TYPE SPECIES *Stomatella imbricata* Lamarck, 1822.

Cotton, 1957, states that Anton, 1839, designated *Stomatella auricula* Lamarck, 1818, as the type of *Stomatella* Lamarck, 1816. This would necessitate a new genus for *imbricata* Lamarck, 1822, and he proposed *Granata* for this purpose. However, we find on examining Anton that he actually lists four species under *Stomatella*; these he divides into two sections: (a) *imbricata* which he states are 'Sigaretus artige' and he lists *imbricata* Lamarck and *rubra* Lamarck here; (b) *lutea* which are 'Haliotisartige' and the species given are *nigra* and *lutea* = *auricula* Lamarck. Nowhere does he indicate a type species and the grouping suggests that he considered the *imbricata* group the primary section with the *lutea*-like species included. Therefore we consider the erection of *Granata* incorrect and believe it must be considered an absolute synonym of *Stomatella* Lamarck.

Imbricated Pearl Shell

Stomatella imbricata Lamarck, 1822

(Figure 75)

Shell creamy white, variegated with scattered dots of reddish brown; solid, ear-shaped. Depressed spire slightly elevated. Whorls rounded. Sculpture consisting of numerous, close, almost equally spaced, finely scaled, spiral riblets. Aperture large, entire, lined with a silvery and iridescent nacre; columella broad, slightly concave. Size: 1½ inches.

Figure 74

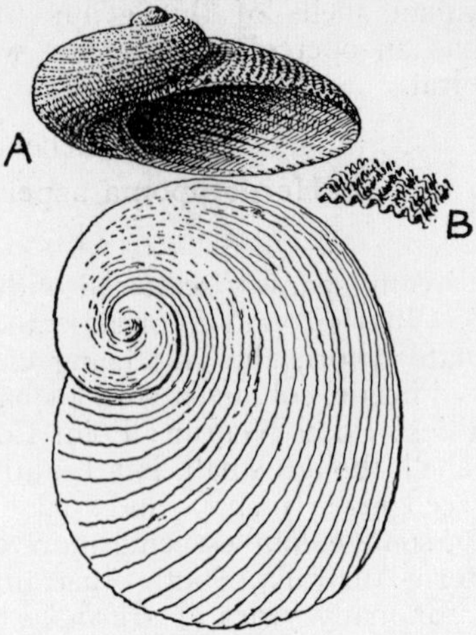

Figure 75

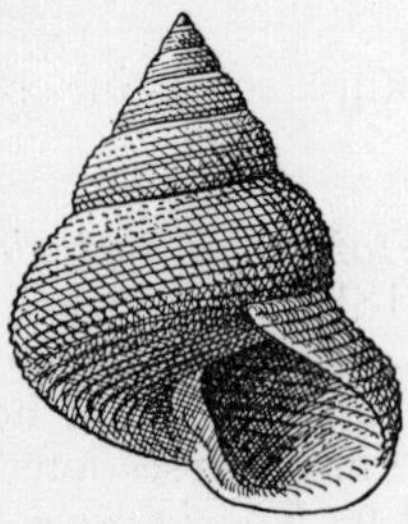

Figure 76

Figure 77

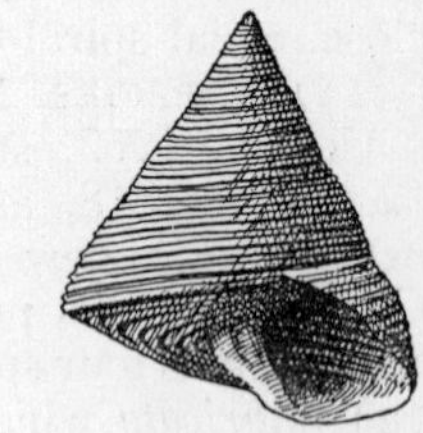

Figure 78

74. *Herpetopoma aspersa* × 2
75. *Stomatella imbricata* × 1½
    A Anterior elevation × 1
    B Sculpture (enlarged)
76. *Calliostoma allporti* × 2
77. *Calliostoma hedleyi* × 2
78. *Calliostoma legrandi* × 2

Localities: Mornington; Sorrento; Portsea; Queenscliff; Western Port. Found adhering to the under-surface of rocks; recorded from all the states of Australia.

Calliostoma Swainson, 1840

Shell imperforate conical, rather thin. Whorls smooth, spirally ridged or granular, the last angulated at the periphery. Aperture quadrangular; columella simple, usually ending anteriorly in a slight tooth.

TYPE SPECIES *Trochus conulus* Linné, 1767.

The Australian members of this genus have been divided into a number of groups which, for convenience, the present authors prefer to regard as subgenera.

Allport's Top Shell

Calliostoma (Fautor) allporti (Tenison Woods, 1875)

(Figure 76)

Shell small, cream, fairly solid, conical, imperforate. Whorls eight, convex, sculptured with ten to eleven concentric, evenly spaced, granulated riblets; granules rounded, separated by oblique, longitudinal striae; sutures well impressed. Body-whorl somewhat rounded at the periphery; base convex, ornamented with about sixteen faintly granular lirae. Interior nacreous. Aperture subquadrate; outer lip acute, crenulate; operculum horny. Size: ½ inch.

Localities: Inverloch; Western Port, dredged alive 6-8 fathoms between Phillip and French Islands.

As with the other representatives of the genus this cream shell possesses a beautiful nacreous interior. It bears some resemblance to *hedleyi* Pritchard & Gatliff but is immediately separable by its more elongate form and stronger granulose ornament. At the last-named locality found in association with *legrandi* (Tenison Woods) and *hedleyi*. Recorded also from Tasmania.

zietzi Verco, 1905; another Victorian species from Western Port appearing also in South Australia; it is somewhat similar in shape to *allporti* but much smaller and without the strong granulated sculpture. Size: ⅛ inch.

Hedley's Top Shell

Calliostoma (Fautor) hedleyi Pritchard & Gatliff, 1902

(Figure 77)

Shell small, yellowish brown with reddish brown spots; solid, conical, imperforate. Whorls eight, convex, embryonic whorl smooth, the two succeeding whorls showing a latticed sculpture, the remainder ornamented with concentric, irregularly granulose riblets of varying size; about eight to nine on the penultimate whorl. Body-whorl angulated at the periphery and with twelve to fifteen red-spotted riblets on the convex base. Interior nacreous. Aperture subquadrate; outer lip sharp, crenulated; operculum horny. Size: ¾ inch.

Localities: San Remo; Inverloch; Airey's Inlet; Lorne; Portland; dredged alive in 6-8 fathoms between Phillip and French Islands, Western Port.

Washed up on the shores and in living condition found in association with *legrandi* (Tenison Woods), this exquisitely sculptured little shell is easily recognized by its conical shape, concentric, granulose riblets, brilliant nacreous interior and subquadrate aperture. It cannot be confused with any Victorian form. Found also in Tasmania. Named in honour of the late Charles Hedley, Conchologist, Australian Museum, Sydney.

Legrand's Top Shell

Calliostoma (Fautor) legrandi (Tenison Woods, 1875)
(Figure 78)

Shell, small, flesh coloured, solid, conical, imperforate. Whorls about six, somewhat flattened, sculptured with numerous, regular, concentric riblets (about nine on the penultimate whorl) frequently with a fine spiral thread in the interspaces, last whorl angulate, base almost flat and spirally lirate. Interior nacreous. Aperture subquadrate, outer lip sharp; operculum horny. Size: ½ inch.

Localities: Cast up on the shores of Western Port; and dredged alive in 6-8 fathoms between Phillip and French Islands; off Lakes Entrance.

A remarkably consistent shell with a rich nacreous interior, and in sculpture much resembling the reversed or sinistral *incerta* (Reeve) from the same locality. In each of these species there is an absence of granulations on the riblets so characteristic of several other similarly shaped Victorian forms, including *hedleyi* Pritchard & Gatliff from the same area. Recorded also from Bass Strait and Tasmania.

compta (A. Adams, 1854); Western Port.

Jewelled Top Shell

Calliostoma (Salsipotens) armillata (Wood, 1828)
(Figure 79)

Shell large, solid, conical, imperforate, surface shining; pinkish with irregular light brown maculations; apical whorls frequently with a rich, bluish green, metallic lustre. Spire approximately two-thirds the length of shell. Whorls nine, sculptured with numerous, regular concentric, granulose lirae of which there are seven on the penultimate whorl. Body-whorl angulated at the periphery with a double-beaded ridge; base spirally ornamented with numerous, strongly beaded ridges. Interior iridescent. Aperture subquadrate through which the spiral lirae are plainly evident; outer lip sharp, operculum horny. Size: 1½ inches.

Localities: Western Port; Inverloch; Anderson's Inlet; Warrnambool; Portland.

A very handsome shell readily recognized. Its shining surface, conical shape, concentric granulose ornament and pearly interior are associated features with which to identify the species. In contour much resembling *Astele subcarinatum* Swainson, but the absence of an umbilicus immediately separates it. Formerly known under the name of *meyeri* Philippi, 1848. Occurs also in Tasmania, New South Wales, South Australia, Western Australia.

Southern Top Shell

Calliostoma (Salsipotens) australe (Broderip, 1835)
(Figure 80)

Shell conical, rather thin, imperforate, shining, creamy white, irregularly maculated with rust brown. Spire about two-thirds the length of shell. Whorls eight to ten, flat, sculptured with strong granulose concentric ridges, the penultimate whorl possessing six, the lowest one stronger and somewhat crenate. Body-whorl prominently carinated; base flat, ornamented with ten strongly grained spiral ridges. Interior iridescent. Aperture subquadrate, through which the spiral ridges are easily discernible; outer lip acute. Size: 1 inch.

Localities: Shoreham; West Head, Western Port.

A beautifully grained species in which the spiral ridges are alternately coloured purple rose. It is a form of infrequent occurrence in Victoria. Recorded from Tasmania, King Island, Bass Strait, South Australia, and Western Australia where it is more common. For many years known under the name of *nobile* (Philippi, 1846).

Reversed or Left-hand Top Shell

Calliostoma (Sinutor) incerta (Reeve, 1863)
(Figure 81)

Shell creamy white occasionally pink tinted, small, conical, sinistral. Whorls obliquely convex, body-whorl obtusely angled at the base. Surface sculptured with close, granulated, concentric ridges. Interior iridescent. Aperture rounded; outer lip acute; operculum horny. Size: ¼ inch.

Localities: San Remo; Western Port; off Wilson's Promontory.

This is a unique little shell. The author in his original description remarked: 'its normal growth is probably dextral'. Such is not the case; it is definitely sinistral, and the writers know of no dextral specimen having been recorded. Occasionally cast up on the beaches, it is dredged alive between Phillip and French Islands at 8 fathoms. Beyond Victoria it occurs in the waters of Tasmania and South Australia.

Astele Swainson, 1854

Shell large, trochiform, solid, with a wide umbilicus. Whorls flat, sculptured with concentric lirae.

TYPE SPECIES *Astele subcarinatum* Swainson, 1854.

Umbilicated Top Shell

Astele subcarinatum Swainson, 1854
(Figure 82)

Shell large, conical or trochiform, solid, broadly umbilicated, fawny pink with irregular splashes of reddish brown. Whorls seven, flat, sculptured with several unequal concentric lirae, the posterior sutural one being slightly more elevated and evenly granulated, the interspaces ornamented with fine spiral striae. Interior nacreous. Aperture broader than high; outer lip

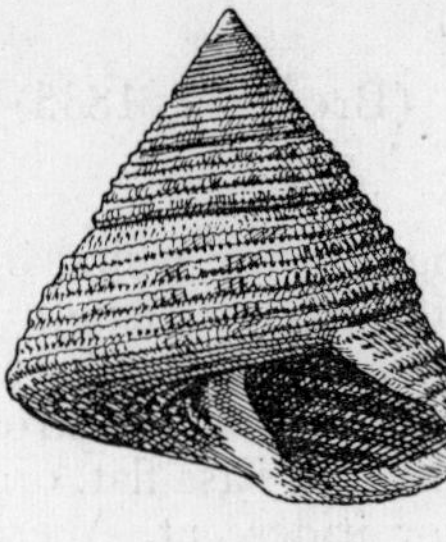

Figure 79

Figure 80

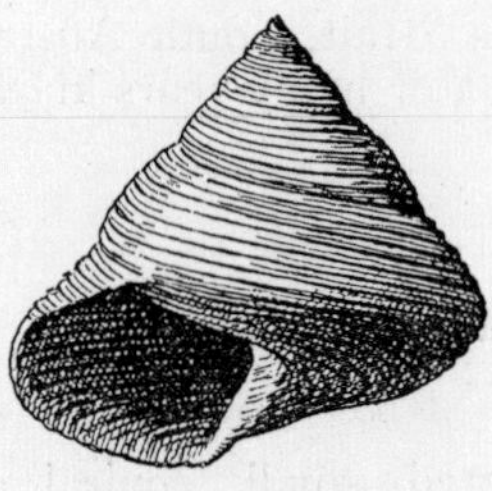

Figure 81

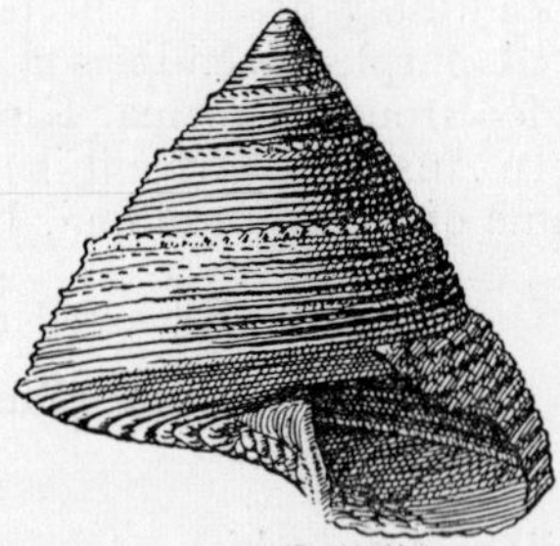

Figure 82

Figure 83

Figure 84

79. *Calliostoma armillata* × 1
80. *Calliostoma australe* × 1⅓
81. *Calliostoma incerta* × 2
82. *Astele subcarinatum* × 1
83. *Astelena scitula* × 3
84. *Gibbula coxi* × 3

acute; umbilicus wide, exposing all the volutions right to the apex, and encircled with three to four rows of granules; base strongly lirate; operculum horny. Size: 1½ inches.

Localities: Western Port; San Remo; Anderson's Inlet; Bass Strait.

In general outline this species recalls *Calliostoma armillata* (Wood) but is immediately separated by its broad, funnel-shaped umbilicus. Distribution from New South Wales, Tasmania, through Bass Strait to South Australia and Western Australia.

Astelena Iredale, 1924

Shell small, thin, trochiform. Whorls convex with concentric sculpture; umbilicus narrow, funnel-shaped.

TYPE SPECIES *Zizyphinus scitulus* A. Adams, 1854.

Elegant or Neat Top Shell

Astelena scitula (A. Adams, 1854)
(Figure 83)

Shell small, conical, yellowish white with brown maculations and two rows of reddish brown spots just above the sutures. Spire approximately half the length of shell. Whorls seven, convex, sculptured with fine concentric ridges and numerous longitudinal microscopic striae. Interior pearly. Aperture rounded; outer lip acute; umbilicus narrow, funnel-shaped. Size: ½ inch.

Locality: Mallacoota.

Although this species is smaller than *Astele subcarinatum* Swainson, possible confusion may arise with the juvenile of that form; distinctive features are at once evident in the rounder whorls, finer ornament, and smaller umbilicus. Quite common on the shores of New South Wales.

Gibbula Risso, 1826

Shell more or less tumid beneath the suture, conical, high as broad, umbilicated, spirally striate; usually conspicuously marked with reds and browns.

TYPE SPECIES *Trochus magus* Linné, 1758.

Gibbula was originally erected for a European species and although these do not as a rule show close affinities with Australian temperate genera, in the present case the differences do not appear to be marked, and we have retained this generic name for the most closely allied of the Victorian species.

Cox's Ridge Top Shell

Gibbula (Notogibbula) coxi Angas, 1867
(Figure 84)

Shell small, solid, orbicularly conical, narrowly umbilicate; whitish with splashes of olive green, pink or brownish red, sometimes faintly red spotted. Spire about half the length of shell. Whorls five, prominently bicarinate and sculptured with numerous, fine, concentric ridges crossed by extremely fine, microscopic, longitudinal striae. Interior pearly. Aperture somewhat rounded; outer lip entire; columella much curved. Size: ½ inch.

Localities: Western Port; off Portsea, Port Phillip.

This shell is easily recognized by its globosely conoid shape, bicarinate whorls and very narrow umbilicus. Occasionally washed up on the shores, in life found on seaweed. Distribution: Tasmania, New South Wales and South Australia.

preissiana (Philippi, 1848); from Point Roadknight, is a close ally; slightly smaller; whitish with reddish brown lines, and spots on the keels; whorls five, bearing two strong keels, base spirally lirate, the lirae larger at the umbilical area; its more conical shape and stronger basal spirals are sufficient features to distinguish the two species.

lehmanni (Menke, 1843); from Point Roadknight, is another representative of the genus in Victoria and may be confused with *coxi* Angas, 1867, but its rose tinted and slightly stronger concentric ridges are features which will easily separate the two forms.

Under this genus several small forms have been recorded from Victoria but distinctive features are evident enough to warrant their removal to other genera as follows:

Cantharidella Pilsbry, 1889 (*Gibbula picturata* A. Adams & Angas, 1864).

picturata (A. Adams & Angas, 1864); a superb species from Mallacoota. Small, conoid, narrowly umbilicate; extremely variable in colour, may be greenish, olive, yellow, or pink, flamed with purple or reddish brown and sometimes with pink or purple dots and with a rich pearly interior. Whorls five to six sculptured with concentric lirae generally five on the penultimate whorls. A common species in New South Wales; in life located on seaweed.

tiberiana (Crosse, 1863); shell about ⅕ inch, fairly common on the shores and dredged alive on seaweed off Point Cook, Port Phillip, and off Rhyll, Western Port. Colour yellowish olive or pink, prettily painted with golden and reddish brown maculations. Interior nacreous. Whorls flatly convex, sculptured with fine encircling striae.

Minopa Iredale, 1924 (*Fossarina legrandi* Petterd, 1879).

legrandi (Petterd, 1879); a common species about ⅛ inch; distributed along the whole coastline; it is at once recognized by its creamy colour, dark apex and reddish brown sagittate markings, the whole surface golden polished and with a rich nacreous interior and minute umbilicus.

tasmanica Petterd, 1879; from Bass Strait, about ⅛ inch; creamy with pink lines and maculations. Whorls convex and sculptured with numerous, fine concentric striae.

brazieri (Angas, 1871); Port Phillip.

petterdi (Crosse, 1870); Port Phillip.

Eurytrochus Fischer, 1880 (*Gibbula danieli* Crosse, 1862).

strangei (A. Adams, 1853); from Mallacoota. A small umbilicated form about ⅓ inch globosely conical; olive with brownish or reddish brown maculations, the spiral ribs showing small brown spots. Whorls five, convex and ornamented with strong concentric ridges, four appearing on the body-whorl. Interior nacreous. Aperture subquadrate, denticulate within; base spirally lirate; umbilical area and outer lip whitish articulated with pinkish brown.

Cantharidus Montfort, 1810

Shell rather thin, ovate-pointed. Whorls striated, or smooth; columella rather straight, simple, not toothed.

TYPE SPECIES *Trochus iris* Gmelin, 1791.

Beautiful Cantharidus

Cantharidus pulcherrimus (Wood, 1828)
(Figure 85)

Shell acutely conical, thick, solid. Whorls almost flat, ornamented with broad, concentric ribs, of which there are five on the penultimate whorl. Colour purplish crimson, with narrow, white marks crossing the ribs. Aperture ovate, interior iridescent, lip edges with rich crimson, more evident on the columella. Size: ¾ inch.

Localities: Point Nepean; Queenscliff; Western Port.

The specific name is appropriate, for it is indeed a beautiful shell with rich colouring and encircling sculpture. Allied to and associated with *ramburi* (Crosse), it is rather common in rock-pools and washed up on beaches. Dead shells become very worn and lose much of their charm; living specimens may be obtained on seaweeds. From *ramburi,* it is distinguished by its more rounded whorls and stronger sculpture. Recorded from Western Australia and South Australia.

Rambur's Cantharidus

Cantharidus ramburi (Crosse, 1864)
(Figure 86)

Shell generally crimson, with narrow, whitish flames on the upper surface. Spire long. Whorls slightly concave below the sutures. Whole surface finely spirally lirate. Aperture oval, iridescent within. Size: ¾ inch.

Localities: Portsea; Point Nepean; Point Lonsdale; Western Port.

Though small, it is nevertheless an ornate form. It is quite common on the beaches and in rock-pools, living examples being obtained on floating seaweeds. A frequent shell in South and Western Australia.

Phasianotrochus Fischer, 1885

Shell, thick, solid, polished, elongated, oval-pointed. Aperture ovate, longer than broad; columella usually with a tooth-like projection in the middle.

TYPE SPECIES *Trochus eximius* Perry, 1811.

The members of this genus usually live on or among algae from near low tide to several fathoms deep and are given the vernacular name Seaweed Shells. With their vividly coloured and polished surface and rich, iridescent, nacreous inner layer they lend themselves for use in decoration and many of the species are sought after for necklaces, etc. They are used both in the natural state and after removal of the outer shell layer to reveal the brilliantly coloured nacre.

Figure 85

Figure 86

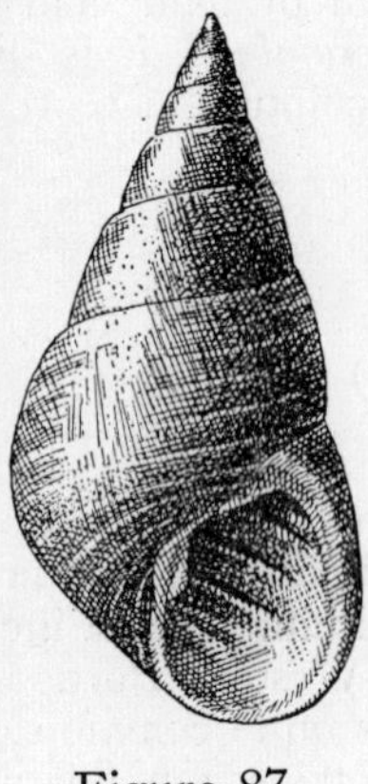

Figure 87

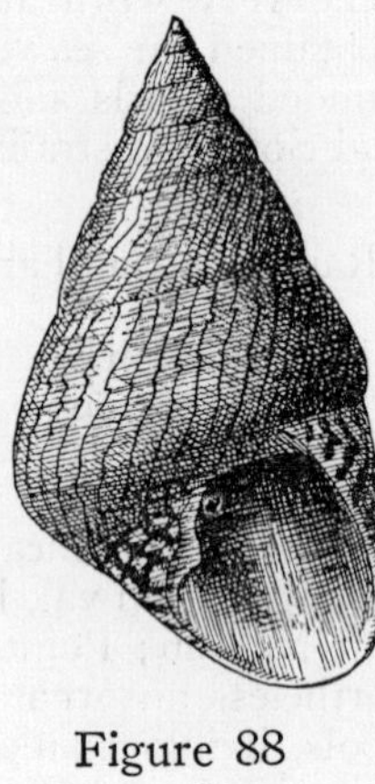

Figure 88

Figure 89

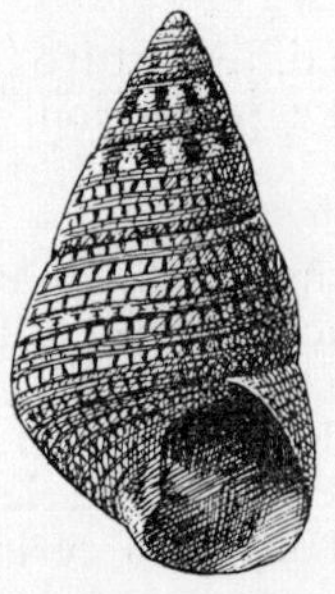

Figure 90

85. *Cantharidus pulcherrimus* $\times 2\frac{2}{3}$
86. *Cantharidus ramburi* $\times 3$
87. *Phasianotrochus eximius* $\times 1\frac{1}{3}$
88. *Phasianotrochus apicinus* $\times 2$
89. *Phasianotrochus bellulus* $\times 3$
90. *Phasianotrochus irisodontes* $\times 3$

Choice Seaweed Shell

Phasianotrochus eximius (Perry, 1811)
(Figure 87)

Shell elongately conical, polished, solid; fawn or rose, with light or dark, narrow encircling lines, and white, longitudinal, linear markings or stripes. Surface microscopically, spirally striated. Spire conical. Whorls flatly convex; apex acute. Aperture ovate, brilliantly iridescent within; columella vertical, with a single tooth below the centre. Size: 1½ inches.

Localities: Point Nepean; Point Lonsdale; Western Port.

The surface of the shell is frequently ornamented with beautiful markings and in this respect shows extreme variation. The interior is vividly iridescent. Occurs in New South Wales, Tasmania, South Australia, and Western Australia.

Striped-base Kelp Shell

Phasianotrochus apicinus (Menke, 1843)
(Figure 88)

Shell small, elongately conical, shining; grey or olive yellow with numerous, narrow, longitudinal reddish brown lines. Spire acute, barely twice the length of shell. Whorls eight, flatly convex, bearing fine, concentric, microscopic striae. Body-whorl obtusely angulated. Interior pearly. Aperture ovate; columella almost centrally denticulate; outer lip and columella margin with fairly close, white markings; base spirally lirate. Size: 1 inch.

Localities: Portland; Port Fairy.

With its characteristic shape, peculiar white basal markings, obliquely descending reddish brown lines, and cerulean blue pearly interior it should be easily recognized. Recorded also from South Australia and Western Australia.

Necklace or Elegant Kelp Shell

Phasianotrochus bellulus (Dunker, 1845)
(Figure 89)

Shell small, solid, elongately conical, highly enamelled; grey, pinkish or brown with encircling narrow orange red bands (two on the penultimate and six to eight on the body-whorl); numerous creamy white longitudinal curved lines arranged in pairs between the spiral bands. Spire acute, barely two-thirds the length of the shell. Whorls eight, almost flat. Interior pearly. Aperture ovate; outer lip strong; columella toothed a little below the centre. Size: ½ inch.

Locality: Portland.

In living condition located amongst smaller weeds, and possessing a very characteristic style of painting, the pairs of longitudinal lines often resolving themselves into the figure 8, a feature so constant it serves as a useful recognition mark; this and the highly enamelled surface should be quite sufficient to identify the species. Occurs also in Tasmania and South Australia where it appears to be extremely common.

Rainbow Kelp Shell

Phasianotrochus irisodontes (Quoy & Gaimard, 1834)
(Figure 90)

Shell small, fairly solid, elongately conical, polished; exhibiting an endless variety of colour patterns in green, olive green, brown, pink or yellow, sometimes with concentric white-spotted lines and reddish brown, longitudinal lines arranged in pairs. Spire slightly more than half the length of shell. Whorls eight, flatly rounded and sculptured with fine, concentric microscopic striae. Body-whorl angulated at the periphery. Interior vividly iridescent. Aperture subquadrate; outer lip acute. Size: ½ inch.

Localities: Apollo Bay; Port Fairy; Portland.

When taken alive the Rainbow Kelp Shell is a most attractive species, but in shore specimens discoloration takes place. It has been confused with *rutilus* (A. Adams) but is immediately separable by its more elongate form. Its occurrence is also noted in South Australia and Western Australia.

Pink-tipped Kelp Shell

Phasianotrochus rutilus (A. Adams, 1851)
(Figure 91)

Shell small, fawn, green or olive green with numerous, oblique, reddish brown longitudinal lines, the apical whorls frequently with white or olive spots. Spire slightly more than half the length of shell. Whorls seven, flatly rounded, sculptured with numerous, fine, concentric striae which are much coarser at the base. Body-whorl obtusely angulated at the anterior third. Interior vividly iridescent. Aperture ovate; outer lip strong; columella denticulate just below the centre. Size: ½ inch.

Locality: Western Port.

A solid species frequently washed up on the shore along the whole coast line. In life found on floating seaweeds. Its squat form and characteristic colour patterns make it easy to recognize. Occurs also in Tasmania and South Australia.

nitidulus (Philippi, 1849); Portland.

Bankivia Krauss, 1848

Shell imperforate, elongated, narrow, conical thin, but slightly pearly. Aperture small; columella slightly twisted, subtruncate towards the base.

TYPE SPECIES *Phasianella fasciata* Menke, 1830.

Members of this genus like the preceding are used for ornamentation in place of beads.

Banded or Silver Kelp Shell

Bankivia fasciata (Menke, 1830)
(Figure 92)

Shell small, elongately conical, shining, without epidermis, exhibiting extreme variation in colour, may be unicolored, white, pink, grey, purple, brown, or almost black, or striped, banded or with longitudinal, undulating lines in an infinity of patterns. Whorls about nine, flatly convex, smooth, sutures impressed. Interior not pearly. Aperture subquadrate, through which the colour bands are clearly seen; outer lip sharp. Size: ¾ inch.

Localities: Portland; Warrnambool; Anderson's Inlet; Kilcunda; Gippsland coast.

The sole representative of the genus in Victoria but widely distributed throughout the state. Washed up on sandy beaches or living on seaweeds at low water. Very thin, fragile, and in colour one of our most variable shells. For many years known under the name of *varians* Beck, 1848. Beyond Victoria it occurs in New South Wales, Tasmania, through Bass Strait to South Australia.

Leiopyrga H. & A. Adams, 1863 (*Cantharidus lineolaris* Gould, 1861).

Two shells belonging to this allied genus may possibly be confused with the last species, but are easily distinguished by their much rounder whorls.

lineolaris (Gould, 1861); from Portland and Gabo Island. A narrowly perforate shell of seven rounded whorls; creamy white with longitudinal wavy markings, and an interrupted band of spots at the anterior fourth of each whorl; ½ inch. This species was known for many years in Victoria as *picturata* H. & A. Adams, 1863.

octona Tate, 1891; a broader shell, from Western Port, Frankston and Portland. It is narrowly perforate, thin, shining; white with longitudinal, zigzag, reddish brown or bluish markings. Whorls eight, convex, ornamented with numerous equidistant concentric striae, a feature immediately separating it from *lineolaris*. Size: ⅓ inch.

Thalotia Gray, 1840

Shell imperforate, elevated conical, thick, solid, granulated or spirally ribbed, periphery rounded or obtusely angular. Aperture small, ovate; outer lip thick, crenulated within; columella toothed at base.

TYPE SPECIES *Monodonta conica* Gray, 1827.

Conical Top Shell

Thalotia conica (Gray, 1827)

(Figure 93)

Shell small, solid, conical, imperforate; colour somewhat variable, may be pinkish or light grey, striped with reddish brown, apical whorls more or less crimson, more evident in fresh examples. Spire slightly more than half the length of shell. Whorls seven, flattened; surface sculptured with evenly spaced beaded ridges about six on the penultimate whorl. Body-whorl obtusely angulate at the periphery, base with fine granulose, spiral ridges. Interior pearly. Aperture subquadrate, through which the concentric sculpture is plainly seen; outer lip strong; columella denticulate. Size: ¾ inch.

Locality: Port Albert.

A rather strong species, fairly common on the Gippsland coast. Its granular surface and purple red spots are consistent characters by which to identify the species. Distribution from New South Wales, Tasmania, through Bass Strait to South Australia and Western Australia.

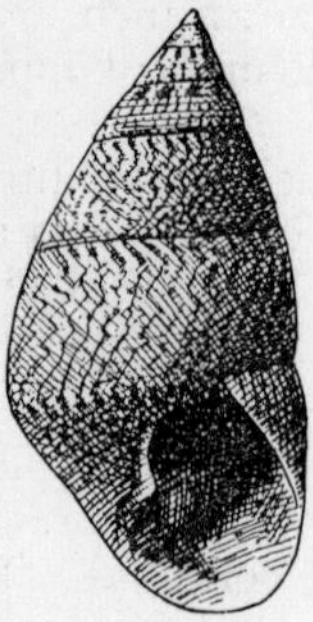

Figure 91

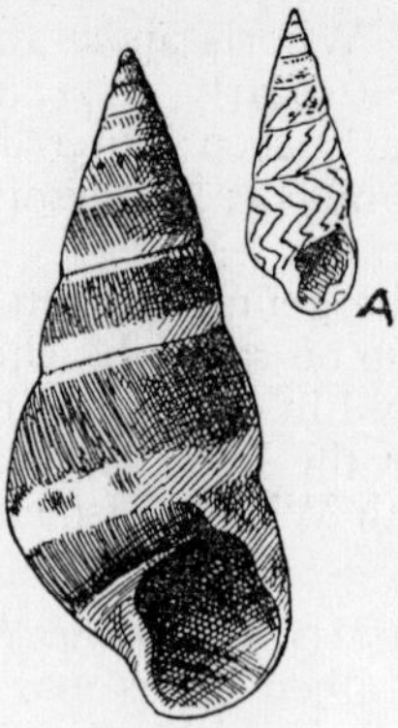

Figure 92

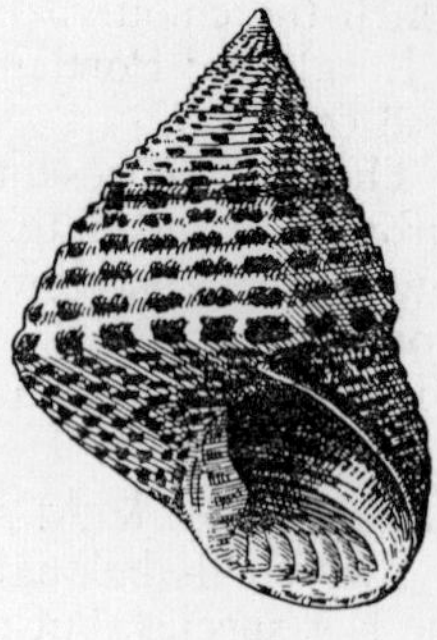

Figure 93

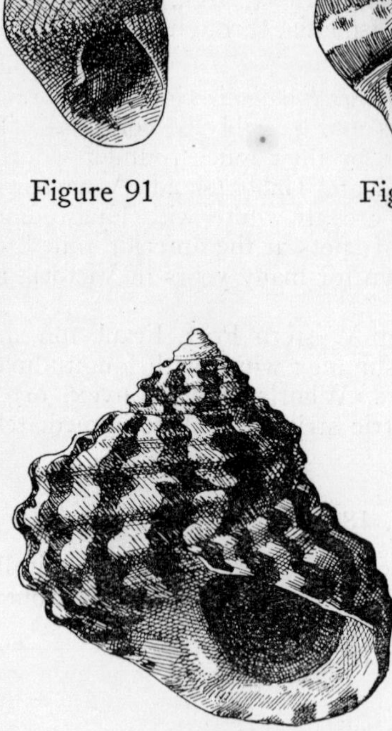

Figure 94

Figure 95

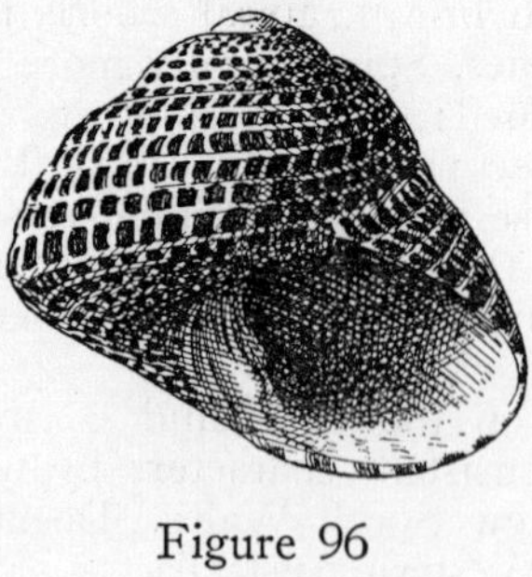

Figure 96

91. *Phasianotrochus rutilus* × 2
92. *Bankivia fasciata* × 2
    A Variation in colour pattern × 1
93. *Thalotia conica* × 2
94. *Austrocochlea constricta* × 2
95. *Austrocochlea concamerata* × 2
96. *Austrocochlea adelaidae* × 2

Austrocochlea Fischer, 1885

Shell turbinate; conical, thick, solid; columella only slightly toothed at the base, not notched; outer lip lirate within.

TYPE SPECIES *Monodonta constricta* Lamarck 1822.

Species now contained in this genus were originally grouped in *Monodonta* but their weaker sculpture and the lack of a strong tooth and notch on the columella are considered grounds for their separation.

Ribbed Top Shell

Austrocochlea constricta (Lamarck, 1822)

(Figure 94)

Shell whitish or greyish, sometimes marked with light or dark radial, zigzag lines or stripes; thick, solid, turbinate. Surface ornamented with strong spiral ribs, of which there are three on the penultimate whorl. Spire conical; apex acute. Aperture rounded, lirate and pearly within; columella short, with a single tooth at the fore part; operculum horny. Size: 1 inch.

Localities: Mornington; Point Nepean; Point Lonsdale; Queenscliff; Western Port, and coast generally.

This Top Shell may be located in great numbers on the surface of rocks; it is one of our commonest species, exhibiting considerable variation in colour and ornament. Some examples are almost colourless, while others are conspicuously longitudinally striped. In addition to Victoria it is common in Western Australia, South Australia, New South Wales and Tasmania.

For many years this species has been recorded as *Austrocochlea obtusa* (Dillwyn, 1817) but recent work by one of us (Macpherson, 1961) has shown this to be a misidentification, and it must now revert to the above name.

Some authors regard the unicoloured and striped forms as separate species, but Mrs Jean Carter (personal communication), who is in process of making a revision of the genus, has been unable to find any anatomical differences between animals with shells of varying colour patterns. Lamarck's original specimens were the unicoloured form so common on the open rock platforms of Victoria and Tasmania, and if the name is to be restricted, it must apply to this form, the striped colour variety being *zebra* (Menke, 1829).

Usually the latter is more common in slightly more sheltered habitats such as bays and inlets in southern waters, though it is common on even the most exposed platforms in New South Wales.

porcata (A. Adams, 1851); applied to the striped stunted salt-marsh variant.

Wavy Top Shell

Austrocochlea concamerata (Wood, 1828)

(Figure 95)

Shell black with yellow spots or longitudinal stripes; conic-globose, thick, solid. Spire very short, apex acute. Ornamented with spiral ribs. Interior white, edged within with black and yellow. Aperture large; columella white, toothed at the base. Size: 1 inch.

Localities: Point Nepean; Queenscliff; Western Port, and coast generally.

This species is decidedly flatter than *Austrocochlea constricta,* with much weaker, concentric ribs, and with the yellow colour more or less evident. Found living in association with *constricta,* but prefers to hide in cracks or under ledges. It also occurs in Queensland, New South Wales, Tasmania, South Australia and Western Australia.

Adelaide Top Shell

Austrocochlea adelaidae (Philippi, 1849)
(Figure 96)

Shell brownish, longitudinally marked with narrow, irregular lines, producing a somewhat tessellated appearance; conical, turbinate, solid, whorls slightly rounded. Sculpture consisting of narrow, spiral grooves, about five on the penultimate whorl and seven on the base. Interior iridescent; outer lip sharp edged with dark green; umbilico-columella area green; operculum brown, horny. Size: 1 inch.

Localities: Mornington; Portsea; Point Nepean; Point Lonsdale; Western Port.

This is a fairly frequent species, but less abundant than *odontis.* It is readily separated by its stronger sculpture and peculiar coloration. A yellowish tip becomes very evident in worn examples. This species is often encrusted with serpulae and is common in Western Australia, South Australia and Tasmania. This and the following species were placed in the genus *Diloma* but Mrs Carter considers their anatomy does not warrant separation from *Austrocochlea.*

Checkered Top Shell

Austrocochlea odontis (Wood, 1828)
(Figure 97)

Shell bluish black, dotted throughout with yellowish white; conic-globose. Whorls finely spirally striate. Outer lip thin, interior green, iridescent; columella not toothed; umbilico-columella area rich green; operculum brown, horny. Size: 1 inch.

Localities: Portsea; Point Nepean; Queenscliff; Western Port.

This Top Shell is frequent in rock-pools at low water. The characteristic spotting provides easy recognition. It also occurs in South Australia and Tasmania.

Clanculus Montfort, 1810

Shell conical, conoidal or turbinate; generally granose-lirate all over. Columella spirally twisted, forming a false umbilicus, plicate throughout and terminating in a multidental varix; outer lip usually dentate within.

TYPE SPECIES *Trochus pharaonius* Linné, 1758.

The Australian *Clanculus* have been divided into a number of subgenera on slight differences in the pattern of the granose-lirate sculpture and the shape and conformation of the teeth on the columella.

Wavy Clanculus

Clanculus (Macroclanculus) undatus (Lamarck, 1816)
(Figure 98)

Shell brown, marked with darker brown, longitudinal stripes, which are sometimes broken into large spots; depressed conoid. Whorls flatly convex, the base also somewhat flattened. Sculpture consisting of numerous, finely granulated spiral ribs. Aperture oblique; outer lip thickened, plicate within. Umbilicus large. Size: 1½ inches.

Localities: Sorrento; Point Nepean; Point Lonsdale; Western Port.

This is the largest species of this genus and not likely to be confused with any other Australian form. It possesses a very large umbilicus, and the peculiar stripes and maculations are very characteristic. Recorded also from South Australia, New South Wales and Tasmania.

Mauger's Clanculus

Clanculus (Macroclanculus) maugeri (Wood, 1828)
(Figure 99)

Shell reddish brown, conical, thick, solid. Spire approximately two-thirds the length of shell. Whorls eight, flatly convex, sutures barely evident. Sculpture consisting of strong, concentric, granulose ridges of varying strength, the body-whorl prominently keeled, with seven rows of granules above the keel and about fifteen at the base which are finer and more closely beaded. Interior pearly. Aperture subquadrate; outer lip thick, crenulate within; columella curved, somewhat wrinkled, with a small tooth at the base. Size: 1 inch.

Locality: Victoria.

A rather strong species separated from any Victorian *Clanculus* by its more conical form and white dotted basal granulations. The Victorian record is by Tenison Woods with no specific locality given. In living condition the species is located on seaweeds. It occurs in New South Wales and Tasmania.

Flamed Top Shell

Clanculus (Euriclanculus) flagellatus (Philippi, 1848)
(Figure 100)

Shell small, conoid, umbilicate, white with irregular, reddish brown maculations. Spire about half the length of shell. Whorls five, convex, sculptured with strong, granulose ridges (six on the penultimate whorl) with interspaces approximately of equal width. Body-whorl strongly carinated with much finer granules on the base than above the keel. Base almost white with a few scattered brownish spots close to keel. Interior pearly white. Aperture subquadrate; outer lip strong, denticulate within; columella oblique with a rather prominent tooth below. Size: ¾ inch.

Localities: Portland; Lorne; Western Port.

This species is of common occurrence along the whole coastline, and easily recognized by its depressedly conical shape and strong granulated surface. Occurs in Tasmania, South Australia and Western Australia.

Figure 97

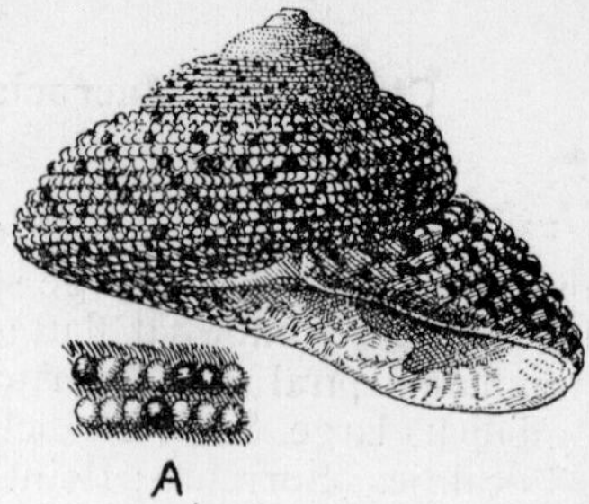

Figure 98

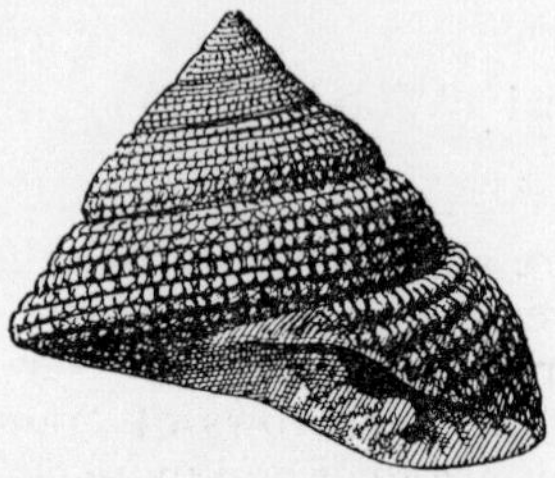

Figure 99

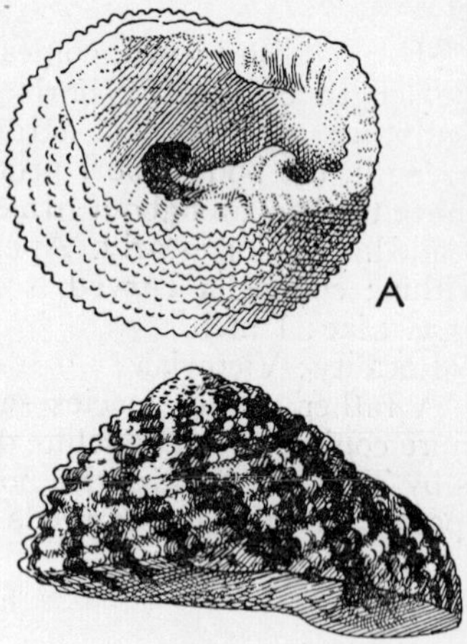

Figure 100

Figure 101

97. *Austrocochlea odontis* × 2
98. *Clanculus undatus* × 1⅓
    A Sculpture (enlarged)
99. *Clanculus maugeri* × 1⅓
100. *Clanculus flagellatus* × 2
    A Ventral view
101. *Clanculus limbatus* × 2

floridus (Philippi, 1848); the Florid Clanculus from Tellaburga Island, is a species of similar contour but with finer granulations; to regard it as *flagellatus* in miniature would not be out of place.

personatus (Philippi, 1846); another species, from Lorne, San Remo, and other localities, which may be confused with *flagellatus* but features which separate it are its white and rose tints, taller spire, stronger granulated surface and narrower umbilicus.

Keeled Top Shell

Clanculus (Euriclanculus) limbatus (Quoy & Gaimard, 1834)

(Figure 101)

Shell small, solid, depressedly conical, umbilicate; white, yellowish, or pink, with irregular brown maculations. Spire about half the length of shell. Whorls five to six, slightly convex, ornamented with fairly strong concentric, granulated ribs, five appearing on the penultimate whorl. Body-whorl prominently keeled at the periphery, base slightly convex, with much finer granose spirals. Interior pearly white. Aperture subquadrate; outer lip thickened; columella oblique, bearing a strong tooth. Size: ¾ inch.

Localities: Port Fairy; Lorne; Sorrento; Cape Schanck; Western Port.

A very characteristic form, of frequent occurrence along our shores. It somewhat resembles *flagellatus* Philippi but is more conical and has flatter whorls; occurs also in Tasmania and South Australia. *brunneus* A. Adams, 1851, is a synonym of this species.

aloysii Tenison Woods, 1878; a smaller shell from Port Phillip, Western Port and off Wilson's Promontory. It is cast up as a common species in shell sand; similar in shape to the last, it is strongly keeled and whitish in colour, variegated with lines and spots of dark brown; sometimes almost black, and with a finely granulated surface.

In Victoria the genus is further represented by the three following species all under ½ inch.

(*Mesoclanculus*) *ochroleucus* (Philippi, 1853); from Western Port, a fawn flatly conoid shell, sculptured with numerous, fine concentric granose riblets; umbilicus narrow and with its margin dentate.

(*Mesoclanculus*) *plebejus* (Philippi, 1851); Plebeian Clanculus, the commonest species of the genus in Victoria obtained in countless numbers on the whole coastline. In life located in rock-pools or under stones at low tide. It has a wide distribution, occurring in New South Wales, Tasmania, through Bass Strait to South Australia and Western Australia.

(*Isoclanculus*) *dunkeri* (Koch, 1943); a deep red or brown shell from Western Port. Conoid, thick, solid, umbilicus shallow; whorls five, flatly rounded, sculptured with concentric granose ribs, three on the penultimate whorl; base with four to five strongly granulated spirals. Interior pearly white; outer lip thickened, denticulate within. Found also in Tasmania and South Australia.

(*Euclanculus*) *leucomphalus* Verco, 1905; off Lakes Entrance.

Ethminolia Iredale, 1924

Shell depressed, trochoid in shape, widely umbilicate; texture thin. Whorls medially angulate, and with strong square shoulder.

TYPE SPECIES *Ethminolia probabilis* Iredale, 1924.

Depressed Top Shell

Ethminolia tasmanica (Tenison Woods, 1877)

(Figure 102)

Shell small, very variable in colour, may be pink, bluish yellow or grey but normally whitish, the upper portion beautifully maculated with olive brown and zoned with white, pink or greyish spots; under surface generally lighter; widely umbilicate, thin, fragile, depressedly conical. Whorls five, rounded, slightly shouldered near the sutures, presenting a terraced appearance to the shell. Sculptured with numerous concentric striae which are crossed by oblique, longitudinal striae cancellating the whole surface, visible only with the aid of lens. Body-whorl obtusely angled at the periphery. Interior iridescent. Aperture rounded, outer lip acute; umbilicus broad, exposing all the volutions; operculum horny. Size: ½ inch.

Localities: Port Phillip; Western Port.

In shell sand frequently cast up on the shores in more or less fragmentary condition. It is a rather fragile shell but easily recognized by its depressed spire, thin texture, fine cancellated sculpture, and peculiar colouring; occurs also in Tasmania.

probabilis Iredale, 1924; off Lakes Entrance.

Spectamen Iredale, 1924 (*Trochus philippensis* Watson, 1881).

philippensis (Watson, 1881); another Victorian species but of rare occurrence, obtained by the *Challenger* in 33 fathoms (sand) off Port Phillip. It is cream with reddish brown zigzag markings, smaller, a little more conical and with a much wider umbilicus. Found also in Tasmania and New South Wales.

Nanula Thiele, 1921 (*Margarita tasmanica* Petterd, 1877).

tasmanica (Petterd, 1877); Port Phillip.

Minolops Iredale, 1929 (*Minolops emendata* Iredale, 1929).

emendata Iredale, 1929; off Lakes Entrance.

Gena Gray, 1847

Shell flat, subspiral, oblong with a very large body whorl. Spire minute. Sublateral, surface smooth or striated. Aperture very large, nearly as long as the shell, pearly within.

Type Species *Stomatella nigra* Quoy & Gaimard, 1834.

This genus resembles *Haliotis* but lacks the characteristic series of perforations. The animal which is too large for the shell is extremely active and often on being disturbed will shed the hind portion of the very large foot.

False Ear Shell

Gena impertusa (Burrow, 1825)

(Figure 103)

Shell fragile, oblong, spire depressed, auriform, outer edge somewhat straightened but generally with a slight sinus; margins arched, so that when placed on a flat surface the shell is supported by its extremities. Whole surface sculptured with very fine, concentric striae, cancellated by finer

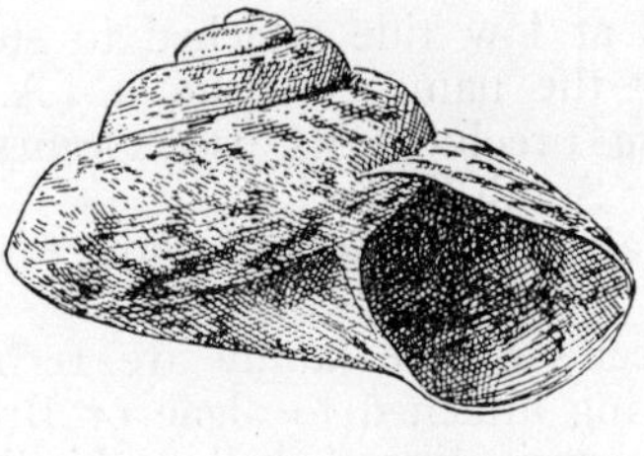

Figure 102

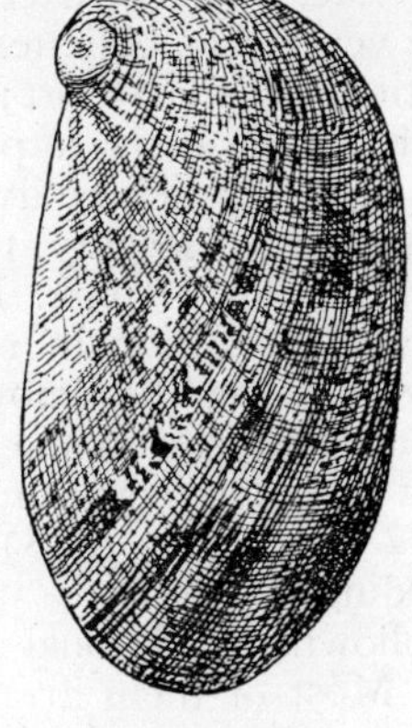

Figure 103

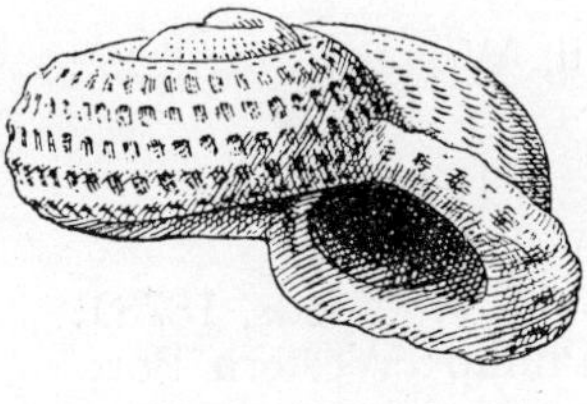

Figure 104

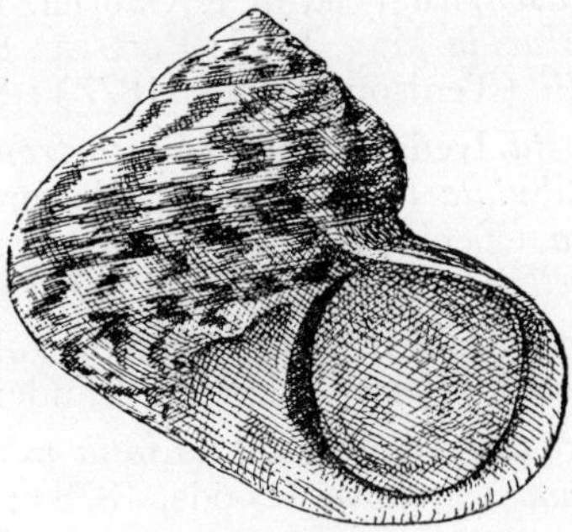

Figure 105

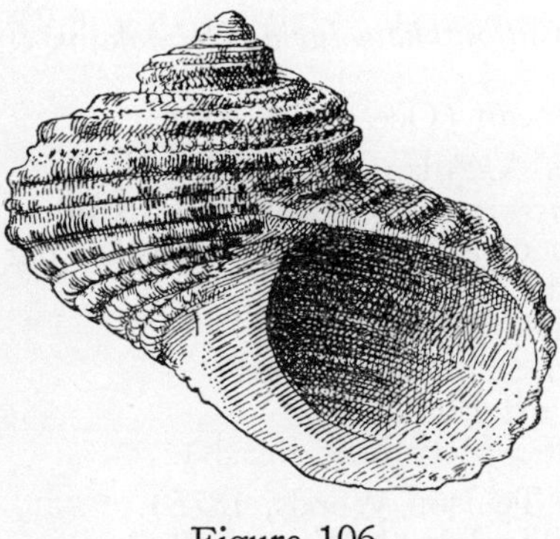

Figure 106

102. *Ethminolia tasmanica* × 3
103. *Gena impertusa* × 2
104. *Munditia australis* × 3
105. *Subninella undulata* × 1
106. *Subninella gruneri* × 1½

oblique striae. Interior nacreous with silvery, green and pink reflections. Aperture very wide, oblique. Size: 1 inch.

Localities: Western Port; Portsea.

Though small, it is a particularly beautiful shell displaying extreme variation in colour. Three living examples adhering to a rock were dredged off the last-named locality, two with a rich pinkish brown and one a deep brown with bands of pink. Found occasionally washed up on the beaches, but in living condition in rock-pools at low tide attached to stones and other shells. Generally known under the name of *strigosa* (A. Adams, 1850), but the prior name *impertusa* precludes its use. Occurs also in Tasmania and more commonly in New South Wales.

auricula (Lamarck, 1818); Port Fairy. A consistently larger shell than the preceding.

The following additional small species of the family are recorded for Victoria. Most of them are found living attached to algae or Bryozoa in deeper water. Sifting and microscopic examination of shell sand is the easiest and most profitable method of collection if living material is not required.

Cirsonella Angas, 1877 (*Cirsonella australis* Angas, 1877 = *Cyclostrema weldii* Tenison Woods, 1877).

microscopica (Gatliff & Gabriel, 1910); Shoreham.
translucida May, 1915; Portsea; Port Phillip; Shoreham; Western Port.
weldii (Tenison Woods, 1877); Sorrento; Port Phillip; Western Port.

Lissotesta Iredale, 1915 (*Cyclostrema micra* Tenison Woods, 1877).

contabulata (Tate, 1899); Western Port.
micra (Tenison Woods, 1877); off Rhyll, Western Port.
porcellana (Tate & May, 1900); Flinders.

Lodderia Tate, 1899 (*Liotia lodderae* Petterd, 1884).

lodderae (Petterd, 1884); Flinders; Otway coast.

Lodderena Iredale, 1924 (*Liotia minima* Tenison Woods, 1878).

minima (Tenison Woods, 1878); Port Phillip; Western Port.

Elachorbis Iredale, 1915 (*Cyclostrema tatei* Angas, 1876).

caperatum (Tate, 1899); Lakes Entrance.
delectabile (Tate, 1899); Point Nepean (Ocean Beach).
harriettae (Petterd, 1884); Western Port; Port Phillip.
homalon (Verco, 1907); off Wilson's Promontory.

Callomphala A. Adams & Angas, 1864 (*Callomphala lucida* A. Adams & Angas, 1864).

lucida Adams & Angas, 1864; Point Nepean (Ocean Beach).

Crossea A. Adams, 1865 (*Crossea miranda* A. Adams, 1865).

Iredale, 1924 erected *Crosseola* with *concinna* as type for the Australian species on the grounds of their possession of a multispiral operculum but this feature is common to the original genus.

cancellata Tenison Woods, 1878; off Rhyll, Western Port.
carinata Hedley, 1903; off Ninety Mile Beach (40 fathoms).
concinna Angas, 1867; Western Port; Port Phillip.
naticoides Hedley, 1907 (40 fathoms off Ninety Mile Beach).

Dolicrossea Iredale, 1924 (*Crossea labiata* Tenison Woods, 1875).

labiata (Tenison Woods, 1875); off Phillip Island, Western Port.

Family CYCLOSTREMATIIDAE

Cyclostremas

Shell small, depressed or rounded, smooth or sculptured, umbilicated, colourless, not nacreous; operculum horny, with several whorls.

Zalipais Iredale, 1915 (*Cyclostrema lissa* Suter, 1908).
inscripta (Tate, 1899); Portsea; Port Phillip.

Brookula Iredale, 1912 (*Brookula stibarochila* Iredale, 1912).
angeli (Tenison Woods, 1876); Western Port.
crebrisculpta (Tate, 1899); off Rhyll, Western Port.
densilaminata (Verco, 1907); off Wilson's Promontory.
johnstoni (Beddome, 1882); off Rhyll, Western Port.
nepeanensis (Gatliff, 1906); Port Phillip Heads.

Liotella Iredale, 1915 (*Liotia polypleura* Hedley, 1904).
annulata (Tenison Woods, 1878); Port Phillip; Western Port.
kilcundae (Gatliff & Gabriel, 1914); Kilcunda.
petalifera (Hedley & May, 1908); Port Albert; off Wilson's Promontory.
vercoi (Gatliff & Gabriel, 1914); off Wilson's Promontory.

Cithna A. Adams, 1863 (*Cithna globosa* A. Adams, 1863).
angulata Hedley, 1907; off Wilson's Promontory.
flexuosa (Gould, 1861); Western Port; Port Phillip.

Family TURBINIDAE

Turbos

Shell turbinate or top-shaped and generally heavy, with a thick lining of iridescent nacre. The mouth is closed with an operculum, usually calcareous, and the shells have an umbilicus. The animal is herbivorous.

Munditia Finlay, 1926

Shell planorbid, sculpture consisting of radial and spiral ribs forming a lattice; umbilicus very wide; lip thickened; operculum horny, multi-spiral.
TYPE SPECIES *Liotina tryphenensis* Powell, 1926.
These shells were originally placed in the fossil genus *Liotina* from which they show a number of differences.

Southern Liotina

Munditia australis (Kiener, 1839)

(Figure 104)

Shell small, white, somewhat turbinate. Whorls rounded. Sculpture consisting of concentric ridges and longitudinal striae, giving the shell a somewhat cancellated appearance. Umbilicus fairly wide and deep, through which the apex may be plainly seen. Interior white. Aperture circular, peristome strong, varicose; operculum circular, horny. Size: ½ inch.
Localities: Port Phillip; Western Port; Sorrento; Lorne; Portland.

This is a heavily ornamented species easily identified by its characteristic cancellate sculpture, circular aperture, and strong outer lip. Well distributed along the Victorian coast and found also in Tasmania and South Australia.

Several smaller forms appear in Victoria, frequently in shell sand cast up on the beaches.

tasmanica (Tenison Woods, 1876); the Tasmanian Liotina, a smaller species from Western Port about ⅛ inch. It is strongly ornamented with nodulose, encircling keels and more numerous closely set longitudinal striae, features which immediately separate it from the above. It is also recorded from South Australia and New South Wales.

subquadrata (Tenison Woods, 1878); from Western Port and Apollo Bay; a species slightly smaller than *tasmanica* and with fewer longitudinal striae. Occurring also in Tasmania and South Australia.

hedleyi (Pritchard & Gatliff, 1899); a minute form from Flinders (about ⅛ inch), the body-whorl of which is angulated by the uppermost of three concentric keels, and crossed by about twenty longitudinal riblets, showing from an apical aspect a definite serration of the keel. Recorded also from South Australia.

mayana (Tate, 1899); exact locality unknown.

A number of other small species allied to *Munditia* are included in the family.

Charisma Hedley, 1915 (*Charisma compacta* Hedley, 1915).
arenacea (Pritchard & Gatliff, 1902); off Phillip Island, Western Port.
josephi (Tenison Woods, 1876); Western Port.

Argalista Iredale, 1915 (*Cyclostrema fluctuata* Hutton, 1883).
rosea (Tenison Woods, 1876); Sorrento, Port Phillip; Western Port.

Subninella Thiele, 1929

Shell wide, roundly turbinate, with a low spire. Sculptured with spiral lines. Umbilicus open, encircled with a white callus which may partially or completely close it; mouth oblique, rounded; operculum smooth, sculptured with indefinite concentric ridges.

TYPE SPECIES *Turbo undulata* Solander, 1786.

Cotton, 1939, proposed the genus *Euninella* for *Turbo gruneri* Philippi but we consider that the differences are specific rather than generic.

Warrener or Wavy Turbo

Subninella undulata (Solander, 1786)

(Figure 105)

Shell wide, roundly turbinate, with a low spire. Sculptured with low spiral lines which tend to disappear on the body-whorl. Colour bluish green, longitudinally marked with white zigzag streaks. Covered with a brown epidermis. Umbilicus open, encircled with a white callus; mouth oblique; round operculum calcareous, solid white, smooth, with strong indefinite concentric ridges which make a thick convex central area. Size: 2 inches.

Localities: A common species in the lower littoral of all rocky platforms.

It is a frequent species in aboriginal kitchen middens. Its peculiar shape, wave-like markings, and unusually large umbilicus, will suffice to identify the species. It is sometimes eaten and provides, at least to some, a very great delicacy.

Gruner's Turbo or Turban Shell

Subninella gruneri (Philippi, 1846)
(Figure 106)

Shell pink, mottled with reddish brown and white, orbicular, conoid. Umbilicus closed by a white callus. Spire short, less than half the length of shell. Whorls rounded, angulated near the sutures, concentrically sculptured with alternate thick and thin granulose cords. Interior iridescent, with pink and green reflections. Aperture circular; columella white; outer lip acute; operculum smooth, white, with indefinite concentric ridges and a flattened rim near the edge. Size: 2 inches.

Localities: Sorrento; Cape Schanck; Flinders; San Remo; Anderson's Inlet; Lorne.

An ornate species easily recognized by its pinkish colour and concentrically ridged sculpture. In fresh examples the internal iridescence, a feature almost absent on beach-worn specimens, is very beautiful. It occurs also in Tasmania, South Australia and Western Australia.

Ninella Gray, 1950

Shell wide, turbinate, sculpture consists of spiral ridges cut by oblique folds, and a row of strong knots under the suture. Umbilicus wide, open, enclosed with a white callus; operculum sculptured with a spiral ridge set with fine spines which enclose a central depression.

TYPE SPECIES *Turbo torquata* Gmelin, 1791.

Heavy Turbo or Turban Shell

Ninella torquata (Gmelin, 1791)
(Figure 107)

Shell large, solid, greyish green, orbicular, strongly umbilicate. Spire short, much less than half the length of shell. Sculpture consisting of oblique, close, longitudinal, lamellose striae and irregular, coarse concentric ridges; sutures canaliculate, with a row of nodules on the adjacent ridge. Interior whitish, slightly iridescent. Aperture circular; columella white, perforated by the wide umbilicus; operculum with four whorls, internally brown, flat; nucleus situated about one-third the diameter from the inner margin; externally white, excavated at the centre and with two strong concentric ribs; almost the whole surface strongly granulated. Size: 3 inches.

Localities: Anderson's Inlet; Port Fairy.

A very coarsely sculptured shell with a wide umbilicus, easily recognized by its lamellose, longitudinal striae, bold concentric ridges, and the extraordinary operculum.

This species has been recorded as a living Victorian shell but intensive collecting in its typical habitat over recent years has failed to produce a single specimen. It is now considered that the Victorian records are from specimens washed out of the Quaternary beds. Shells in these beds are very

Figure 107

Figure 107A

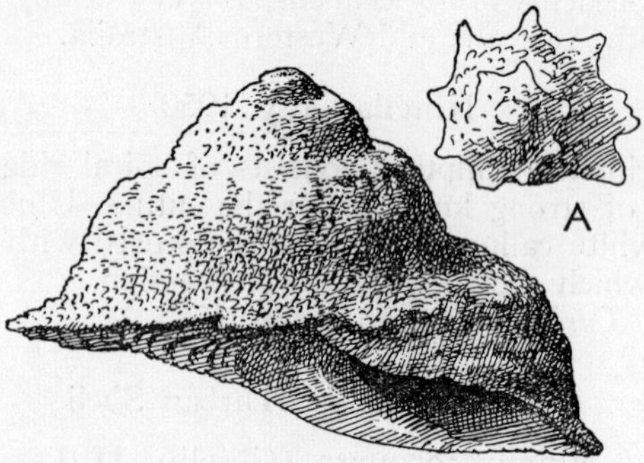

A

Figure 108

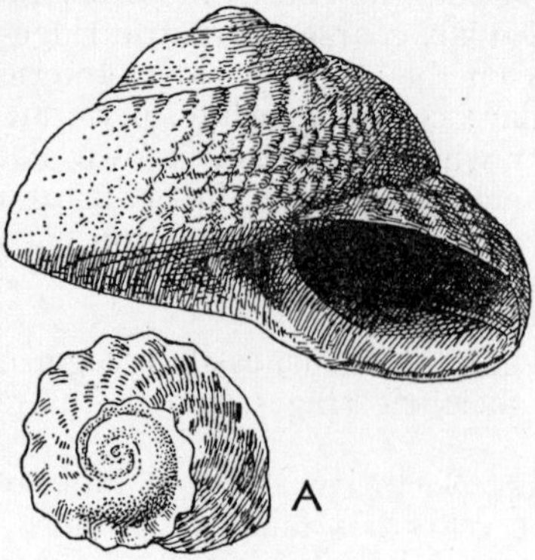

A

Figure 109

107. *Ninella torquata* $\times \frac{2}{3}$
107A. Operculum
108. *Bellastraea kesteveni* × 2
A Dorsal view
109. *Micrastraea aurea* × 3
A Dorsal view

well preserved, some even having the operculum in position. It is a common shell in New South Wales, South Australia and Western Australia; in the latter state it shows some differences in form and is called *Ninella whitleyi* Iredale, 1949. It has been known for many years as *Turbo stamineus* Martyn, 1784, a name not acceptable to the International Commission on Zoological Nomenclature.

Bellastraea Iredale, 1924

Shell conical, sculptured with thick spiral lines, squarish on the edge from which project more or less folded lamellae; underside slightly convex. Mouth opening very oblique, umbilicus covered; operculum white, smooth with one rib.

TYPE SPECIES *Bellastraea kesteveni* Iredale, 1924.

Star Shell

Bellastraea kesteveni Iredale, 1924

(Figure 108)

Shell small, conical, colour very variable, may be fawn, brown, golden, grey; imperforate. Spire about half the length of shell. Whorls not rounded, at the lower edge flattened out into a flange or rim, this feature being very pronounced on the body-whorl. Sculpture consisting of irregular, longitudinal striae just below the sutures, and numerous, concentric, more or less imbricating ridges very distinct at the slightly convex base. Interior nacreous. Aperture transversely ovate, outer lip acute, channelled at the outer angle; columella very broad; operculum solid, calcareous. Size: 1 inch.

Localities: Cape Schanck; Warrnambool; off Wilson's Promontory (dredged); Western Port (dredged between Phillip and French Islands).

In living condition these Stars are easily identified by the conspicuous flange at the lower edge of the whorls, and the concentric, imbricating ridges, features almost absent in water-worn beach specimens. These animals are vegetable feeders and the shells sometimes become encrusted with *Lithothamnion* and species of Bryozoa. The name Taut Shells is often applied. Recorded also from Tasmania, New South Wales, South Australia and Western Australia. Previously known in Victoria as *Astraea fimbriata* (Lamarck, 1822), a preoccupied name.

sirius (Gould, 1849); a closely related species. It is higher, more conical and flat based and has an oval solid operculum. Found in eastern Victoria, and a common shell on New South Wales rock platforms where it clusters in groups near low tide.

Micrastraea Cotton, 1930

Shell small, depressed, solid, imperforate. Aperture small oblique; columella callus spread over umbilical region; operculum spirally ribbed, having a granular surface.

TYPE SPECIES *Trochus aureus* Jonas, 1844.

Golden Star

Micrastraea aurea (Jonas, 1844)

(Figure 109)

Shell small, solid, yellowish or golden, depressedly conical, imperforate. Spire about half the length of shell. Whorls ornamented with diagonal ribs (which occasionally divide towards the periphery) crossed by concentric lirae, the point of crossing more or less producing beads giving the surface a granular appearance, more evident at the base. Interior nacreous. Aperture transversely oval; outer lip sharp; columella broad; operculum solid, calcareous. Size: ¾ inch.

Localities: Port Phillip; Western Port; Lorne.

This is a rather attractive shell identified by its peculiar oblique, bifurcating, longitudinal ribs, distinct granulated base, and nacreous interior. Consistency of colour is not apparent, olive and vivid green examples being of frequent occurrence. It is a common species on the whole Victorian coastline and occurs in Tasmania, South Australia and Western Australia.

Phasianella Lamarck, 1804

Shell elongate, oval, or subglobose, polished, without epidermis, variegated with bright colours; imperforate. Outer lip simple, thin; operculum heavy, calcareous, externally convex, white.

Type Species *Buccinum australis* Gmelin, 1788.

The genus, though widespread, is numerically not very strong in species, but under suitable conditions there may be a large congregation of individuals of a single species. The most favoured habitat of Australian species seems to be bays and inlets where rocks and sand meet and there is a strong growth of algae on which they feed.

Pheasant Shell or Painted Lady

Phasianella australis (Gmelin, 1788)

(Figure 110)

Shell large, elongated, rather thin, smooth, with a produced conical spire. Whorls about eight, slightly convex. Aperture pyriform; outer lip thin; operculum white, calcareous, solid. Size: 3 inches. Specimens are frequently obtained exceeding this dimension, more particularly in South Australian waters.

Localities: Sorrento; Western Port; Portland.

For brilliancy of colouring and diversity of painting, few of our Victorian shells can be compared with this species, and it certainly joins the company of our most beautiful Australian shells. The colour pattern is extremely variable, being richly marked with lines and waves of various delicate tints, among the more frequent being purplish rose, pink, cream, and red. This is the largest species of the genus and, in addition to the above localities, is found in South Australia, Western Australia and Tasmania.

Swollen Pheasant Shell

Phasianella ventricosa Swainson, 1822

(Figure 111)

Shell large, solid, ovate, spire short, conical. Shining, pale flesh-colour, variously marked with a series of arrow-shaped, square, or oblong blotches, or longitudinally striped with various coloured festoons. Whorls five to six, very convex. Interior white. Aperture ovate; operculum calcareous. Size: 1½ inches.

Localities: Point Nepean; Queenscliff; Western Port.

In Port Phillip, it is not so frequent as the previous species, but is fairly abundant in Western Port, and is recorded from all the states in Australia. While not so large as *australis* (Gmelin), it exhibits the same diversity of colour pattern and certainly ranks with that species as one of our foremost shells. Some specimens are unicoloured, while a well-defined variety, *zebra* Reeve, 1868, is easily recognized by its oblique, longitudinal, very conspicuous broad bands of chestnut red and yellow; another variety, *reticulata* Reeve, 1862, is closely, undulately painted throughout with brown lines and flesh-coloured flames beneath the sutures.

This shell has been known by the name *perdix* (Wood, 1828), but the above prior name takes precedence.

Iredale, 1924, erected the subgenus *Mimelenchus* for this species but we agree with Thiele that it is inseparable from *Phasianella s.s.*

Variegated Pheasant Shell

Phasianella variegata Lamarck, 1822

(Figure 112)

Shell ovately conical, polished; colour extremely variable, may be flesh tinted or pink, flamed just below the sutures with dark pink, brown or almost black, the whole shell fairly regularly encircled with fine reddish brown lines about twelve to fifteen on the body-whorl, interrupted by white dots, the white occasionally predominating. Spire about half the length of shell. Whorls five, smooth, convex. Interior whitish. Aperture ovate, widely rounded anteriorly; outer lip entire; columella curved, somewhat calloused posteriorly. Size: 1 inch.

Localities: Port Phillip; Western Port; Portland.

The Variegated Pheasant, though small, is a handsome species and like its congeners, exhibits great variation in colour pattern. However, the distinctive features which separate it from *australis* and *ventricosa* are its smaller size and elongate form. It occurs also in Tasmania.

There are two other small species allied to *Phasianella* recorded from Victoria:

Gabrielona Iredale, 1917 (*Phasianella nepeanensis* Gatliff & Gabriel, 1908).
nepeanensis (Gatliff & Gabriel, 1908); Point Nepean.

Pellax Finlay, 1926 (*Phasianella huttoni* Pilsbry, 1888).
rosea (Angas, 1867); the Rosy Pheasant Shell, found often at Western Port, Cape Schanck and Cape Otway. It is a little species (⅜ inch); thick, shining, uniformly rose throughout with longitudinal markings of similar

Figure 110

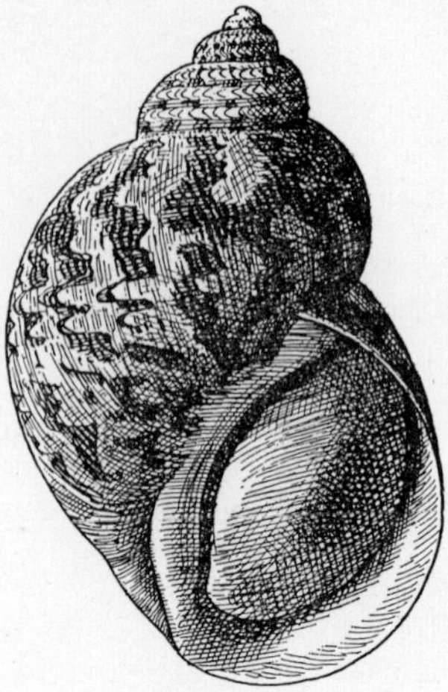

Figure 111

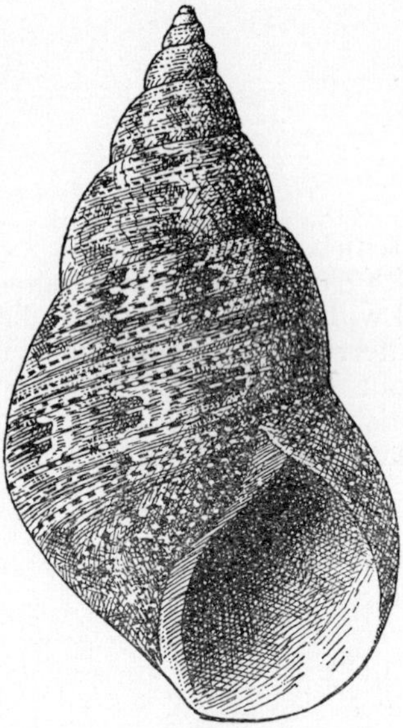

Figure 112

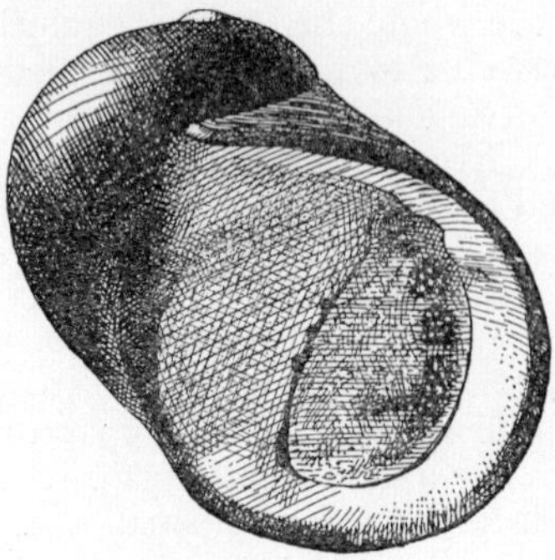

Figure 113

110. *Phasianella australis* × 1
111. *Phasianella ventricosa* × 1⅓
112. *Phasianella variegata* × 4
113. *Melanerita melanotragus* × 2

but darker coloration. Aperture ovate; columella white. Whorls four, rounded. Occuring also in New South Wales and Tasmania. This is quite a common species in shell sand along the coast, its uniform rose tint serving as a useful recognition mark.

Family PHENACOLEPADIDAE

Sugar Limpets

Shell symmetrical, cap-shaped, or conical with the backward-pointing apex near the posterior margin; without an operculum. Muscle scar horseshoe-shaped, opening directed anteriorly.

Phenacolepas Pilsbry, 1891 (*Scutella crenulata* Broderip, 1834).
calva (Verco, 1906); off Ninety Mile Beach (40 fathoms).

Superfamily COCCULINACEA

Little Cap Shells

Deep-water, limpet-like animals with cap-shaped shells, the apex directed posteriorly and somewhat asymmetrically but mostly situated in the mid-line. Muscle scar horseshoe-shaped, opening anterior. This superfamily is divided into two families with similar shell characters, but anatomical differences warrant their separation.

Family COCCULINIDAE

Nucleus of apex spiral and usually deciduous.

Notocrater Finlay, 1927 (*Cocculina craticulata* Suter, 1908).
tasmanica (Pilsbry, 1895); Shoreham, Western Port.

Family LEPETELLIDAE

Cocculinella Thiele, 1909 (*Acmaea minutissima* Smith, 1904).
coercita (Hedley, 1907); exact locality unknown.

Family NERITIDAE

Nerites

Shell thick, porcellanous, semi-globose with a small spire. Aperture semi-lunate, closed by a limy articulated operculum.

Nerites are principally inhabitants of warmer shores, living on the upper surface of rocks between tide marks. They move about at night feeding on seaweed.

Melanerita von Martens, 1889

Shell more or less globose, usually distinctly ribbed; dominant colour black; moderately small with flat columella surface which as a rule is granulated. Outer edge denticulate within; operculum symmetrically granulated.

TYPE SPECIES *Nerita melanotragus* Smith, 1884.

Black Nerite or Crow

Melanerita melanotragus (A. E. Smith, 1884)
(Figure 113)

Shell black, occasionally bluish black, shining, semi-globose. Spire low. Whorls rounded. Sculpture consisting of uneven, narrow, shallow, spiral grooves, crossed by fine, irregularly spaced growth-striae. Aperture large, semi-lunar, white or bluish white within; outer lip solid, entire, sharp, black at the edge, dentate; columella concave, minutely toothed; inner lip resolving itself into a broad, white, shining callus; operculum calcareous, externally granulate, with an appendage by which it is locked under the edge of the columella. Size: 1 inch.

Localities: Portsea; Point Nepean; Queenscliff.

This is a very common species, and the only representative of the genus in Victoria. The present species is recorded from all the states of Australia and also New Zealand.

Order MESOGASTROPODA

Family LITTORINIDAE

Periwinkles

Spiral shells of few whorls without an umbilicus and having a flattened columella; operculum horny and more or less spiral. Shore living snails usually found at and above high-tide mark.

Problitora Iredale, 1931

Shell thin, smooth, horny, with a large body-whorl; imperforate. Aperture large, round, thin lipped.

TYPE SPECIES *Amauropsis moerchi* A. Adams & Angas, 1863.

Authors have had some difficulty over placing members of this genus and have mistakenly put them in *Amauropsis,* a genus of Naticidae, and in *Pellilitorina,* an Antarctic genus of Littorinidae from which they differ in shape and the hairy periostracum.

Globular Sand Snail

Problitora globula (Angas, 1880)
(Figure 114)

Shell small, light brown, thin, elongately ovate, globose. Spire about one-third the length of shell. Whorls well rounded, sutures canaliculate; not umbilicated. Interior whitish. Aperture pyriform; outer lip thin; columella slightly curved. Size: ½ inch.

Localities: Dredged between Phillip and French Islands (5-6 fathoms); San Remo.

A shell of infrequent occurrence and unknown to the writers from any other Victorian locality. Nothing approaches it in our waters and it should be easily recognized by its thin texture, light brown colour, and turbinate form. Occurs also in South Australia.

Laevilitorina Pfeffer, 1886 (*Littorina caliginosa* Gould, 1849).
mariae (Tenison Woods, 1875); Western Port; Port Phillip.

Melarapha Cristifordi & Jan, 1832

Shell rather thin, conical, with pointed spire and inflated body whorl. Usually sculptured with fine encircling lirae and often variegated with light colours. Mouth ovoid, outer lip thin, columella flattened. Operculum horny.

TYPE SPECIES *Paludina glabrata* Pfeffer, 1821 = *Littorina neritoides* Linné, 1758.

Iredale, 1912, introduced the name *Melaraphe* Menke, 1828, into the Australasian literature for this group of shells and it has been adopted generally since. However, Guiler, 1958, claimed it was not acceptable as the name was invalid and the species should not be separate from *Littorina s.s.* Ferussac, 1822. McMichael, 1959, has reviewed all the evidence and has shown that the species do warrant separation and in this we concur. However, the original introduction is invalid and must be replaced by the next available alternative which is the above name.

Banded Periwinkle or Australwink

Melarapha unifasciata (Gray, 1826)

(Figure 115)

Blue grey shell with a brown apex. A darker band encircling the middle of the body-whorl and continuing up the spiral just above the suture. Columella and aperture chestnut, the latter with a white band near the base. Sculpture of faint encircling grooves. Size: ½ inch.

Localities: Mornington; Portsea; Point Nepean; Queenscliff; Western Port, and all open coast rock platforms.

This is a pretty little blue-tinted species, exceedingly common and generally found in clusters on the rocks at and above high-water mark. The species is very widely distributed, being found in every state in Australia.

This shell was formerly known as *Littorina mauritiana* (Lamarck, 1822), a New Zealand shell which lacks the spiral lirae.

Checked Australwink

Melarapha praetermissa May, 1908

(Figure 116)

Shell white with brown markings frequently zigzag; small, globosely turbinate. Spire very sharp approximately one-third the length of shell. Whorls much rounded, irregularly, concentrically grooved and with growth-lines visible but only under lens. Body-whorl faintly keeled at about the anterior third. Interior reddish brown with a narrow white band at the base. Aperture ovate; outer lip sharp; operculum horny, semicircular, dark brown, thin. Size: ½ inch.

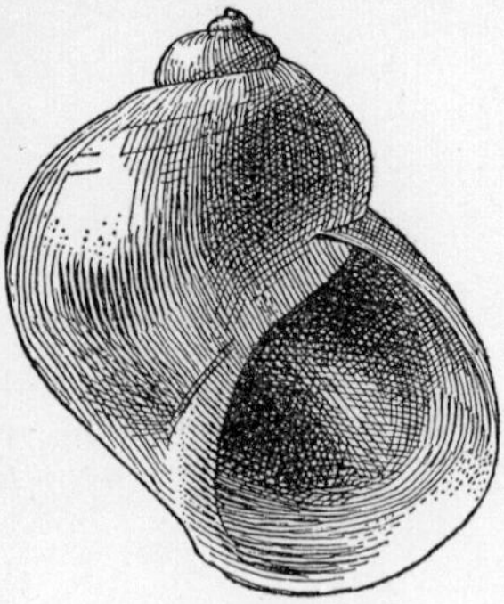
Figure 114

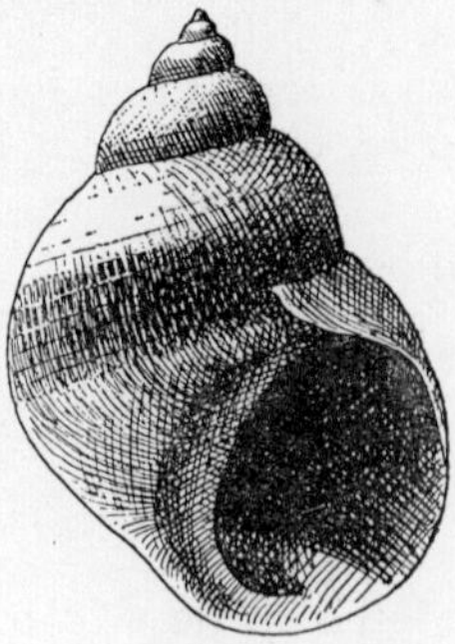
Figure 115

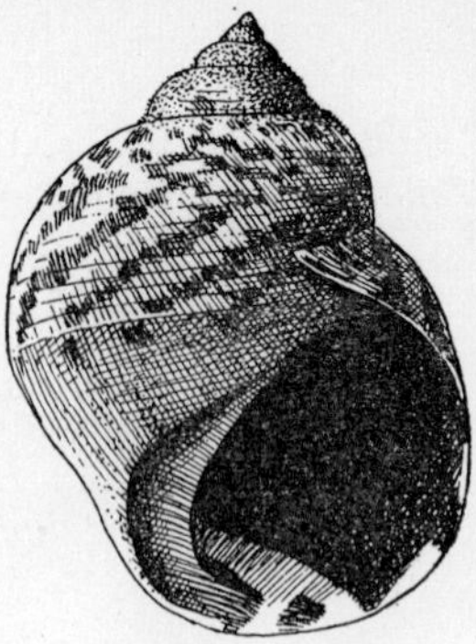
Figure 116

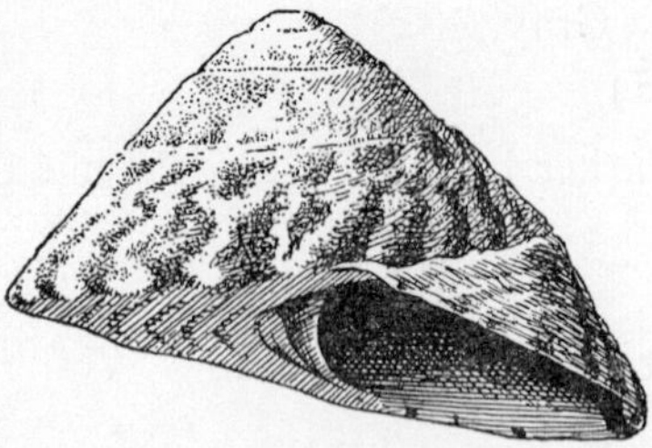
Figure 117

Figure 118

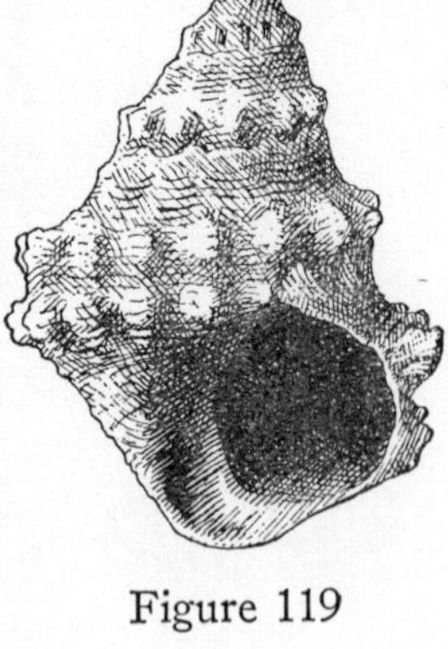
Figure 119

114. *Problitora globula* × 6
115. *Melarapha unifasciata* × 3
116. *Melarapha praetermissa* × 3
117. *Bembicium nanum* × 3
118. *Bembicium auratum* × 3
119. *Nodilittorina pyramidalis* × 3

Localities: Western Port; Williamstown back beach; Lorne; Port Albert; common throughout Victorian shores.

This small Australwink is somewhat similar in shape to *unifasciata* Gray but more globose, and lacks the bluish tint so characteristic of that species. The habits of these forms are much alike being found in clusters on the rock surfaces at and above high water mark. Occurs also in Tasmania.

This shell was once thought to be the same as *novaezelandiae* (Reeve, 1857), but is now separated.

paludinella (Reeve, 1857); occurs in New South Wales, Victoria and Tasmania, living in the upper littoral of rock platforms among and above the barnacle *Chamaesipho columna*. These tiny dark shells are so small that they are often either overlooked or dismissed as the young of one of the other species. Hedley confused this species with *Ampullaria tasmanica* Tenison Woods, 1876, a Tasmanian fluviatile species without the flared lip.

Bembicium Philippi, 1846

Shell trochiform, imperforate, with a flat base, last whorl somewhat keeled; often sculptured with spiral lines or nodules; mouth oblique; outer lip thin; operculum horny.

TYPE SPECIES *Trochus melanostoma* Gmelin, 1789.

Risella Gray, 1847, is an absolute synonym of this genus and was used previously by Australian conchologists for these shells.

Shells of this genus show considerable variation in shape and colour, and this has led, in the past, to a number of species being described. However, recent work by Mrs Hilda Anderson shows that there are three species in south-eastern Australia, and that all other names are synonyms of one or other of these.

Striped-mouth Conniwink

Bembicium nanum (Lamarck, 1822)
(Figure 117)

Shell small, light brown, top-shaped; angulated, with a flattish concave base; imperforate; last whorl somewhat keeled. Whorls flatly convex with oblique zebra-like dark bands, surface more or less corroded. Mouth shining, with the brown stripes showing on the sharp outer lip; operculum horny. Size: ½ inch.

Localities: Lorne; Port Phillip; rocky parts of coast generally. Located on the upper littoral of exposed rock platforms in Queensland, New South Wales, Tasmania and South Australia as well as Victoria.

Gold-mouthed Conniwink

Bembicium auratum (Quoy & Gaimard, 1834)
(Figure 118)

Shell small, yellow or brown, more or less articulated with bluish grey but extremely variable in colour and pattern; trochiform. Spire approximately two-thirds the length of shell. Whorls rugose, flattish, overlapping. Body-whorl possessing a strongly tuberculated keel at the anterior third;

base flat or sometimes concave. Interior dark chocolate. Aperture transversely ovate, outer lip acute; operculum horny. Size: ¾ inch.

Localities: Port Phillip; Western Port; Port Albert.

The rugose sculpture and dark mouth readily distinguish this shell from the following species.

Found living more often in quieter waters of sheltered muddy inlets and estuaries, occurring also in Queensland, New South Wales, Tasmania and South Australia.

melanostomum (Gmelin, 1791); inhabits sheltered bays, inlets and mangrove swamps. It is smaller and less strongly keeled than the above species with a dark outer lip and the columella tinged with orange.

Nodilittorina von Martens, 1897

Shell small, tuberculate, imperforate. Mouth smooth within; columella broad.

TYPE SPECIES *Littorina pyramidalis* (Quoy & Gaimard, 1833).

Members of this genus were included once under *Tectarius* from which they differ in shape and sculpture. There has been confusion on the genotype but R. Tucker Abbott, 1954, clearly outlined the evidence and selected the above species.

Tubercled Noddiwink

Nodilittorina pyramidalis (Quoy & Gaimard, 1833)

(Figure 119)

Shell small, bluish grey, turbinate, tuberculate. Spire acuminate, slightly less than half the length of shell. Whorls sculptured with numerous, concentric ridges crossing the irregular lines of growth. On the body-whorl appear two very prominent nodulose keels almost centrally situated, the nodules being fairly evenly spaced. Interior reddish brown, shining. Aperture ovate; outer lip acute; columella curved, somewhat calloused; operculum thin, horny. Size: ½ inch.

Locality: Mallacoota.

Found living gregariously on the surface of rocks at and above high-water mark. Its turbinate shape and strong, evenly spaced tubercles provide features to readily identify the species. Occurs commonly on the New South Wales coast and through the Indo-Pacific region.

Previously known as *tuberculata* (Menke, 1828), a western Atlantic species.

Family RISSOIDAE

Rissoids

Shells mostly small, conical, cylindrical or turreted, smooth or variously sculptured. Aperture round or ovate sometimes with an anterior sinus; operculum horny with or without an inner process.

This is a large family of small shells, the members of which show considerable variation in shape and sculpture. The species have been grouped into a number of genera by the various authors and several are mentioned below. However, because of their small size no attempt is made here to discuss the family in detail. Comprehensive papers on the Australian forms are Laseron (1950) and Cotton (1944). The following genera and species are recorded from Victoria:

Notosetia Iredale, 1915 (*Barleeia neozelanica* Suter, 1898).
atropurpurea (Frauenfeld, 1867); Rhyll.
muratensis Cotton, 1944; Western Port.
nitens (Frauenfeld, 1867); Rhyll.
pellucida (Tate & May, 1900); Anglesea.
pertranslucida (May, 1912); Kilcunda.
purpureostoma May, 1919; Western Port.
simillina (May, 1915); off Ninety Mile Beach (40 fathoms).

Pelecydium Fischer, 1885 (*Pelecydium venustulum* Folin, 1871).
Hedley, 1904, erected the genus *Epigrus* with *Rissoina cylindracea* Woods, 1878, as type species for the Australian shells of this form. Recent authors have realized their relationship to *Pelecydium* and *Epigrus* is now placed in its synonymy.
apicilata (Gatliff & Gabriel, 1913); Port Albert; Western Port (6 to 8 fathoms).
badius (Petterd, 1884); Western Port; Kilcunda.
cylindraceus (Tenison Woods, 1878); off Wilson's Promontory.
dissimilis (Watson, 1886); Port Fairy.

Rissopsis Garrett, 1873 (*Rissopsis typica* Garrett, 1873).
Cotton, 1944, erected *Eusetia* with *Rissopsis expansa* Powell, 1930, as type species and included the Australian species previously consigned to the above genus. We, however, consider the species closer to *typica* than to *expansa* and prefer to retain *Rissopsis.*
brevis May, 1919; Bass Strait.
bulimnoides Tate & May, 1900; Port Albert.
consobrina Tate & May, 1900; Ocean Beach, Point Nepean.
maccoyi Tenison Woods, 1876; Flinders; Western Port.

Lironoba Iredale, 1915 (*Rissoa suteri* Hedley, 1904).
agnewi (Tenison Woods, 1876); Portsea, Port Phillip.
archensis (May, 1913); 45 miles east of Lakes Entrance.
imbrex (Hedley, 1908); San Remo.
layardi (Petterd, 1884); exact locality unknown.
schoutanica (May, 1912); off Ninety Mile Beach (40 fathoms).
tenisoni (Tate, 1899); a small, white, turreted shell about 1/12 inch. Whorls seven, rounded, sculptured with strong, elevated, spiral ridges, five on the body-whorl then three to apex. A very common species and distributed on the whole coastline; it is found also in New South Wales, Tasmania and South Australia. Originally described as *Rissoa* (*Cingulina*) *australis* Tenison Woods, 1877, the name was preoccupied by *Rissoa* (*Oina*) *australis* Sowerby, 1876, another species.

Pisinna Monterosato, 1884 (*Rissoa punctulum* Philippi, 1836).
Iredale, 1915, erected *Estea,* with *Rissoa zosterophila* Webster, 1905, as type species, for the Australasian forms, but recent authors place it in the synonymy of the above genus.
approxima (Petterd, 1884); Western Port.

bicolor (Petterd, 1884); Portsea.

columnaria (Hedley & May 1908); off Wilson's Promontory.

frenchiensis (Gatliff & Gabriel, 1908); Kilcunda, Western Port; Cape Schanck; Sorrento.

incidata (Frauenfeld, 1867); Western Port; Kilcunda. Cotton, 1944, renamed the South Australian shell previously placed in this species *erma,* but Victorian shells are definitely conspecific with the New South Wales species.

janjucensis (Gatliff & Gabriel, 1913); Jan Juc; Western Port (8 to 10 fathoms).

kershawi (Tenison Woods, 1877); off Rhyll, Western Port; a small light brown pupaeform shell. Whorls six, convex and ornamented with close, longitudinal, slightly oblique ribs; apex blunt. Aperture rounded. Size: about ½ inch. Locality: Western Port, and occurring also in Tasmania.

microcosta (May, 1919); off Wilson's Promontory.

obeliscus (May, 1915); San Remo; off Ninety Mile Beach (40 fathoms).

olivacea (Frauenfeld, 1867); Sorrento; Kilcunda.

pyramidata (Hedley, 1903); exact locality not recorded.

rubicunda (Tate & May, 1900); Western Port.

subfusca (Hutton, 1873): between Phillip and French Islands, Western Port; Portsea, Port Phillip.

tasmanica (Tenison Woods, 1875): Port Fairy.

tumida (Tenison Woods, 1875); Western Port.

Microdryas Laseron, 1950 (*Cingula australiae* Frauenfeld, 1867).

australiae (Frauenfeld, 1867); Kilcunda.

iravadioides (Gatliff & Gabriel, 1913); off Wilson's Promontory; Western Port (8 to 10 fathoms).

Subestea Cotton, 1944 (*Alvania seminodosa* May, 1915).

flindersi (Tenison Woods, 1876); exact locality not recorded.

selebrosa (Frauenfeld, 1867); Western Port; Kilcunda.

Scrobs Watson, 1886 (*Rissoa jacksoni* Brazier, 1894).

jacksoni (Brazier, 1894); Kilcunda; San Remo.

Notoscrobs Powell, 1927 (*Notoscrobs ornata* Powell, 1927).

petterdi (Brazier, 1894); a minute, brown, smooth species, elongately ovate. Whorls four and a half, rounded. Aperture oval. Size: 1/25 inch. Common in Port Phillip and Western Port and appearing also in New South Wales and Tasmania.

triangulus (May, 1915); exact locality not recorded.

Botelloides Strand, 1926 (*Onoba bassianus* Hedley, 1911).

bassianus (Hedley, 1911); Port Albert.

Merelina Iredale, 1915 (*Rissoa cheilostoma* Tenison Woods, 1876).

cheilostoma (Tenison Woods, 1876); a small, elongated, turreted, solid, yellowish shell, ⅛ inch, with a conspicuously rounded aperture. Whorls seven. Sculptured with longitudinal and concentric ribs giving it a clathrate appearance. Obtained in Port Phillip, Western Port, and occurs also in New South Wales, Tasmania and South Australia.

apicilirata (Tate & May, 1901); Western Port.

gracilis (Angas, 1877) ; Ocean Beach, Flinders.
hulliana (Tate, 1893) ; Port Phillip; Western Port.
strangei (Brazier, 1894) ; Kilcunda; Western Port.

Linemera Finlay, 1924 (*Linemera interrupta* Finlay, 1924).
filocincta (Hedley & Petterd, 1906) ; off Ninety Mile Beach (40 fathoms).
suprasculpta (May, 1915) ; exact locality not recorded.

Turboella Leach, 1847 (*Turbo parva* Da Costa, 1826).

Iredale, 1915, erected *Haurakia* for the Australasian shells of this type with *Rissoa hamiltoni* Suter, 1898, as type species. Recent authors have included it in the synonymy of the above genus.

descrepans (Tate & May, 1900) ; Shoreham.
liddelliana (Hedley, 1907) ; off Wilson's Promontory.
supracostata (May, 1919) ; exact locality unknown.

Rissoina d'Orbigny, 1840

Shell turreted, mostly colourless; spirally ribbed. Aperture half rounded, produced dorsally; columella oblique, aperture edge thickened dorsally with a sinus; operculum solid, inside with an oblique process on the nucleus.

TYPE SPECIES *Rissoina inca* d'Orbigny, 1840.

d'Orbigny's Rissoina

Rissoina d'orbignyi (A. Adams, 1851)
(Figure 120)

Shell small, solid, white or cream, elongated. Whorls about eight, flattish, sculptured with closely set, longitudinal ribs a feature more evident on the spire, and fine microscopic, concentric striae. Interior white, shining. Aperture elongately ovate; outer lip thickened. Size: ½ inch.

Localities: Port Phillip; Western Port.

A very common species washed up on the beaches and distributed on the whole coastline; in life located under stones often encrusted with serpulae. It is readily recognized by its elongate shape, solid texture, longitudinal sculpture and expanded, thick, outer lip.

elegantula Angas, 1880; size: ¼ inch; a very small, milky white, elongated shell possessing eight whorls sculptured with regular, close longitudinal riblets, the interspaces ornamented with extremely fine microscopic striae, visible only with the aid of a lens. Like the above species, it possesses a thick outer lip; it is, however, easily separated by its smaller size and much finer ornament.

Locality: Port Phillip; Western Port; Kilcunda; Barwon Heads.

fasciata A. Adams, 1851; locality not recorded.
gertrudis Tenison Woods, 1875; exact locality not recorded.
lintea Hedley & May, 1908; Bass Strait.
nivea A. Adams, 1851; exact locality not recorded.
rhyllensis Gatliff & Gabriel, 1908; San Remo; off Rhyll, Western Port; Ocean Beach, Point Nepean.
spirata (Sowerby, 1824) ; exact locality not recorded.
variegata Angas, 1867; Point Roadknight.

Stiva Hedley, 1904 (*Stiva ferruginea* Hedley, 1904).

This genus is very similar to *Rissoina s.s.* but its consistently larger size and stronger ribbing warrant its distinction.

ferruginea (Hedley, 1904); off Cape Everard (65 fathoms).

royana (Iredale, 1924); one of the largest members of the family in Victoria. Size: ¾ inch. Obtained in ten fathoms off Gabo Island and off Lakes Entrance. The author described it as awl-shaped, apex blunt; mouth ovate, slightly channelled, anteriorly. Colour white marbled with orange, forming a subsutural band in many cases; the apical whorls uniform orange. The ornament consists of longitudinal ribs about thirty-two on the penultimate whorl, the interstices being wider and latticed with very fine scratched lines.

Rissolina Gould, 1861 (*Rissoina elegantissima* d'Orbigny, 1840).

angasi (Pease, 1872); widespread on coast.

crassa (Angas, 1871); Kilcunda; Western Port.

Anabathron Frauenfeld, 1867 (*Anabathron contabulatum* Frauenfeld, 1867).

contabulatum Frauenfeld, 1867; Flinders; Cowes; Sorrento.

Eatoniella Dall, 1876 (*Eatonia kerguelenensis* Smith, 1875).

Iredale, 1915, erected *Dardanula* to replace *Dardania* Hutton, 1882 (*Dardania olivacea* Hutton, 1882); a genus which Suter had already relegated to the synonymy of *Eatoniella* an opinion which both Thiele and Wenz uphold and in which we concur.

aurantiocincta (May, 1915); off Wilson's Promontory.

dubitabilis (Tate, 1899); off Rhyll, Western Port.

erratica (May, 1912); off Wilson's Promontory.

flammea (Frauenfeld, 1867); Portsea, Port Phillip; Western Port.

melanchroma (Tate, 1899); Western Port; Port Phillip.

Skenella Pfeffer, 1886 (*Helix serpuloides* Montagu, 1808).

bruniensis (Beddome, 1882); exact locality not recorded.

Tatea Tenison Woods, 1879 (*Tatea huonensis* Tenison Woods, 1875 = *Diala rufilabris* A. Adams, 1862).

rufilabris (A. Adams, 1862); from Frankston creek; Portland; Western Port; Tarraville; Lakes Entrance. A small elongated form, ⅙ inch, horn-coloured with dark brown epidermis. Whorls eight, flat. Spire acuminate. Aperture ovate; operculum horny. Professor Tate commenting on the habitats of this species remarks 'usually on submerged weeds, or sheltering under stones at the margin in watercourses subject to tidal influences, though occasionally in fresh water, but rarely in waters wholly saline. It is a characteristic estuarine shell, but is present in some inland brackish waters; is abundant throughout southern Australia and Tasmania, and extends to the Clarence River in New South Wales.'

Many of the minute forms of Rissoidae appear in shell sand in countless numbers; in life they are located mostly in seaweed both in shallow and deep water. It is suggested that collection may be simplified by obtaining seaweed from rock-pools and under stones and immersing it in fresh water when the little shells will be seen to drop off. It is obvious that to deal satisfactorily with the group much material from different localities is necessary but the student will be rewarded for his efforts in a collection of minutae possessing a great variety of colour, shape and ornament.

Family ASSIMINIDAE

Assiminids

Shell usually less than ½ inch, mostly ovate either high or flattened. Mouth round or ovate, mostly without an operculum; operculum horny as a rule, sometimes calcareous.

Assiminea Fleming, 1828 (*Assiminea grayana* Fleming, 1828).

Thiele, 1927, divided the genus into a number of subgenera and sections *Metassiminea,* with *Assiminea philippinea* Boettger, 1893, as type species, was erected to include the Philippines and Australian species.

brazieri (Tenison Woods, 1876); from Frankston; Western Port; Anderson's Inlet; Port Albert. A small species ⅛ inch, conoid, cream with one broad, brown encircling band; whorls flatly rounded, angulated at the periphery; operculum horny.

tasmanica Tenison Woods, 1876; from Western Port; Port Albert. A small form ⅙ inch, somewhat resembling the above but with no colour band.

Family TORNIDAE (ADEORBIDAE)

Tornids

Shell small, mostly with a low spiral, seldom round or conical; colourless; smooth or variously sculptured. Mouth oblique, round or ovate; operculum thin, horny, spiral.

Cochliolepas Stimpson, 1858 (*Cochliolepas parasitica* Stimpson, 1858).

Hedley, 1913, elected *Naricava* with *Adeorbis angasi* A. Adams, 1863, as type species but later authors have shown that it must be replaced by the older genus.

angasi (A. Adams, 1863); Portsea.

kimberi (Verco, 1907); Shoreham.

vincentiana (Angas, 1880); Portsea; San Remo.

Pseudoliotia Tate, 1898 (*Cyclostrema micans* A. Adams, 1850).

micans (A. Adams, 1850); Port Phillip; Western Port.

Family SKENEOPSIDAE

Skeneops

Shell very small, with a low or flattened spire, whorls rounded, with an impressed suture, umbilicus wide. Mouth round; operculum thin with a central nucleus and a slowly enclosing spiral.

Microdiscula Thiele, 1912 (*Microdiscula vanhoffeni* Thiele, 1912).

charopa (Tate, 1899); off Rhyll, Western Port.

Family RISSOELLIDAE

Rissoellids

Shell small, mostly translucent, thin, umbilicate, round or ovate, smooth. Mouth ovate; operculum thin, nucleus near the middle of the spiral.

Jeffreysiella Thiele, 1912 (*Jeffreysiella notabilis* Thiele, 1912).

Iredale, 1912, erected *Heterorissoa* with *H. secunda* Iredale, 1912, as type for the same group of shells, but as Thiele's name has priority, it must stand.

wilfredi (Gatliff & Gabriel, 1911); Ocean Beach, Point Nepean.

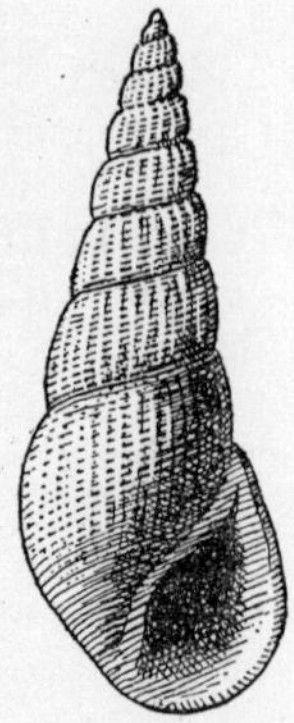

Figure 120

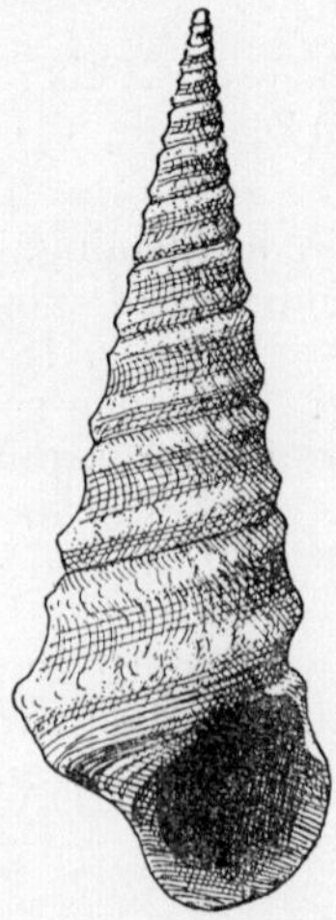

Figure 121

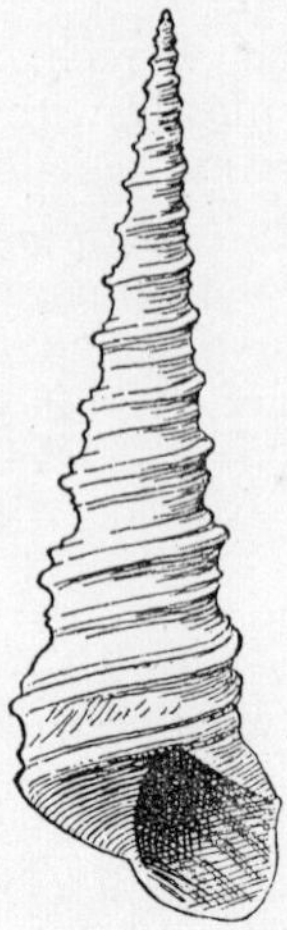

Figure 122

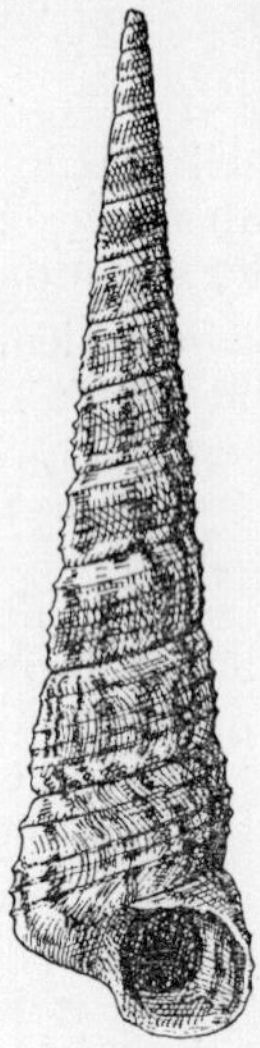

Figure 123

Figure 124

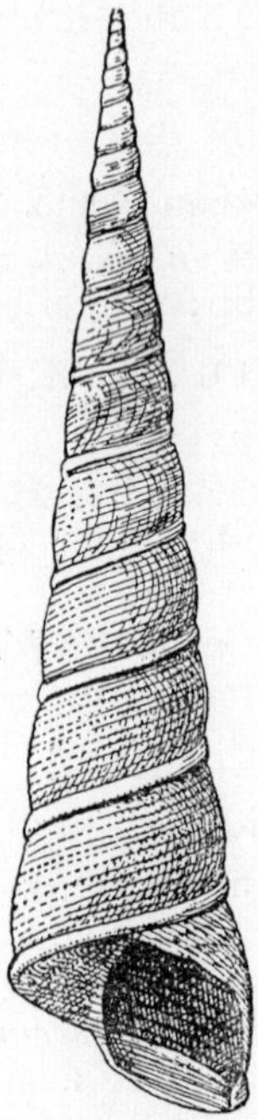

Figure 125

120. *Rissoina d'orbignyi* × 4
121. *Ctenocolpus australis* × 3
122. *Stiracolpus atkinsoni* × 3
123. *Gazameda gunni* × 1⅓
124. *Gazameda iredalei* × 1⅓
125. *Gazameda subsquamosa* × 2

Family TURRITELLIDAE

Turritellas or Screw Shells

Elongated many-whorled shells with a rounded mouth, the outer lip usually thin and sharp; the operculum is horny. A family of sand-dwelling molluscs.

Ctenocolpus Iredale, 1925

Turritellas with the whorls bearing two smooth or knotted keels.
TYPE SPECIES *Turritella australis* Lamarck, 1822.

Southern Granulated Screw Shell

Ctenocolpus australis (Lamarck, 1822)
(Figure 121)

Shell small, strong; cream or yellowish with whorls and base brown banded; acuminately turreted. Whorls about twelve, concentrically sculptured with a prominent granulose keel at the anterior third and one of lesser strength just below the sutures which are deeply impressed. The longitudinal ornament consisting of numerous fine, undulating striae. Interior cream. Aperture rounded; outer lip acute, with a fairly deep sinus centrally situated. Size: 1 inch.

Localities: Western Port; Port Albert and Gippsland coast generally.

A sturdy little species, its strongly granulated character immediately separating it from any member of the family in Victoria. *Turritella granulifera* Tenison Woods, 1875, is a synonym recording its existence in Tasmanian waters.

diffidens Iredale, 1925; a shell of similar shape but without the characteristic granulations; its type locality is Gabo Island.

Stiracolpus Finlay, 1927

Turritella having an apex of one whorl without a keel, the sculpture consisting of three subequal spiral cords and a few weak interstitial riblets.
TYPE SPECIES *Turritella symmetrica* Hutton, 1873.

Atkinson's Screw Shell

Stiracolpus atkinsoni (Tate & May, 1900)
(Figure 122)

Shell small, white, cream, or light brown; acuminate, turreted. Whorls twelve to thirteen, carinated, keels strong and of varying strength, on the penultimate whorl four appear, the peripheral one being much the largest and pronounced, a slight band appearing more or less at the sutures. Interior creamy white. Aperture rounded, outer lip sharp with a central rather deep sinus. Size: 1 inch.

Locality: Bass Strait.

The vernacular name Screw is particularly applicable to this charming little shell which is of rare occurrence in Victoria. A fragile species with much character easily identified by its acuminate form and prominently keeled whorls. Recorded also from Tasmania where it was for some time known as *Turritella tasmanica* Tenison Woods, 1877 (non Reeve).

Gazameda Iredale, 1924

Turritella with long spire, sinuate mouth; long, obtuse, glossy and almost Scaphelloid protoconch; operculum simple. The animal is ovoviviparous.
TYPE SPECIES *Turritella gunni* Reeve, 1849.

Gunn's Screw Shell

Gazameda gunni (Reeve, 1849)

(Figure 123)

Shell cream, elongately acuminated, turreted. Whorls sixteen to twenty, narrow, flatly concave, the centres of which are concentrically grooved; sutures deep. Aperture rounded, lip thin; base pinkish; operculum horny. Size: 2 inches.

Locality: Point Nepean; off Lakes Entrance.

This is the largest and strongest of our Screw Shells, easily recognized. This species is more frequently seen on eastern Victorian shores, and is also recorded from New South Wales and Tasmania.

Turritella philippensis Watson, 1881; regarded as a synonym of this species.

Latticed Screw Shell

Gazameda iredalei Finlay, 1927

(Figure 124)

Shell cream or light brown, obscurely marked with spots of reddish brown; keels, paler, narrow, lanceolately acuminate. Whorls sixteen to eighteen, bicarinated on the lower half and further ornamented with fine concentric lirae, and numerous close, longitudinal, somewhat undulating striae. Sutures well defined. Interior creamy. Aperture rounded, outer lip acute. Size: 1½ inch.

Locality: South-west Victorian coast.

This species was known as *Turritella clathrata* Kiener, 1844, but preoccupation of the specific name necessitated a change.

This graceful, acuminated Screw Shell with its two keeled feature is easily recognized and cannot be confused with any Victorian form. A species not of common occurrence in our waters and known only to the writers from the above locality. Recorded also from Tasmania.

Acute or Sharp Screw Shell

Gazameda subsquamosa (Dunker, 1871)

(Figure 125)

Shell cream or light brown, acutely lanceolately turreted. Whorls fifteen to eighteen, flattish or flatly convex, earlier ones smooth, later whorls encircled with seven or eight slightly elevated, irregular ridges and a well-developed keel at the base of each whorl close to suture. Further sculptured with numerous, fine, longitudinal, undulating lamellate striae. Interior cream. Aperture rounded, outer lip sharp, with a rather deep sinus. Size: 1½ inches.

Locality: Cowes; dredged between Phillip and French Islands (8 fathoms).

An acutely lanceolate species, somewhat resembling *gunni* (Reeve) but easily separated by its flatly convex whorls and more numerous concentric ridges. *Turritella lamellosa* Watson, 1881, and *T. acuta* Tenison Woods, 1875, are synonyms. It occurs in the waters of Tasmania and New South Wales.

decoramen Iredale, 1936; off Lakes Entrance.

Colpospira Donald, 1900 (*Turritella runcinata* Watson, 1881).

guilleaumei Iredale, 1924; off Lakes Entrance.

quadrata (Donald, 1900); from Bass Strait. A small shell somewhat pyramidal in shape; colour cream with brownish maculations and darker spots on the two encircling ridges situated a little above and below the sutures. Whorls twelve to fourteen, the latter ones showing slight concavity between the ridges, the interspaces being ornamented with fine concentric striae. Interior light brown. Aperture subquadrate; outer lip thin with a well-developed sinus. Size: ¾ inch.

sinuata (Reeve, 1849); has been united with the above species but it is distinguished by its narrower and deeper sinus and well-developed encircling ridges.

smithiana Donald, 1900; San Remo.

runcinata (Watson, 1881); from off East Moncoeur Island in sand and shell at 38 fathoms, is yellowish more or less tinged with and speckled with brown and white; sculpture consisting of rounded, unequal, irregular, interrupted concentric threads, many stronger than the rest. Size: 1¼ inch. *Turritella accisa* Watson, 1881, is a synonym of this species.

Kimberia Cotton & Woods, 1935 (*Turritella kimberi* Verco, 1908).

microscopica (May, 1910); off Ninety Mile Beach (40 fathoms).

Family MATHILDIDAE
Mathilds

Shell conical with a high spire, a smooth sinistral embryonic shell, and following whorls with spiral threads, mostly crossed with longitudinal striae. Mouth small, angular, outer lip thin, columella smooth; operculum thin, with many narrow whorls.

Opimilda Iredale, 1929 (*Mathilda decorata* Hedley, 1903).

decorata (Hedley, 1903); off Lakes Entrance.

Glyptozaria Iredale, 1924 (*Turritella opulenta* Hedley, 1907).

euglypta (Iredale, 1929); off Cape Everard (65 fathoms).

opulenta (Hedley, 1907); off Wilson's Promontory.

Family SOLARIIDAE
Staircase Shells

Shell conical or disc-shaped with an open umbilicus. Whorls spirally sculptured. Aperture circular or angular and closed by a horny operculum.

Philippia Gray, 1847

Shell discoidal, subconical, smooth aperture subquadrate; umbilicus wide, the margin crenulated; operculum horny, flat orbicular, many-whorled.

TYPE SPECIES *Solarium lutea* Lamarck, 1822.

Yellow Sundial Shell

Philippia lutea (Lamarck, 1822)

(Figure 126)

Shell small, obtusely conical, thin, smooth. Whorls five, flatly convex; yellowish, encircled at the periphery of body-whorl with two white, regular, reddish beaded keels. Interior white. Aperture rounded; umbilicus small, deep, with white crenulations; outer lip thin; base flattish; operculum flat, horny. Size: ½ inch.

Localities: Portsea; Western Port; Barwon Heads; Lorne; Portland.

This little Sundial is the only representative of the genus in Victoria and easily recognized by its flattish form and the outstanding, symmetrical red beaded keels. Occasionally specimens are cast up on the beaches; they are not littoral in their habits and like the larger tropical species apparently prefer deeper waters. For many years known under the genus *Solarium* Lamarck, 1799. Occurs also in New South Wales, Tasmania through Bass Strait to South Australia.

Family VERMETIDAE

Worm Shells

Shell irregularly coiled, fused to substratum, septate. Animal vermiform. Operculum simple, horny, becoming reduced or lost.

Serpulorbis Sasso, 1827

Shell tubular, irregularly twisted, attached. Aperture circular, margin thin continuous; operculum absent.

TYPE SPECIES *Serpulorbis polyphragma* Sasso, 1827.

New Holland Worm Shell

Serpulorbis sipho (Lamarck, 1818)

(Figure 127)

Shell light brown, agglomerated, tubular, irregularly twisted; sculpture consisting of rugose growth-striae, crossed by numerous, unequal, longitudinal riblets; the latter sculpture appears to predominate, but frequently the surface is irregularly reticulated by the growth-marks. Aperture rounded. Size: 6 inches.

Localities: St Kilda; Frankston; Portsea; Altona; and common in Western Port.

A distinctive shell, often in large colonies and generally attached to stones and other shells. It is frequently found on our shores, but the better examples are obtained by dredging. These shells resemble Serpulae in general appear-

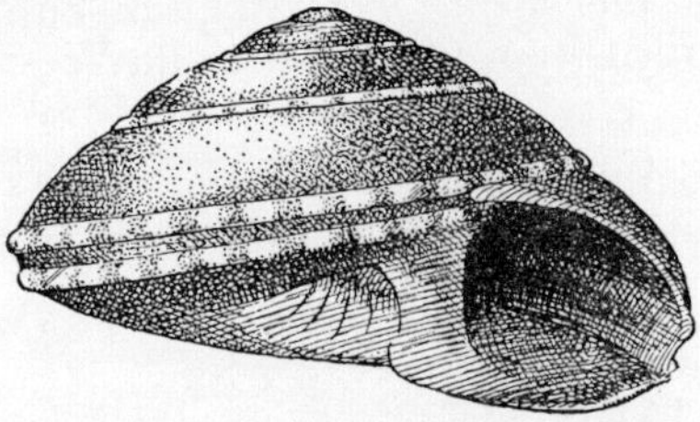

Figure 126

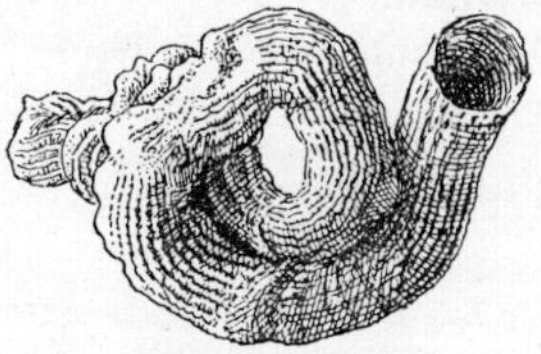

Figure 127

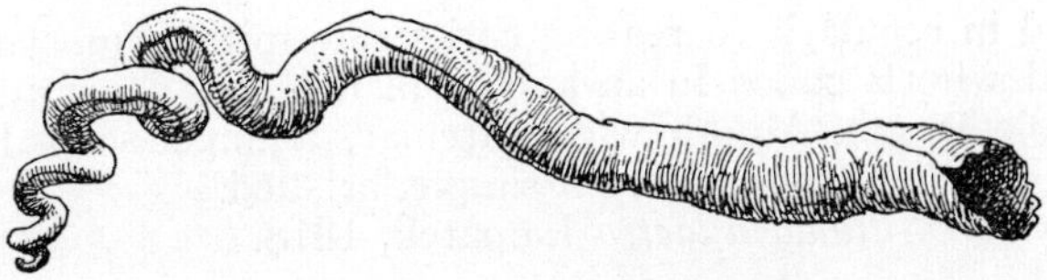

Figure 128

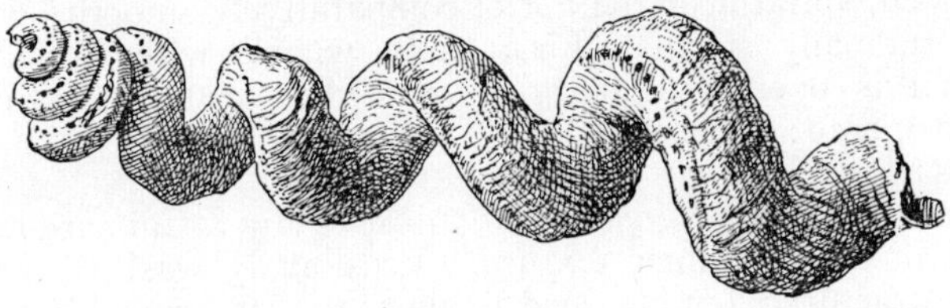

Figure 129

126. *Philippia lutea* × 4⅓
127. *Serpulorbis sipho* × 1
128. *Pyxipoma weldii* × 1⅓
129. *Siliquaria australis* × ⅔

ance, although regularly spiral near the apex. They appear on the whole Victorian coastline and are recorded from Queensland and Western Australia.

Hedleyi (Finlay, 1927); San Remo; Cowes.

Vermicularia flava Verco, 1907; this is the tube of a Polycheate worm, not a mollusc.

Family SILIQUARIIDAE

Slit Worm Shells

Shell loosely spiral, embedded in substratum, non-septate. Animal never vermiform. Operculum always well developed, horny, bristly.

Pyxipoma Mörch, 1860

Shell coiled in regularly increasing corkscrew spiral; embryonic shell one and one-third whorls coiled in an almost plane spiral; longitudinal fissure open its whole length. Mouth wide, circular, trumpet-shaped; operculum elevated to form a tall dome, margin simple, bristled.

TYPE SPECIES *Siliquaria lactea* Lamarck, 1818.

Weld's Worm Shell

Pyxipoma weldii (Tenison Woods, 1875)

(Figure 128)

Shell white, yellowish or very light pink; thin, fragile, tubular, irregularly loosely twisted, spiral near the apex; comparatively smooth but under lens finely longitudinally striated. First three whorls with the fissure filled, afterwards it is open with margins somewhat acute. Aperture circular; margin sharp. Size: 3 inches.

Localities: Port Phillip; Western Port.

Somewhat resembling *Siliquaria australis* (Quoy & Gaimard), but smaller in size and of much thinner texture. Occasionally washed up on beaches more or less fractured, but obtained living at the last named locality between Phillip and French Islands, 8 fathoms, in colonies embedded in sponges. Masses not infrequently measure 6 to 8 inches in diameter. Beyond Victoria it occurs in New South Wales, Tasmania and South Australia.

Siliquaria Bruguière, 1792

Shell tubular, cylindrical, irregularly twisted; apex subspiral. Whorls with, on the dorsal side, a longitudinal fissure modified into a series of holes extending from the apex to the mouth. Aperture circular; operculum horny, cylindrical.

TYPE SPECIES *Serpula anguina* Linné, 1758.

There has been some confusion as to whether *Siliquaria* or *Tenagodus* Guettard, 1770, is the name to be used for the type genus of this family, but in a personal communication Dr John Morton advises the continued use of *Siliquaria* for the present.

Southern Worm Shell

Siliquaria australis (Quoy & Gaimard, 1834)

(Figure 129)

Shell creamy white, large, tubular, thick, irregularly twisted, spiral near apex. Sculpture consisting of fine longitudinal striae on the inner side, rugose and transversely fissured on the outer side, with a longitudinal fissure radiating from the apex and proceeding through all the whorls. Aperture circular. Size: 8 inches.

Locality: Point Nepean.

Frequently washed up on beaches. In life these shells are found in colonies generally embedded in sponges. On one of her fishing cruises, the trawler *Endeavour* obtained in Bass Strait an immense assemblage measuring six feet long by one foot. The species occurs in Tasmania and South Australia.

Magilia Velain, 1877 (*Magilia scapuliformis* Velain, 1877).
caperata Tate & May, 1900; Anglesea.

Family CAECIDAE

False Tooth Shells

Shell small tubular, regular, minute, with fugitive spiral nucleus. Mouth circular; operculum horny, with a central nucleus.

Caecum Fleming, 1831 (*Dentalium trachea* Montagu, 1803).
amputatum Hedley, 1893; Ocean Beach, Point Nepean.

Pedumicra Iredale & Laseron, 1957 (*Strebloceras cygnicollis* Hedley, 1904).
cygnicollis Hedley, 1904; Port Albert.

Family PLANAXIDAE

Clusterwinks

Shell of medium size, thick, ovate, not umbilicated and often covered with a strong periostracum. Mouth ovate with a narrow sinus, grooved in the aperture; operculum horny.

Hinea Gray, 1847

Shell smooth, covered with a brownish yellow epidermis. Whorls flattened. Outer lip thickened, grooved internally.

TYPE SPECIES *Buccinum brasiliana* Lamarck, 1822.

Yellow-coated Clusterwink

Hinea brasiliana (Lamarck, 1822)

(Figure 130)

Shell small, ovately conical, creamy, covered by a minutely pilose, brown epidermis, generally absent on the earlier whorls. Spire about half the length of shell. Whorls flatly convex. Interior white. Aperture oval, white, outer lip much rounded, strengthened inside by a thick, white, shining callus, denticulate within; anterior canal very short; operculum horny. Size: ¾ inch.

Locality: Otway coast, between Ryan's Pen and Moonlight Head.

The genus is represented by one species in Victoria easily recognized by its conically oval shape, thickened outer lip and brownish epidermis which is mostly worn off on the earlier whorls. Occurs frequently in New South Wales. For many years known under the name of *Planaxis mollis* Sowerby, 1822.

Family POTAMIDIDAE

Mud Whelks

Shell thick, many-whorled, turreted. Aperture with an anterior sinus, outer lip more or less produced; operculum roundish, spiral, with a central nucleus.

Pyrazus Montfort, 1810

Shell turreted, many-whorled. Whorls rugose, transversely sulcate and granular, varices few, indistinct. Aperture with a short canal in front; columella with an oblique, spiral callosity in the middle; outer lip dilated and thickened, united at the fore-part to the inner lip.

TYPE SPECIES *Strombus palustris* Linné, 1767.

Club Mud Whelk

Pyrazus ebeninus (Bruguière, 1792)

(Figure 131)

Shell large, dark brown, turreted. Spire elongate approximately two-thirds the length of shell, whorls of which are rounded but somewhat angulated at about the centre. Body-whorl more conspicuously nodosely angulated but angle nearer to suture. Sculptured with oblique, longitudinal ribs and numerous somewhat irregular concentric ridges. Interior brown, white towards the outer lip which is strong, sinuate posteriorly and margined with dark brown. Aperture almost circular, anterior canal short, rather broad; columella curved, inner lip reflected; operculum brown, flat. Size: 3 inches.

Localities: Mallacoota; Lakes Entrance; Wingan Inlet.

A remarkable, dark brown turreted shell, the only species of the genus in Victoria and easily recognized. The nearest ally is *Velacumantus australis* (Quoy & Gaimard), but immedaitely separated by its smaller size. This shell will aways possess a sentimental interest to Australian naturalists since it is one of the half dozen shells collected by Captain Cook at Botany Bay in April, 1770.

A species which appears in countless numbers in many parts of New South Wales and Queensland, it inhabits estuaries of rivers, mud-flats and mangrove-swamps near the sea. Extremely gregarious and after receding tides their crossing and recrossing tracks form amazing patterns. It is not infrequent to see young oysters adhering to these Mud Whelks. The animal was used as an article of food by the aborigines, many examples appearing in kitchen middens. Toondah was the name given to the shell by aborigines of Moreton Bay, Queensland. It is recorded also from Tasmania. For many years known as *Potamides ebeninus* Bruguière.

Velacumantus Iredale, 1936

Shell medium sized, solid, sculptured with spiral cords and longitudinal costae. Aperture compressed; anterior canal short; inner lip not produced over the body-whorl.

TYPE SPECIES *Cerithium australis* (Quoy & Gaimard, 1834).

Southern Mud Whelk

Velacumantus australis (Quoy & Gaimard, 1834)

(Figure 132)

Shell brownish black, turreted, many-whorled. Whorls somewhat rounded, ornamented with numerous spiral cords and broad, distant, longitudinal costae. Sutures deeply impressed. Aperture ovate, terminating in a very short canal; outer lip sharp, denticulated by the spiral sculpture; interior dark brown, grooved. Size: 1½ inches.

Localities: Williamstown; Altona.

These molluscs are quite common, inhabiting estuaries of rivers, mangrove-swamps and mud-flats near the sea. In the present species the colour is fairly constant, but as regards sculpture is subject to considerable variation, some examples being more strongly costate. The genus is represented in Victoria by one species and should be easily recognized. It possesses a wide distribution, with records from South Australia, New South Wales, Queensland and Tasmania. This species was for many years known as *Batillaria australis* (Quoy & Gaimard).

Zeacumantus Finlay, 1926

Shell small, many-whorled, sculpture by rows of granules. Aperture oval; outer lip expanded, thin; anterior canal pronounced.

TYPE SPECIES *Cerithidea subcarinata* Sowerby, 1885.

Common Mud Whelk

Zeacumantus diemenensis (Quoy & Gaimard, 1834)

(Figure 133)

Shell greyish brown, small, acuminate. Spire approximately three-quarters the length of shell. Whorls are flatly convex and granulated in four concentric rows of decussating sculpture. Body-whorl exhibiting the same feature except at the lower portion where the granules become less conspicuous. Interior brown, polished. Aperture ovate; outer lip sharp; anterior canal short, fairly wide. Size: 1 inch.

Localities: Western Port; Port Fairy; Mallacoota.

A very acuminate shell and easy to recognize by its strong granulate ornament. In living condition, generally located on weedy sandbanks. During a stay, in 1912, at Port Fairy, F. Chapman, then Palaeontologist to the

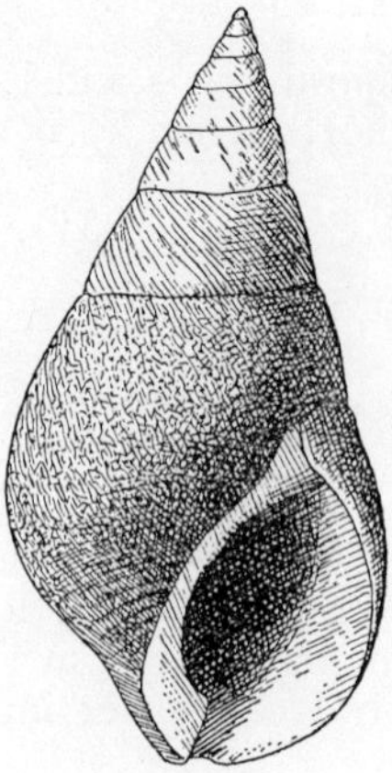

Figure 130

Figure 131

Figure 132

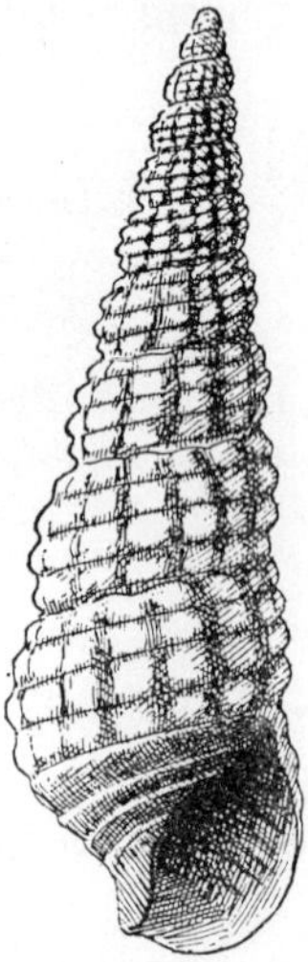

Figure 133

130. *Hinea brasiliana* × 3
131. *Pyrazus ebeninus* × 1
132. *Velacumantus australis* × 2
133. *Zeacumantus diemenensis* × 3

National Museum, Victoria, and C.J.G. were amazed at the tracks made by this curious mollusc. Chapman sent to the *Victorian Naturalist* the following note: 'we noticed in searching the lagoon to the east of the township, a stretch of muddy sand left in a moist condition by the retreating tide which had evidently ebbed away a few hours before. The patch under observation measured about 20 yards by 1 yard, and was covered with a network of shallow grooves about 3 mm. in diameter, and about 1·5 mm. in the centre. There were many other patches of a like nature in the surrounding area. At the terminus of each long trail, and partly ensconced under a tumulus of sand, lay a living Creeper. The mollusc in each case was slowly working its way along as it fed on the organic particles, amongst which, no doubt, Foraminifera were included, for they were common in the sand. In its progression the mollusc tumbled the damp sand to each side, the result of which was to form a fairly smooth groove, the edges of which were raised slightly above the general surface of the sand. From our observations at this spot it is probable that there were many hundreds of these little Gastropods at work.' Found also in Tasmania and South Australia.

The shells grouped in this species show considerable variation in size and sculpture and were originally divided into several species of which *Turritella cerithium* Quoy & Gaimard, 1834, was one. Pritchard & Gatliff (1900 B) suggest that these apparently different forms are really male and females of the same species.

Family CERITHIIDAE
Creepers

Elongated many-whorled shells which have a small oblique aperture with a short anterior canal. The operculum is horny. These molluscs live mostly in shallow water on sea grass and algae.

Diala A. Adams, 1861

Shell elongate, ovate, thin and translucent, usually with white or brown lines or flecks. Protoconch glossy, the following whorls smooth, nodulose or angular, sometimes with a row of knots. Aperture uniform, above angular; columella straight or somewhat oblique, sometimes with a slight sinus at the end.

TYPE SPECIES *Diala varia* A. Adams, 1861.

Spotted Diala

Diala monile (A. Adams, 1862)
(Figure 134)

Shell small, elongately conical, smooth, shining; white, encircled with a reddish brown spotted band just above the sutures, the spots being squarish and spaces between approximately of equal width. Spire slightly more than half the length of shell. Whorls flatly convex, a little angulated on the body-whorl. Interior whitish, through which the spots are plainly visible. Aperture oval; outer lip sharp. Size: ¼ inch.

Localities: Common in Port Phillip particularly on the western shores, such as Portarlington, Altona; Western Port; Portland; Port Albert.

An extremely pretty shell at once recognized by its concentric spotted band of colour, a consistent feature. Occurs also in New South Wales, Tasmania and South Australia.

lauta A. Adams, 1862; from Port Phillip and Western Port. About ¼ inch and frequently associated with the above. A small, smooth, elongated conical species, creamy, girdled throughout with reddish brown spotted lines, this feature showing considerable variation with some specimens exhibiting much heavier bands.

pagodula (A. Adams, 1862); of similar size from Port Phillip and Port Albert. Creamy with a brownish reticulated pattern throughout, whorls flattened and noduled just above the sutures.

picta A. Adams, 1861; from Western Port, similar to the last species but smaller and without the sutural nodules.

imbricata A. Adams, 1862; Western Port.

magna Tate, 1891; Port Phillip Bay (deep water).

phasianella Angas, 1867; Western Port.

pulchra A. Adams, 1862; Port Phillip; Western Port.

semistriata (Philippi, 1836); Western Port; Kilcunda. This species was misidentified as *varia* A. Adams, 1861.

translucida Hedley, 1905; off Phillip Island, Western Port (7 fathoms).

Cacozeliana Strand, 1928

Shell turreted, many-whorled, granular, often with irregular varices. Aperture with a slight canal in front; inner lip simple; outer lip acute not reflexed or expanded; operculum is paucispiral.

TYPE SPECIES *Cerithium lacertinum* Gould, 1859.

These shells were oginially grouped in *Bittium* Gray, 1847, a European genus from which they differ in the few whorled operculum. Iredale erected *Cacozelia* for this Southern hemisphere group, but his name is preoccupied and Strand replaced it with the above name.

Granulated Creeper
Cacozeliana granaria (Kiener, 1842)
(Figure 135)

Shell small, brown or reddish brown but extremely variable in colour; in dead shells may be yellow white, cream; elongate. Spire about four-fifths the length of shell, the whorls of which are flatly rounded and granulated, bearing a few irregular, feeble varices. Sculptured with numerous, longitudinal ribs crossed by concentric ridges, the point of crossing forming well-defined beads. This granulated ornament is continued on the body-whorl except at the anterior end where the spirals are less conspicuous and more or less without granules. Interior reddish brown, polished. Aperture ovate; outer lip acute; anterior canal short, wide; operculum brown, horny. Size: ¾ inch.

Localities: Western Port; very common along the whole coastline.

Found alive on weedy sandbanks. A species somewhat approaching *Zeacumantus diemenensis* (Quoy & Gaimard, 1834) but separated by its more cylindrical shape, smaller size and finer granulated sculpture; widely distributed, being recorded from New South Wales, Tasmania and South Australia.

icarus Boyle, 1880; Flinders; San Remo; Portsea.

Eubittium Cotton, 1937 (*Bittium lawleyanum* Crosse, 1863).
lawleyanum (Crosse, 1863); an elongate form much smaller in size, dark brown and sculptured with numerous concentric striae.
insculptum Reeve, 1865; Western Port; Port Phillip; Corner Inlet (on weed banks).
Batillariella Thiele, 1929 (*Bittium estuarina* Tate, 1893).
estuarina (Tate, 1893); Western Port; Port Phillip. A rather stout form, reddish brown, covered by a bluish black epidermis; longitudinally costate and with several concentric ridges to each whorl.

Hypotrochus Cotton, 1932

Shell thin, subtranslucent. Whorls bicarinate, ventricose, longitudinally plicate, beset with varices and spirally striate. Aperture produced anteriorly into a short, narrow canal; protoconch very small, horny, smooth, of two and a half whorls.

TYPE SPECIES *Cerithium monachus* Crosse & Fischer, 1864.

Species of this genus were originally grouped in *Plesiotrochus,* a genus of fossils.

Monk Shell
Hypotrochus monachus (Crosse & Fischer, 1864)
(Figure 136)

Shell fragile, acuminate, cream or light brown, covered with reddish brown markings between the ribs. Spire approximately three-quarters the length of shell. Whorls obtusely keeled at the lower third and sculptured by longitudinal flatly rounded ribs and numerous concentric microscopic striae; sutures deeply impressed. Body-whorl bicarinate, the lower keel conspicuously spotted; whorls bearing about fourteen to eighteen prominent varices. Interior light brown through which the darker markings are plainly visible. Aperture rounded; outer lip sharp; anterior canal very short and wide. Size: ¼ inch.

Localities: Point Nepean; Western Port; Lorne; Portland; Port Albert.

This is a very characteristic shell and should be easily recognized by its peculiar, acuminate shape, prominent varices, and microscopic striae. Found also in Tasmania, South Australia, Western Australia.

Ataxocerithium Tate, 1894

Shell elongate, turreted. The aperture is almost complete posteriorly. There are no varices and there is no posterior canal. The outer lip is expanded anteriorly and laterally projecting, and behind it the whorl is peculiarly contracted. Columella lamellated and elevated, contiguous with the outer lip and forming a tubular canal of moderate length, shortly but abruptly bent to the left and backward at the tip.

TYPE SPECIES *Cerithium serotinum* Crosse & Fischer, 1864.

Square-mouthed Creeper
Ataxocerithium serotinum (A. Adams, 1855)
(Figure 137)

Shell small, elongate, white, cream, pale brown or light pink. Spire slightly more than three-quarters the length of shell, the whorls of which are some-

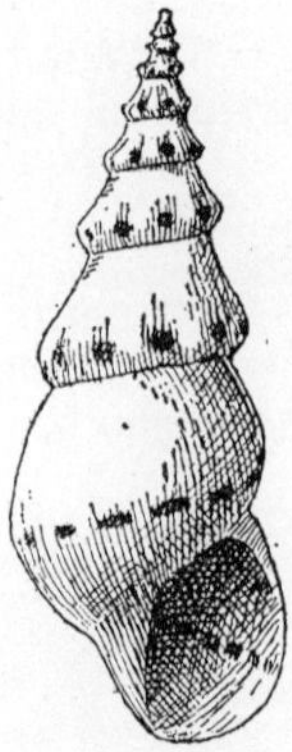

Figure 134

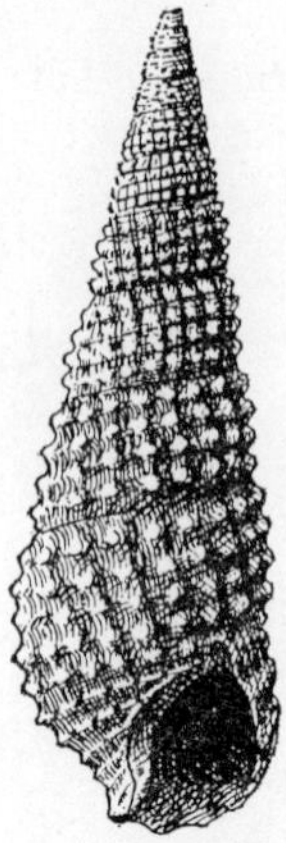

Figure 135

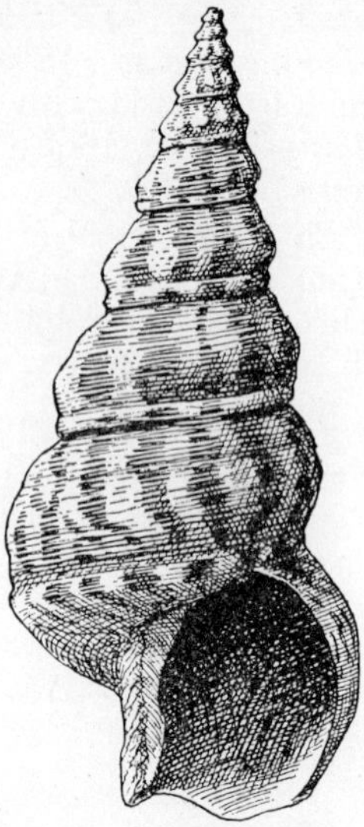

Figure 136

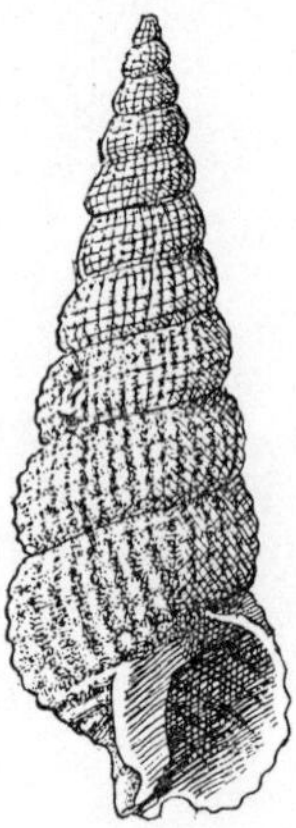

Figure 137

Figure 138

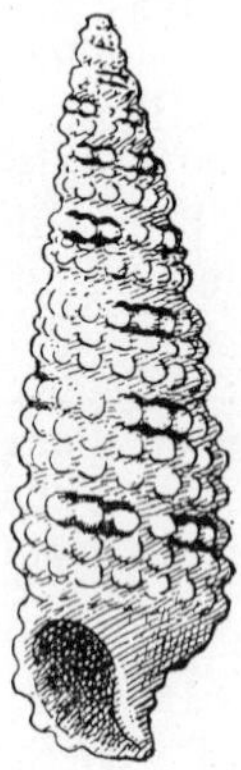

Figure 139

134. *Diala monile* × 8
135. *Cacozeliana granaria* × 3
136. *Hypotrochus monachus* × 3
137. *Ataxocerithium serotinum* × 2⅔
138. *Seila crocea* × 6
139. *Notosinister maculosa* × 6

what rounded and cancellated by longitudinal riblets and concentric ridges; this sculpture is less conspicuous towards the anterior end; sutures deeply impressed. Interior cream. Aperture ovate; outer lip sharp, crenulated; columella reflected; anterior canal somewhat contracted, short. Size: ¾ inch.

Localities: Port Phillip; Western Port; Portland; dredged off Wilson's Promontory.

The shell is not of frequent occurrence on our shores and because of the fineness of its outer lip usually comes in a more or less damaged condition. The characteristic cancellate ornament will serve as a useful recognition mark and the species cannot be confused with any other shell in Victoria. Distribution: New South Wales, Tasmania through Bass Strait to South Australia.

applenum Iredale, 1936; off Lakes Entrance.

scruposum Iredale, 1936; off Lakes Entrance.

Family CERITHIOPSIDAE

Little Creepers

Small, cylindrical, many-whorled shells with a small opening and nearly straight canal. The surface is usually sculptured with beads and most species live in deep water. The members of this family are all small and many are minute, therefore they have an interest for the specialist only. Those wishing to know more of the Australian species should consult the systematic lists of the various states.

Seila Adams, 1861

Shell elongate, conical with flat whorls, sutures often indistinguishable. Aperture with a thin outer margin; the columella flat, broad, short, the base smooth and excavate. Protoconch varied, but with the nucleus unfolded. The sculpture consists of three spiral keels with a fourth on the body-whorl, the surface of the keels smooth and unbroken, the transverse sculpture confined to minute striations between the keels.

Type Species *Triforis dextroversus* A. Adams & Reeve, 1850.

Recent authors have separated the Australian species here listed under *Seila* as *Notoseila* Finlay, 1926 (*Cerithiopsis terebelloides* Hutton, 1873). But as Finlay's separation was mainly on the protoconch which shows considerable variation in the Australian species we prefer to retain Victorian species in *Seila* until further work clarifies the group.

Yellow Creeper

Seila crocea (Angas, 1871)

(Figure 138)

Shell small, elongate, orange. Whorls barely rounded, sculptured by three closely set concentric ridges, the narrow interstices of which are longitudinally, microscopically striated. Interior orange. Aperture ovate; outer lip sharp; anterior canal short, wide. Size: ⅓ inch.

Localities: Port Phillip; Western Port; Portland; dredged off Wilson's Promontory.

It is found also in New South Wales, Tasmania and South Australia.

albosutura (Tenison Woods, 1876) ; from Western Port and Port Phillip. A narrower purplish brown shell which was originally described as having a lighter coloured band at the suture, but this feature is absent in Victorian specimens.

insignis (May, 1911) ; off Ninety Mile Beach (40 fathoms).

marmorata Tate, 1893; exact locality not recorded.

Joculator Hedley, 1909 (*Cerithiopsis ridiculus* Watson, 1886).

cessicius (Hedley, 1905) ; Western Port; Port Phillip.

Seilarex Iredale, 1924 (*Seila attenuatus* Hedley, 1899).

attenuatus (Hedley, 1900) ; Ocean Beach, Point Nepean; Western Port (6 fathoms).

turritelliformis (Angas, 1877) ; misidentified as the above species.

Tubercliopsis Laseron, 1956 (*Tubercliopsis capricornia* Laseron, 1956).

septapila Laseron, 1951; Torquay.

Specula Finlay, 1926 (*Cerithiopsis styliformis* Suter, 1908).

turbonilloides (Tenison Woods, 1878) ; Western Port.

Zaclys Finlay, 1926 (*Cerithiopsis sarissa* Murdoch, 1905).

angasi (Semper, 1874) ; Western Port.

dannevigi (Hedley, 1911) ; San Remo; off Ninety Mile Beach (40 fathoms).

semilaevis (Tenison Woods, 1877) ; San Remo.

Family TRIPHORIDAE

Sinistral Creepers

Small elongate, left-handed snails, the shell is many-whorled and the surface is usually beaded. The small aperture has a short canal.

This is an interesting group of molluscs easily separated by the sinistral form—the aperture being to the left of the central axis, but specifically somewhat tantalizing as the shells are small, and more or less granular, with this characteristic, although distinctive, at times showing little difference. Furthermore some species exhibit much variation and it requires close examination and often the use of the microscope to detect the distinctions on which the species are founded.

Twenty species are recorded from Victoria with many more throughout Australia. In a handbook of this description it is unnecessary to report on every one in detail. However, the following species will serve to represent the family and if further study is desired it is suggested much information would be gained on reference to papers by Hedley (1902) ; Verco (1909) ; Cotton and Godfrey (1931) ; Laseron (1954 C) ; Laseron (1958).

Notosinister Finlay, 1926

Triphoras with typical sculpture of three rows of gemmules, and a simple aperture in which the posterior canal is reduced to a subsutural slit or sinus; anterior canal short and open; columella pillar broad, the anterior margin of the aperture sometimes expanded into a narrow fold or spur which may reach or partially cross the columella. The protoconch begins as a smooth dome, a keel developing on the second or third whorl; white transverse sculpture may appear as minute threads. The whole protoconch may have from two to five whorls, and adult sculpture appears quite abruptly.

TYPE SPECIES *Triphora fascelina* Suter, 1908.

When Finlay created *Notosinister* he divided it into two subgenera, *Cautor* and *Teretriphora,* based on the protoconch. Cotton and Godfrey in their review of South Australian species accepted these subgenera as of generic rank. However, Laseron disagrees with this and, with the exception of one species, groups all such Australian shells in *Notosinster s.l.*

Splashed Sinistral Creeper

Notosinister maculosa (Hedley, 1903)
(Figure 139)

Shell small, sinistral, elongate, solid; colour white normally irregularly maculated with chocolate or reddish brown immediately under the well-defined sutures. Whorls nine to ten, sculptured with three rows of somewhat flattened tubercles but reduced to two rows in the earlier whorls; the tubercles are fairly evenly spaced, the intervals being of similar width. Interior white. Aperture roundly oval; outer lip thin; anterior canal short, somewhat tubular. Size: ¼ inch.

Localities: San Remo; Flinders; Shoreham.

cinerea (Hedley, 1903); off Rhyll. Laseron, 1958, places this species in his genus *Aclophora* (*A. robusta* Laseron, 1958) but we consider the division is insufficient to warrant generic separation from *Notosinister.*

albovittata (Hedley, 1903); Shoreham.
ampulla (Hedley, 1903); Portsea.
angasi (Crosse & Fischer, 1865); Western Port.
armillata (Verco, 1909); Ocean Beach, Point Nepean; San Remo.
festiva (A. Adams, 1851); Portsea; Anderson's Inlet.
granifera (Brazier, 1894); Flinders.
innotabilis (Hedley, 1903); off Wilson's Promontory.
jacksonensis Laseron, 1954; Torquay.
labiata (A. Adams, 1851); Shoreham.
leuca (Verco, 1909); Western Port.
mammillata (Verco, 1909); Portsea.
nigrofusca (A. Adams, 1851); a misidentification of the species below.
nocturna (Hedley, 1903); San Remo.
pfeifferi (Crosse & Fischer, 1865); off Lakes Entrance; Western Port; Sorrento.
regina (Hedley, 1903); West Head, Flinders.
robusta Laseron, 1954; Portarlington.
spina (Verco, 1909); off Wilson's Promontory.

Eutriphora Cotton & Godfrey, 1931 (*Triphora cana,* Verco, 1909).
cana (Verco, 1909); Portsea.

Isitriphora Cotton & Godfrey, 1931 (*Triforis tasmanica* Tenison Woods, 1876).
disjuncta (Verco, 1909); between Phillip and French Islands, Western Port.

tasmanica (Tenison Woods, 1876); Western Port; Sorrento; Kilcunda.
tasmanica nivea (Verco, 1909); Torquay.

Teretriphora Finlay, 1927 (*Triphora huttoni* Suter, 1908).
gemmegens (Verco, 1909); between Phillip and French Islands, Western Port.

Several other species were originally placed in this genus, but we concur with Laseron in their transfer to *Notosinister*.

Family SCALIDAE

Scalas or Wentle-Traps

Shell spiral, consisting of a number of rounded ribbed whorls, usually white and polished. The mouth round with a reflected outer lip which becomes a new rib-like varix when the mollusc adds to the shell. The mouth is closed with a horny operculum. The animals are carnivorous and when molested emit a purple fluid.

Granuliscala de Boury, 1909

Shell medium, elongately conical. Whorls a little flattened, longitudinal sculpture marked on early whorls, becoming obsolete on latter, interstices minutely striate and roughened so as to appear granulose; whorls many, ribs eight to a whorl. Mouth broadly oval, not complete nor varicose; inner lip reflected, forming a pseudo-canaliculate basal tip with the outer lip.

TYPE SPECIES *Scala granosa* Quoy & Gaimard, 1834.

Granulated Wentle-Trap

Granuliscala granosa (Quoy & Gaimard, 1834)

(Figure 140)

Shell cream, elongated. Whorls slightly convex, sometimes varicose, crenulated at the suture; surface minutely granulated. Aperture rounded; inner lip thick. Size: 1½ inches.

Localities: Frankston; Dromana; Portsea; Queenscliff; also Western Port.

Washed up on the beaches, this strong little shell is fairly common. It is readily distinguished by the presence of the varices and the characteristic granulated sculpture. It also occurs on the coasts of South Australia, Western Australia and Tasmania.

Opalia H. & A. Adams, 1853

Medium carinate scaloid shells; turreted, rather straight whorls; imperforate, boldly distinctly ribbed, ribs not continuous. Mouth small, oval; base flattened; ten ribs to a whorl, interstices microscopically scratched.

TYPE SPECIES *Scalaria australis* Lamarck, 1822.

Southern Wentle-Trap

Opalia australis (Lamarck, 1822)

(Figure 141)

Shell white, elongated, turreted. Whorls numerous, united, convex, ornamented with rather thick, straight, longitudinal ribs, the interstices of which are microscopically, radially and concentrically sculptured; last whorl encircled below with a definite thick ridge. Aperture rounded; lip reflected; operculum horny. Size: 1 inch.

Localities: Frankston; Dromana; Portsea; Point Lonsdale; common in Western Port.

This species is pure white and its elegance of form and delicate sculpture affords easy recognition. This mollusc is obtained under stones at low water, and enjoys a wide range of distribution, being recorded from Western Australia, South Australia, New South Wales and Tasmania.

Cirostrema Mörch, 1852 (*Scalaria varicosa* Lamarck, 1819).

Iredale, 1936, erected *Plastiscala* with *Scala morchi* Angas as type for these small shells, but we do not consider the variations from *varicosa* Lamarck, the type species of *Cirostrema*, are of more than specific value.

morchi (Angas, 1871); from Western Port. A white, solid, acuminate little shell with nine well-rounded whorls, cancellated with longitudinal ribs and numerous concentric ridges; base rounded. Size: ¼ inch.

invalida (Verco, 1906); from Western Port. A rather thin, elongate form with about eight regularly convex whorls which are almost smooth but under the lens show fine, microscopic longitudinal and concentric striae; it bears a close resemblance to *morchi* but has finer sculpture. Size: ⅜ inch.

Dannevigena Iredale, 1936 (*Dannevigena martyr* Iredale, 1936).

martyr Iredale, 1936; south of Gabo Island (115-45 fathoms).

Limiscala de Boury, 1909

Shell umbilicated, high, top-shaped. Whorls strongly convex, with numerous narrow lamellae and with fine and closely set spiral lines. Aperture oviform.

Type Species *Scala lyria* Sowerby, 1834.

Delicate Wentle-Trap

Limiscala helicornua Iredale, 1936

(Figure 142)

Shell small, cream with two or three brown bands, the anterior being darker; broad, imperforate. Whorls much rounded, sculptured by numerous, longitudinal, closely set lamellae—about twenty-five on the body-whorl between which will be seen, with the aid of lens, fine, longitudinal microscopic striae. Interior cream. Aperture oval, through which the bands and lamellae are plainly visible; outer lip entire. Size: ½ inch.

Localities: Port Albert; Western Port.

A squat, cream, brown-banded species, ornamented with numerous longitudinal lamellae. These associated features give distinction to the species and it cannot be confused with any Victorian form. For many years known as *Scala tenella* Hutton, 1885, a New Zealand shell.

acanthopleura (Verco, 1906); Port Albert.

platypleura (Verco, 1906); Western Port.

The two species above were placed in *Folaceiscala* de Boury, 1912 (*Scalaria dubia* Sowerby, 1844) by Cotton & Godfrey, 1938, but this genus is now reduced to a synonym of *Limiscala*.

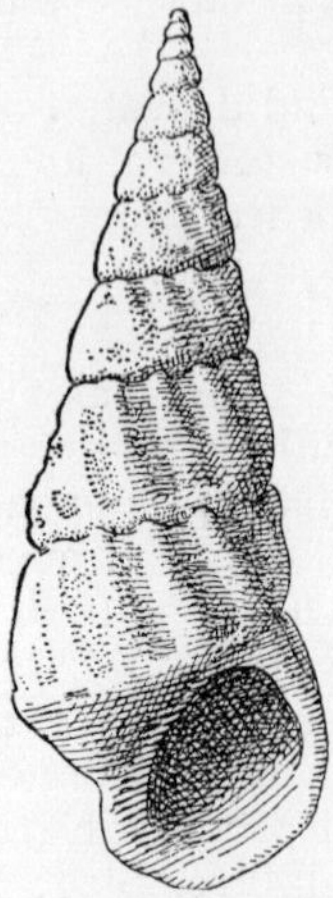

Figure 140

Figure 141

Figure 143

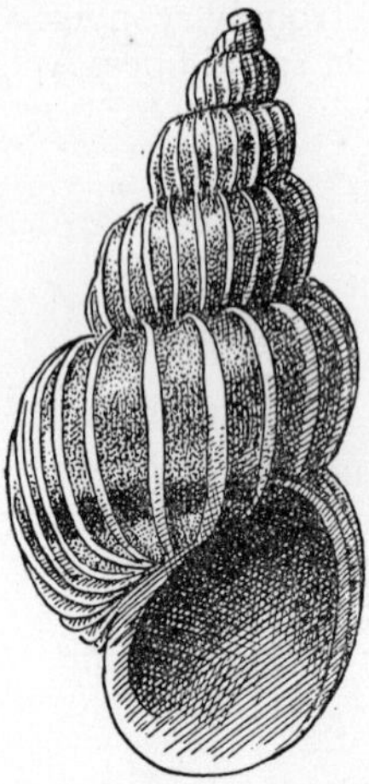

Figure 142

Figure 144

140. *Granuliscala granosa* × 1½
141. *Opalia australis* × 3
142. *Limiscala helicornua* × 6
143. *Clathrus jukesiana* × 2
144. *Clathrus minora* × 3

Clathrus Oken, 1815

Moderately thick; whorls united; longitudinal ribs numerous. Aperture suboval; umbilicus covered by the inner lip, no basal rib.

TYPE SPECIES *Scalaria communis* Lamarck, 1845.

de Boury, 1909, erected *Acutiscala* with *Scalaria philippinarum* Sowerby, 1844, as type and later authors included allied Australian species. However, Thiele, 1929, does not consider that sufficient deviation from *Clathrus* is shown to warrant the acceptance of *Acutiscala* and in this we concur.

Jukes' Wentle-Trap

Clathrus jukesiana (Forbes, 1852)

(Figure 143)

Shell small, white, elongated, acuminate, imperforate. Whorls nine, rounded, sculptured with numerous, nearly equal, longitudinal lamellae about twenty on the body-whorl; sutures deeply impressed. Interior white. Aperture ovate; outer lip strong, reflected; operculum horny. Size: 1 inch.

Localities: Portsea; Port Phillip; Western Port; Portland; Port Albert.

A very graceful shell, widely distributed on the whole coastline. After careful consideration of figures, descriptions and specimens, we have come to the conclusion that *Acutiscala faba* Iredale, 1936, is a synonym of this species and that *Scalaria delicatula* Crosse & Fischer, 1864, also falls into its synonymy. Beyond the state it is found in New South Wales and Tasmania, South Australia and Western Australia.

ampacta (Iredale, 1936); this has a very similar appearance but is more elongate with fewer lamellae (approximately fifteen) on the body-whorl.

coreta (Iredale, 1936); Western Port.

Shining Wentle-Trap

Clathrus minora (Iredale, 1936)

(Figure 144)

Shell small, white, somewhat shining, elongate. Spire approximately three-quarters the length of shell. Whorls much rounded, bearing distant, rather erect lamellae which are slightly produced and aculeated near the suture, nine appearing on the body-whorl, interspaces showing fine, microscopic, concentric and longitudinal striae. Interior white. Aperture slightly ovate; outer lip broad. Size: 1 inch.

Locality: Western Port.

Occasionally cast up on the beaches and dredged alive on sand and shell in eight to ten fathoms between Phillip and French Islands. Even in worn condition the shell should be easily identified by its beautiful ornament, the aculeated lamellae at the sutural area presenting an outstanding feature; occurs also in Tasmania and New South Wales.

This species was originally confused with *Scalaria philippinarum* Sowerby, 1844.

Parviscala de Boury, 1887 (*Scalaria algeriana* Weinkauff, 1866).

tacita (Iredale, 1936); Latticed Wentle-Trap from Western Port, an elongate, dull white species slightly smaller than *minora* with about eight to ten whorls, well rounded but possessing less erect lamellae which are somewhat folded back, eight on the body-whorl; the interspaces delicately sculptured with concentric and longitudinal fine microscopic striae, visible only under the lens, giving the surface a latticed appearance. Interior white. Aperture almost round; the shell should not be confused with the above, being easily separated by its less prominent lamellae and the fine latticed ornament of the interstices; occurs also in New South Wales. This species has been placed in *Laeviscala* which is now considered a synonym of *Parviscala*.

This species was previously called *Scala aculeata* Sowerby, 1844, but is a Northern hemisphere species from which ours shows marked deviation.

Propescala Cotton & Godfrey, 1932 (*Scala translucida* Gatliff, 1906).

translucida (Gatliff, 1906); from Portsea and Western Port. A small, white, fairly solid shell of eight well-rounded whorls and blunt apex; sculptured with strong, round, longitudinal ribs about fifteen on the penultimate whorl, the interstices being ornamented with fine concentric striae. Size: ⅛ inch.

Narvaliscala Iredale, 1936 (*Narvaliscala dorysa* Iredale, 1936).

dorysa Iredale, 1936; Gabo Island (150-200 fathoms).

Family IANTHINIDAE
Violet Snails

Delicate spiral shells with a short spire usually lavender or purple and without an operculum. They are pelagic molluscs living on the surface of the open sea.

Ianthina Röding, 1798

Shell subglobose, violet, spiral, thin, translucent. Spire short. Whorls slightly angulated, nucleus sinstral minute. Aperture large, quadrangular, inner lip reflexed; columella tortuous; outer lip thin, notched or sinuated in the middle.

TYPE SPECIES *Ianthina nitens* Menke, 1828 = *Ianthina prolongata* Blainville, 1822.

The snail-like animal with forked tentacles and small foot is pelagic, living in colonies in the open sea. They float on the surface of the sea with the aid of a gelatinous frothy raft secreted by the foot. In some species the eggs are attached to the under side of the float until they hatch but in others the young are born alive. Their main food is *Velella* and other pelagic jellyfish whose colouring and habit is similar.

The animal exudes a violet fluid on being molested, which no doubt serves as a defence mechanism.

Common Violet or Purple Sea Snail

Ianthina janthina (Linné, 1758)
(Figure 145)

Shell large, thin, fragile, turbinate, variable in shape from trochoid to globose. Whorls sharp to somewhat roundish triangular shape which gives a triangular to roundish triangular aperture. Surface of the shell striated; the first one and a half whorls may have sharp regular striae which pass into a fainter irregular striation. Colour violet which may be lighter on upper portion of spire and below the sutures, and there may or may not be a whitish band round the columella. Columella straight with a slight twist; outer lip sharp, slightly sinuate. Size: 1½ inches.

Localities: Gabo Island; Mallacoota; Portland.

Ianthina because of their colour and delicacy have always attracted interest and this species because of its extreme variability in shape, size and colour has received a great many different names. A few of the names that have been recorded in the Victorian literature are *communis* Lamarck, 1799; *smithae* Reeve, 1858; *balteata* Reeve, 1858. The animal is viviparous.

pallida Thompson, 1841; from Portland, has a globose shell, and widely expanded aperture with a sinus above the mid-line from which a slight continuous concentric keel can be seen on all whorls. This species was recorded as *striolata* A. Adams & Reeve, 1850.

Little Violet or Purple Sea Snail

Ianthina exigua (Lamarck, 1816)
(Figure 146)

Shell small, thin, fragile, colour violaceous with whitish band below the sutures, conically turbinate. Spire about half the length of shell. Whorls rounded, slightly obtusely angular about the middle. Sculptured with fine, somewhat irregular, oblique, radial striae. Interior light violet. Aperture somewhat ovate; outer lip very thin possessing a deep sharply angular sinus almost centrally situated; columella slightly reflected. Size: ½ inch.

Locality: Curdie's Inlet.

This is the smallest Violet Sea Snail in Victoria, easy to recognize by its radially sculptured whorls and deep angular sinus.

Family ACLIDIDAE
Aclids

Shell very small, colourless, mostly umbilicate, high turreted, smooth or sculptured. Whorls more or less numerous. Mouth ovate; outer lip sharp, not continuous; operculum horny, spiral increasing rapidly.

Coenaculum Iredale, 1924 (*Scalaria minutulum* Tate & May, 1900).

minutulum (Tate & May, 1900); Portsea; Port Phillip; Shoreham, Western Port.

Family MELANELLIDAE
Urchin Shells

Small, high-spired shells, usually polished and white with an oval aperture and slightly thickened outer lip. The spire is often slightly bent; operculum horny, ovate, subspiral. Parasitic on marine animals, particularly Echinoderms.

There has been and still is a great deal of disagreement and confusion amongst authors over the definition and naming of the various genera of this family. Laseron, 1955, made a revision of the New South Wales species and erected a number of new genera to replace those in use, his main grounds for doing this being that they had been erected for extra-Australian species, and therefore could not be closely related to our fauna.

Very little is known of the animals and habitats of the Australian members of this family, and any classification suggested is likely to have grave faults. Therefore, we prefer to retain a few broad divisions until such time as adequate material is at hand for a detailed review.

Melanella Bowdich, 1822

Shell elongated, white, smooth, polished. Spire produced, many whorled, frequently with an interrupted varix on one side; apex acute. Aperture oval, pointed behind; inner lip reflected over the pillar; axis imperforate; outer lip thickened internally.

TYPE SPECIES *Melanella dufresnii* Bowdich, 1822.

Large Urchin Shell

Melanella augur (Angas, 1865)

(Figure 147)

Shell opaque white, smooth, shining, solid, elongated. Spire about three-quarters length of shell; apex acute. Whorls ten, rather flat. Interior white. Aperture somewhat pyriform; inner lip a little reflected over the slightly curved columella; outer lip thickened. Size: ½ inch.

Localities: Port Phillip; Western Port.

The largest *Melanella* in Victoria, easily recognized by its size and strong texture. Recorded also from South Australia, New South Wales and Tasmania.

inflata (Tate & May, 1900); the Inflated Urchin, found on starfish at Flinders and other parts along the coast, a smaller species (¼ inch) broader in form with the last whorl much inflated.

immaculata (Pritchard & Gatliff, 1900); another smaller species from Shoreham, Frankston and Port Albert, is rarely seen but should appear after diligent search through shell-grit. About five whorls and a fairly well-marked, though small, apex. Size: ⅛ inch.

commensalis Tate, 1898; Flinders; Western Port.

coxi Pilsbry, 1899; Shoreham; Port Albert.

indiscreta (Tate, 1898); Flinders; Shoreham; San Remo.

mucronata Reeve, 1866; Kilcunda; Western Port; Port Phillip; off Ninety Mile Beach.

orthopleura Tate, 1898; Port Campbell.

petterdi (Beddome, 1882); Flinders.

schoutanica (May, 1915); Portsea.

tenisoni Tryon, 1886; off Wilson's Promontory; Port Albert; Western Port; Port Phillip.

tryoni Tate & May, 1900; Frankston; Dromana.

victoriae Gatliff & Gabriel, 1914; off Wilson's Promontory.

Strombiformis da Costa, 1778 (*Strombiformis glaber* da Costa, 1778).

acutissima (Reeve, 1866); Western Port; Port Phillip; Port Albert.

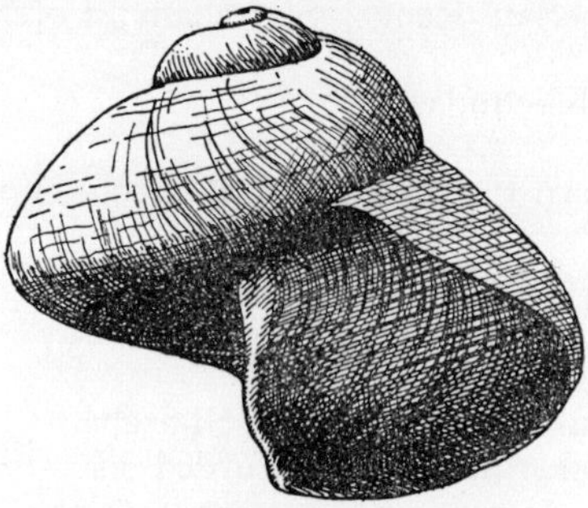

Figure 145

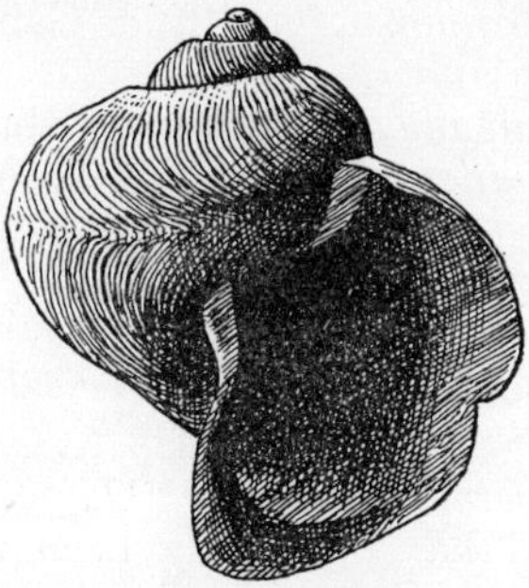

Figure 146

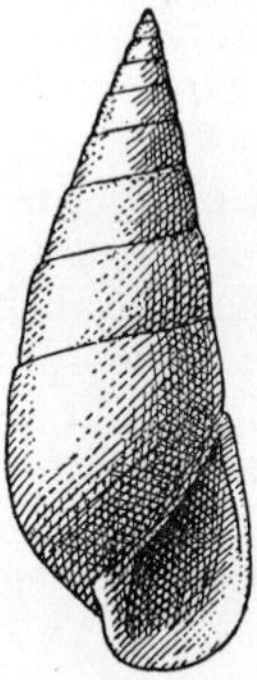

Figure 147

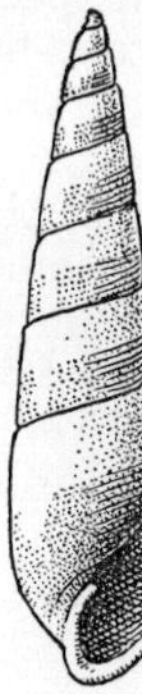

Figure 148

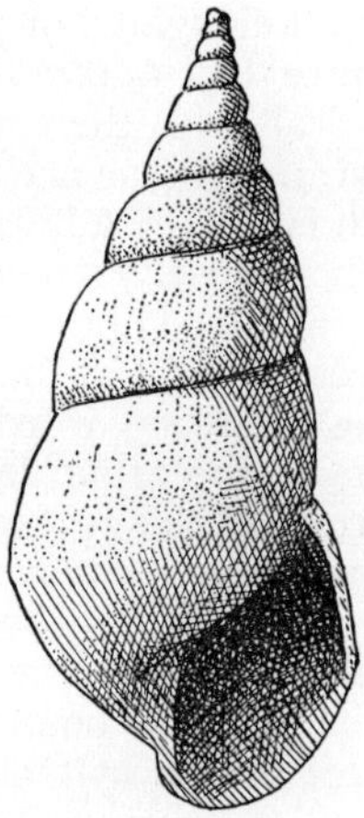

Figure 149

Figure 150

145. *Ianthina janthina* × 1⅓
146. *Ianthina exigua* × 3
147. *Melanella augur* × 3
148. *Syrnola tincta* × 8
149. *Stilifer petterdi* × 4
150. *Larinopsis turbinata* × 2

joshuana (Gatliff & Gabriel, 1910); San Remo; Shoreham; Portsea; Portland.

kilcundae (Gatliff & Gabriel, 1914); Kilcunda.

lodderae Hedley, 1903; Western Port.

perexigua (Tate & May, 1900); Western Port; off Wilson's Promontory; Port Albert; Portland.

styliformis (Gatliff & Gabriel, 1914); off Wilson's Promontory.

topaziaca (Hedley, 1908); Western Port; Port Albert.

Stilapex Iredale, 1925 (*Stilapex lactarius* Iredale, 1925).

lactarius Iredale, 1925; off Cape Everard (65 fathoms); Shoreham. This shell has previously been incorrectly identified and recorded as *Stilifer brazieri* Angas, 1877.

Family PYRAMIDELLIDAE

Pyramid Shells

Small elongated or pyramidal shells with many flattened whorls. The oval aperture has a thin outer lip and plicated columella; operculum horny, subspiral, with the columella margin sinuated.

Pyramidellidae are ectoparasites, each species feeding on a particular host which is usually a bivalve mollusc or a tube-building polychaete worm. They attach themselves to the host by an oval anchor, pierce the body wall and suck the blood and soft tissue.

Fretter and Graham (1949) summarize their recent work on this family in the following words: 'a consideration of the structure and reproductive habits of the group leads to the conclusion that they are opisthobranchiate gastropods'. They further suggest that the family should be placed among the primitive members of that subclass. From the above conclusions it seems likely that the preceding closely related families Melanellidae and Stiliferidae should also be transferred to Opisthobranchia.

The family contains many small species which have been divided in several genera by the different authors.

Syrnola A. Adams, 1860

Typically elongated shells with flattened whorls, smooth and porcellanous in texture, often coloured or with brown bands; the protoconch exsert and tilted heterotrophe; the columella fold single, generally prominent, but occasionally nearly obsolete and only visible inside the aperture.

TYPE SPECIES *Syrnola gracillima* A. Adams, 1860.

Shining Syrnola

Syrnola tincta Angas, 1871

(Figure 148)

Shell subulate, rather solid, smooth, shining, whitish, irregularly banded and marked with brown. Whorls ten and a half, nearly flat; sutures deeply impressed. Aperture ovate; outer lip thin; columella plait rather prominent. Size: ¼ inch.

Localities: Western Port; Port Phillip; Barwon Heads.

This is one of the common species of our coast and is found both on ocean and sheltered bay beaches. It is also recorded from South Australia, Tasmania and New South Wales.

aurantiaca (Angas, 1867); Shoreham; San Remo.
bifasciata Tenison Woods, 1875; Western Port; Port Phillip.
jonesiana (Tate, 1898); Flinders.
manifesta Hedley, 1912; definite locality not recorded.
victoriae (Gatliff & Gabriel, 1911); Port Albert.

Puposyrnola (Cossmann, 1912 (*Auricula acicula* Lamarck, 1822).
harrisoni (Tate & May, 1900); Portsea.
petterdi Gatliff, 1900; Western Port.
tasmanica (Tenison Woods, 1877); off Lakes Entrance.

Agatha A. Adams, 1860 (*Pyramidella* [*Agatha*] *virgo* A. Adams, 1860).
angasi (Tryon, 1886); Western Port.
australis (Angas, 1871); Port Phillip; Western Port.
laevis (Angas, 1867); Dromana; Kilcunda.
metcalfei Pritchard & Gatliff, 1900; Flinders.
simplex (Angas, 1871); between Phillip and French Islands, Western Port.

Odostomia Fleming, 1813 (*Turbo plicata* Montagu, 1803).
deplexa Tate & May, 1900; Flinders.
occultidens May, 1915; Portsea; Rhyll.

Egilia Dall & Bartsch, 1904 (*Parthenia lacunata* Carpenter, 1857).
mayii (Tate, 1898); Portsea, Port Phillip.

Linopyrga Laws, 1941 (*Odostomia rugata* Hutton, 1886).
brevis (Pritchard & Gatliff, 1900); off Rhyll.
nugatoria (Hedley, 1903); off Ninety Mile Beach (40 fathoms).
portseaensis (Gatliff & Gabriel, 1911); Portsea.

Miralda A. Adams, 1864 (*Parthenia diadema* A. Adams, 1864).
suprasculpta (Tenison Woods, 1877); Portsea, Port Phillip; Rhyll.

Cinctiuga Laseron, 1951 (*Cingulina diaphana* Verco, 1906).
diaphana Verco, 1906; Ocean Beach, Point Nepean.

Turbonilla Risso, 1826 (*Turbo lactea* Linné, 1758).
beddomei (Petterd, 1884); Western Port.
scalpidens (Watson, 1886); exact locality not recorded.

Chemnitzia d'Orbigny, 1893 (*Melania campanellae* Philippi, 1836).
acicularis (A. Adams, 1853); Portsea; Western Port; Port Albert.
hofmani Angas, 1877; Port Phillip; Western Port; Barwon Heads; Anderson's Inlet.
mariae Tenison Woods, 1876; Port Phillip; Western Port.

Pyrgiscus Philippi, 1841 (*Melania rufa* Philippi, 1841).
fusca (A. Adams, 1853); the Brown Turbonilla, a slender form from Portland, Port Phillip and Western Port. About ¼ inch, possessing eight whorls delicately sculptured with close, longitudinal riblets which become obsolete at the base; this feature, with the rich brown colour, is sufficient to readily identify it.
varicifera (Tate, 1898); off Lakes Entrance.

Pyrgiscilla Law, 1937 (*Turbonilla* [*Strioturbonilla*] *chattonensis* Marwick, 1929).
tiara (May, 1910); off Ninety Mile Beach (40 fathoms).

Eulimella Jeffreys, 1947 (*Eulima macandrewi* Forbes, 1846).
moniliformis Hedley & Musson, 1891; Swan Bay, Port Fairy.
turrita (Petterd, 1884); Portsea; Swan Bay.

Cingulina A. Adams, 1860 (*Cingulina circinata* A. Adams, 1860).
magna Gatliff & Gabriel, 1900; Shoreham.
rhyllensis Gatliff & Gabriel, 1900; between Phillip and French Islands, Western Port.
spina (Crosse & Fischer, 1864); Port Phillip; Western Port.

Oscilla A. Adams, 1867 (*Oscilla cingulata* A. Adams, 1867).
tasmanica (Tenison Woods, 1876); Portsea; San Remo; Rhyll.

Aclis Loven, 1846 (*Turbo nitidissima* Montagu, 1803).
pellucida Gatliff & Gabriel, 1911; between Phillip and French Islands, Western Port; Port Albert.

Pseudorissoina Tate & May, 1900 (*Stilifer tasmanica* Tenison Woods, 1876).
tasmanica (Tenison Woods, 1876); Portsea; Flinders; San Remo.

Family STILIFERIDAE
Stilifers

Shell thin, subulate or subglobose. Aperture entire; columella simple.

Although displaying very little character these shells are an interesting group; show much variation. They have no operculum and occasionally the apex is a little out of perpendicular. The animals possess a linguiform foot with an anterior lobe and are parasitic or commensal in habits on Echinoderms, burrowing into them. When removed and placed in water, they show much movement extending the tongue-shaped foot using it as an exploring agent. It is thought that the species living on the external surface of their host are probably commensal rather than truly parasitic. A search among the larger starfish especially may be profitable to the collector.

Stilifer Broderip, 1832

Shell subulate or subglobose, thin pellucid, smooth, polished, many-whorled; apex of spire produced and styliform with a sinistral nucleus. Aperture subovate, pointed behind, rounded and entire in front; inner lip smooth arcuated; outer lip thin, simple.

TYPE SPECIES *Stilifer astericola* Broderip, 1832.

Petterd's Stilifer or Urchin Shell or Starfish Shell

Stilifer petterdi Tate & May, 1900
(Figure 149)

Shell opaque white, tumidly pyramidal, polished. Spire approximately two-thirds the length of shell. Whorls rounded and separated by well-impressed sutures, ultimate whorl very convex and somewhat angled about the centre. Sculpture consisting of feeble, irregular, concentric threads with longitudional growth-lines plainly visible. Interior white, shining. Aperture ovately pyriform; outer lip thin; columella much curved. Size: ½ inch.

Locality: Western Port (dredged 6 to 8 fathoms).

The largest Stilifer in Victoria; a very thin shell of outstanding shape easily recognized by its robust form, the shell being known for many years as *robusta* Petterd, 1884, a prior application of this name precluding its use. Also found in New South Wales, Tasmania and South Australia.

lodderae Petterd, 1884, Lodder's Stilifer, its nearest ally in Victoria, is a white shell from Flinders and Port Phillip, similar in size and shape but more elongate, and recorded also from New South Wales, Tasmania and South Australia.

brunneus Tate, 1888, the Brown Stilifer, from Shoreham and parasitic on the big Sea Urchin *Heliocidaris tuberculata* Lamarck (invariably on the periproct); a small (⅕ inch) dark brown form with an extremely short spire acuminately produced, the spire and ultimate whorls sculptured concentrically and longitudinally with microscopic striae; found also in South Australia.

auricula Hedley, 1907; Ocean Beach, Point Nepean.

Family FOSSARIDAE

Fossarids

Shell flattened or ovate, conical, usually umbilicated and with a prominent protoconch, as a rule spirally sculptured. Body-whorl mostly large; mouth oblique; columella elongated, connected to outer lip; operculum thin, horny with an upstanding nucleus and open spiral.

Larinopsis Gatliff & Gabriel, 1916

Shell fragile, turbinate, umbilicate. Whorls rapidly increasing in size; peristome continuous, slightly reflexed epidermis olivaceous operculum horny, angular with nucleus intramarginal. The animal is ovoviviparous and marine.

TYPE SPECIES *Larina turbinata* Gatliff & Gabriel, 1909.

Turbinate Larinopsis

Larinopsis turbinata (Gatliff & Gabriel, 1909)

(Figure 150)

Shell fragile, turbinate, white, umbilicate, semi-transparent. Whorls six including the minute apex of two whorls which are brown and appear to be smooth, but under the microscope are seen to be densely spirally pitted. Whorls ventricose, rapidly increasing, the last and penultimate comprising nearly the whole of the shell. Sculpture faint, under lens it is densely irregularly, spirally striate, the striae being somewhat undulating, and crossed by growth-lines of similar strength. Aperture circular, slightly reflexed; umbilicus small, deep, almost hidden by the reflexion of the lip. Interior vitreous; operculum horny. Shell covered with an olivaceous epidermis. Size: ¾ inch.

Locality: Dredged in 5 fathoms between Phillip and French Islands, Western Port.

One of our rarest univalves; the type rests in the National Museum of Victoria collection; the few other specimens known, all from the type locality, are more or less fragmentary. It is a remarkable species, very fragile, and its whiteness and peculiar turbinate shape provide features sufficient to identify the shell. Great difficulty was experienced in obtaining the radula with unsatisfactory results and during the process the animal was seen to possess about 400 embryonic shells so minute it is estimated thousands could rest on a threepenny piece. One example was found on the north coast of Tasmania. Twenty years later a second species appeared trawled in 65 fathoms off Jervis Bay, New South Wales, and is known under the name of *ostensus* Iredale, 1929, a more elongated shell with deeper sutures.

Family VANIKORIDAE

Vanikorids

Shell umbilicate, with a small, smooth embryonic shell and large round body-whorl; colourless, with spiral sculpture and more or less strong sloping ribbing. Mouth large, sloping; columella edge weakly curved; outer lip strongly curved, not thickened; operculum thin, with a terminal nucleus, weakly spiral.

Vanikoro Quoy & Gaimard, 1832 (*Sigaretus cancellata* Lamarck, 1819).
quoyiana A. Adams, 1853; Hobson's Bay.

Gray proposed the name *Merria* for these shells on the grounds that *Vanikoro* was a vernacular name; however, it has been accepted by modern authors.

Family HIPPONICIDAE

Bonnet Limpets

Shell cap or cup-shaped, smooth or ribbed; embryonic shell spiral; no operculum.

Cheilea Modeer, 1793

Shell cup-shaped, apex posterior, embryonic shell spiral. Surface differently sculptured. Mouth wide, roundish; a grooved appendix is attached to the inside of the apex and hangs into the body cavity of the shell.

TYPE SPECIES *Patella equestris* Linné, 1766.

Cup and Saucer Limpet

Cheilea flindersi Cotton, 1935

(Figure 151)

Shell white or cream, large, conical, texture rather thin, very irregularly circular in outline; apex subcentral, slightly curved, inclined anteriorly; from apex to posterior end fairly convex; anterior slope either straight or slightly concave. Surface displaying rather strong, irregular, concentric lines of growth crossed by inequidistant radial striae or ridges. Interior white, polished, with a delicate, calcareous half cup-shaped process open on posterior side and extending from the apex into the interior of the shell; margin slightly crenulated. Size: 2 inches.

Localities: Flinders; San Remo.

A most curious shell easily identified by its conical shape and the internal calcareous plate. It adheres to rocks and the under surface of stones partly buried in the sand. The vernacular name of Horse Hoof is sometimes applied to these shells. Occurs also in South Australia.

Hipponyx Defrance, 1819

Shell cap or cup-shaped, with spiral embryonic shell, apex usually near posterior edge; sculpture varied, no internal appendage.

Type Species *Hipponyx cornucopiae* Defrance, 1819.

In recent years a number of names have been used for this genus. They are *Amalthea* Schumacher, 1817, which is preoccupied (see Iredale, 1937, p. 252); *Hipponyx* Defrance, 1819; and *Sabia* Gray, 1847. Iredale rejected *Hipponyx* because it was erected for a fossil which, however, is obviously very similar to *Amalthea conica* Schumacher, 1817, the type of *Sabia,* and we do not consider that there is sufficient evidence to warrant the division of the fossil and living species.

Conical Horse Hoof or Bonnet Limpet

Hipponyx conicus (Schumacher, 1817)

(Figure 152)

Shell brown, solid, conical; apex posteriorly situated, directed backwards. Surface of shell sculptured with rather broad flatly rounded ribs with narrow interspaces, adult forms very often developing a rugose, wrinkled appearance. Interior white, polished, bordered brown, margins crenulated; muscle scar horseshoe-shaped. Size: ¾ inch.

Localities: Western Port; Port Phillip; Lakes Entrance; Mallacoota.

A common shore-living species located under stones at low water and easily recognized by its characteristic bonnet shape, brownish and coarse radial sculpture. The animal by means of its foot excavates a shallow cavity in the surface of another shell or stone to which it attaches itself, and like the Slipper Limpets are found frequently living on one another and often on rocks in clusters of eight to twelve or more. Bryozoa and *Serpula* appear in many cases attached to the species. For many years known under the name of *Hipponyx australis* (Lamarck, 1819). Found also in Tasmania and South Australia.

Antisabia Iredale, 1937 (*Hipponyx foliacea* Quoy & Gaimard, 1835).

foliacea (Quoy & Gaimard, 1835); the Foliaceous Bonnet Limpet is another species of the family in Victoria, about ½ inch and characterized by the concentric, close imbricating laminae and distinct radiating striae presenting a surface resembling the tiles of a roof, invested throughout by a pilose epidermis. No difficulty should be experienced in separating the two forms. Previously known as *Hipponyx antiquatus* (Linné, 1758). Found also in New South Wales, Tasmania and South Australia.

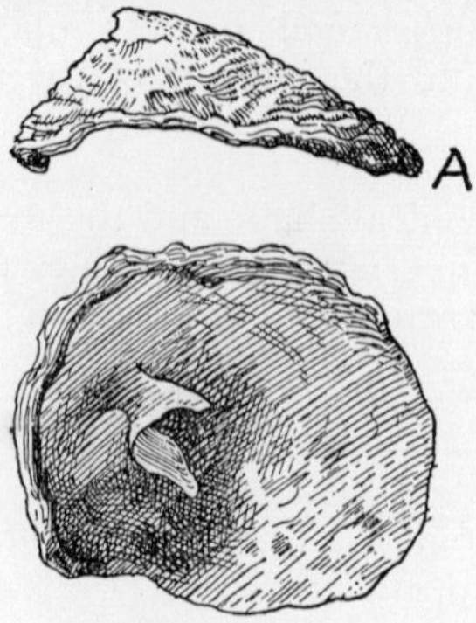

Figure 151

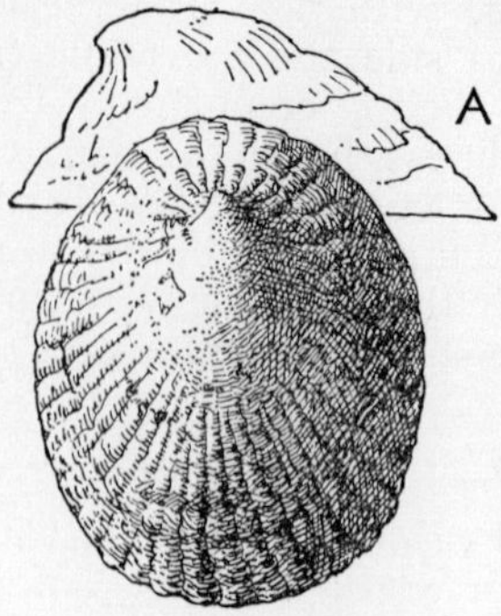

Figure 152

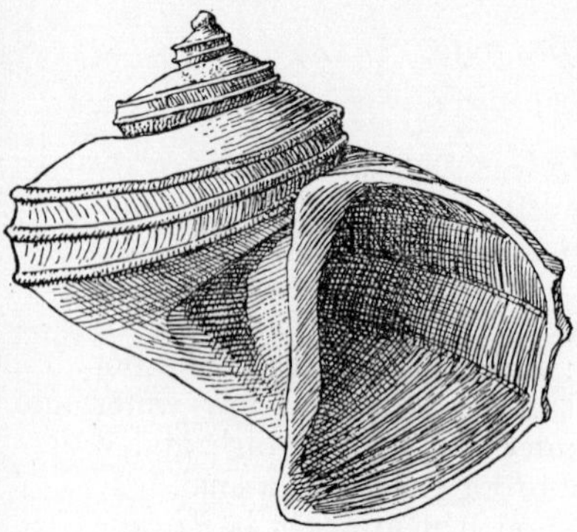

Figure 153

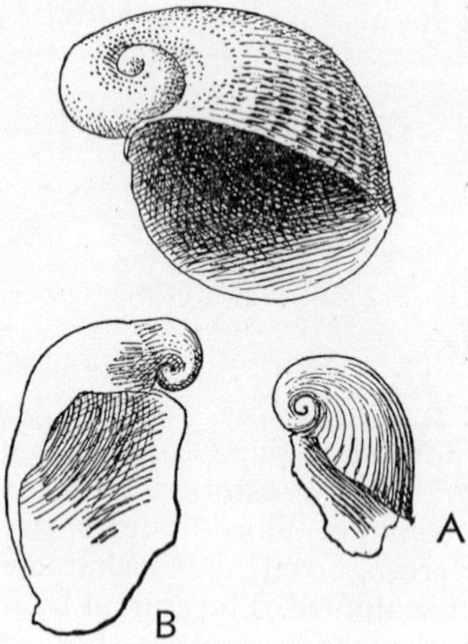

Figure 154

151. *Cheilea flindersi* × ⅔
A Lateral elevation

152. *Hipponyx conicus* × 1⅓
A Lateral elevation

153. *Zelippistes blainvilleanus* × 3

154. *Capulus violaceus* × 12
A Anterior elevation × 4 B Lateral elevation × 4

Family TRICHOTROPIDAE
False Turban Shells

Medium-sized thin shells, turban shaped and carinated. The epidermis is fringed on the ridges. The operculum is horny.

Zelippistes Finlay, 1926

Shell small, transparent, deeply umbilicated. Spire very small. Whorls quickly increasing with three cross-barred encircling cords. Aperture wide D-shaped.

TYPE SPECIES *Separatista benhami* Suter, 1902.

Blainville's Lippistes

Zelippistes blainvilleanus (Petit, 1851)
(Figure 153)

Shell thin, fragile, turbinate, creamy, invested with a light brown, scaly epidermis. Spire slightly less than half the length of shell. Whorls rounded, carinated, those of the spire bicarinate, the body-whorl tricarinate, and more or less separated from the spire-whorls. The whole surface sculptured with fine concentric, microscopic striae crossed by close, faint growth-lines. Interior cream. Aperture semicircular; outer lip acute; columella slightly curved; umbilicus very wide, with the whorls visible to the apex; operculum horny. Size: ½ inch.

Locality: Western Port, dredged between Phillip and French Islands (8 fathoms).

A fragile, remarkably shaped shell, easily recognized by its turbinate form and with its wide and deep umbilicus. Few specimens appear to be washed up but it should occur on the shores of Western Port. *Trichotropis gabrieli* Pritchard & Gatliff, 1898, is now regarded as a synonym of the above and the authors in relegating this to synonymy remarked: 'When we described *T. gabrieli* it was from a single specimen, which has only two encircling carinae on the body-whorl. Other shells have been since dredged in the same locality, and in every instance the body-whorl is tricarinate. Examination of the type shows that the growth of the penultimate whorl has been interfered with, and apparently this accounts for its only having two carinae on the body-whorl. We are now constrained to regard it as an abnormal form of *L. blainvilleanus*.'

Family CAPULIDAE
Cap Limpets

Cap-shaped shells with a spirally recurved posterior apex. Aperture rounded; muscle scar horseshoe-shaped.

Capulus Montfort, 1810

Shell conical; surface striated, usually covered with a horny epidermis; apex inclined posteriorly, recurved, subspiral. Aperture dilated.

TYPE SPECIES *Patella hungarica* Linné, 1758.

Violet Cap Limpet

Capulus violaceus Angas, 1867

(Figure 154)

Shell small, colour light violet, much elevated, cap-shaped, base oblong-ovate; apex free, recurved, inclined to the right, posteriorly situated. Sculptured with very fine radial striae, and feeble lines of growth visible only with the aid of lens. Interior violet tinted; outer margin acute. Size: ½ inch.

Locality: Dredged in Western Port.

The animals of *Capulus* apparently are possessed of limited locomotion and like *Hipponyx* are found adhering to shells and stones, adapting themselves to the objects on which they live, forming shallow excavations. Beach specimens occur commonly in New South Wales, and the species is recorded also from Tasmania through Bass Strait to South Australia.

devotus Hedley, 1904, Devout Cap Limpet; dredged in 40 fathoms off the Ninety Mile Beach, a much smaller form about ⅕ inch, white; more rounded and with concentric, microscopical hair lines, features which immediately distinguish it from the above. This occurs also in New South Wales, Tasmania and South Australia.

Icuncula Iredale, 1924 (*Cingulina torcularis* Tenison Woods, 1878).
torcularis (Tenison Woods, 1878); off Lakes Entrance.
zodiacus (Hedley, 1907); off Wilson's Promontory.

Sirius Hedley, 1900 (*Raulinia badius* Tenison Woods, 1878).
badius (Tenison Woods, 1878); San Remo.

Family CALYPTRAEIDAE

Shelf Limpets

Circular or irregularly shaped limpet-like animals, with a plate or cup-like process on the inner side of the shell. These cup-saucer limpets are found living attached to other shells from shallow to deep water.

Sigapatella Lesson, 1831

Shell spiral, vaguely sculptured, with strong periostracum. Last whorl large. Ventral edge of aperture wide, slightly concave.

TYPE SPECIES *Calyptraea novaezelandiae* Lesson, 1830.

Shelf Limpet

Sigapatella calyptraeformis (Lamarck, 1822)

(Figure 155)

Shell small, depressed conical, white or cream with apex sometimes lavender or violet. Whorls slightly rounded, smooth, marked by lines of growth. The whole surface invested with a thin fawn, somewhat fibrous epidermis. Interior white, polished, sometimes stained with violet. Aperture fairly deep, elongately oval, margin sharp; basal plate pure white with fine growth-lines plainly visible. Size: 1 inch.

Localities: Flinders; dredged off Wilson's Promontory; Point Cook.

This is a striking little shell easily recognized by its depressed conical shape and internal basal plate or shelf. The species enjoys a rather wide distribution along the Victorian coast and in living condition is located attached to stones and on the outer and inner surfaces of larger shells; after heavy weather the species appears in quantity on the shores of Port Phillip. Found also in New South Wales, Tasmania, through Bass Strait to South Australia and Western Australia.

scutum (Lesson, 1830); a mistaken identification of *C. hedleyi* below.

Crepidula Lamarck, 1799 (*Patella fornicata* Linné, 1758).

In recent years the genus *Crypta* Humphrey, 1794, has been used for this genus, but it must be rejected as its author has been ruled non-binomial.

aculeata (Gmelin, 1791); the Prickly Slipper Limpet, from Port Phillip but not frequently occurring in Victoria. It is easily separated from the last by its radiating prickly or spinose ridges, a feature showing much variation in development of spines. About one inch. Yellowish or brown, interiorly chocolate coloured and possessing a white basal plate or septum. Found also in New South Wales where it appears to be common.

Clypeola Gray, 1867 (*Sigapatella tenuis* Gray, 1867).

Finlay, 1927, erected *Zegalerus,* with *tenuis* as type. As it is an absolute synonym, it must give way to Gray's prior name.

hedleyi (Smith, 1915); a smaller, light brown form from Western Port, Lakes Entrance and Port Albert. It is ⅜ inch, thin in texture with whorls less rounded and apex almost central. This species was known in Victoria as *Sigapatella tenuis* Gray, 1867, a New Zealand species.

Zeacrypta Finlay, 1926

Shell arched, elongate, smooth, apex terminal, small, septate, edge straight.

TYPE SPECIES *Calyptraea monoxyla* Lesson, 1830.

These slipper limpets live upon or in the mouths of shells, their hosts are usually dead but at times they may attach themselves to living animals of other or their own species.

Slipper Limpet

Zeacrypta immersa (Angas, 1865)
(Figure 156)

Shell brown, oblong-ovate, fairly smooth, convex, sculptured by faint concentric lines of growth; apex posteriorly situated, overlapping the margin; covered by a thin horny epidermis. Interior chocolate colour, polished, the posterior almost half covered by a flattish or slightly convex basal plate the edge of which is nearly straight. Size: 1 inch.

Localities: Port Phillip; Western Port; dredged 38 fathoms off East Moncoeur Island.

Two species of *Zeacrypta* were originally recorded from Victoria, the above species and *unguiformis* (Lamarck, 1822). However, the latter is a west American species and it is now recognized that the supposed two species which were separated on form and colour are synonymous. The dark convex form (*immersa*) inhabits the external curve of a shell whereas the

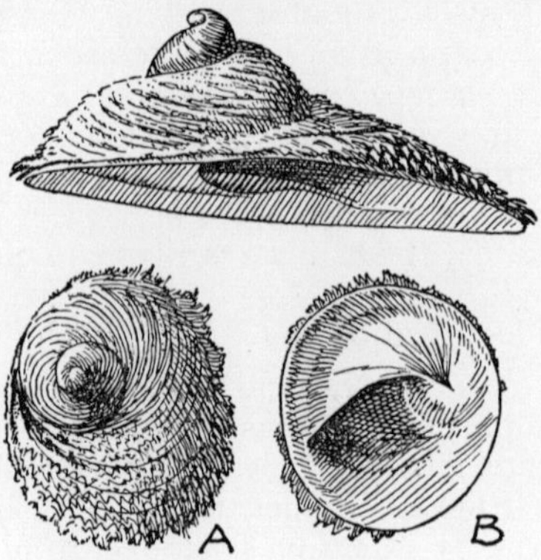

Figure 155

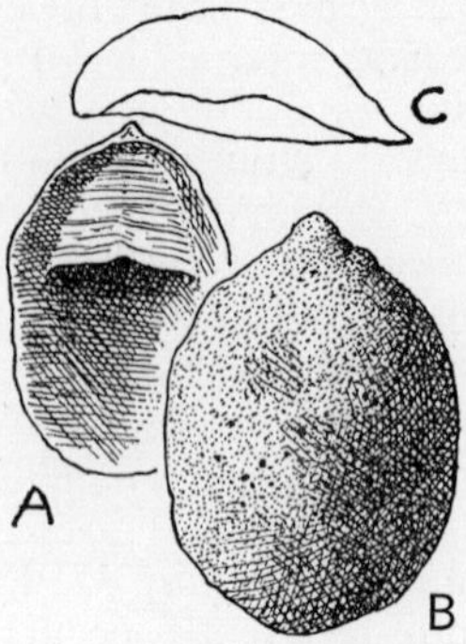

Figure 156

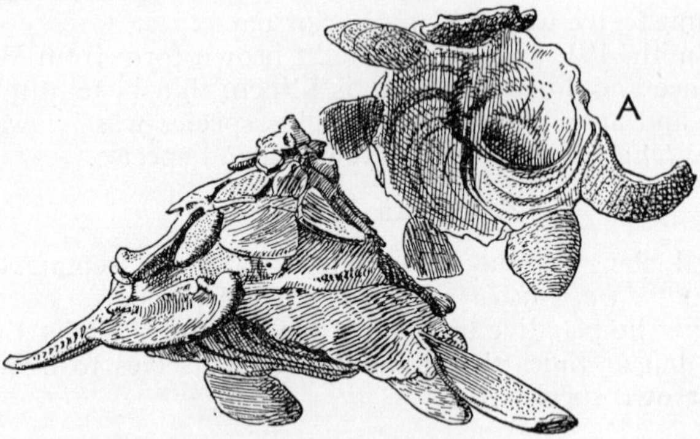

Figure 157

155. *Sigapatella calyptraeformis* × 2
A Dorsal view × 1 B Ventral view × 1

156. *Zeacrypta immersa*
A Anterior view × 1 B Dorsal view × 1
C Lateral elevation

157. *Xenophora peroniana* × 1
A Anterior view × ⅔

concave-shaped and lighter coloured shell (*unguiformis*) lives on the inner (mouth) curve of the host and assumes the shape and colour of this surface. Recorded also from New South Wales, Tasmania and South Australia.

Family XENOPHORIDAE

Carrier Shells

Shell a flattened conical spiral, umbilicated; solders shells and stones to its exterior surface. Operculum large, horny subannular.

Xenophora Fischer, 1807

Shell trochiform, concave beneath, irregular. Whorls flat, more or less concealed by fragments of shells and stone. Spire depressed. Aperture wide; umbilicus small.

TYPE SPECIES *Turbo trochiformis* Born, 1778.

Peronian Carrier Shell or Shell Collector

Xenophora peroniana (Iredale, 1929)
(Figure 157)

Shell whitish, conical, trochiform; apex small but prominent with about three semitransparent coils, succeeded by six to seven rapidly increasing, roundish, irregular whorls all of which are minutely sculptured with closely set, obliquely spiral lirae crossed by fine lines of growth. The whole surface of shell is almost completely covered with extraneous bodies, such as small pebbles, Bryozoa, and fragments of various shells such as Dentalium, Glycymeris, Ostrea, Venerids. The base of the body-whorl is concave. Aperture transversely oval. Size: 2 inches.

Locality: 20 fathoms off Lakes Entrance; Cape Conran.

The excrescences are soldered or incorporated with the limy material of the shell by the animal, which in this species shows a definite preference for fragments of bivalves, the concavities of which are almost without exception placed uppermost so as to gain a better grip and not to retard locomotion. This animal provides a wonderful example of protection or camouflage, and living as it does on a seabed of pebbles and fragments of shells, its enemies could easily be deluded.

Family PTEROTRACHEIDAE

Winged-foot Slugs

Pelagic, elongated, cylindrical, translucent animals with a ventral fin and a tail fin used for swimming; without a shell. Gill external, posterior.

Firoloida Le Sueur, 1817 (*Firoloida demarestia* Le Sueur, 1817).
demarestia Le Sueur, 1817; Port Albert.

Family NATICIDAE

Sand Snails

Shell globular or oval with a short spire. Aperture entire, semi-lunar, outer lip sharp; the columella lip with a callus more or less reflected over the umbilicus; operculum calcareous or horny.

The animal has a very large foot on which it moves quickly, leaving a distinct track as it seeks bivalves buried in the sand or mud. Sand snails are carnivorous, piercing with their radula the shells of bivalves in the thin umbonal area and sucking the victim's flesh. The egg-case is peculiar, being a circular collar formed of sand grains amongst which the eggs are arranged.

Polinices Montfort, 1810

Shell ovate or subovate, solid, smooth, usually without epidermis. Spire small, acute. Aperture semicircular; inner lip oblique, thickened, calloused; umbilicus narrow, deep; columella adherent to, and spirally contorted in, the umbilicus; operculum large, horny, simple.

TYPE SPECIES *Nerita mamilla* Linné, 1758.

Charles Hedley transferred these shells to the prior genus *Uber* Humphrey, 1797, but the International Commission on Zoological Nomenclature has ruled this work non-binominal and the name must now revert to *Polinices.*

Conical Sand Snail

Polinices conicus (Lamarck, 1822)

(Figure 158)

Elongated conical-ovate, a little compressed and whitish or orange tinted at the suture; balance of shell fulvous ash colour, fading to white at the base; columella callus heavy, almost filling the umbilicus—which with the callus, is more or less tinged with chestnut colour. Size: 1½ inches.

Localities: Frequent in all parts of Port Phillip; Norman Bay; Apollo Bay; Portland.

Readily distinguished by its conical form. Its range is wide, being recorded from Tasmania and the whole of the southern Australian coastline.

Finlay & Marwick, 1937, erected the subgenus *Conuber* with this species as type for the southern and eastern Australian shells. They state 'it differs from *Polinices s.s.* in its consistently high conic shape, the course of its growth-lines and the peculiar way in which the parietal callus ends abruptly, leaving exposed a narrow umbilicus and half of the funicle.'

The following species are also included in this subgenus: *sordidus, aulacoglossa,* and *ayresi.*

Sordid Sand Snail

Polinices sordidus (Swainson, 1821)

(Figure 159)

Shell lead grey, conically ovate, rather solid, narrowly, deeply umbilicated. Whorls faintly, diagonally striated. Aperture ovate; columella callosity tinted with orange brown. Size: 1½ inches.

Locality: Port Phillip; Western Port; Portland.

Distinguished by its more globose form which with the lead grey of the shell makes it easily identified. These animals are located most frequently in muddy sand at low tide. This prior name takes the place of the familiar *Polinices plumbeus* (Lamarck, 1822).

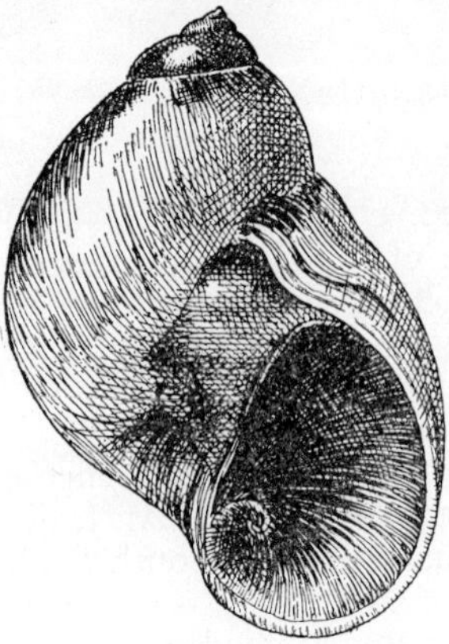

Figure 158

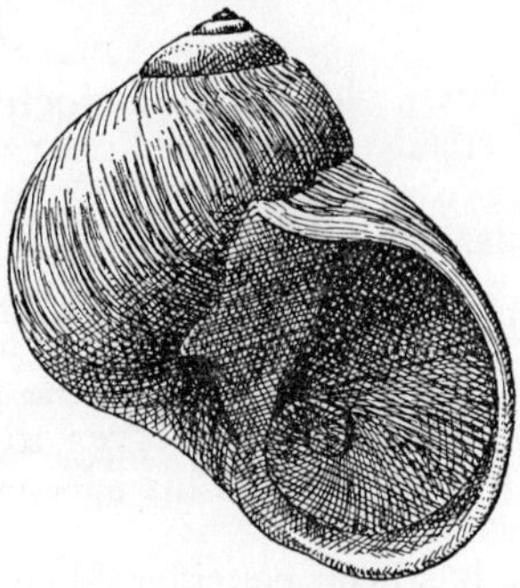

Figure 159

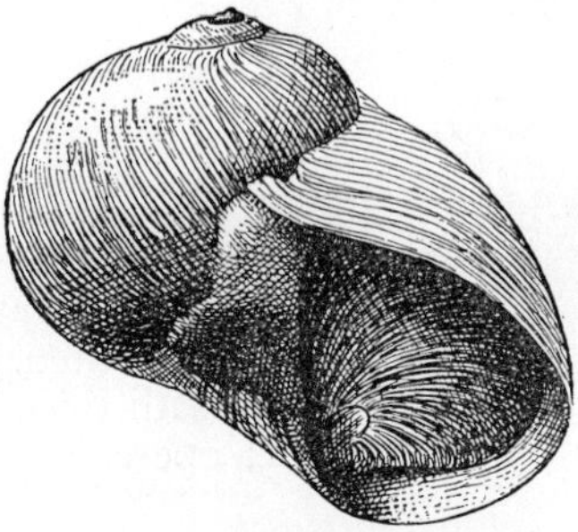

Figure 160

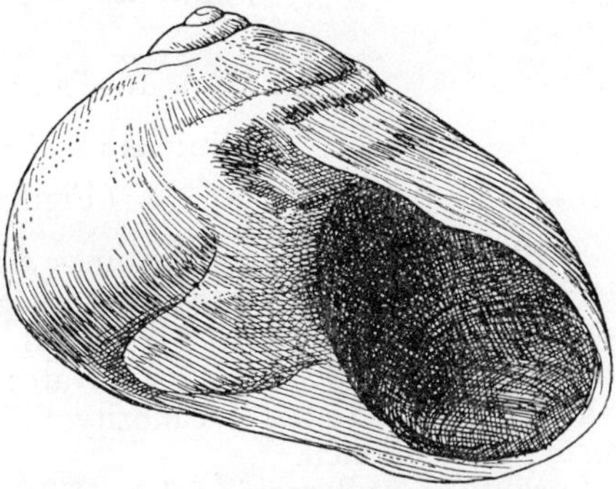

Figure 161

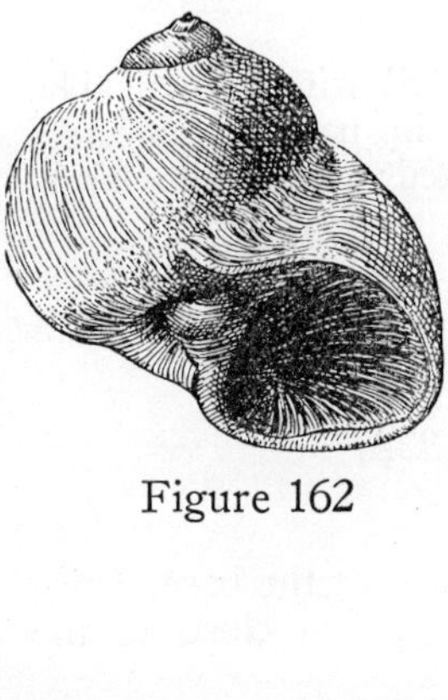

Figure 162

158. *Polinices conicus* × 1⅓
159. *Polinices sordidus* × 1
160. *Polinices aulacoglossa* × 1
161. *Polinices incei* × 2
162. *Mammilla melastoma* × 1

Common Sand Snail

Polinices aulacoglossa (Pilsbry & Vanatta, 1908)
(Figure 160)

Shell fawn, globose, indistinctly spirally engraved. Umbilicus wide, faintly spirally striate; the umbilical callosity the centre of which is transversely deeply grooved is reflected, partly concealing the upper part of the umbilicus. Aperture semi-lunar; outer lip thin; operculum horny. Size: 1½ inches.

Localities: Western Port; Waratah Bay; Port Phillip.

The callus is darker than the remainder of the shell and the presence of the transverse groove serves to distinguish it from the other Australian species of the genus. This species occurs in Queensland, New South Wales and Tasmania.

There has been considerable confusion over the name applied to this shell which was originally called *Polinices didymus* Chemnitz, 1781. Also, it was misidentified as *Natica tasmanica* Tenison Woods, 1875, another but indeterminate species.

ayresi Gabriel, 1961; 18 miles east of Lakes Entrance (5-15 fathoms).

Ince's Sand Snail

Polinices incei (Philippi, 1851)
(Figure 161)

Shell white, shining, sometimes tinted with pale violet, and a narrow reddish brown band near the sutures; solid. Spire short. Whorls flatly convex, base flat. Surface of shell microscopically decussately striated. Interior white. Aperture lunar-ovate; outer lip acute; columella with thick callus; a broad rounded callosity filling in the umbilical area; operculum horny. Size: 1 inch.

Localities: Anderson's Inlet; Wilson's Promontory.

A solid, depressed shell with a flattened base, features sufficient to distinguish it from any Victorian Sand Snail. Occurs also in South Australia and New South Wales.

Mammilla Schumacher, 1817

Shell oval, conical, relatively thin; white or fasciated with brown. The whorls oblique, rapidly enlarging. Mouth oblong, inner lip narrow, reflected, usually brown and blackish; umbilicus not funiculated; operculum large, brown, horny, filling the aperture.

TYPE SPECIES *Natica melanostoma* Gmelin, 1791.

Black-mouthed Sand Snail

Mammilla melastoma (Swainson, 1822)
(Figure 162)

Shell bluish grey or fawn, solid, ovate, rather flattened at the base. Spire very short. Whorls faintly excavated at the upper portion close to the sutures. The whole surface sculptured with fine oblique, radial striae, growth-lines plainly visible. Interior dark reddish brown. Aperture oblong, ovate; outer lip acute; columella calloused, callosity orange red almost covering the umbilical area, in some cases entirely so; operculum horny. Size: 1½ inches.

Localities: Lakes Entrance; Wingan Inlet.

The above species, apparently confined to the eastern portion of the state, is easily distinguished by the characteristic rich orange red callosity almost or entirely filling the umbilicus. Beyond Victoria it is found in New South Wales and Queensland.

Sigaretotrema Sacco, 1890

Spire very small, blunt; last whorl large, arched above; umbilicus wide, open, round edged. Mouth semicircular; operculum thin, horn coloured.

TYPE SPECIES *Sigaretus michaudi* Michelotti, 1847.

Iredale, 1924, erected *Propesinum* with *Natica umbilicata* Quoy & Gaimard, 1834, as type for these Australian shells.

Umbilicated Sand Snail
Sigaretotrema umbilicata (Quoy & Gaimard, 1833)
(Figure 163)

Shell cream, axially streaked with chestnut brown, a white band beneath the suture and another interrupting the colour markings at the centre of body-whorl; covered by a thin yellowish epidermis. Spire very short. Sculpture consisting of fine growth-lines which are crossed by fine spiral striae, giving it a decussated appearance. Aperture oval; umbilicus large, deep. Size: 1 inch.

Localities: Point Nepean; Point Lonsdale; Western Port.

This Sand Snail, with its characteristic colouring and large umbilicus, is a species easy to recognize. It is fairly constant in shape, but some examples are more globose, while others show a distinct flatness; sometimes there are rows of encircling brown spots (variety *picta* Reeve, 1864), while others are uniformly white. Crawling about, these molluscs may be located on the sand at low water, and dredged at shallow depths. Found on the Victorian coast generally, with an extension of range to South Australia, New South Wales and Tasmania.

nitida (Reeve, 1864); Kilcunda.

Ectosinum Iredale, 1931

Shell flattened, much broader than high, protoconch flat, glossy, two-whorled, succeeded by two rapidly increasing adult whorls with fine spiral sculpture and fine growth-lines. Operculum absent.

TYPE SPECIES *Ectosinum pauloconvexum* Iredale, 1931.

Zoned Sinum
Ectosinum zonale (Quoy & Gaimard, 1833)
(Figure 164)

Shell creamy white, covered by a thin yellowish epidermis; ear-shaped. Spire small, depressed. Whorls radially striated and concentrically, densely, finely wrinkled; obsoletely red zoned. Interior white and polished. Aperture very wide. Size: 1 inch.

Localities: Portsea; Queenscliff.

The zone of colour which characterizes the shell is but slightly indicated. The animal is much larger than the shell. In living conditions, it may be located on muddy sand flats or dredged in fairly deep water. It is a frequent species at Western Port, and is also recorded from South Australia and Western Australia.

Friginatica Hedley, 1916 (*Natica beddomei* Johnston, 1884).

beddomei (Johnston, 1884); Western Port; off Lakes Entrance. About ½ inch, white with a deeply excavated suture and deep umbilicus.

There has been considerable confusion over the placing of the small calcareous operculate members of this family.

Naricarius Dumeril, 1895 (*Nerita canrena* Linné, 1758).

Operculum with numerous spiral grooves; funicle broad and heavy.

subcostata (Tenison Woods, 1878); off Phillip Island, Western Port. About ¼ inch; with well-defined radiating grooves just below the sutures; umbilicus deep; operculum spirally grooved. Iredale, 1936, suggested that this species should be separated subgenerically from *Naricarius* and proposed the name *Quantonatica* for its reception.

Tanea Marwick, 1931 (*Natica zelandica* Quoy & Gaimard, 1834).

Operculum with two weak marginal grooves; funicle broad and heavy. Some authors have failed to see the outside groove when examining the operculum of the following species and so have mistakenly placed them in *Notocochlis* Powell, 1933 (*Cochlis migratoria* Powell, 1927), which has a single groove.

sagittata Menke, 1843; Port Phillip; Western Port. About ½ inch; with a row of sagittate markings just below the sutures; callus partly covering the umbilical area.

luculenta (Iredale, 1929); 18 miles east of Lakes Entrance (5-15 fathoms).

Proxiuber Powell, 1933 (*Lunatia australis* Hutton, 1878).

shorehami (Pritchard & Gatliff, 1900); Shoreham. About ⅜ inch; with slight radiating grooves close to sutures; umbilicus covered with a pad almost filling the umbilical area.

Tasmatica Finlay & Marwick, 1937 (*Natica schoutanica* May, 1912).

schoutanica (May, 1912); Bass Strait. About ⅝ inch; yellowish white, irregularly netted with broken zigzag lines of chestnut more distinct on the upper part of whorl; umbilicus fairly deep.

Family LAMELLARIIDAE

Lamellarias

Very thin, fragile, inflated shells with a large and flairing aperture and a low spire. There is no operculum. The shell is completely hidden by the mantle of the large fleshy animal which is often brightly coloured.

Lamellaria Montagu, 1815

Animal slug-like, with the shell completely covered by the thick mantle. Shell spiral, thin pellucid. Spire tumid, small. Aperture very large.

TYPE SPECIES *Helix perspicua* Linné, 1767.

Ophione Hidden Bubble Shell

Lamellaria ophione Gray, 1849

(Figure 166)

Shell white, thin, fragile, oblong-elongate. Surface showing fine lines of growth and covered with an extremely thin epidermis which is shining and slightly iridescent. Interior vitreous. Aperture oblique, ovate; outer lip very thin with the epidermis sometimes overlapping; columella much curved; apex plainly visible through the aperture. Size: ½ inch.

Localities: Dredged alive in Western Port; and off Portsea, Port Phillip.

The shell is internal being completely concealed by the smooth mantle of the animal. Occurs also in Tasmania and South Australia. There has been some confusion regarding the genus into which this species should be placed, but examination of the radula shows it to be *Lamellaria* and not *Marseniopsis*.

Mysticoncha Allan, 1936

Shell completely internal, thin, large, globose. Spire small, few-whorled. Body-whorl large; whitish buff. The animal is large and globose; the thick fleshy mantle projects beyond the body so that the foot and head are hidden from above.

TYPE SPECIES *Lamellaria wilsoni* Smith, 1886.

The single species of this genus was originally placed in *Lamellaria* from which it differs in both animal and shell characters.

Wilson's Hidden Bubble Shell

Mysticoncha wilsoni (Smith, 1886)

(Figure 165)

Shell internal, large, white, thin, fragile, elongately ovate, covered by a very thin, membranous epidermis. Spire short, apex obtuse. Whorls three, rapidly increasing in size. Body-whorl large, well rounded. Sculpture consisting of fine longitudinal microscopic striae. Interior white, shining. Aperture ovately pyriform, outer lip tenuous, epidermis slightly overlapping; columella much curved. Size: 1½ inches.

Localities: Port Phillip (dredged); Portland.

A remarkable internal shell and extremely fragile, requiring great care in handling. From the above description no difficulty should be experienced in identifying the species. The animal is large and almost pleurobranch-like in appearance; mantle thick and fleshy, smooth and much larger than the body over which it arches almost hiding the head and foot. The ground colour is orange yellow with a brown black network of conspicuous lines forming irregular five- or six-sided areas in the centre of which are rounded blotches of the same colour. Basedow, 1905, described a similar animal from

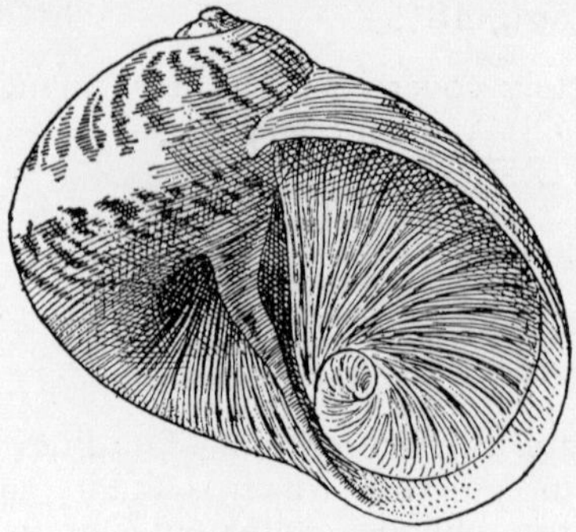

Figure 163

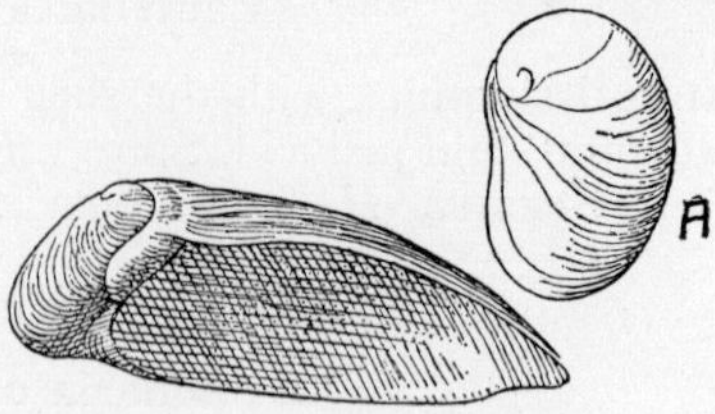

Figure 164

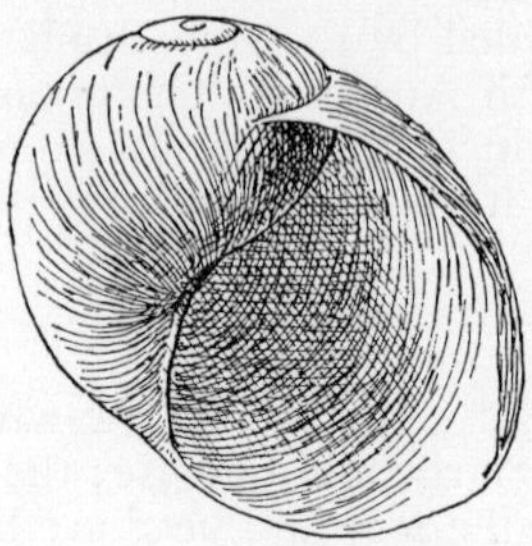

Figure 165

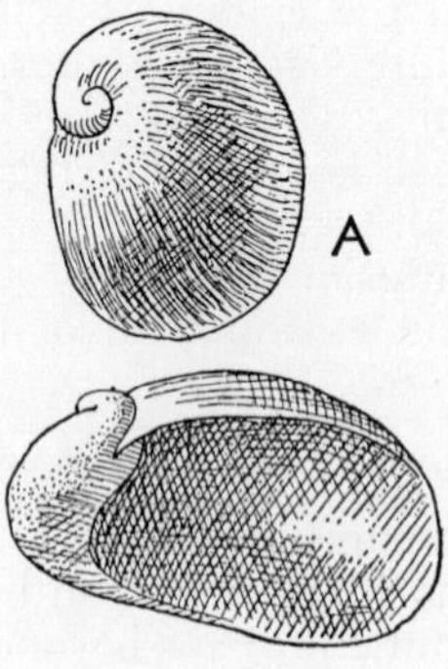

Figure 166

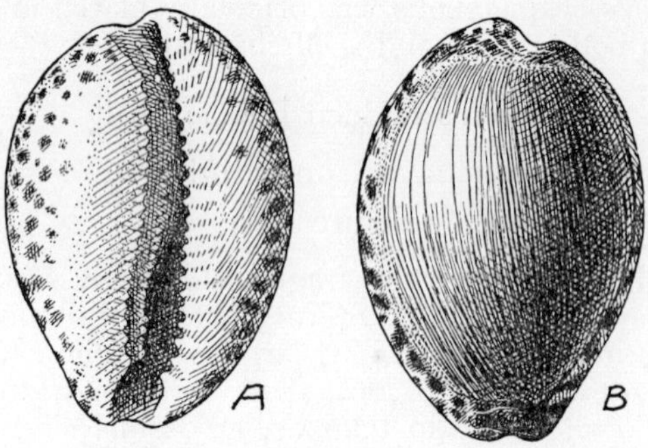

Figure 167

163. *Sigaretotrema umbilicata* × 2
164. *Ectosinum zonale* × 1⅓
 A Dorsal view × ⅔
165. *Mysticoncha wilsoni* × 1⅓
166. *Lamellaria ophione* × 2⅔
 A Dorsal view
167. *Notocypraea angustata* × 1⅓
 A Ventral view B Dorsal view

South Australia as *Caledoniella contusiformis* and Joyce Allan has placed it in the synonymy of *wilsoni* with which we concur.

Family CYPRAEIDAE

Cowries

Shell ventricose, convolute, covered with shining enamel. Spire concealed by the last whorl which is large. Aperture long and narrow with a short canal at each end; outer lip inflected and both lips crenulated. There is no operculum. The young shell has a thin outer lip and prominent spire and can easily be mistaken for a thin-shelled Olive.

Notocypraea Schilder, 1927

Shell small, subcylindrical, pyriform, smooth, not umbilicated, without an elevated spire. Colour cream with a pattern of darker peppering or maculations.

TYPE SPECIES *Cypraea piperita* Gray, 1825.

The genus *Notocypraea* is confined to southern Australia where a number of forms occur. These forms which show variation in colour and shape have been variously grouped by different authors, some preferring to consider them a single species with colour varieties, while others have separated them into a number of species and subspecies.

Brown Cowry

Notocypraea angustata (Gmelin, 1791)

(Figure 167)

Shell ovate, smooth, shining, reddish brown to plum colour or rarely pale brown on the dorsum. The base convex, white; the margin slightly thickened, rounded, white with small round brownish black spots which coalesce to form larger spots at the posterior and anterior extremities. The aperture rather large; outer lip rounded with strong white teeth sloping into the aperture; inner lip slightly thickened with small teeth which do not extend on to the base. Size: 1 inch.

Localities: Portsea; Sorrento; Western Port.

Cotton & Godfrey, 1932, proposed a new name *verconis* for this shell on the grounds that the type locality of Gmelin's shell was South Africa. We can find nothing to substantiate this, and as the figure quoted by Gmelin is similar to this southern Australian shell, we prefer to retain the older well-established name.

J. R. Griffiths has been working on a revision of this genus for the past year and he has supplied us with the following notes on the species recorded from Victoria. The chief features which distinguish the Victorian species of *Notocypraea* are the colour and markings of the dorsum, and the shape and teeth of the fossula. Other differences exist between the species, but they are to a great extent masked by the considerable variation within each species, and are therefore only of small diagnostic value. These main distinctions are themselves subject to variation, and occasional specimens are found which cannot be identified even by these means; in these cases the shape of the radula is the only satisfactory criterion.

angustata (Gmelin, 1791); the dorsum is deep brown, even in slightly worn shells. Western and central Victorian specimens only rarely show any trace of transverse dorsal bands, but beach specimens from eastern Victoria are more distinctly banded, the interrupted brown bands being wider than those of *comptoni*. The columellar teeth are large, and less numerous than those of most of the other species. The fossula is crossed by a few regular parallel teeth. Sometimes, however, they are formed only at the top and bottom, with the resultant trough running at a constant depth along the full length of the fossula, and extending into the columellar sulcus. On the average, this species is larger, broader and taller than other *Notocypraea*.

comptoni (Gray, 1847); Portland; Port Phillip, Western Port (Figure 168). This species usually has three or four uninterrupted dorsal bands, the central two being the most prominent, and the rear one least distinct. In some live specimens the deep brown of the top layer of the shell conceals the bands, but they show readily as soon as the shell becomes worn or faded. The bands are rarely more than one millimetre wide, and usually remain separate from each other over the whole width of the shell. The fossular teeth are smaller than those of *angustata,* and always exhibit a depression half-way down; the depression is deepest in front, and tapers up so that it is shallow where it meets the columellar sulcus. The lateral spots are usually smaller than those of *angustata*, but more numerous than those of *piperita*. *Cypraea* (*Notocypraea*) *trenberthi* Trenberth, 1961, is a synonym, see Schilder, *Cowry* 1 (2), 1961, p. 4.

declivis (Sowerby, 1870); Portland (Figure 169). Fully adult specimens are easily recognized by the dorsal pattern, which consists of a pale milky layer densely covered by hundreds of small brown specks. On immature and worn shells, where these characters are not present, dorsal bands somewhat similar to those of *comptoni* are visible. *declivis* differs from *comptoni* by its general colour being sepia rather than reddish brown, and by the more prominent pairs of terminal patches, the rear pair of which is the larger.

The close relationship between the two species is shown by some similarity in the fossular arrangement, though the teeth of *declivis* do not usually extend so far into the shell, and the fossular depression is less noticeable.

piperita (Gray, 1825); off Lakes Entrance; Western Port; Port Phillip; Portland (Figure 170). The dorsal colour is usually paler than that of the three preceding species. The four dorsal bands are more prominent and wider than those of *comptoni,* and in most shells the central two join into one very wide band on top or on the right of the shell. The front and rear bands, which like the central pair are usually interrupted and not continuous, are in most shells clearly visible. The dorsum is sometimes spotted (piperate), and even occasionally reticulate, the colour of the marks being medium dark brown. The teeth in the region of the fossula are smaller than those of *angustata,* and more regular than those of *comptoni*; they are longitudinally depressed in the centre, the depression continuing to form a marked columellar sulcus running the whole length of the shell. The lower part of the fossular teeth sometimes bends out to form a marked protrusion into the shell. Occasional albino specimens have been found in most species of *Notocypraea*, but those of *piperita* usually bear at least a faint trace of bands on the columellar side of the dorsum.

wilkinsi (Griffiths, 1959); Flinders; dredged Western Port. This is easily distinguished from the other Victorian species by its dorsal colour, which is pale gold or pale yellow. Furthermore, there is no trace of dorsal bands, either on the dorsum or the sides. The fossular arrangements are sometimes reminiscent of *piperita*, but there is considerable variation, and insufficient specimens are known to permit description of its usual form. The teeth of *wilkinsi* are, however, the more numerous. Its shells are usually without lateral spots; when present they are small. In proportion to length, shells of *wilkinsi* are lighter in weight than the shallow-water specimens of the other species.

molleri (Iredale, 1931); the range in Victoria is from the eastern border to off Lakes Entrance. Found in sponges at about 60 fathoms. Shell ovate, 20 to 30 mm. long, with a tall, wide and swollen dorsum; right side calloused. Dorsal colour usually pale flesh, though occasionally white. There are four fairly wide interrupted light brown dorsal bands; the central ones sometimes extend completely across the dorsum, while the end ones usually reach only to the top. On some shells the bands are only present well down on the left side, or are even absent altogether. There are usually two pale brown anterior terminal blotches. The labial spots, which are usually but not always large in size, number about ten; the few shells with the columellar side spotted have fewer but equally large spots. In mature shells the base is white. The aperture is wide throughout, considerably bent at the rear, and almost unconstricted in front. The teeth are fairly big; on the labial side they run half-way over the base, while on the columellar side they extend only just beyond the aperture. The fossula is deep, wide and long; it is formed of ridges of teeth on top and bottom, somewhat similar to many specimens of *N. angustata*; however, the lower edge goes down more deeply into the shell than the lower edge of the columellar sulcus. The lower edge of the fossula protrudes slightly out into the shell.

Umbilia Jousseaume, 1884

Large, elongate-oval, smooth shells with produced extremities; deep umbilicus, uncalloused margins and comparatively small teeth.

TYPE SPECIES *Cypraea umbilicata* Sowerby, 1825 (non Dillwyn, 1823) = *Umbilia hesitata* Iredale, 1916.

Umbilicated Cowry

Umbilia hesitata Iredale, 1916
(Figure 171)

Shell solid, pyriformly ovate, ventricose, highly enamelled, whitish, besmeared with chestnut, and thickly spotted with a similar but darker coloration; base white, sides with more numerous spots more or less of a violet tinge. Aperture somewhat twisted, teeth well developed. Spire deeply umbilicated. Size: 4 inches.

Localities: Cape Schanck; Gippsland coast; dredged Bass Strait.

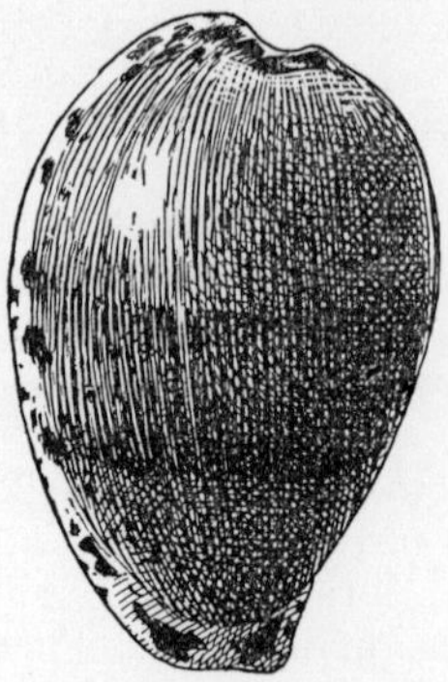

Figure 168

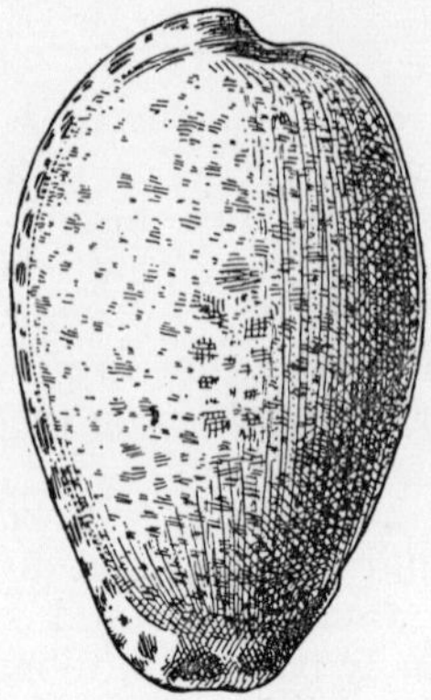

Figure 169

Figure 170

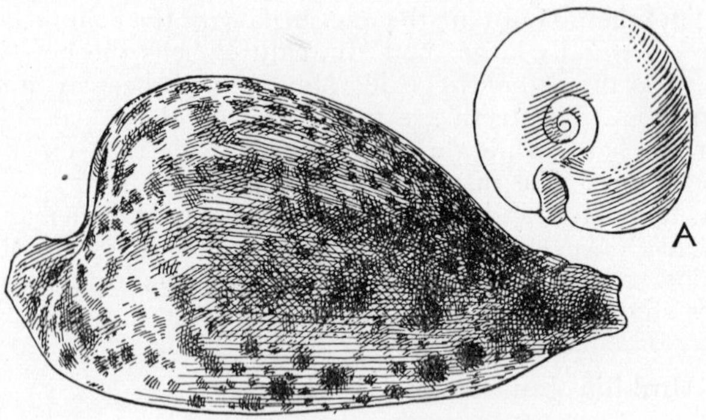

Figure 171

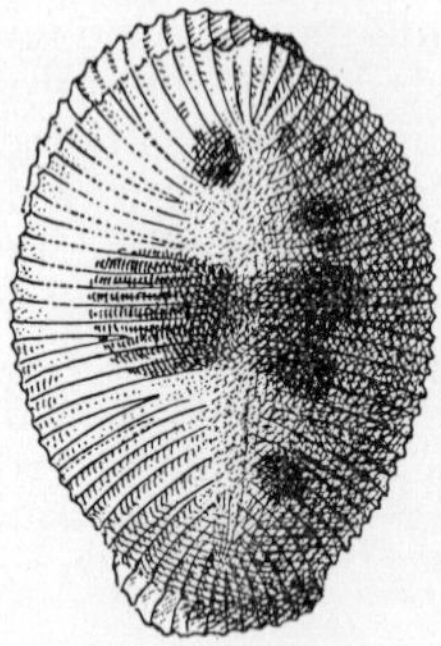

Figure 172

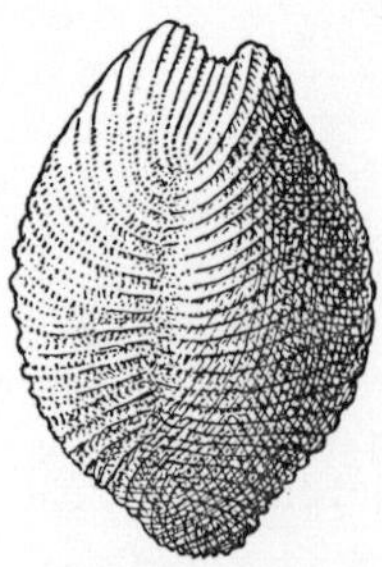

Figure 173

168. *Notocypraea comptoni* × 2
169. *Notocypraea declivis* × 2
170. *Notocypraea piperita* × 2
171. *Umbilia hesitata* × ⅔
   A Posterior elevation × 4/9
172. *Ellatrivia merces* × 3
173. *Ellatrivia oryza* × 3

South-eastern Australia appears to be the metropolis of this handsome cowry; worn specimens occasionally wash up on the beaches but living examples are generally secured through trawling operations, the principal source of supply. An extremely solid and heavy species. Much variation in colour exists and series of specimens can be graded from the typical dark chestnut through to light shells with few markings. Albinos sometimes appear and are known as *howelli* Iredale, 1931. A smaller (3 inches) dark form with a distinct ventral blotch is called *beddomei* Schilder, 1930. Known for many years as *Cypraea umbilicata* Sowerby, 1825, a preoccupied name, this shell was first obtained in quantity by the trawler *Endeavour* while dredging in Bass Strait. It also occurs off the southern New South Wales and Tasmanian coasts.

Ellatrivia Iredale, 1931

Shell small, thin, sculptured with fine ribs; white blotched with light brown.

TYPE SPECIES *Triviella merces* Iredale, 1924.

Southern Ribbed Cowry

Ellatrivia merces (Iredale, 1924)

(Figure 172)

Shell small, fragile, ovate, slightly narrowed anteriorly. Colour pinkish white with a few brown spots of varying size on the back, the extremities being rose tinted. Apex fairly conspicuous. The surface is ornamented by transverse ribs throughout, interrupted by an impressed median line indicating where the mantle lobes of the animal meet. Aperture narrow, curved at the extremities. Size: ½ inch.

Localities: Common at southern end of Port Phillip Bay; Lorne; and coast generally.

This species enjoys a rather wide distribution. In addition to the entire Victorian coastline, it occurs in South Australia, Western Australia, New South Wales, Tasmania and New Zealand. A species readily identified by its peculiar ribbed sculpture. As in the case of *Notocypraea piperita* this form may be obtained hiding in clumps of Bryozoa. Better opportunities for procuring them in living condition are offered when dredging is undertaken.

This shell was originally known as *Trivia australis* Lamarck, 1822, but this name is preoccupied by Schroëter, 1804.

Rice-grain Cowry

Ellatrivia oryza (Lamarck, 1810)

(Figure 173)

Shell small, snow white, oblong-ovate, transversely regularly finely ribbed throughout with a well-defined linear dorsal impression. Spire concealed. Aperture narrow, curved at the extremities. Size: ⅓ inch.

Locality: Port Phillip dredged in 5 fathoms.

This small, pure white shell is easily separated from *merces* Iredale by its smaller size, more ovate form, and in the absence of the pinkish coloration and dorsal spots. We are aware of one record of this species in Victoria (MacGillivray, 1852, p. 365).

Family CASSIDIDAE

Helmet Shells

Shell solid, subglobular with a short spire and large body-whorl. Whorls sometimes varicose. Aperture terminating in a short anterior canal, columella calloused and plicated; outer lip may be thickened and reflected, usually dentate within. The small horny operculum does not fill the aperture.

Helmet Shells are a fairly large assemblage deriving their name from an apparent likeness to the helmet of the ancients, to which the large tropic species were likened. These large forms are a well-known article of commerce as it is from them that shell cameos, etc., are carved. The smaller, more rounded species, with widened apertures, which include our Victorian representatives, are known as False Helmets.

Antephalium Iredale, 1927

Small, ovate shells, with strong sculpture. Mouth narrow; outer lip recurved, not usually strongly variced; columella reflected so as to leave only a minute false umbilicus.

TYPE SPECIES *Cassis semigranosa* Lamarck, 1822.

Half-grained Helmet

Antephalium semigranosum (Lamarck, 1822)

(Figure 174)

Shell creamy white, ovate. Spire acuminate, less than half the length of body-whorl, apex somewhat papillary. Whorls rounded, sculptured with oblique radial threads, and at the middle of body-whorl faintly, concentrically, grooved, a feature becoming more pronounced at the anterior end; on the upper portion of whorl it is further ornamented by four conspicuous, encircling rows of grains. Aperture pyriform, very smooth within; anterior canal short, much recurved; columellar callus finely plicated; operculum elongated, horny. Size: 1½ inches.

Localities: Mornington; Sorrento; Portsea; Queenscliff.

The rows of grains are extremely characteristic, serving as a useful recognition mark, and students should experience no difficulty in identifying this very attractive species. As regards colour, variation is evident, some specimens being much darker while others are distinctly pink tinted.

Located on sandbanks, the animals are extremely active, preying on bivalve molluscs; excellent living examples have been obtained during night collecting at Portsea, while cast-up specimens are not uncommon at many points on the Victorian coast. Beyond this state it is recorded from Tasmania, South Australia and Western Australia.

adcocki (Sowerby, 1896); a smaller, white shell, 1¼ inches, readily distinguished by the rows of small bright red brown square spots on the spire. The sculpture of impressed spiral lines overridden by strong radials on the shoulder is not as obvious as in *semigranosum*. It is a rare shell but has been recorded from south-west Australia, South Australia and Bass Strait, Victoria.

sinuosum (Verco, 1904); the smallest of the Victorian Helmets (1 inch) and is easily distinguished by its colour pattern of five spiral rows of faint square yellowish red spots on a white ground. Taken in 20-25 fathoms off Lakes Entrance. It was originally described from South Australia.

Xenogalea Iredale, 1927

Medium-sized shell characterized by its open mouth, rarely denticulate within, and by its short spire; juvenile ornament spirally striate; adult smooth, sometimes with a nodulous or keeled shoulder.

TYPE SPECIES *Cassis pyrum* Lamarck, 1822.

Members of this genus were mistakenly placed in *Xenophalium* Iredale, 1927, by Gatliff & Gabriel, 1931, and by Macpherson & Chapple, 1951.

Pear Helmet

Xenogalea pyrum (Lamarck, 1822)

(Figure 175)

Shell fawn, with bands of brownish wavy spots; strong, moderately large, ovately globose. Spire low, the earlier whorls of which are spirally striated. Whorls convex, more or less angulated round the upper portion, generally nodulose. The shell is adorned with a row of nodules on the angle of the shoulder, with occasionally a second row a little lower down. The middle of the body-whorl is almost smooth, but towards the base about six to eight shallow grooves appear. Aperture large, very smooth within, and terminated by a sharply recurved canal; columella bearing several small folds; outer lip thickened, reflected and banded with purplish brown on the outer edge; inner lip spreading as a white plate beyond the columella. Size: 4 inches.

Localities: Sorrento; Portsea; Queenscliff.

This is a species showing considerable variability and is sometimes comparatively thin, but more frequently much thickened. With hermit crabs in possession these shells are frequently trapped in crayfish pots, and the deeper waters of Bass Strait are noted for beautifully coloured specimens. The species has a wide distribution, being recorded from Western Australia, South Australia, New South Wales, Tasmania and New Zealand.

Spotted Helmet

Xenogalea spectabilis Iredale, 1929

(Figure 176)

Shell creamy pink with five to six brownish bands, extending to the outer lip where they are darker; large, globose, elongately ovate. Spire short about one-quarter the length of shell. Body-whorl fairly smooth except for numerous definite lines of growth and more or less nodulose shoulder ridges.

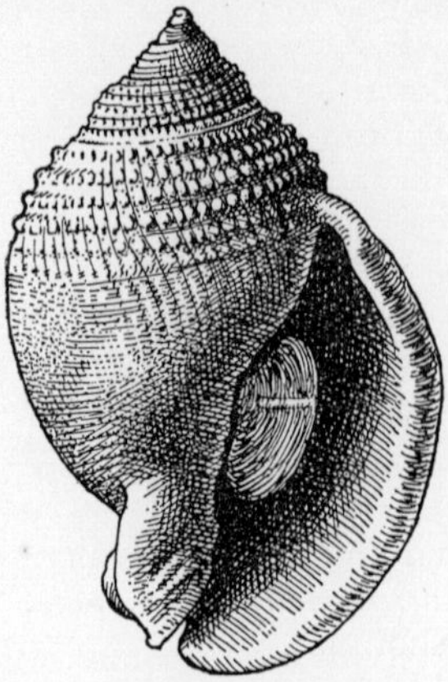

Figure 174

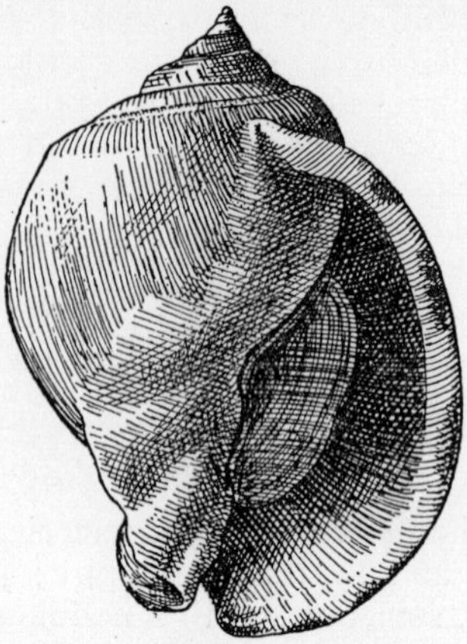

Figure 175

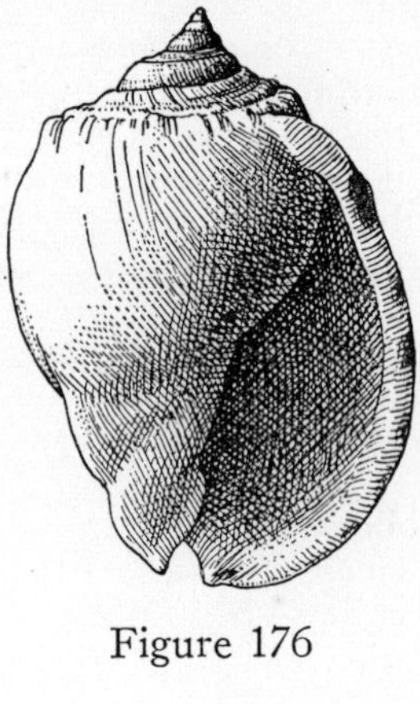

Figure 176

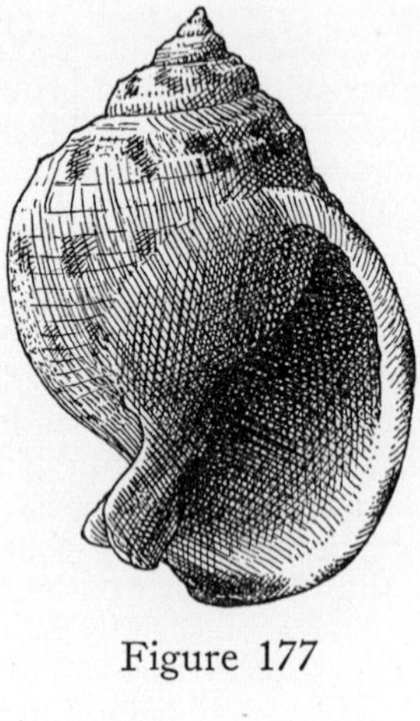

Figure 177

174. *Antephalium semigranosum* × 1
175. *Xenogalea pyrum* × 1
176. *Xenogalea spectabilis* × ⅔
177. *Xenogalea thomsoni* × ⅔

Interior cream, smooth. Aperture fairly wide; umbilicus almost hidden by the reflection of the inner lip; columella internally lirate; outer lip reflected, denticulate within; anterior canal short, recurved. Size: 4 inches.

Localities: Queenscliff; Ninety Mile Beach.

A handsome shell easily recognized by its ovate form and the nodulous shoulder ridges. It may be mistaken for *pyrum* but is heavier and with the above feature is readily separated. Occurs also in New South Wales.

Thomson's Helmet

Xenogalea thomsoni (Brazier, 1875)

(Figure 177)

Shell thin, somewhat fragile, globose. Spire attenuate, less than half the length of body-whorl, the whorls of which are angularly shouldered and sculptured with feeble, irregular, spiral ridges, crossed by fine, slanting radial threads; colour pinkish fawn with interrupted rows of reddish brown spots, one on the earlier whorls and three on the body-whorl, the darkest on the shoulder. Body-whorl shouldered, marked by a distinct ridge with a definite depression separating this from another angular ridge. Middle of body-whorl is comparatively smooth except for irregular growth-lines, but towards the base about six to eight shallow grooves appear. Aperture fairly wide; outer lip slightly reflected and banded with purplish brown on the outer edge; internally the outer lip is smooth but slightly denticulate in the adult stage; inner lip spreading as a white plate beyond the columella; canal short, recurved. Size: 2-3 inches.

Locality: off Cape Everard.

This is easily recognized by its consistent pinkish fawn colour and regular rows of darker spots. Though similar in shape to *pyrum* Lamarck it is definitely more fragile. By no means common in Victoria but frequently appears on the New South Wales coast down to 70 fathoms. Occurs also in northern Tasmania.

Labiate or Lipped Helmet

Xenogalea labiata (Perry, 1811)

(Figure 178)

Shell strong, ovate, shining. Spire comparatively short and without varices; earlier apical whorls with extremely fine spiral threads which soon vanish, the shell becoming quite smooth showing only faint lines of growth. Colour rather distinctive, generally bluish purple, splashed or flamed with brownish purple, the body-whorl with three or four encircling, interrupted bands of darker coloration, still darker on the outer lip which is denticulate within. Aperture narrowly ovate; canal short. Interior purplish. Size: 2½ inches.

Localities: Waratah Bay; San Remo; Bass Strait.

A distinctively coloured species, the darkest of our Victorian Helmets, sometimes known under the name of Agate Helmet. Iredale in the records of the Australian Museum remarks: 'I secured a shell at Port Fairy, Victoria, in 1923 which may be the Victorian representative, but it is larger, less

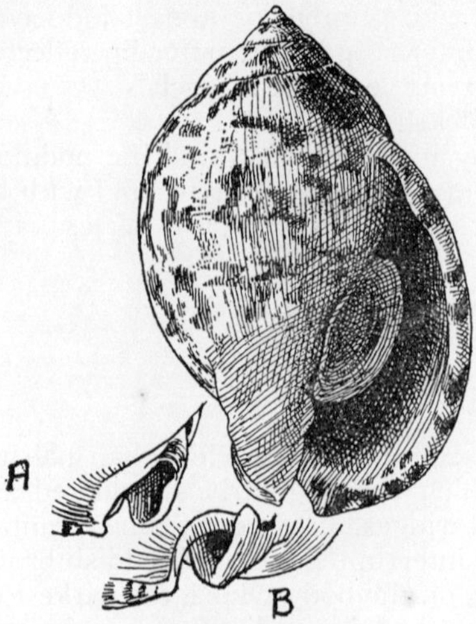

Figure 178

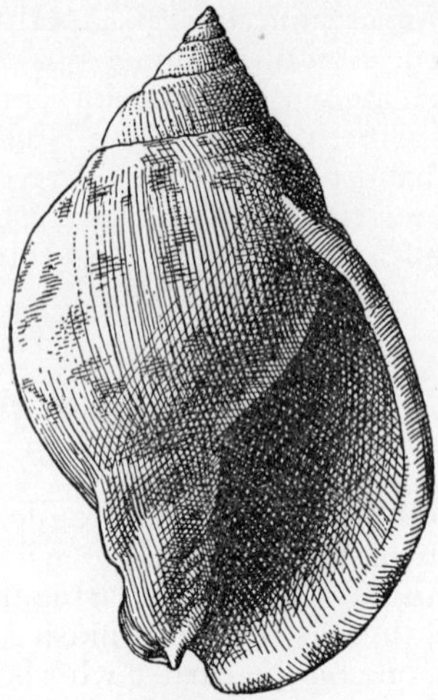

Figure 179

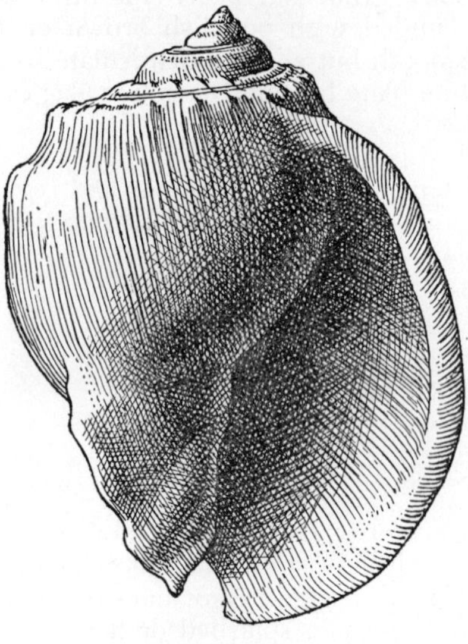

Figure 180

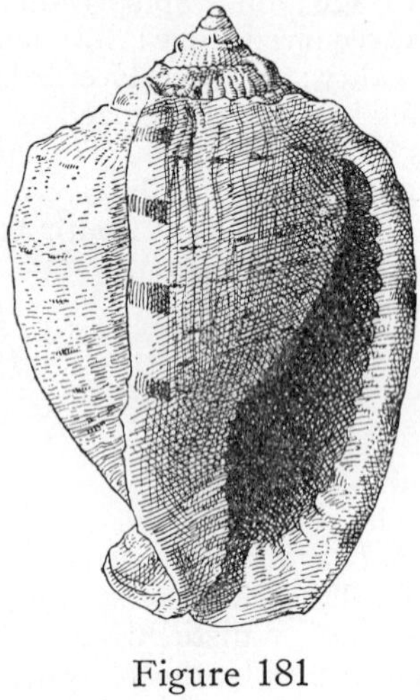

Figure 181

178. *Xenogalea labiata* × ⅔
 A Dorsal view of anterior notch
 B Anterior view of anterior notch

179. *Xenogalea stadialis* × ⅔

180. *Xenogalea nivea* × 1

181. *Hypocassis bicarinata* × 1

brightly coloured, and lacks the distinctive colour scheme of the typical form.' Another specimen apparently similar to this appears in the collection of C.J.G. A very common shell all along the New South Wales coast, and is recorded from south Queensland and Norfolk Island.

Smooth Helmet

Xenogalea stadialis (Hedley, 1914)
(Figure 179)

Shell large, globose, elongately ovate; colour pinkish with four to five interrupted rows or bands of brownish blotches extending to the outer lip, the colours of which are much darker; earlier whorls decussated with fine concentric threads which are crossed by oblique lines of growth, this feature disappearing on the penultimate and body-whorls; here the shell is quite smooth except for numerous, irregular growth-lines. Interior pinkish white; inner lip with a definite plate of callus; columella showing several irregular plications. Aperture fairly wide; outer lip reflected, not denticulate within. Size: 3 to 4 inches.

Locality: Dredged Bass Strait.

This is a very striking species, at one time regarded as a deep-water variety of *pyrum* Lamarck. Trawling operations on the continental shelf of New South Wales, however, have provided numerous specimens for study. Consistent features are sufficient to rank the shell as a distinct species and not a variety. It certainly shows some similarity to *pyrum* but is immediately separated by its large size, absence of angles and tubercles on the shoulder, absence of spiral grooves at the base and lack of denticles on the outer lip. Found also in Tasmania.

White Helmet

Xenogalea nivea (Brazier, 1872)
(Figure 180)

Shell white, occasionally slightly banded, globose. Spire very short about one-quarter the length of shell. Body-whorl fairly smooth except for two rows of nodules on the shoulder; longitudinal growth-lines are plainly visible. Interior white, polished; umbilicus very small almost hidden by the reflection of inner lip; outer lip reflected, with about six distant, brown blotches thereon; anterior canal short, recurved. Size: 3 inches.

Locality: Portland.

This is a shapely, pure white shell, easily recognized by the double row of tubercles at the shoulder of the body-whorl. It occurs also in Tasmania where it has been regarded as a variety of *pyrum*.

This shell has been mistakenly identified as *paucirugis* Menke, 1843, a south-western Australian species which has not yet been recorded from as far east as Victoria.

Hypocassis Iredale, 1927

Medium to large shells with many varices. Spire very short; whorls strongly shouldered and nodulose. Mouth narrow; outer lip thickened, folded back, internally faintly denticulate; columella nearly straight, widely reflected over the body-whorl; false umbilicus narrow but open, umbilical cavity present.

TYPE SPECIES *Cassis bicarinata decresensis* Hedley, 1923.

Bicarinate Helmet

Hypocassis bicarinata (Jonas, 1839)

(Figure 181)

Shell large, colour pinkish violet, thick, triangularly ovate, ventricose. Spire very short, apex papillate. Whorls with single varices. Surface of shell sculptured with irregular, longitudinal plications and at the base crossed by fine spiral striae giving it at this area a somewhat wrinkled appearance. At the upper portion of the body-whorl will be seen two or three more or less distinct nodulose ridges, the lowest almost entirely absent; the whole surface showing eight to ten interrupted concentric bands of brownish colour, sometimes appearing as faint lines. Interior white or cream. Aperture narrow; outer lip thickened, faintly denticulate within; columella almost straight, reflected over body-whorl and bearing about six plications. Size: 4-5 inches.

Locality: Portland.

The only representative of the genus in Victoria, its nodulose character makes recognition easy. This ornate species like all the Helmets is predatory, living in sandy localities where bivalves abound and upon which it preys. It is rare in Victoria, but common in South and Western Australia.

bicarinata decresensis (Hedley, 1923); a small strongly sculptured form to which some of the smaller specimens taken in Victoria would be referred.

Family CYMATIIDAE

Rock Whelks or Tritons

Shell stout, fusiform, with a short anterior canal, and varices which are either few and irregularly placed or form a continuous row crossing the whorls on opposite sides. The number of varices does not exceed two to each whorl. Horny operculum angular with subspiral or subcentral nucleus.

Argobuccinum Bruguière, 1792

Shell thick with encircling sculpture, varices more or less strong. Spire elevated. Anterior canal short, posterior absent. Mouth edge toothed.

TYPE SPECIES *Murex argus* Gmelin, 1791.

Finlay, 1926, erected the genus *Gondwanula* for the New Zealand species *Ranella tumidum* Dunker, 1862, in the mistaken idea that *Argobuccinum* dated only from Mörch, 1852, and was thus preoccupied. Therefore *Gondwanula* falls into the synonymy of *Argobuccinum*.

Flag Triton

Argobuccinum vexillum (Sowerby, 1835)

(Figure 182)

Shell large, fusiformly ovate, colour fawn with ridges of brown and dark brown. Spire short about one-third the length of shell. Whorls with eight broad, flatly rounded varices. The whole shell invested with a light brown, velvety epidermis. On the whorls of the spire appear four to five distinct dark brown, concentric ridges, crossed by longitudinal ribs and at the point of crossing small tubercles are formed; this feature more or less evident on the body-whorl, the longitudinal ribs in some specimens being almost obsolete; between the concentric ribs of the body-whorl will be seen parallel ribs of lesser strength and generally lighter in colour. Interior white. Aperture roundly ovate; outer lip expanded strengthened by a very broad varix; columella slightly curved, with a prominent tooth at the posterior end; anterior canal short; operculum light brown, horny. Size: 2 to 4 inches.

Localities: Portland; Port Fairy; Western Port.

A rather distinctive shell, readily recognized by its fusiform shape and concentric, brown ribs, with varices coloured similarly to the rest of the shell. Much variation exists as regards the number and strength of tubercles, the body-whorl in some cases being almost devoid of such. Occurs also in Tasmania and South Australia.

Bass Triton

Argobuccinum bassi (Angas, 1869)

(Figure 183)

Shell light brown, ovately fusiform, solid. Spire approximately one-third the length of shell; apex obtuse. Whorls with about six rounded varices and sculptured throughout with close irregular concentric, flatly rounded ridges, beaded here and there, more particularly towards the anterior end. Body-whorl slightly angled near the suture, the angle bearing about nine rather prominent tubercles. Interior slightly violet tinted. Aperture ovate; outer lip white, thickened, strongly denticulated within; operculum horny. Size: 1 inch.

Localities: Corner Inlet; Flinders; dredged Western Port.

This is a strong shell, easy to recognize by its squat form and concentric sculpture. When taken alive the shell is frequently encrusted with a species of Bryozoa. Occurs also in Tasmania.

Austrosassia Finlay, 1931

Shell solid, finely spirally striate. Spire about half length of shell which is covered by a thick brownish periostracum; approximately two varices to each whorl; turbinate, with globose rapidly increasing whorls; usually wider than high, the tip minute and planorbid, set almost symmetrically on the shell, polished, white and shining with no trace of a horny envelope, differentiated from normal shell only by inception of adult sculpture.

TYPE SPECIES *Septa parkinsoniana* Perry, 1811.

Iredale placed the above species in *Austrotriton* Cossman, 1893, but Finlay showed that it differed from the type of the latter genus in the form of the protoconch.

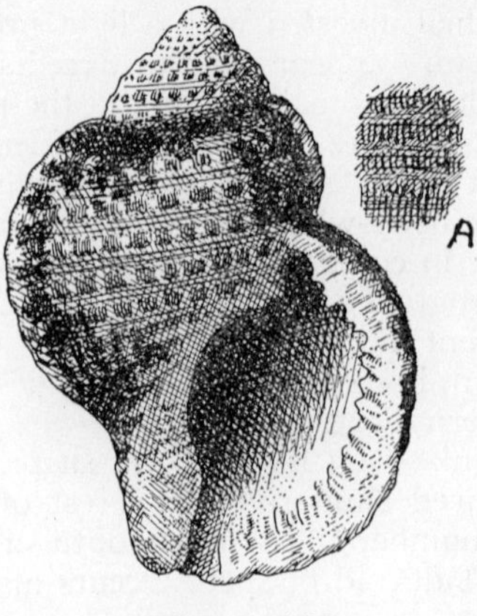

Figure 182

Figure 183

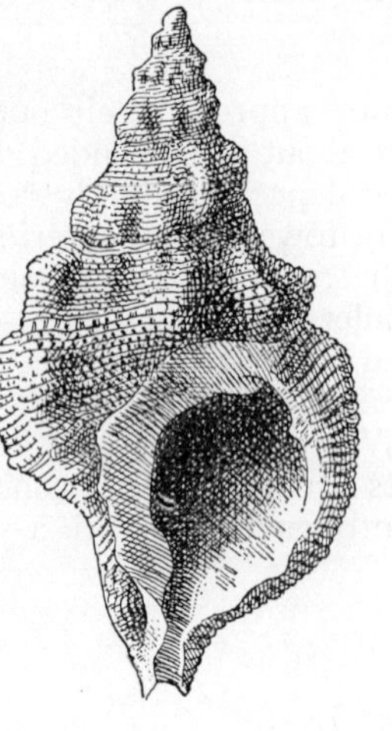

Figure 184

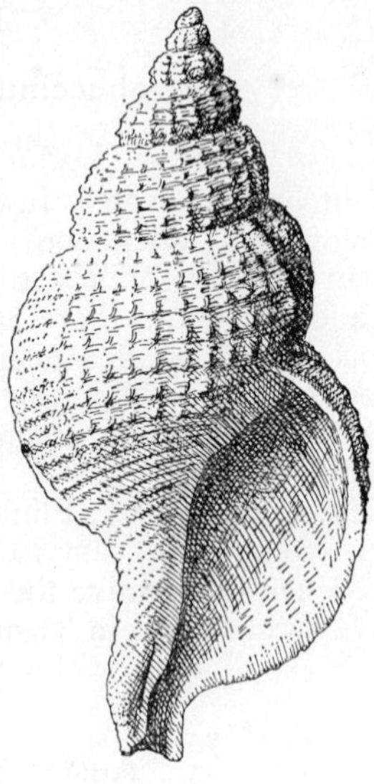

Figure 185

182. *Argobuccinum vexillum* $\times \frac{2}{3}$
A Sculpture (enlarged)
183. *Argobuccinum bassi* $\times 1$
184. *Austrosassia parkinsoniana* $\times 2$
185. *Fusitriton retiolus* $\times \frac{2}{3}$

Parkinson's Whelk

Austrosassia parkinsoniana (Perry, 1811)
(Figure 184)

Shell solid, somewhat broadly fusiform, yellowish brown, the spiral striae articulated with darker brown and white clearly visible under lens, the whole surface invested with a thin, brownish epidermis. Spire about half the length of shell. Whorls with six to seven strong, rounded varices between which will be seen four to five elongated, rounded knobs. Sculptured throughout with equidistant concentric, granulose striae. Interior white. Aperture ovate, outer lip fortified with a strong varix, denticulated within; columella much curved; anterior canal narrowly open. Size: 1½ inches.

Localities: Conran Point; eastern Victoria.

A distinctive species, easily recognized by prominent knobs on the body-whorl and the close, even, granulose concentric striae. An extremely frequent shell on the beaches in New South Wales.

Iredale, 1924, separated deeper water shells as *parkinsoniana basilicus*. He considers they differ consistently in shape, being narrower, longer spired, and longer canalled. Having examined a series we find the variation is not consistent and that a series from deep water cannot be separated.

Fusitriton Cossman, 1903

Shell fairly thin, varices small, canal long. Sculpture of encircling and longitudinal ribs forming a reticulate pattern; periostracum thick.

TYPE SPECIES *Triton cancellatum* Lamarck, 1815.

Netted Triton

Fusitriton retiolus (Hedley, 1914)
(Figure 185)

Shell large, thin, ovately fusiform, colour white or creamy covered by a fawn epidermis. Spire acuminate, approximately one-third the length of shell. Whorls inflated, bearing thirteen to fourteen rounded varices, and reticulated throughout by concentric and longitudinal striae, small tubercles being produced at the point of crossing. Interior white and olivaceous towards the outer lip which is thin but strengthened by an outside rounded varix. Aperture ovate; columella curved, furnished at the posterior end with a prominent elongated tooth; anterior canal fairly long, well open, and slightly upturned. Size: 5 inches.

Locality: South of Gabo Island (100 fathoms).

A shapely shell and easily identified by its beautiful reticulate sculpture. It is recorded from the New South Wales coast at 50-100 fathoms between Green Cape and Gabo Island.

This shell was originally placed in the genus *Argobuccinum* by the author.

Cymatiella Iredale, 1924

Shell small, varices irregular, strong. Inner mouth edge toothed; edge of columella free.

Type Species *Triton quoyi* Reeve, 1844 = *Triton verrucosa* Reeve, 1844 (page priority).

Members of this genus were originally included in the genus *Personella* Conrad, 1865.

Warted Sand Whelk

Cymatiella verrucosa (Reeve, 1844)

(Figure 186)

Shell solid, creamy, invested with a thin epidermis of light brown; fusiformly oblong. Spire approximately half the length of shell. Whorls rounded bearing eight to ten varices. Sculpture consisting of concentric ridges and striae, with more prominent longitudinal ribs, noduled at the point of crossing and giving the surface a somewhat cancellated appearance. Interior white. Aperture ovate; outer lip strong, denticulated within; columella curved, scarcely wrinkled; anterior canal short; operculum horny. Size: ¾ inch.

Localities: Lorne; Sorrento; dredged in Western Port.

One of our commonest shells, it is a solid, little species and with its coarse sculpture and conspicuous varices should be easily identified. A sinistral example appears in the collection of C.J.G. Recorded also from New South Wales, Tasmania and South Australia.

Lesueur's Sand Whelk

Cymatiella lesueuri Iredale, 1929

(Figure 187)

Shell solid, white, sometimes with a pink apex, covered with a thick brown epidermis, fusiformly oblong. Spire about half the length of shell. Whorls rounded with about eight to ten varices. Ornamented with fairly evenly spaced concentric ridges and more prominent longitudinal ribs, noduled at the point of crossing. Interior white. Aperture ovate; outer lip strong, strengthened with a stout varix, denticulated within; columella curved, with four, short, pronounced ridges at the anterior end, canal short, well open. Size: ¾ inch.

Localities: Port Phillip; Western Port; Airey's Inlet; Lorne.

This species bears much resemblance to *verrucosa* (Reeve) but is smaller and may be distinguished by its definitely finer sculpture, and although not a consistent feature, the pink apex serves as a useful recognition mark. This species has for many years been incorrectly called *Personella eburnea* Reeve, 1844, which is not an Australian shell.

Two other Victorian members of the genus recorded from Port Fairy appear less frequently.

columnaria (Hedley & May, 1908); of similar sculpture but more elongate.

gaimardi Iredale, 1929; also more elongate with the nodules more pronounced and altogether more strongly sculptured (Figure 188).

Ratifusus Iredale, 1929

Shell elongate, oblong with distinct varices. High spire, faint spiral sculpture; short canal.

Type Species *Ratifusus adjunctus* Iredale, 1929.

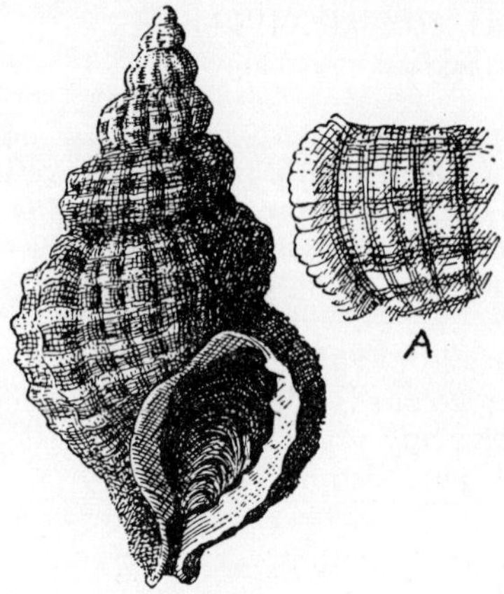

Figure 186

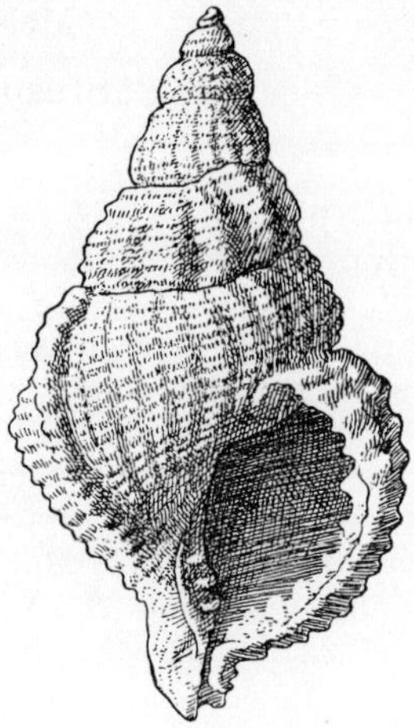

Figure 187

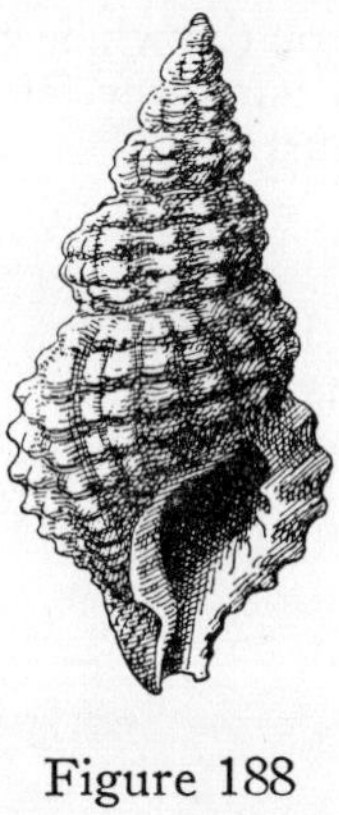

Figure 188

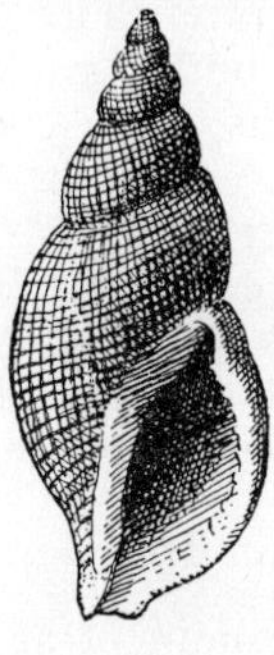

Figure 189

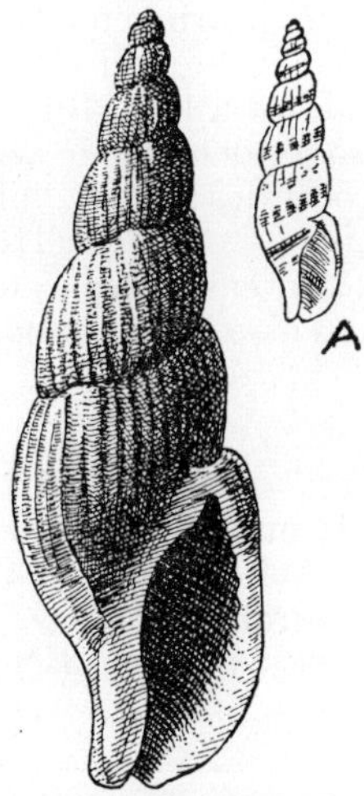

Figure 190

186. *Cymatiella verrucosa* × 2
 A Sculpture (enlarged)
187. *Cymatiella lesueuri* × 3
188. *Cymatiella gaimardi* × 2⅔
189. *Ratifusus mestayerae* × 1⅓
190. *Ratifusus bednalli* × 2⅔
 A Variation in colour pattern × 1

Mestayer's Truncated Spindle

Ratifusus mestayerae (Iredale, 1914)
(Figure 189)

Shell elongately oblong, solid; normally brown. Spire almost half the length of shell. The whole surface ornamented with a pronounced reticulate sculpture. Interior white. Aperture oblong-ovate; outer lip thickened by a definite varix; denticulate within; anterior canal wide. Size: 1 inch.

Localities: Port Phillip; Western Port; Anderson's Inlet; Lorne; Airey's Inlet; Warrnambool.

This is a common species on the whole coastline, easily recognized by its strong reticulate sculpture. Consistent colour is not apparent as specimens may be light brown, cream or white. For many years known under the name of *Pisania reticulata* A. Adams, 1854. Found also in Tasmania.

Bednall's Truncated Spindle

Ratifusus bednalli (Brazier, 1875)
(Figure 190)

Shell fusiformly elongate, solid; colour white, brown or cream with interrupted bands of brown. Spire more than half the length of shell. Whorls bearing ten, obtusely rounded varices and sculptured with longitudinal striae latticed by numerous, finer, concentric striae visible only with the aid of a lens. Interior brown or white. Aperture ovately oblong; outer lip solid, denticulate within. Size: ¾ inch.

Localities: Port Phillip; Western Port; Inverloch.

Like *mestayerae* (Iredale) it is of frequent occurrence on the whole coastline and easily separated from that species by its much narrower form. It occurs also in Tasmania, New South Wales and South Australia.

Mayena Iredale, 1917

Shell ovate conical; varices prominent at about each half whorl. Protoconch smooth, adult whorls keeled, with the keel emphasized by an ornamentation of nodules. Canal short, open, recurved; operculum oval, nucleus subcentral.

TYPE SPECIES *Biplex australasia* Perry, 1811.

Australian Rock Whelk

Mayena australasia (Perry, 1811)
(Figure 191)

Shell brown, fusiform, solid. Spire slightly less than half the length of shell, apex decollated. Whorls with twelve to fourteen rounded varices which are irregularly banded with reddish brown and white, the whole surface covered with a fawn-coloured velvety epidermis. Sculpture consisting of numerous, irregular, concentric ridges crossed by fine microscopic striae, visible only under the lens, giving the shell a cancellate appearance. Whorls angulated, the angle armed with a row of prominent tubercles. Interior white, highly polished. Aperture roundly ovate; anterior canal narrowly open; outer lip fortified with a strong varix, inside with two rows of teeth separated by a groove corresponding with the elevation of the outside varix; columella curved showing at the posterior end a very pronounced tooth and a few wrinkles anteriorly; operculum brown, horny. Size: 2-3 inches.

Localities: Portland; Gippsland coast.

A shell with outstanding characters, very solid in texture and with a row of tubercles on each whorl; this, in addition to the white interior, and teeth within the mouth are features more than sufficient to recognize the shell. In life generally located at low tide in crevices of rocks. It occurs also in New South Wales, Tasmania and South Australia.

A variety with spire slightly more elongated which may be regarded as a deep-water form is known as *australasia benthicola* Iredale, 1929.

In the trawl, a perfect example of this variety 5 inches long and in living condition was obtained in Bass Strait and is now in the National Museum of Victoria collection.

Cabestana Röding, 1798

Shell heavy, umbilicate, with a few, strong, undulate, concentric ridges and unifying varices. Mouth oval, outer lip strongly toothed, columellar callus smooth, canal short.

Type Species *Triton cutaceum* Gmelin, 1891.

Iredale, 1936, considered that the Australian species should be separated from *Cabestana* and erected the genus *Cymatilesta* for them. We do not consider this is justified and prefer to retain *Cabestana*.

Spengler's Rock Whelk

Cabestana spengleri (Perry, 1811)
(Figure 192)

Shell large, yellowish brown, very solid, somewhat fusiform, spirally costate and radially finely ribbed. Spire elevated, conical, the early whorls finely reticulated. Whorls angulated at the upper part, tuberculated on the angle. The whole shell covered by a thin, horny epidermis. Aperture oval, produced anteriorly into a short, open, slightly recurved canal; internally white; outer lip much expanded; inner lip white, smooth; operculum horny. Size: 6 inches.

Localities: Sorrento; Portsea; Point Lonsdale; Western Port.

Dredging frequently provides large and well-developed specimens, examples up to seven inches in length being not uncommon. This is a strongly sculptured species.

These molluscs are occasionally found on our shores in the act of ovipositing in rock-pools at low water. The egg-mass, containing numerous capsules, is circular, and has been likened to a peach split down the centre, the cavity answering to that occupied by the peach stone. Each 'nest' is attached to rock or sometimes to the concave side of an oyster valve. The capsules, each about half an inch in length, and containing the eggs are closely packed and attached to the inner wall of the rather tough envelope. In fresh condition

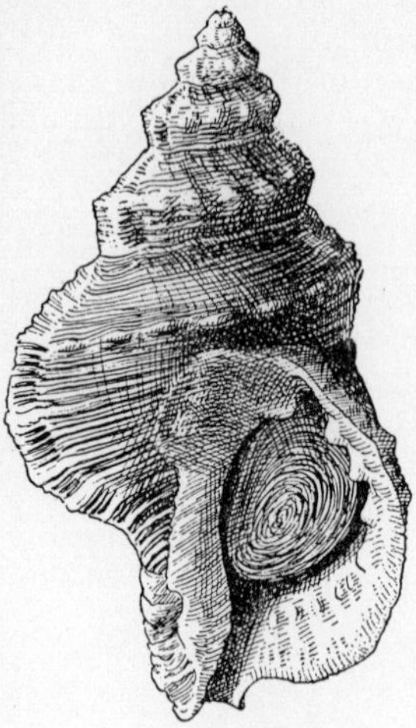

Figure 191

Figure 192

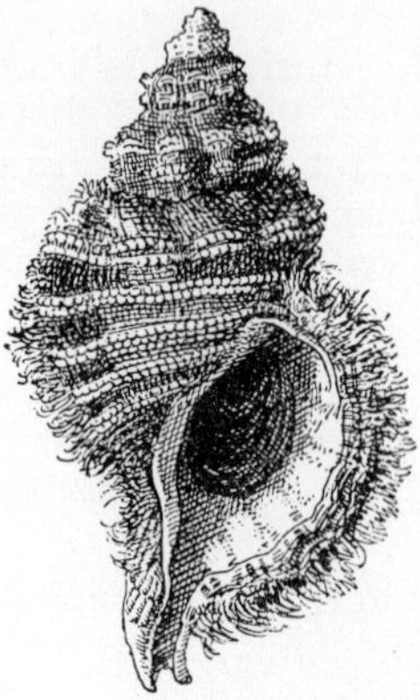

Figure 193

191. *Mayena australasia* × ⅔
192. *Cabestana spengleri* × ⅔
193. *Cabestana waterhousei* × ⅔

these 'nests' are of a yellowish colour. Besides Victoria the species is found in Tasmania, Queensland, New South Wales and South Australia.

Iredale, 1929, in a review of the family, stated that the Victorian shell is more obese and should bear the name *spengleri barthelmyi* (Bernardi, 1857). In this we cannot concur, as considerable variation in size and shape occurs in specimens from one locality. Iredale, 1929, also suggested that the deep-water members of the species which are more elongate in form should be separated as subspecies *spengleri procerum*.

boltenianum (A. Adams, 1954); a small (1½ to 2 inches high) broad form of *Cabestana* which Hedley and originally Iredale considered a dwarf form of *spengleri*. However, recent authors including Iredale now make it a separate species. Its distribution is south-eastern Australia and New Zealand.

Waterhouse's Rock Whelk

Cabestana waterhousei (A. Adams & Angas, 1864)
(Figure 193)

Shell medium size, yellowish brown, solid, broadly spindle-shaped; spirally costate and radially finely ribbed, spiral costae in pairs, the radial ribs giving them a nodulose appearance. Spire elevated, conical, two indistinct varices on each whorl. Periostracum brown, thick, with short bristles on the varices areas. Aperture oval, white; outer lip broad, flattened, the spiral costae produced on the interior; operculum horny. Size: 4 inches.

Localities: Sorrento; Lorne; Portland; dredged off Lakes Entrance.

This species is distinguished from *spengleri* by its smaller size, more scaly and hairy periostracum, and the finer, more separated spiral costae. It also occurs in Tasmania, New South Wales, Western Australia and South Australia where it is a common intertidal species.

Iredale, 1929, has separated the elongate, deep-water form of this species as *waterhousei frigidulum*.

Ploughed Rock Whelk

Cabestana tabulata (Menke, 1843)
(Figure 194)

Shell somewhat triangularly fusiform, whitish; variegated with light brown and blue and covered with a brownish scaly epidermis. Spire less than half the length of shell. Whorls, with two varices, angled near the sutures, the sloping area more or less flat, angles slightly knobbed. Sculptured with close, concentric ridges and crossed by fine longitudinal striae giving the surface a somewhat cancellated appearance. Interior white, with purple tinted grooves. Aperture roundly ovate; columella white and wrinkled; outer lip denticulate within. Size: 2 inches.

Localities: Portland; Port Fairy; San Remo.

This ornate species is of infrequent occurrence on our shores. The concentric sculpture and strikingly peculiar flat area near the sutures provide features to recognize the species readily. Found in South Australia, Western Australia and rather commonly in New South Wales.

For many years these shells were known as *Cymatium exaratum* (Reeve, 1844), but Menke's name has priority. Iredale, 1936, erected the genus *Cabestanimorpha* for this species but admitted that it was not unlike *Triton cutaceum* the type of *Cabestana*. We prefer to retain it in this genus until shown that it should be removed.

Cymatona Iredale, 1929 (*Nassarius kampyla* Watson, 1883).

kampyla (Watson, 1883); a deep-water representative of the family from 40 fathoms off Ninety Mile Beach. A species about 1 inch, with spire approximately half the length of shell; whorls ornamented with longitudinal riblets and irregular concentric ridges, body-whorl nodosely angled at the upper portion close to suture; aperture ovate, outer lip denticulate within; canal rather long, slightly upturned.

Monoplex Perry, 1811

Shell large, moderately elongate, canal of medium length. Whorls nodosely ribbed. Outer lip thick, plicated internally.

TYPE SPECIES *Monoplex australasiae* Perry, 1811.

Southern Rock Whelk

Monoplex australasiae Perry, 1811

(Figure 195)

Shell large, solid, broadly fusiform; light brown, outer lip and columella grooves variegated with dark brown, and covered with a thin, scaly epidermis with hair-like appendages. Spire about half the length of shell. Whorls possessing two to three varices and sculptured with prominent nodulous concentric ribs, two on the spire whorls and about ten to twelve on the body-whorl, interstices showing fine concentric striae, the whole surface crossed by fine longitudinal striae. Interior white. Aperture ovate, deeply grooved within, outer lip strong, anterior canal short, fairly open; columella curved, with narrow plaits and wrinkles; operculum horny. Size: 6 inches.

Locality: Portland Bay.

The second largest species of the family in Victoria, easily recognized by the prominent nodulous, concentric ribs and variegated aperture. It has a wide distribution, with records from New South Wales, South Australia, Western Australia and Tasmania. At the above-named locality taken alive, and in crayfish pots with hermit-crabs in possession.

Originally all shells of similar type to this were grouped in the one species *Cymatium parthenopium* von Salis, 1793.

Charonia Gistel, 1848

Shell large, spiral-shaped. Spire elevated; protoconch smooth, glossy, rounded; varices prominent, approximately one to each whorl. Body-whorl large. Mouth oval, canalled above, with a tooth on the upper edge of the columella; outer lip strongly denticulate within; basal canal short; truncated, recurved; operculum thick, horny.

TYPE SPECIES *Murex tritonis* Linné, 1758.

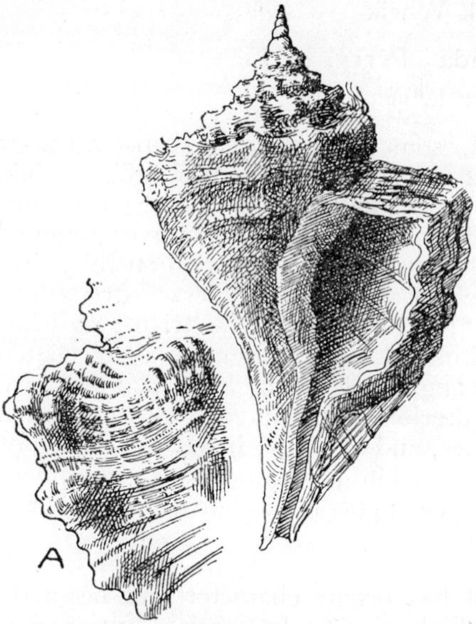

Figure 194

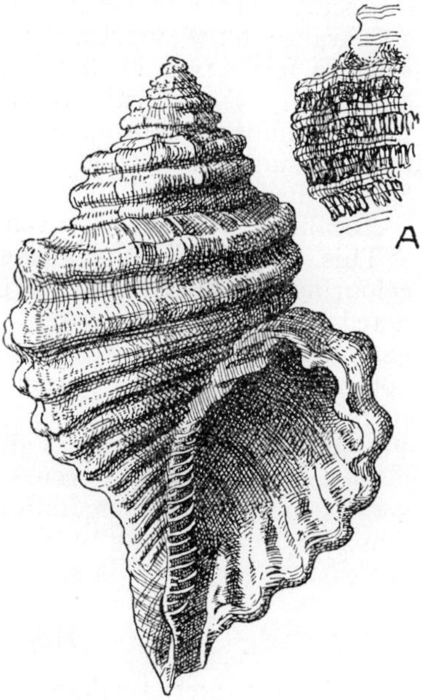

Figure 195

Figure 196

194. *Cabestana tabulata* × 2
  A Sculpture (enlarged)
195. *Monoplex australasiae* × ⅔
  A Sculpture (enlarged)
196. *Charonia rubicunda* × ½

Red Rock Whelk

Charonia rubicunda (Perry, 1811)

(Figure 196)

Shell large, solid, ovately conical, somewhat trumpet-shaped; colour yellowish brown, variegated with light and dark brown, apex delicate pink, the whole surface covered with a thin, scaly, brown epidermis. Spire acuminated, less than half the length of shell. Whorls with nine to eleven rounded varices. Sculptured with strong, nodulose, concentric ribs, two on the spire-whorls and about eight on the body-whorl and several of lesser strength at the anterior end; intervening spaces finely, concentrically striate; crossing this sculpture are numerous fine longitudinal striae presenting a cancellated appearance to the surface. Whorls are angulated, the angles armed with two rows of fairly prominent tubercles. Interior violet tinted. Aperture ovate, outer lip expanded with brown denticles, and arranged in groups of two or three; columella with numerous, narrow, white plaits and a very prominent tooth of similar colour on the posterior end; operculum horny, dark brown. Size: 6 inches.

Localities: Portland; off Cape Howe.

This shining, trumpet-shaped shell has ornate character and beautiful colouring, associated features giving distinction to the species. Variation is noted in the colouring, some examples being much lighter with less variegation. Occasionally washed up on the shores, sometimes trapped in crayfish pots with hermit crabs as inhabitants. In living condition, like the common Rock Whelk *Cabestana spengleri* it is located in crevices of rock at low tide. Found also in New South Wales, Tasmania and South Australia.

euclia Hedley, 1914; occurs in deeper water and differs from the above species in its more elongate form and lighter colouring. Is found in South Australia through Bass Strait to New South Wales. Iredale, 1929, separated the eastern form of this species as *euclia instructa.*

Negyrina Iredale, 1929

Shell moderately small. Spire high, with numerous small encircling ridges and with a knotted sculpture. Columella smooth or with upper part wrinkled; mouth oval; operculum with apical nucleus.

TYPE SPECIES *Triton subdistorta* Lamarck, 1822.

Thiele, 1929, erected a new section *Charoniella* close to *Charonia* for *subdistorta,* but Iredale, 1929, had anticipated Thiele and *Charoniella* is an absolute synonym of *Negyrina.* The species had previously been located in *Eugyrina* Dall, 1904.

Distorted Rock Whelk

Negyrina subdistorta (Lamarck, 1822)

(Figure 197)

Shell solid, ovately conical, somewhat distorted; white or cream and more often than not, mottled with reddish brown, with a narrow white band encircling the centre of body-whorl, the whole surface invested with a thin horny epidermis. Spire approximately half the length of shell. Whorls with

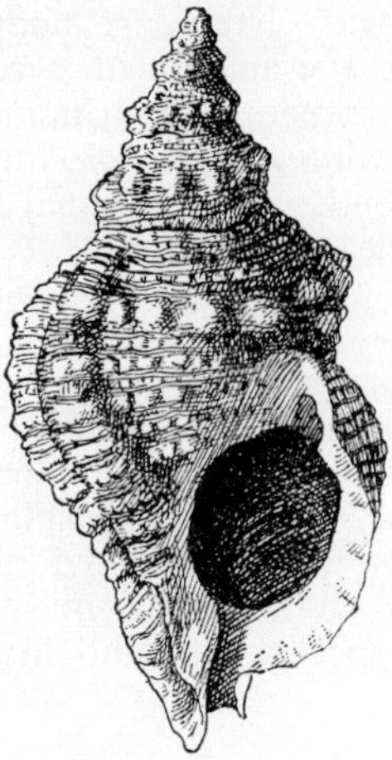

Figure 197

Figure 198

197. *Negyrina subdistorta* × 1
198. *Tonna cerevisina* × ⅔

seven to nine varices, sculptured throughout with irregular, nodulose, concentric ridges and angulated at the upper part, the angle with a row of prominent tubercles; the whole surface further ornamented with numerous longitudinal, fine microscopic striae. Interior white, smooth. Aperture roundly oval, outer lip acute; columella fairly smooth; anterior canal short, narrowly open; operculum horny. Size: 2-3 inches.

Localities: Lakes Entrance; off East Moncoeur Island, Bass Strait; Airey's Inlet; Lorne; Western Port.

A well-chosen name for this peculiar shell, a form frequently washed up on the beaches. With the consistent distorted feature and the encircling nodulose ridges identification is rendered easy. From the last-named locality, between Phillip and French Islands in 8 fathoms, excellent examples are obtained in living condition. One of these was taken in the act of ovipositing on a valve of the Fan Shell *Chlamys asperrimus* Lamarck, the mass containing twenty-three eggs. Beyond Victoria the species is found in New South Wales, Tasmania and South Australia.

Family TONNIDAE
Tun Shells

Shell large, thin, inflated, whorls ornamented with encircling ribs. Spire short, turbinate. Body-whorl very large. Aperture with an oblique notch in front; operculum present in juvenile stage but absent in adult except in the genus *Oocorys*.

Tonna Brunnick, 1772

Shell medium to large, thin, sculptured with strong spiral ridges, axial sculpture of fine growth-lines only, except on post-embryonic whorls which may be reticulated; callus reflected over the umbilicus. Aperture large; outer lip usually thin, not reflected.

Type Species *Buccinum galea* Linné, 1758.

There has been some doubt as to the correct name for this group of molluscs as *Tonna* was originally described without a species list. However, the name has been in use for a long period and, in 1946, was accepted by the International Commission on Zoological Nomenclature in place of the less well known *Dolium* Lamarck, 1801.

Beer Barrel Tun

Tonna cerevisina Hedley, 1919
(Figure 198)

Shell large, thin, globose, light brown, occasionally with three or four spiral, paler bands of colour. Whorls five in addition to a three-coiled horny protoconch. The surface of shell adorned with more or less equally spaced, flatly rounded concentric ribs separated by furrows of slightly lesser width, seventeen to eighteen appearing on the ultimate whorl; colour bands frequently, irregularly spotted with shades of chocolate, a feature restricted to the ribs. Aperture semi-lunate; outer lip simple; canal short, widely open. Interior smooth, shining, corrugated by the external ribbing, colour bands and spots discernible from within. Size: 6 inches.

Localities: Marlo, 3 miles off shore, 12 fathoms; Lakes Entrance, 6 miles off shore, 20 fathoms.

This is the first record of this shell in Victorian waters. Tun shells are among the largest of our Gastropods and the above species extends its range all along the eastern Australian coast; a huge specimen of this Beer Barrel from New South Wales was recorded by Hedley as—weight one pound two ounces, capacity four and a quarter pints, with a length of almost ten inches. Variation exists as regards colour.

tetracotula Hedley, 1919; from New South Wales, is perhaps the only species to be compared with it, and it differs in its more globose form and lack of smaller intermediate spiral ribs.

Order NEOGASTROPODA

Family MURICIDAE

Murex or Rock Dwellers

Shell spiral, turreted, with an anterior canal. The whorls thickened by varices or nodules at each rest period. Operculum horny with subapical or lateral and marginal nucleus.

Muricidae is divided into two subfamilies:

Muricinae with three or more varices, which may be nodulous, foliated or spinose on each whorl, and a well-marked canal.

Purpurinae without varices, but tuberculate and the anterior canal very short. Also the columella is characteristically flattened, e.g. *Dicathais*.

Subfamily MURICINAE

Torvamurex Iredale, 1936

Shells of medium size. Spire approximately same length as body-whorl. Each whorl with three varices, bearing a number of straight, folded fronds, the frond nearest the suture longer than remainder. Anterior canal half length of body-whorl.

TYPE SPECIES *Triplex denudata* Perry, 1811.

Fronded Murex

Torvamurex denudatus (Perry, 1811)
(Figure 199)

Shell pinkish brown, fusiform. Spire acuminately turreted, barely half the length of shell. Whorls seven to eight, angulated, with three well-developed, more or less frondose varices between which appear, normally, two prominent tubercles. Sculpture consisting of numerous, concentric ridges and striae. Interior white, shining. Aperture slightly ovate; outer lip denticulated; anterior canal narrow, recurved; operculum brown, horny. Size: 2 inches.

Localities: Port Melbourne; Port Phillip Heads; Airey's Inlet; off East Moncoeur Island, 38 fathoms; Western Port.

A species easily recognized by its fusiform shape and frondose varices, characteristics which easily distinguish it from other Victorian forms. In 6 to 8 fathoms at the last-named locality between Phillip and French Islands

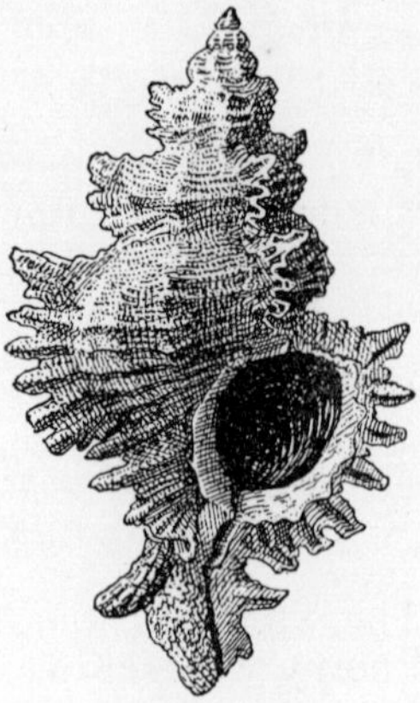

Figure 199

Figure 200

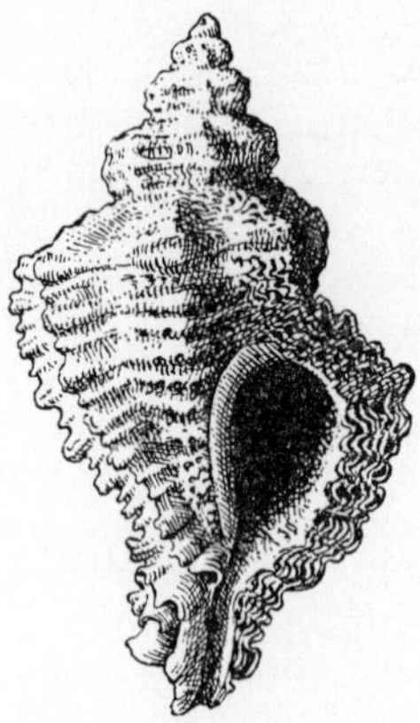

Figure 201

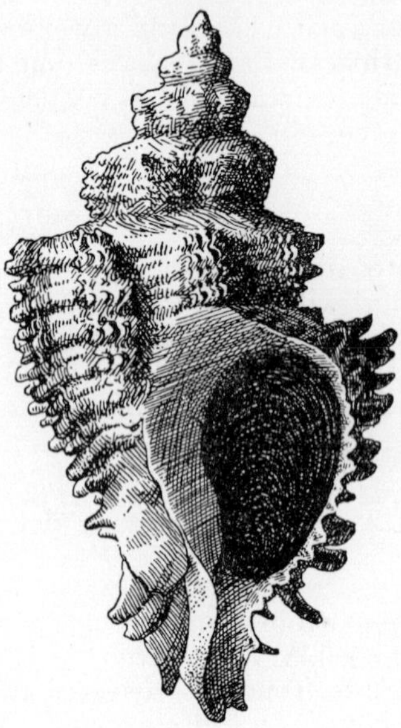

Figure 202

199. *Torvamurex denudatus* × 1
200. *Torvamurex damicornis* × 1⅓
201. *Murexsul brazieri* × 4
202. *Murexsul umbilicatus* × 3

examples have been procured in living condition in association with *Pterynotus triformis* (Reeve) and *P. angasi* (Crosse). The bitubercled feature between the varices is not consistent as specimens occasionally occur with one or three tubercles. The shell frequently appears in collections under the name of *Murex palmiferus* Sowerby, 1840, which, however, is a synonym. Occurs in New South Wales, Tasmania, through Bass Strait to South Australia.

Long-horned Murex

Torvamurex damicornis (Hedley, 1903)

(Figure 200)

Shell cream, tinged with pink, more noticeable at apex and on spines. Whorls rounded. Three varices on each whorl, each varix possessing, at the shoulder, a prominent upcurved, broad, hollowed spine with the tip generally bifid; below these are three to four shorter spines of varying length. Further ornament is seen in the close concentric threads. Aperture almost round; outer lip extending beyond the varix and denticulate within; canal rather long and much upcurved. Size: 2 inches.

Locality: South-west of Gabo Island, 30 fathoms; off Lakes Entrance.

Apparently a rare species and cannot be confused with any other Victorian shell. Beyond the state it is recorded from New South Wales (dredged at 50 fathoms off Coogee) and in deep water off the South Australian coast.

Murexsul Iredale, 1915

Shell small. Spire moderately tall. Whorls with many more or less thorny varices. Mouth oval; canal moderately long.

TYPE SPECIES *Murex octogonus* Quoy & Gaimard, 1833.

Brazier's Murex

Murexsul brazieri (Angas, 1877)

(Figure 201)

Shell small, elongately oval, light brown. Spire about one-third the length of shell. Whorls five, rounded, somewhat flattened on the upper portion just below the sutures, resulting in the formation of a nodulose angle. Sculpture consists of strong longitudinal ribs, five appearing on the body-whorl, crossed by several erect ridges, at the point of crossing producing a more or less foliaceous character. Interior brownish purple. Aperture ovate; outer lip crenulated, denticulate within. Size: ½ inch.

Localities: Port Melbourne; Flinders; dredged off Wilson's Promontory.

This is the smallest *Murex* in Victoria and not of frequent occurrence. The nearest ally is, perhaps, the common *umbilicatus* (Tenison Woods) from which it may be separated by its smaller size and the lesser number of ribs on the body-whorl. Its distribution is from New South Wales where it is common, to Tasmania, and through Bass Strait to South Australia.

Umbilicated Murex

Murexsul umbilicatus (Tenison Woods, 1875)

(Figure 202)

Shell yellow or chestnut, solid. Spire raised, almost half the length of shell. Whorls five to six, angulate above, longitudinally, conspicuously plicate (eight on the body-whorl) and further ornamented with strong, nodular, concentric lirae. Interior white; outer lip denticulate within; umbilicus well-defined and margined with rounded imbricated scales. Size: 1 inch.

Localities: Cape Otway; Airey's Inlet; Western Port.

It is a rather strongly sculptured species and easily recognized by its strong, longitudinal plicae. Frequently cast up on the shores at the last-named locality, and dredged alive in 8 fathoms. Occurs also in Tasmania and South Australia.

fimbriatus (Lamarck, 1822); a smaller species from Portland and other Victorian localities, may possibly be confused with the above but is easily distinguished by its finer concentric sculpture.

Pterynotus Swainson, 1833

Spire more or less tall. Whorls with three flat, thin varices. Mouth small, oval; canal moderately long, bent.

TYPE SPECIES *Murex pinnatus* Swainson, 1833.

Angas' Murex

Pterynotus angasi (Crosse, 1863)

(Figure 203)

Shell small, variable in colour, may be creamy yellow, fawn, brown or brown blotched with white, fusiform. Spire composed of six whorls almost half the length of shell. Whorls sculptured with three foliated varices which, at the shoulder angle, give rise to semitubular, upturned spines; between each two varices appear three well formed, large tubercles practically at a level with the spines. The shell is further adorned with a few irregular spiral lirae and longitudinally with fine growth-lines. Aperture oblong with the outer lip slightly advanced beyond the varix; canal moderately long and upturned. Size: 1 inch.

Localities: Off East Moncoeur Island, 38 fathoms; dredged off Altona Bay; San Remo; Flinders; Western Port.

This shapely shell is easily recognized by its foliated varices with the characteristic tubercles between each two. In the warmer waters of New South Wales it frequently occurs in rock-pools and is much more highly coloured than southern representatives.

The range of distribution is further extended from Tasmania, through Bass Strait to South Australia.

Three-shaped Murex

Pterynotus triformis (Reeve, 1845)

(Figure 204)

Shell brown, strong triangularly ovate; three laminate varices, spirally ridged and wrinkled with two tubercles between each. Aperture ovate, sinuated at the upper or posterior end; operculum much darker. Size: 2 inches.

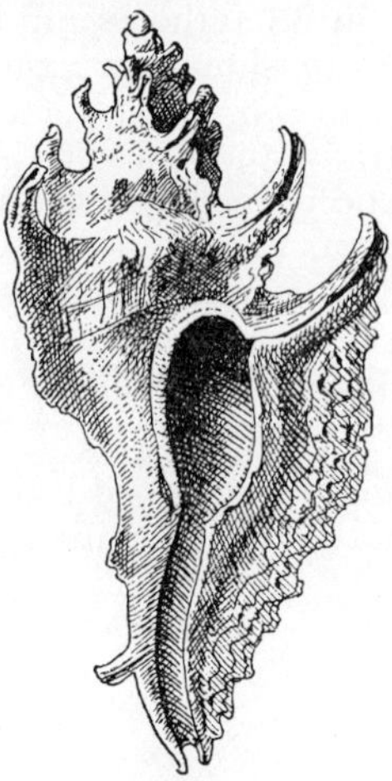

Figure 203

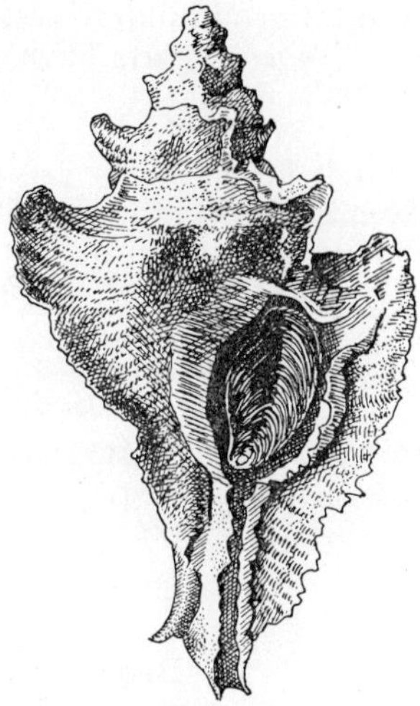

Figure 204

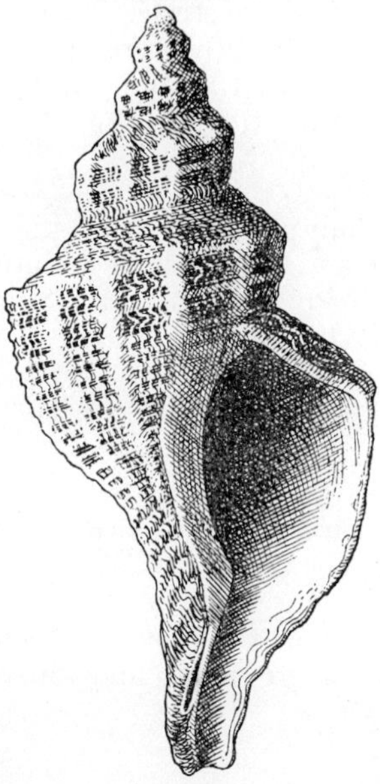

Figure 205

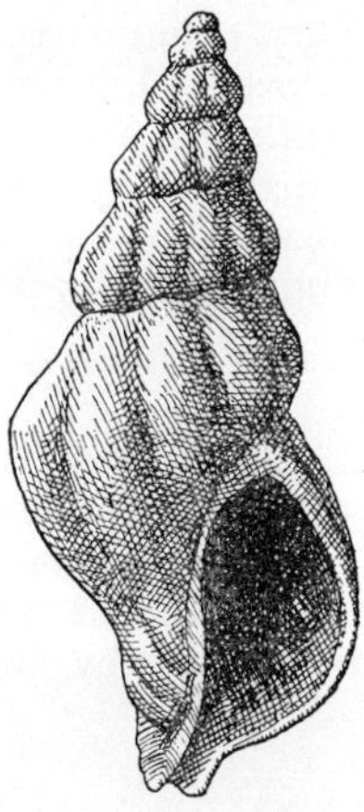

Figure 206

203. *Pterynotus angasi* × 3
204. *Pterynotus triformis* × 1
205. *Bedeva paivae* × 4
206. *Litozamia brazieri* × 4

Localities: Port Phillip; Western Port; East Moncoeur Island.

The *Challenger* reports having obtained it alive in 38 fathoms off East Moncoeur Island, Bass Strait, and the present figure shows an example obtained by dredging in Western Port. In addition to the whole Victorian coastline, it is known from Tasmania, South Australia and Western Australia. Even in the juvenile condition no Victorian species could be confused with this characteristic form. Very little variation exists except in the better developed varices of some specimens, while the intermediate tubercles are sometimes three instead of two.

Bedeva Iredale, 1924

Shell spindle-shaped, whorls angular above. A number of broad rounded varices on each whorl. Mouth oval; canal open, shorter than mouth.

Type Species *Trophon hanleyi* Angas, 1867.

Mussel Drill

Bedeva paivae (Crosse, 1864)

(Figure 205)

Shell cream or brown, solid, fusiform. Sculptured on body-whorl with about eleven, longitudinal ribs and numerous fine lamellose striae. The whole surface crossed by close concentric lirae. Interior generally white but sometimes brown towards the outer lip which is denticulated within. Aperture ovate; anterior canal not very long, slightly upturned; operculum thin, horny. Size: 1 inch.

Localities: Port Phillip; Western Port.

This is a very common species washed up in fair numbers after stormy conditions in Port Phillip Bay. Its fusiform shape and longitudinal sculpture are features sufficient to identify it. Found also in Tasmania and South Australia. Sometimes known under the name of Oyster Drill.

assisi (Tenison Woods, 1876); Port Fairy.

hanleyi (Angas, 1867); Mallacoota.

Litozamia Iredale, 1929

Shell small, spindle-shaped, rounded. Protoconch smooth; varices closely set, forming axial ribs. Mouth oval; canal short.

Type Species *Peristernia rudolphi* Brazier, 1894.

Members of this genus were originally included in the genus *Trophon* but are distinguished by their fusiform shape, short canal and smooth protoconch.

Brazier's Trophon

Litozamia brazieri (Tenison Woods, 1875)

(Figure 206)

Shell small, cream, ovately fusiform. Spire about half the length of shell; apex papillose. Whorls seven, strongly longitudinally ribbed, about six to seven on the body-whorl. Aperture ovate, light brown or green; outer lip denticulate within; anterior canal short, slightly recurved. Size: ½ inch.

Localities: Sandringham; Flinders; Western Port.

A very common species easily detected by its longitudinal sculpture and fusiform shape. Recorded also from Tasmania.

goldsteini (Tenison Woods, 1875); Goldstein's Trophon, Flinders; off Port Phillip Heads. It has longitudinal ribs, distant concentric lirae and possesses a peculiar chalk-like coating. These features are sufficient to distinguish it from the above.

rudolphi (Brazier, 1894); off Lakes Entrance.

Benthoxystus Iredale, 1929 (*Trophon columnarius* Hedley & May, 1908).

petterdi (Crosse, 1870); Port Phillip. A yellowish or brown-banded form of similar dimensions but with a distinct clathrate sculpture, occurring also in Tasmania and New South Wales.

recurvatus (Verco, 1909); off Ninety Mile Beach (40 fathoms).

Enatimene Iredale, 1929 (*Trophon simplex,* Hedley, 1903).

simplex (Hedley, 1903); off Ninety Mile Beach (40 fathoms).

Emozamia Iredale, 1929

Shell short, squat, with a short spire, large open mouth, and short open canal.

TYPE SPECIES *Murex licinus* Hedley & Petterd, 1906.

Broad Emozamia

Emozamia licinus (Hedley & Petterd, 1906)

(Figure 207)

Shell small, light brown, broad. Spire about one-quarter the length of shell. Whorls about six somewhat angular at the shoulder. Sculpture consisting of more or less erect varices, about eight to nine on the ultimate whorl; further ornamented with strong, spiral ribs divided by deep grooves, about twelve to fourteen appearing behind the aperture which is large, light purple and shining; anterior canal fairly wide; operculum horny. Size: ¾ inch.

Locality: Off Lakes Entrance.

This Victorian record is based on an example obtained by Mr J. Walker in 1953. The type, a living example described under *Murex* was dredged at 250 fathoms off Sydney. Its range of distribution has been extended to 100 fathoms off Cape Pillar, southern Tasmania. The peculiar squat form and characteristic sculpture provide means to identify the species readily.

Typhis Montfort, 1810

Shell ovate to oblong, muriciform, with projecting hollow tubes between the spinose varices, the last open, occupied by the exhalant canal. Aperture orbicular, prolonged in front into a closed siphonal canal.

TYPE SPECIES *Murex tubifer* Bruguière, 1792.

Yates' Typhis

Typhis yatesi (Crosse, 1865)

(Figure 208)

Shell cream with a narrow brown band on the shoulders just below the sutures, fusiform. Spire short somewhat pagoda-like, a little less than half

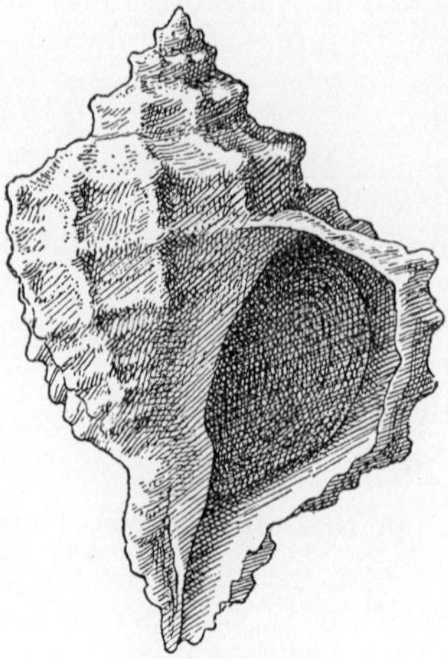

Figure 207

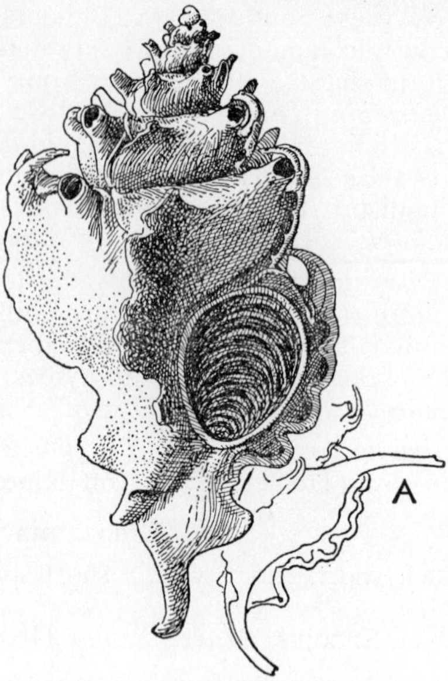

Figure 208

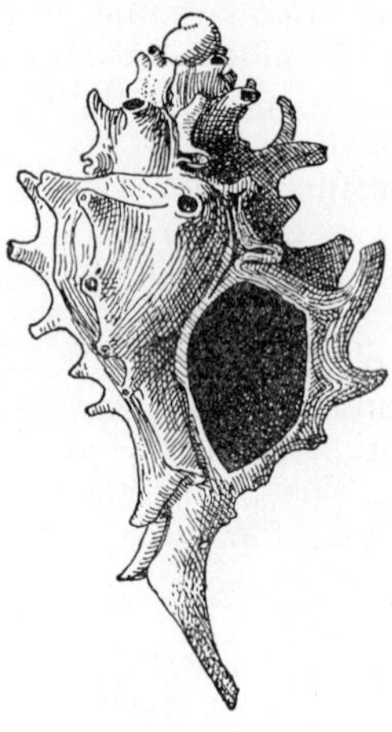

Figure 209

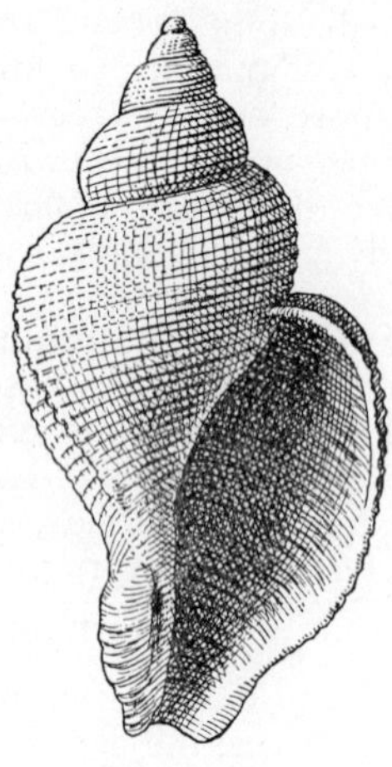

Figure 210

207. *Emozamia licinus* × 3
208. *Typhis yatesi* × 3
A Spines around aperture (enlarged)
209. *Typhis philippensis* × 5
210. *Galfridus eburneus* × 4

the length of shell, apex papillose. Whorls angulated, with foliated varices and a hollow, tubular projection between each varix, the last, the only one open, extremely lengthy. Aperture oval, with anterior canal close and upturned; outer lip slightly in advance of the last varix; operculum ovate, horny with thin, irregular imbricated lamellae. Size: 1 inch.

Locality: Western Port.

This is a remarkable shell and with its characteristic pagoda-like spire should be easily identified. In the adult stage and when taken in living condition the last tubular projection is much produced, a feature often absent in beach-worn examples. Dredged alive in 8 fathoms between Phillip and French Islands, and occurs also in South Australia.

Short-spined Typhis

Typhis philippensis (Watson, 1886)

(Figure 209)

Shell light brown, shining, oblong-fusiform. Spire short, barely half the length of shell, apex papillose. Whorls angulated, variced, bearing hollow tubular projections between the spinose varices. Aperture oval; anterior canal closed, slightly upturned; outer lip a little extended beyond the last varix; operculum ovate, horny and showing thin, irregular imbricated lamellae. Size: 1 inch.

Locality: Off entrance to Port Phillip, 33 fathoms; off Lakes Entrance.

This shell is easily recognized by its characteristic spinose ornament. From Twofold Bay, New South Wales, at 15 to 20 fathoms a variety, a little broader than the Victorian form, occurs. It is known under the name of *interpres* Iredale, 1924.

Cyphonochelus Jousseaume, 1881 (*Typhis arcuatus* Hinds, 1843).

syringianus (Hedley, 1903); a somewhat similar Victorian form. A solid, brown, ovate shell, of smaller dimensions and with the anterior canal less upturned, occurring also in New South Wales and Tasmania.

Galfridus Iredale, 1924

Ovately spindle-shaped, with rounded whorls. Protoconch of two whorls. Spirally lirate, sutures impressed. Mouth large; outer lip thin; canal short.

TYPE SPECIES *Triton speciosus* Angas, 1871.

Petterd's Galfridus

Galfridus eburneus (Petterd, 1884)

(Figure 210)

Shell white, fusiform, whorls four and a half. Sculpture consisting of irregular growth-lines crossed by numerous fine spiral striae, giving the shell a somewhat decussated appearance. Aperture ovate; outer lip thin; canal short. Size: ½ inch.

Locality: Flinders.

A fine white, decussated shell occurring also in Tasmania.

speciosus (Angas, 1871); from Mallacoota, is a common New South Wales form with a coarser sculpture; it is very closely allied and possibly a large series would prove them conspecific.

Maculotriton Dall, 1904

Shell ribbed axially and sulcate spirally. Protoconch of two or three whorls; apex obtuse. Spire elevated, with appressed sutures, varices inconspicuous, two or three only. Canal short.

TYPE SPECIES *Triton bracteatus* Hinds, 1844.

Southern Spotted Triton

Maculotriton australis (Pease, 1872)

(Figure 211)

Shell small, elongately fusiform, solid; colour dark chocolate or reddish brown with an encircling broad, whitish band. Spire slightly more than half the length of shell. Whorls seven, rounded, sculptured by irregular concentric riblets with reddish brown interspaces of lesser width, and broad, prominent, longitudinal ribs about fourteen on the body-whorl. Interior whitish. Aperture elongately oval, through which the encircling band may be clearly seen; outer lip crenulated; anterior canal fairly wide and long. Size: ½ inch.

Locality: Port Fairy.

A characteristic, elongately fusiform species with nothing approaching it in Victoria. Readily identified by its regular concentric sculpture and bold longitudinal ribs. The reddish brown of the interstices presents the shell with a beautiful lined appearance, serving as a useful recognition mark. For many years known under the generic name of *Cantharus*. Recorded also from New South Wales, where it is common, the animal being fairly active at low-water level.

Drupa Röding, 1798

Shell ovate, solid. Spire short. Whorls tubercular or spinose. Aperture narrow, oval; inner lip wrinkled; outer lip with plait-like teeth; canal short.

TYPE SPECIES *Drupa grossularia* Röding, 1798.

The only Victorian member of this genus was for many years placed in the genus *Sistrum* Montfort, 1810, now considered a synonym of *Drupa*.

Rough Castor Bean

Drupa aspera (Lamarck, 1822)

(Figure 212)

Shell white, dark brown between the ribs, obtusely ovate. Spire short, about one-third the length of shell. Sculpture consisting of prominent, longitudinal ribs, eight on body-whorl with numerous concentric ridges and two definite, strongly noduled keels. Interior purplish. Aperture elongately oval; outer lip crenulated, denticulate within; columella fairly straight; anterior canal rather narrow. Size: ¾ inch.

Locality: Lakes Entrance.

A species easily recognized by its bold knobby sculpture and the dark brown blotches between the ribs; these features, with the purplish or lavender interior and strong outer lip, denticulate within, should be ample to identify the species. These shells are generally located on rocky shores wedged in fissures and crevices of boulders, and more often than not are

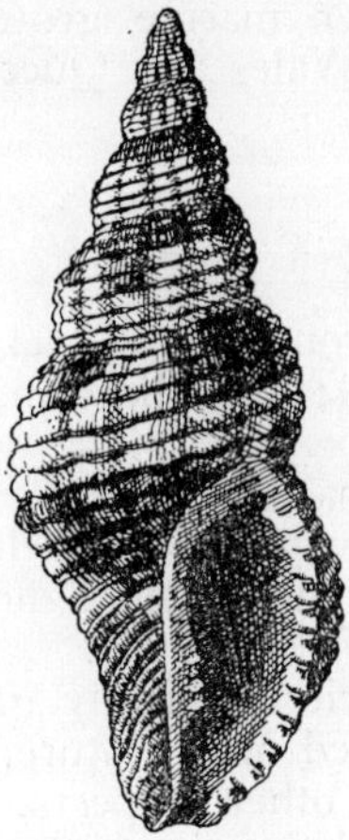

Figure 211

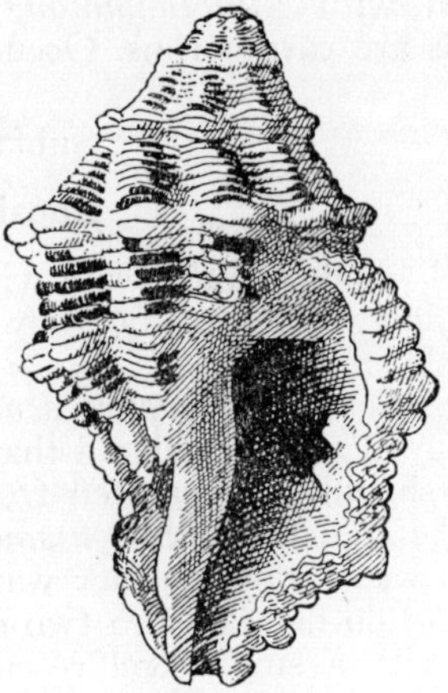

Figure 212

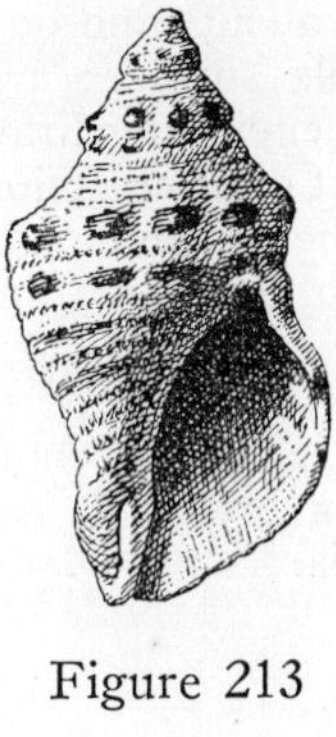

Figure 213

Figure 214

211. *Maculotriton australis* × 4
212. *Drupa aspera* × 3
213. *Lepsiella reticulata* × 2
214. *Lepsiella vinosa* × 3

covered with *Lithothamnion, Serpula* and other marine growths. The animals are carnivorous. Occurs in New South Wales and Queensland.

Subfamily PURPURINAE

Lepsiella Iredale, 1912

Shell small, oblong-ovate with one or more strong, spiral keels on each whorl and several on the body-whorl. Mouth oval; outer lip thin, crenate; canal short.

TYPE SPECIES *Purpura scobina* Quoy & Gaimard, 1833.

Finlay, 1928, considered that the shells with two or more keels on each whorl should be separated from *Lepsiella s.s.* and erected a new genus *Lepsithais* (with *Polytropa squamata* Hutton, 1878, as type).

This we do not consider warranted as the species originally grouped in *Lepsiella* do not fall into two distinct groups based on sculpture. We find shells with a single keel resemble *Lepsithais* in other respects, and vice versa; in fact a series of species show a gradation in sculpture from *Lepsiella* to *Lepsithais*. Therefore we have retained all the Victorian species in the one genus *Lepsiella*.

Reticulated Lepsiella or Purple

Lepsiella reticulata (Blainville, 1832)

(Figure 213)

Shell small, light brown, solid, fusiform. Spire slightly less than half the length of shell. Sculpture consisting of radial ribs about nine on the body-whorl, and fine longitudinal striae over the whole surface; a somewhat reticulate character is produced by the crossing of encircling lirae and four to six carinae, the lowest least pronounced; the point of intersection is rather strongly nodulose, each nodule much darker than the remainder of the shell. Interior yellowish. Aperture ovate; outer lip sharp, lirate within; anterior canal fairly open. Size: ½ inch.

Localities: Port Phillip; Western Port; Lorne.

An exceedingly common species on our shores; it may be confused with *vinosa* (Lamarck) but is easily separated by its brownish colour, and nodosely carinated whorls. Occurs also in Tasmania.

Wine-mouthed Lepsiella

Lepsiella vinosa (Lamarck, 1822)

(Figure 214)

Shell greenish white, acuminately oval. Spire less than half the length of shell. Whorls six, angulated. Sculpture consisting of several irregular, concentric broad, rounded lirae, the whole surface crossed by numerous very close, scaly intricate lamellae. Interior brownish purple, very polished. Aperture somewhat ovate; outer lip slightly crenulated. Size: ½ inch.

Localities: Sandringham; Flinders; Lorne.

An extremely common species all along the coast on rocks between tide marks. Like all the Muricidae, it is carnivorous, and is usually found in association with *Brachidontes* and *Galeolaria* on which it feeds. Readily recognized by its characteristic shape and the longitudinal lamellae, a feature easily detected with the aid of lens.

propinqua (Tenison Woods, 1876); another form which can only be spoken of in a varietal sense, is a little lighter in colour and with a finer sculpture. Both recorded from Tasmania, South Australia, New South Wales and Western Australia.

Flinders' Lepsiella

Lepsiella flindersi (A. Adams & Angas, 1863)
(Figure 215)

Shell greenish white, acuminately oval. Spire conical, about one-third the length of shell. Whorls sculptured with several irregular, concentric ridges. The entire surface crossed by numerous, very fine, scaly, imbricate lamellae difficult to discern in worn specimens. Body-whorl with a distinct somewhat nodulose angle at the upper portion close to suture. Interior violet brown. Aperture elongately ovate, outer lip coarsely crenulated by the outside sculpture, denticulate within; columella fairly straight; anterior canal narrow. Size: 1 inch.

Locality: Port Fairy.

Somewhat approaches *vinosa* Lamarck but is larger and with the concentric sculpture not so close. Recorded also from South Australia and south Western Australia.

Agnewia Tenison Woods, 1877

Shell small. Spire tall; whorls rounded with longitudinal and spiral ribs forming a lattice. Mouth small, elliptical; canal short.

Type Species *Purpura tritoniformis* Blainville, 1832.

Triton Purple

Agnewia tritoniformis (Blainville, 1832)
(Figure 216)

Shell creamy yellow with narrow interrupted, brown, bands about seven to eight on the body-whorl, solid, fusiform. Spire less than half the length of shell. Sculptured with about twelve equidistant, somewhat nodose, longitudinal ribs, separated by intervening spaces of similar width; the whole surface crossed by irregular, concentric striae. Interior creamy white, brown banded towards the outer lip. Aperture ovate, with a very short, straight, open, anterior canal; outer lip sharp, crenulated externally. Size: 1 inch.

Localities: Port Phillip; Western Port; Lorne; Apollo Bay.

Cast up on the shores this is a common species, generally distributed along the whole coastline. In life located in pools at low tide down to 50 fathoms. Under the lens this proves to be a beautifully sculptured shell, easily recognized by its longitudinal and spiral ornament, giving it a latticed appearance. Occurs also in New South Wales, Tasmania and New Zealand.

Figure 215

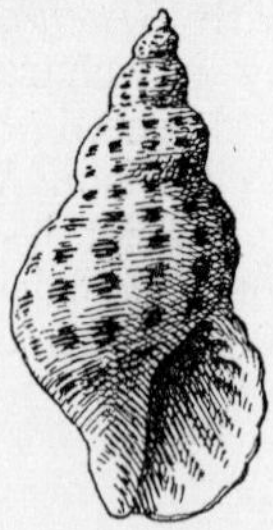

Figure 216

Figure 217

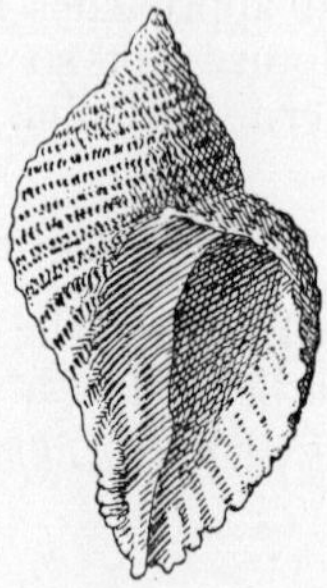

Figure 218

215. *Lepsiella flindersi* × 1⅓
216. *Agnewia tritoniformis* × 1⅓
217. *Dicathais textilosa* × ⅔
218. *Dicathais baileyana* × 1

Dicathais Iredale, 1936

Shell heavy. Short spired, sculptured with flat, spiral ribs. Body-whorl very large, rounded. Mouth oval; anterior canal short; outer lip dentate-lirate; columella flattened thaid-like.

TYPE SPECIES *Buccinum orbita* Gmelin, 1791.

Dog Winkle

Dicathais textilosa (Lamarck, 1822)

(Figure 217)

Shell white or cream, solid, ovate-ventricose; obsoletely, irregularly, spirally ribbed throughout; radial sculpture resolving itself into a series of irregular, growth-lamellae; spire short. Aperture large, oval, with a widely open anterior canal, white within; outer lip tinged with orange and grooved within; operculum horny. Size: 3 inches.

Localities: Mornington; Portsea; Point Nepean; Queenscliff; Western Port.

This species was known for many years under the name of *Thais succincta* Martyn, 1784, but as Martyn's names are not acceptable, under the International Rules of Zoological Nomenclature, Lamarck's name has come into use.

Dog Winkles generally live in small clusters in rock pools and crevices below half tide. They are found in the whole of southern Australia but show some differences in form and are distinguished by different specific names. The New South Wales shell has bold encircling ribs with deep grooves between and is called the cart-rut shell *orbita* (Gmelin, 1791). The Western Australian species *aegrota* (Reeve, 1846) lacks the deep grooves and has a nodulose sculpture.

Bailey's Dog Winkle

Dicathais baileyana (Tenison Woods, 1881)

(Figure 218)

Shell dull white or creamy, solid, whorls five. Spire conical, about one-quarter the length of shell. Sculpture consisting of numerous concentric ribs with interspaces of about equal width, and fine longitudinal closely set, undulating lamellae and thick irregularly spaced growth-lines almost like varices. Interior white. Aperture oblong-ovate; outer lip thick, sulcate within; anterior canal a deep groove, elongate; operculum horny. Size: 1 inch.

Localities: Mornington; Flinders; Anderson's Inlet.

Much rarer than the other Victorian member of the genus but located under similar conditions in rock-pools and crevices, and frequently, like *textilosa,* much eroded. A solid species, separated by its finer concentric and longitudinal sculpture. As a means of protection the animals of many species of this group secrete a purple fluid, hence the name Purple Shells which is often applied to these interesting forms. Found also in Tasmania and South Australia.

vector Thornley, 1952; off Lakes Entrance.

Family MAGILIDAE

Coral Dwellers

Shell variable in form, often resembling Muricidae, usually with a small spire and few whorls, and as a rule more or less distinct spiral sculpture. Mouth with a short or moderately long anterior canal; operculum horny with a marginal nucleus, sometimes involuted.

Liniaxis Laseron, 1955

Small Rapana-like shells with sinusigeral protoconch of three to five whorls, the last two with two spiral ridges with transverse striae between. Anterior canal slightly prolonged, the axis adjoining it solid and covered with narrow overlapping plates. Sculpture predominantly spiral, though broad round axial ribs, may be present. The spiral keels are numerous, closely packed and covered with fine, closely spaced scales. Operculum multi-spiral.

TYPE SPECIES *Liniaxis elongata* Laseron, 1955.

Wilson's Coral Shell

Liniaxis wilsoni (Pritchard & Gatliff, 1898)

(Figure 219)

Shell small, solid, ovately fusiform, colour light brown. Spire a little less than half the length of shell; spire-whorls centrally angulate. Body-whorl rather strongly angulated at about the upper fourth of the whorl. Sculptured throughout by strongly developed longitudinal costae advancing right to the sutures, intervening spaces approximately of equal width to the costae. The surface is further ornamented by fairly regular concentric striae bearing close, erect, forwardly projecting scales. Interior pink, white towards the acute, finely crenulated outer lip. Aperture ovate; columella fairly straight; anterior canal short, broad. Umbilicus small, margined by a strong ridge; operculum horny. Size: 1 inch.

Localities: San Remo; Point Lonsdale; Portland; off Bass Strait cable.

The species was described in 1890 from an immature example dredged off Port Phillip Heads by Bracebridge Wilson; since then odd examples have appeared, as indicated, at the above localities but is by no means a common species. A living specimen was taken off the rocks at the second locality and it is suggested that diligent search at this area may provide examples. It is a coarsely sculptured species recognized by its longitudinal ribs and scaly concentric striae.

Cotton, 1932, considers the above species a synonym of *nodosa* (Adams, 1853), but if Laseron, 1951, is correct in his idea of *nodosa* then the two species are separate.

sertata Hedley, 1902; off Lakes Entrance. A more delicate and less stout shell than the above species. Size: ¾ inch.

Rhombothais Woolacott, 1954 (*Rhombothais arbutum,* Woolacott, 1954).

mira (Cotton, 1932); a fair example, obtained at Portland, is in the collection of the National Museum of Victoria. It is a white, solid species with very bold longitudinal ribs and rather closely set concentric lirae;

easily separated from *wilsoni* by its more angular form and flatly rounded ribs. Size: 1 inch.

These shells were originally described in the genus *Coralliophila* from which they differ in a number of points.

Tolema Iredale, 1929

Shells biconical. Sculptured with prominent, spiral keels bearing closely packed, elongated scales. Protoconch sinusigeral, of four whorls, first minute, the next bearing incipient spiral keels which become prominent on the third and fourth. Adult whorls five or six, flattened above periphery, rounded below; peripheral keel bearing spines. Mouth pyriform; anterior canal long, straight.

TYPE SPECIES *Purpura sertata* Iredale, 1929 (non Hedley, 1902) = *Tolema australis* Laseron, 1955.

Southern Pagoda

Tolema australis Laseron, 1955

(Figure 220)

Shell large, white, biconical. Whorls angulated, about ten in number including a four-coiled protoconch; the angle, which is situated a little above the periphery, is accentuated by the presence of a strong concentric keel with prominent semitubular scales, about twenty appearing on the body-whorl. The surface of shell is further ornamented above and below the angle with smaller keels bearing scales of a curved semitubular nature. Aperture pyriform, smooth within, corrugated by the external sculpture; columella arched; outer lip spinose; anterior canal long and fairly straight. Size. 1½ inches.

Locality: Off Lakes Entrance (16-20 fathoms).

A charming shell with nothing approaching it in Victoria. Its pagoda-like contour and concentric scaly-ribbed sculpture render it easy to recognize. Sometimes known under the name of Deep-water Purple. It is also found in New South Wales, off Twofold Bay, and in 110 fathoms off Sydney, the type locality.

Family COLUMBELLIDAE

Dove Shells

Shell oval. Spire generally short. Aperture narrow, elongated, terminating in a short anterior canal; columella lip usually dentate; outer lip thick and internally crenulated; operculum small and horny.

Dentimitrella Ludbrook, 1958

Shell small, elongate fusiform with a moderately elevated spire about equal to the body-whorl. Whorls smooth, suture linear, base ribbed; protoconch of two or more whorls, elevated, smooth. Mouth moderately short, narrow; outer lip usually varicose, strongly denticulate within; columella with a flat groove, denticulate usually in the area where the callus passes over the ribs of the base; anterior canal short, rather narrow, oblique.

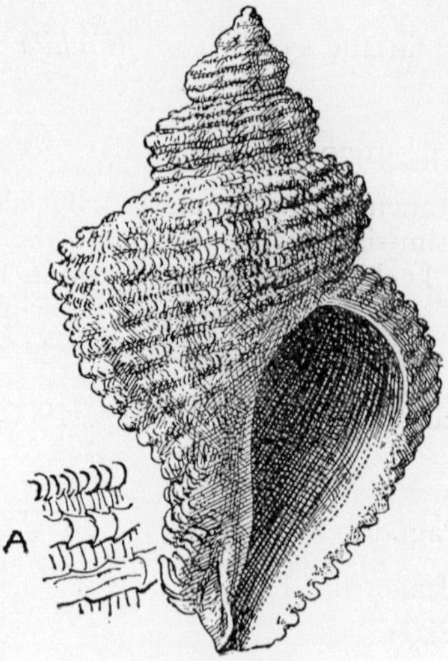

Figure 219

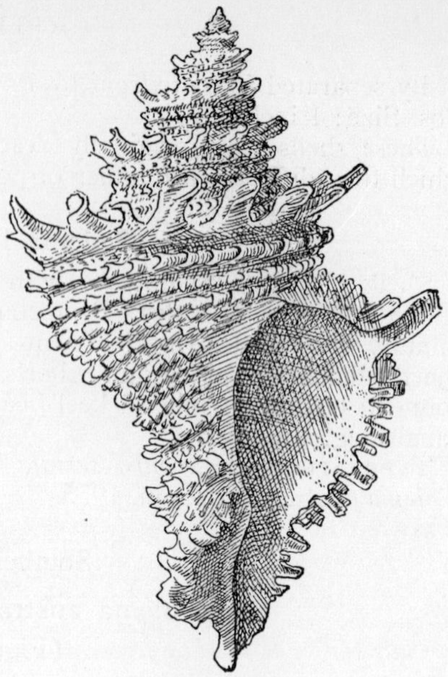

Figure 220

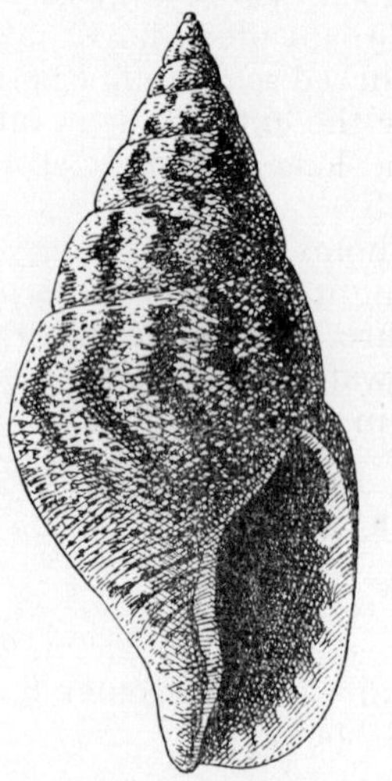

Figure 221

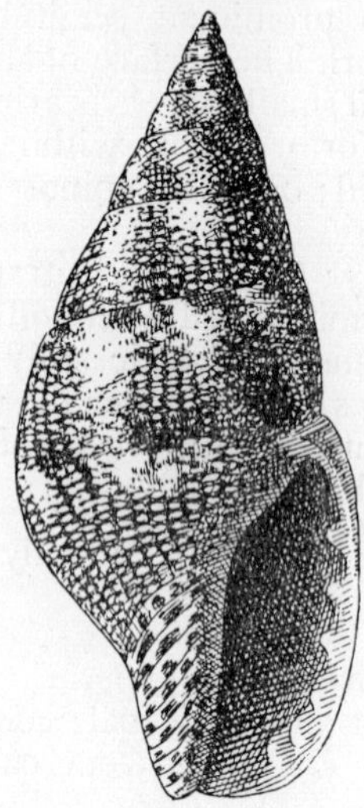

Figure 222

219. *Liniaxis wilsoni* × 2⅔
A Sculpture (enlarged)

220. *Tolema australis* × 2

221. *Dentimitrella semiconvexa* × 3

222. *Dentimitrella pulla* × 4

TYPE SPECIES *Columbella lincolnensis* Reeve, 1859.

The author erected this genus for species, previously included in the New Zealand *Zemitrella* Finlay, 1926 (*Lachesis sulcata* Hutton, 1873) from which they differ in having smooth whorls, narrow aperture and canal and denticulations on the outer lip. Members of the genus have a very similar appearance but show considerable variation in colour pattern even within a species, so that separation is difficult and many colour varieties have been designated as species.

Semiconvex Dove Shell

Dentimitrella semiconvexa (Lamarck, 1822)
(Figure 221)

Shell oblong-ovate, smooth, rather thick, colour extremely variable, may be yellow, green, white, brown with fine markings, or longitudinally flamed with reddish brown, covered by a thin light brown epidermis. Spire half the length of shell. Whorls flatly convex. Sculptured with fine concentric striae. Interior white. Aperture narrow; outer lip thickened, conspicuously denticulated within; anterior canal short, wide; operculum horny. Size: ¾ inch.

Localities: Coast generally.

A very pretty little shell and probably the commonest species of Dove Shell on our shores. Normally found with the reddish brown longitudinal flames but this is far from being a constant feature. In life located crawling on the surface of sand patches in shallow water. Beyond Victoria it occurs in New South Wales, Tasmania and South Australia.

Long Dove Shell

Dentimitrella pulla (Gaskoin, 1851)
(Figure 222)

Shell small, acuminately oblong, brown, with dark brown maculations just above the sutures. Spire approximately half the length of shell. Interior light brown. Aperture elongately oval; outer lip acute, denticulate within; columella white. Size: ½ inch.

Localities: Western Port; Port Phillip; Lorne.

A small species much narrower than *semiconvexa*, it is generally located under stones at low water.

austrina (Gaskoin, 1851); from Western Port, Lorne, and Port Phillip. A squat shell about ½ inch, fawn and with a consistent, broad pinkish band on the body-whorl plainly visible through the aperture which is denticulate within.

lincolnensis (Reeve, 1859); from Western Port. A fairly solid, acuminate shell, whitish, variously marked with reddish brown and chestnut. Size: ½ inch.

menkeana (Reeve, 1859); from Lorne, Western Port. An elongated species; spire narrowly acuminate, fawn with a concentric, interrupted reddish band; aperture oblong-ovate, outer lip thickened, denticulate within. A number of variations occur of which the variety *xavierana* (Tenison

Woods, 1876) is one; in it the dark markings assume the form of 'undulating chestnut longitudinal broad lines of colour which under the lens are sometimes seen to be flecked with white'.

axiaerata (Verco, 1910); off Lakes Entrance.

dictua (Tenison Woods, 1878); a variety of *nubeculata* with a netted pattern.

franklinensis (Gatliff & Gabriel, 1910); Port Franklin; Portsea.

nubeculata (Reeve, 1859); Western Port; Port Phillip; Portland.

nux (Reeve, 1859); a variety of *pulla* (Gaskoin).

tenuis (Gaskoin, 1851); a mistaken identification of *pulla* (Gaskoin).

saccharata (Reeve, 1859); Western Port.

tenebrica (Reeve, 1859); Western Port.

tenisoni (Tryon, 1883); Western Port.

vincta (Tate, 1893); Western Port.

Zella Iredale, 1924 (*Terebra beddomei* Petterd, 1884).

beddomei (Petterd, 1884); Barwon Heads; off Lakes Entrance.

Pseudamycla Pace, 1902 (*Buccinum dermestoideum* Lamarck, 1822).

miltostoma (Tenison Woods, 1876); from Western Port and Port Albert. A small species, whitish with brownish maculations; smooth, tumid, readily recognized by the aperture which is delicately margined with a reddish tint and is denticulate within.

dermestoidea (Lamarck, 1822); from Lorne and Portland, somewhat similar in shape to the previous species, banded with reddish brown but without the characteristic reddish margin of the outer lip.

Truncaria Adams & Reeve, 1850 (*Truncaria filosa* A. Adams & Reeve, 1850).

australis Angas, 1877; Flinders, Western Port; Port Albert.

Macrozafra (Finlay, 1926 (*Clathurella subabnormis* Suter, 1899).

angasi (Brazier, 1871); Western Port; Port Phillip.

atkinsoni (Tenison Woods, 1875); Western Port; Outer Harbour, Geelong.

calva (Verco, 1910); off Wilson's Promontory.

cominelliformis (Tate, 1892); Western Port.

fulgida (Reeve, 1859); Hedley considered this shell to be the same as *angasi* Brazier, a decision with which Gatliff disagreed and we concur with him.

legrandi (Tenison Woods, 1875); San Remo; Ocean Beach, West Head, Flinders.

lurida (Hedley, 1907); San Remo; Portsea; Torquay.

remoensis (Gatliff & Gabriel, 1910); San Remo.

smithi (Angas, 1877); San Remo; Anderson's Inlet.

Retizafra Hedley, 1913 (*Pyrene gemmulifera* Hedley, 1907).

gemmulifera (Hedley, 1907); off Wilson's Promontory.

Antizafra Finlay, 1926 (*Columbella pisaniopsis* Hutton, 1885).

plexa (Hedley, 1901); Bass Strait.

Aesopus Gould, 1860 (*Aesopus japonicus* Gould, 1860).

cassandra (Hedley, 1904); Port Albert.

gatliffi (Verco, 1909); off Inverloch (8 fathoms).

pallidulus (Hedley, 1905) ; Port Albert.
plurisulcatus (Reeve, 1859) ; off Rhyll, Western Port (7 fathoms).

Family BUCCINIDAE
Whelks

Shell ovate, oblong or pear-shaped. Anterior canal moderate or short; columella without folds and plaits; operculum horny with a terminal or lateral nucleus.

Austrosipho Cossmann, 1906

Spindle-shaped shells of large size with a globular, paucispiral protoconch asymmetrically wound, the nucleus being bulbous and lateral.

TYPE SPECIES *Siphonalia roblini* Tate, 1888.

In recent years there has been considerable diversity of opinion amongst authors as to the correct placing of the shells included in this genus. In 1914, Iredale proposed *Verconella,* to replace *Penion* Fischer, a preoccupied name which he had previously (1912) used for the Austro-Neozelanic molluscs classed under *Siphonalia.* Finlay, 1927, showed that these are two distinct series, the New Zealand *Verconella* and an Australian group for which Cossmann's *Austrosipho* is available. Iredale, 1924, proposed *Berylsma,* with *Fusus waitei* Hedley, 1903, as type, for the pale coloured elongate deep-water shells of the series. These shells do not differ structurally from *Austrosipho* and we consider *Berylsma* a synonym of that genus.

In 1929 Iredale stated that there was confusion in connection with the fossils, and therefore he preferred to save further confusion by erecting a new genus *Largisipho* for *Siphonalia oligospira* Tate, 1891 group, with a new subspecies *L. oligospira spectanda* as type. Unfortunately the protoconch of the type was missing but as the shell appears closely related to *oligospira* once again there seems no valid reason to separate it from *Austrosipho.* Therefore, we consider that *Largisipho* also falls into the synonym of that genus.

Grand Whelk

Austrosipho grandis (Gray, 1839)
(Figure 223)

Shell brown, solid, fusiform, spire produced, body-whorl ventricose. Whorls spirally, conspicuously ridged throughout, very frequently with a fine thread in the interstices, angulated round the upper portion, angle tuberculated. Interior white. Aperture ovate, with a widely open canal; operculum horny, pyriform. Size: 5 inches.

Localities: Frankston; Portsea; Sorrento; Western Port.

With hermit crabs in possession, these are frequently caught in crayfish pots. Considerable variation exists in the length of the anterior canal, which is noticeably longer in the deep-water examples. The tubercles in some instances are more or less compressed, while in others they show a stronger development. The species is not confined to Victoria, being recorded from

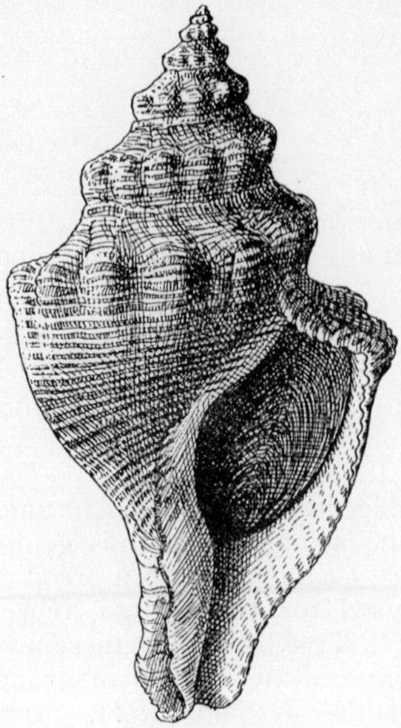

Figure 223

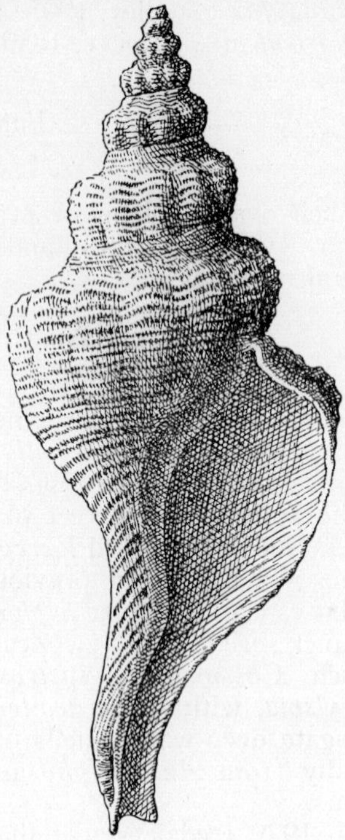

Figure 224

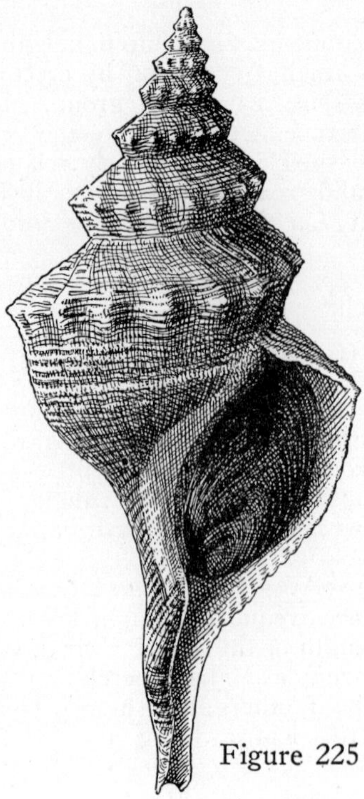

Figure 225

223. *Austrosipho grandis* $\times \frac{2}{3}$
224. *Austrosipho waitei* $\times \frac{2}{3}$
225. *Austrosipho maxima* $\times \frac{1}{2}$

Tasmania, South Australia and Western Australia. The animals are carnivorous.

grandis tasmaniensis (A. Adams & Angas, 1863) ; a paler shell with a short anterior canal and often showing one or more brown bands.

oligospira (Tate, 1891) ; a relatively non-costate, fusiform shell that occurs at Port Fairy, Warrnambool, and is common in South Australia.

Waite's Whelk

Austrosipho waitei (Hedley, 1903)
(Figure 224)

Shell dull white, covered by a scaly, fawn epidermis, fusiform, solid. Spire acuminate, less than half the length of shell. Whorls angled at the shoulder, somewhat tuberculated, sculptured with broad rounded ribs, about fourteen appearing on the body-whorl, these project at the angle gradually vanishing as they approach the base, the whole surface crossed by fairly regular concentric ridges between which appears a fine thread; numerous growth-striae are plainly visible but only with the aid of lens. Interior white, shining. Aperture oblong-ovate; outer lip crenulated by the outside sculpture; lirate within; columella curved; anterior canal elongate, open; operculum horny, centrally grooved. Size: 6 inches.

Locality: Trawled off Gabo Island, Bass Strait.

With its more elongate canal it may be readily distinguished from *grandis*. It also occurs off the New South Wales coast.

Large Whelk

Austrosipho maxima (Tryon, 1881)
(Figure 225)

Shell large, narrowly fusiform, solid. Spire acuminate, much less than half the length of shell; spire-whorls angled at the centre. The body-whorl a little above the flattened periphery, the angles armed with fairly sharp, prominent nodules. Sculptured throughout with irregular, flattened concentric riblets, crossed by numerous, fine longitudinal striae giving the surface a cancellated appearance. Colour fawn with an interrupted band of brown on the angle, and a second faint band at about the centre of body-whorl. Interior white, highly polished. Aperture ovate; outer lip crenulated by the external sculpture, lirate within; columella fairly straight; interior canal broad, open, elongate, slightly upturned; operculum horny, furrowed centrally. Size: 8 inches.

Localities: Port Albert; dredged off Phillip Island.

This large species is occasionally cast up on the shores in more or less worn condition. When trapped in crayfish pots with hermit crabs in possession better examples are often obtained. The species lives in Bass Strait and many excellent specimens have been procured through trawling. As in *grandis* much variation exists in the development of the nodules; in some specimens this feature is much less pronounced. However, the elongate anterior canal, the angulated whorls and cancellate sculpture are characters which identify the species. Beyond Victoria it is found in New South Wales and Tasmania.

Phos Montfort, 1810

Shell elongate, turreted. Spire sharp, elevated. Whorls ornamented with prominent longitudinal costae, and less pronounced spiral threads. Aperture oblong; outer margin lirate within; columella plicate in front, canal short, slightly twisted.

TYPE SPECIES *Murex senticosus* Linné, 1758.

Thorny Phos

Phos senticosus (Linné, 1758)

(Figure 226)

Shell brown, strong, acuminately oblong, turreted. Spire slightly more than half the length of shell. Whorls nine, rounded sculpture consisting of regular, longitudinal ribs, thirteen on body-whorl. The whole surface cancellated with fine, longitudinal and concentric striae. Interior bluish violet. Aperture ovate, outer lip strong with a slight notch near the anterior end; columella with an oblique fold; anterior canal short, fairly wide. Size: 1 inch.

Locality: Gippsland coast; Port Phillip Heads (dredged).

An ornate shell common in tropical waters. Its longitudinal sculpture with the prickly ribs and the sinus at the anterior termination of the outer lip are features to readily identify the species. The animals of this genus are peculiar in that the hind portion of the foot ends in a tapering filament.

Cominella Gray, 1850

Shells with the last whorl somewhat depressed beneath the suture, so that the aperture forms a small groove posteriorly. Outer lip sharp or crenate internally. Usually sculptured with spiral ribs.

TYPE SPECIES *Buccinum testudinea* Martyn, 1784.

Ribbed Cominella

Cominella eburnea (Reeve, 1846)

(Figure 227)

Shell yellowish brown frequently with reddish brown spots or blotches; fusiform; spire turreted. Whorls concavely angled at the upper part, nodose at the angle. Sculpture consisting of broad, rounded radial ribs and close spiral grooves which are continued over the ribs. Aperture light brown, narrowly ovate, terminating in a fairly wide anterior canal, lirate within; outer lip sharp; operculum horny, dark brown. Size: 1½ inches.

Localities: Brighton; Frankston; Dromana; Portsea; Western Port.

This ribbed form, found on sandbanks and in rock-pools, is a common species, particularly in bays and inlets where it replaces *lineolata,* an open coast species. It is also common in South Australia and Tasmania.

Lineated Cominella

Cominella lineolata (Lamarck, 1809)

(Figure 228)

Shell greyish brown, encircled with fairly regular, brown or black lines, which are sometimes interrupted, conically ovate. Spire smooth, rather

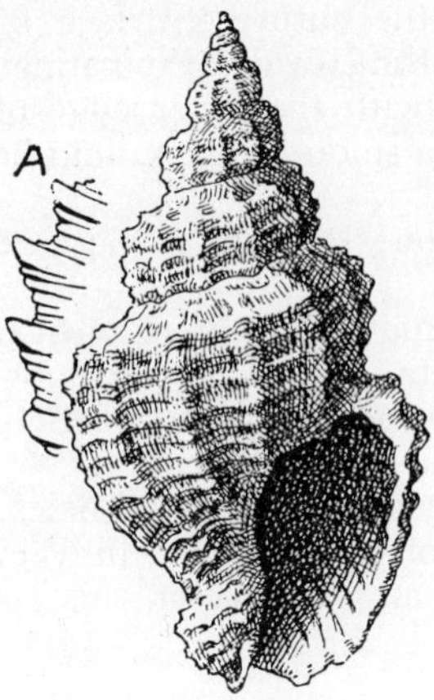

Figure 226

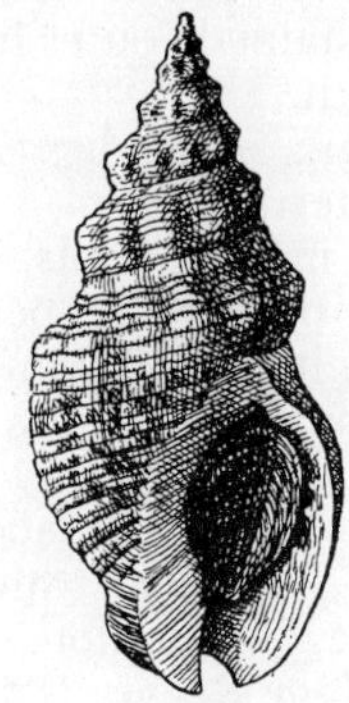

Figure 227

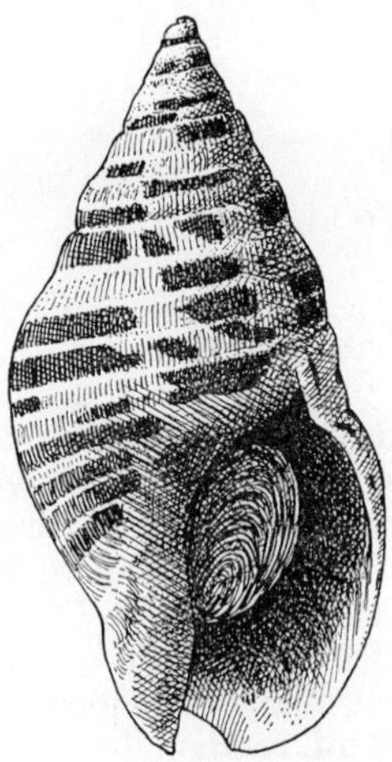

Figure 228

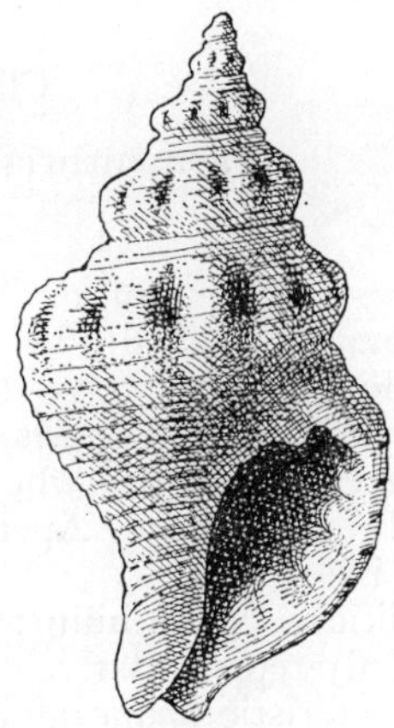

Figure 229

226. *Phos senticosus* × 2
A Sculpture (enlarged)
227. *Cominella eburnea* × 1⅓
228. *Cominella lineolata* × 2
229. *Tasmeuthria clarkei* × 3

elevated. Whorls concavely impressed below the suture. Surface microscopically radiately striated. Aperture lirate within, oval, terminating in a short anterior canal. Interior deeply coloured, with the inner edge of the lip lightly stained; outer lip crenulated; columella smooth; operculum horny. Size: 1 inch.

Localities: St Kilda; Sandringham; Frankston; Portarlington; Queenscliff; Western Port.

Located in rock-pools, this is one of the commonest shells of open ocean rock platform; it also penetrates bays and inlets. Possessing an extensive synonymy, it provides an excellent example for the study of variation, but examination of a series will convince one of their specific identity.

These shells frequently appear more or less perforated by small holes, due probably to attack by sponges. With records of its existence in Western Australia, South Australia, New South Wales and Tasmania, this species has a wide distribution.

acutinodosa Reeve, 1846; locality not recorded.
kingicola Tate & May, 1900; Queenscliff.

Tasmeuthria Iredale, 1925

Small spindle-shaped shell, spire approximately half length of shell. Sculptured with longitudinal ribs and concentric striae. Aperture oval; outer lip thick, denticulate within.

TYPE SPECIES *Siphonalia clarkei* Tenison Woods, 1875.

Clarke's False Cominella

Tasmeuthria clarkei (Tenison Woods, 1875)

(Figure 229

Shell small, cream or light brown, somewhat fusiform, solid. Spire approximately half the length of shell. Whorls six with strong, rounded, longitudinal ribs, twelve on the body-whorl, becoming obsolete at the anterior extremity. Whorls are angled, accentuated by a reddish brown spot between each rib, the whole surface further ornamented with concentric striae. Interior white. Aperture ovate; outer lip thick, denticulate within. Size: ¾ inch.

Localities: Port Phillip; Western Port.

The only representative of the genus in Victoria, it is easily identified by its characteristic shape and the interrupted band of reddish brown spots. Occurs also in Tasmania and South Australia.

Fax Iredale, 1925 (*Phos tabida* Hedley, 1904).
tabida (Hedley, 1904); off Ninety Mile Beach (40 fathoms).

Family NASSIDAE

Dog Whelks

Shell ovate. Spire usually elongate. Base of aperture with a notch; canal short; inner lip calloused; operculum horny with apical nucleus and usually a serrated margin. Extremely active in their movements they are carnivorous, readily boring through the shells of other molluscs and feeding upon them.

Alectrion Montfort, 1810

Shells mostly fairly large, with a high spire and sunken suture. Protoconch partially or entirely fluted. Outer lip not compressed, serrated.

TYPE SPECIES *Buccinum papillosa* Linné, 1767.

The southern Australian representative of this genus has been placed in *Nassarius* from members of which it differs in spire structure and the armature of the mouth.

Lined Nassarius

Alectrion particeps (Hedley, 1915)

(Figure 230)

Shell oblong-ovate, smooth, fawn pink with brownish blotches and concentric reddish brown lines. Spire approximately half the length of shell. Interior white. Aperture ovate outer lip acute. Size: 1 inch.

Localities: Portarlington; Lakes Entrance.

A very pretty shell approaching the northern *glans* Linné, 1767, but the prickles on the anterior portion of the outer lip so characteristic of that species are evanescent in *particeps* Hedley, which is also a more slender species. The author provides an interesting note: 'The animal lives sunk beneath the surface of the sand. Into a pool where no *particeps* were visible, a few crushed shellfish were thrown; a few minutes later a number of individuals, including *particeps,* appeared in various directions, all steadily crawling towards the bait. The animal of *particeps* is particularly bold and active. When lifted out of the water by the shell, the animal twists and kicks with much vigour. The colour of it is cream irregularly splashed with black.' Occurs also in New South Wales and South Australia.

Parcanassa Iredale, 1936

Small stout shells, with a round mouth and toothed outer lip. Spire and body-whorl of approximately equal length.

TYPE SPECIES *Parcanassa burchardi ellana* Iredale, 1936.

Impoverished Nassarius

Parcanassa pauperata (Lamarck, 1822)

(Figure 231)

Shell whitish or cream, with a narrow brown band near the suture of the spire-whorls, and a single brown band at the centre of the body-whorl; ovate, ventricose, radially, granosely ribbed and concentrically striated, a character more evident anteriorly. Body-whorl rather smooth at the back, shining. Aperture small, oval; outer lip varicose, with a deposit of enamel about the columella; operculum horny. Size: ¾ inch.

Localities: Brighton; Frankston; Dromana; Portsea; Portarlington; Western Port.

Figure 230

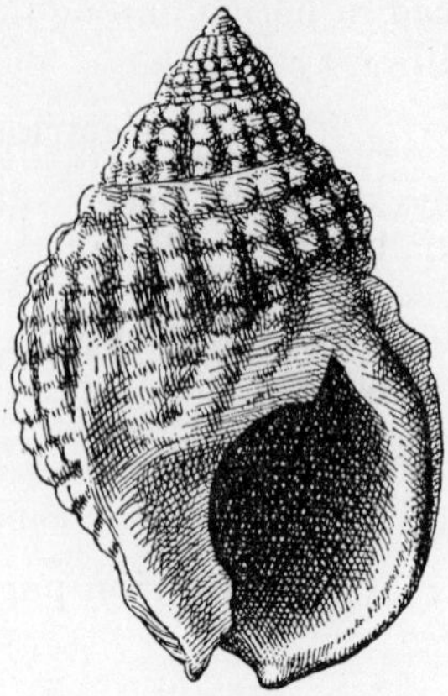

Figure 231

Figure 232

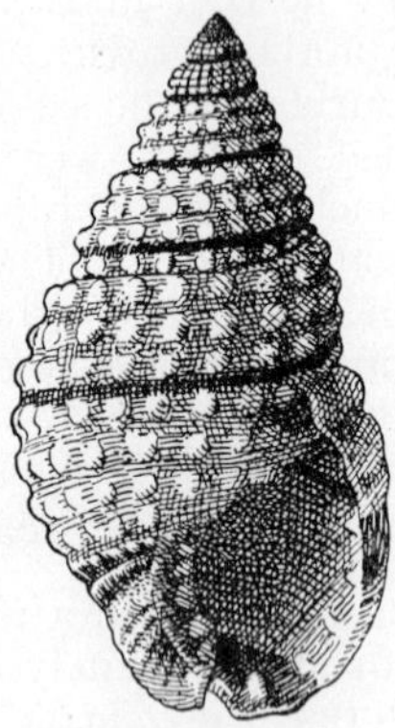

Figure 233

230. *Alectrion particeps* × 2
231. *Parcanassa pauperata* × 3
232. *Parcanassa burchardi* × 4
233. *Niotha pyrrhus* × 2

A similar shell to *Niotha pyrrhus,* but decidedly broader and with only one colour band on the body-whorl; these features with the presence of the enamel near the columella will suffice to distinguish the two species. It is found alive on sandbanks at low tide. Recorded from Western Australia, South Australia, New South Wales and Tasmania.

Blemished Nassarius

Parcanassa burchardi (Philippi, 1851)

(Figure 232)

Shell ovately conical, light brown with a band of dark brown just below the sutures, and on the body-whorl a second band situated at the anterior end. Spire rather short, a little less than half the length of shell. Whorls rounded, sculptured with longitudinal ribs which are somewhat tubercled at the upper portion close to the sutures. Interior white or light brown. Aperture ovate; outer lip thickened; columella curved and with a rather expanded callosity. Size: ½ inch.

Localities: Port Phillip; Western Port; Lakes Entrance.

A common shell unlikely to be confused with any Victorian form, easily detected by its encircling brown bands, tubercled longitudinal ribs and peculiar squat form. Occurs in South Australia and New South Wales. This species was known as *Nassarius labecula* A. Adams, 1851, but the above name has priority. It has also been mistakenly identified as *jonasi* Dunker, 1846, but is easily distinguished from it by the lack of striae between the ribs. Iredale, 1936, named the small New South Wales form subspecies *ellana* and this smaller form occurs in North Arm, Lakes Entrance, and at Rhyll, Western Port.

Tavaniotha Iredale, 1936 (*Nassa optata* Gould, 1850).

optata (Gould, 1850); the Semi-granular Nassarius. This shell was known for many years as *Nassarius semigranosus* Dunker, 1846, a preoccupied name, and is common in Port Phillip, Western Port, and the south-west coast generally; about ½ inch long, somewhat shining, possessing a granular surface to some extent resembling *N. pyrrhus*.

Iredale, 1936, separated the southern Australian shells subspecifically as *munieriana* (Crosse, 1864) on the grounds that they were larger and did not develop teeth on the outer lip. However, we find that a series of Victorian shells have the same size range as Sydney specimens and do develop an armed outer lip.

Reticunassa Iredale, 1936 (*Nassa paupera* Gould, 1850).

compacta (Angas, 1865); brown-banded Nassarius. A common species in Port Phillip, Western Port, Lakes Entrance. It is ornamented with about sixteen prominent longitudinal ribs crossed by concentric ridges between which are eight to ten microscopical concentric threads. Shells usually bear a red brown band just below the periphery but may be white. A smaller shell than the following species with the concentric ridges more closely set and lacking the microscopical threads between.

compacta benthali Gabriel, 1961; off Cape Everard (65 fathoms); off Lakes Entrance (20 fathoms); has more numerous radials and concentrics than the above species giving it a fine latticed appearance.

paupera (Gould, 1850); Lakes Entrance.

Niotha H. & A. Adams, 1853

Shell stout. Spire short. Whorls granulated or cancellated. Inner lip with a callus very large and spreading; outer lip crenate, not variced externally.
TYPE SPECIES *Nassa cumingii* A. Adams, 1851.

Banded Nassarius

Niotha pyrrhus (Menke, 1843)
(Figure 233)

Shell white, with a brown band near the suture of the spire-whorls, and conspicuously two-banded on the body-whorl; ovately conical; surface ornamented with numerous radial riblets which are crossed by spiral ribs, the points of intersection developing into prominent granules. Aperture rather small, ovate, with a short, reflected anterior canal; outer lip dentate; operculum horny. Size: ¾ inch.

Localities: St Kilda; Brighton; Mordialloc; Portsea; Portarlington; Western Port.

Crawling about on sandbanks at low water, these creatures may be obtained in numbers. It is a handsome little species, easily recognized by the encircling colour bands. Widely distributed, with records of its existence in Western Australia, South Australia, Tasmania and New Zealand.

Radulphus Iredale, 1924 (*Radulphus royana* Iredale, 1924).
royana Iredale, 1924; off Lakes Entrance (20 fathoms).

Family FASCIOLARIIDAE

Tulip Shells or Spindle Shells

Shell more or less spindle-shaped, without varices. Outer lip of the aperture not thickened; columella may bear plaits; operculum horny with spiral nucleus.

Dolicholatirus Bellardi, 1881

Shell imperforate, spindle-shaped, slender turreted, with flattened ribs and small spiral ridges. Mouth opening small, elliptical with long straight canal.
TYPE SPECIES *Murex lancea* Gmelin, 1791.

Spicer's Spindle Shell

Dolicholatirus spiceri (Tenison Woods, 1876)
(Figure 234)

Shell uniformly reddish brown, elongately fusiform, solid. Spire approximately half the length of shell. Whorls eight, sculptured with irregular, concentric ridges and crossed by fine longitudinal striae. Interior of similar colour but lighter. Aperture ovate terminated by a long, fairly straight, anterior canal. Size: 1 inch.

Locality: Western Port (dredged in 8 fathoms).

A very elongate, spindle-shaped shell of uniform coloration which with its strong sculpture should be easily identified. Diligent searching along the

Figure 234

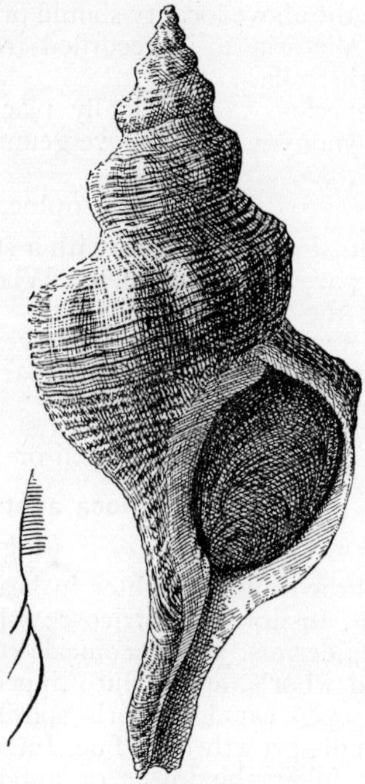

Figure 235

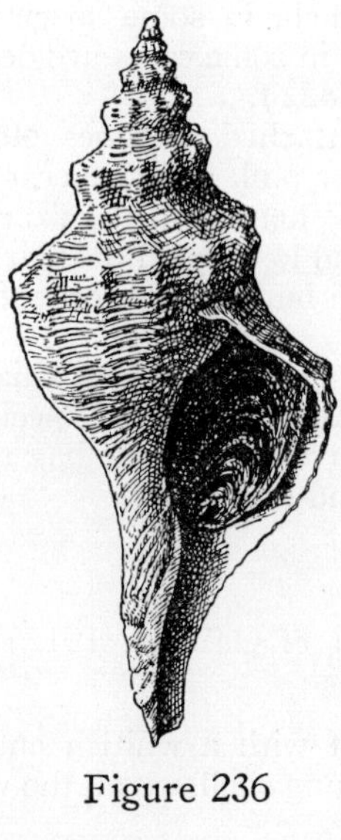

Figure 236

234. *Dolicholatirus spiceri* × 3

235. *Pleuroploca australasia* × ½
A Columella plaits (enlarged)

236. *Pleuroploca australasia bakeri* × ⅔

shores at the above locality should provide specimens of this interesting form. Beyond Victoria it is recorded from New South Wales, Tasmania and South Australia.

This species was originally placed in the genus *Latirofusus* Cossman, 1889, a synonym of the above genus.

Pleuroploca Fischer, 1884

Shell fusiform. Aperture with a straight anterior canal and with plaits on the fore part of the columella. Whorls with a row of more or less strong knobs at the angle.

TYPE SPECIES *Buccinum trapezium* Linné, 1767.

This genus is immediately separable from *Austrosipho* by the folds on the columella.

Tulip Shell or Short Spindle Shell

Pleuroploca australasia (Perry, 1811)

(Figure 235)

Shell brown, much lighter in beach-worn specimens; rather large, solid, elongated, fusiform, ventricose; when alive covered with a shining, dark brown epidermis. Spire conical, consisting of a few rounded or slightly angulated whorls, apex white tipped. Aperture oval, terminating in a long, straight, open canal. Whorls spirally, obtusely ridged throughout, nodose at the angle, growth-lines fine. Interior white. Columella with three oblique plaits or folds, the lower or anterior of which is larger than the rest; operculum horny, pyriform. Size: 6 inches.

Localities: Mordialloc; Frankston; Portsea; Altona. Fine examples are dredged in Port Phillip and Western Port.

Considerable confusion has existed with regard to this species and its fusiform varieties; in Victoria one species is generally recognized. Variation is apparent in the angulosity of the whorls, which, in some specimens, is very much emphasized; furthermore, the nodules in some cases are decidedly more pronounced, variety *coronata* (Lamarck, 1822).

Egg masses of this species are often collected attached to stones, etc.; they are bunches of beautifully coloured, rich cream or pink bell-shaped capsules with fluted edges, half to one inch in length. When fresh, they are most attractive, but unfortunately become shrivelled and lose much of their beauty on drying out. Dozens of capsules may form the bunch, each one of which contains from six to eight embryonic forms.

In addition to Victoria, this species is recorded from Tasmania, New South Wales, South Australia and Western Australia. This species was originally placed in the genus *Fasciolaria* Lamarck, 1799, but has recently been separated because of differences in shape and sculpture.

Baker's Tulip Shell

Pleuroploca australasia bakeri (Gatliff & Gabriel, 1912)

(Figure 236)

Shell similar in colour to the last species but with a whitish encircling band at the periphery, which is plainly seen on young shells on all the whorls;

the rest of the shell is light reddish brown, with a brownish yellow epidermis. The chief points of difference: spire more elongate, longitudinal plications pronounced, spiral sculpture a little stronger, comprising strong riblets with from two to four well-defined intermediate threads. Size: 4 inches.

Localities: Anderson's Inlet; Lakes Entrance; off Cape Howe.

Although showing minor differences, the authors were convinced that the shell was not worthy of specific distinction, regarding it merely as a deep-water variety. It also occurs off the New South Wales coast.

Propefusus Iredale, 1924

Elongate, spindle-shaped. Spire moderately high. Entire shell spirally ribbed and with longitudinal costae becoming obsolete on the body-whorl. Canal moderately long, slightly curved; operculum horny.

TYPE SPECIES *Fusus pyrulatus* Reeve, 1847.

Flame Spindle Shell

Propefusus pyrulatus (Reeve, 1847)

(Figure 237)

Shell white, with brown colour bands on the shoulder and at the anterior end; somewhat stoutly fusiform; in life covered with a horny epidermis. Spire high, turreted. Whorls rounded, ventricose, sculptured throughout with spiral ridges and longitudinal plications. Aperture generally oval, terminating in a long straight canal which is about the same length as the spire; columella straight; outer lip sharp; operculum horny, pyriform. Size: 3 inches.

Localities: South Melbourne; St Kilda; Frankston; Point Nepean; Outer Geelong Harbour; Western Port; Lakes Entrance.

The present form, with its graceful contour and characteristic sculpture, is easily identified. The animal is carnivorous. Not many years ago at the two first-named localities it was not an infrequent shell, but evidently owing to drainage and the dumping of mud at the northern part of the bay, this species is now seldom seen. The *Challenger* dredged it in forty fathoms off East Moncoeur Island, Bass Strait. The channels of Western Port provide examples in excellent condition and in addition to the Victorian coast, it appears on the list of South Australian mollusca.

undulatus (Perry, 1811) has been confused with the above species. It is narrower, smaller and more solidly built than *pyrulatus* with a rounder mouth and a straight, stout canal. It is recorded from Port Fairy, western Victoria; South Australia; Western Australia and Tasmania. Synonyms of this species are *Fusus ustulatus* Reeve, 1847, and *Fusus legrandi* Tenison Woods, 1875.

Ollaphon Iredale, 1929 (*Trophon molorthus* Hedley & May, 1908).
molorthus Hedley & May, 1908; off Lakes Entrance.

Fusus Bruguière, 1792

Elongate spindle-shaped shells with a long, many-whorled spire, and a very long, straight, open canal. Sculptured with spire ribs and longitudinal costae.

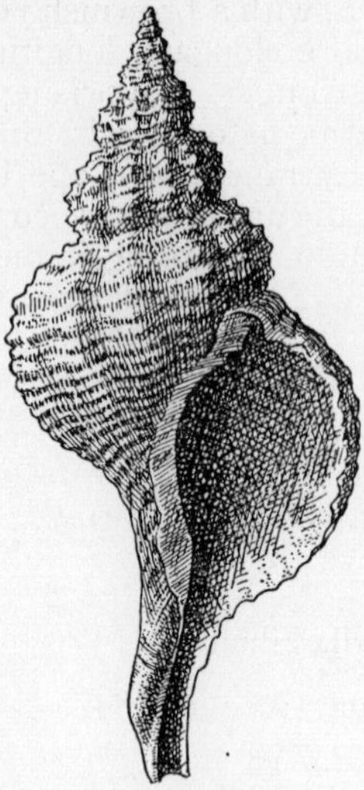

Figure 237

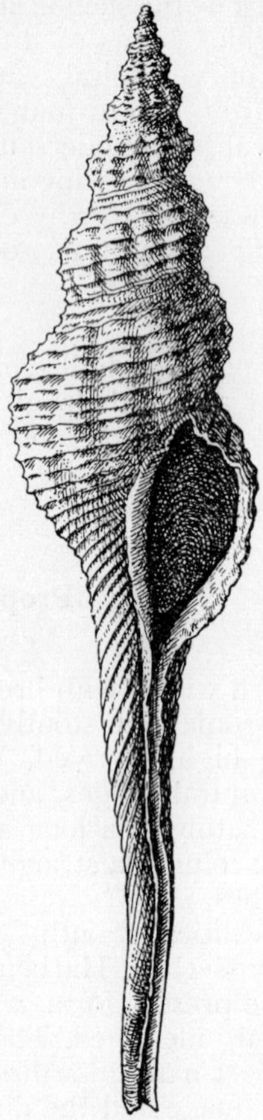

Figure 238

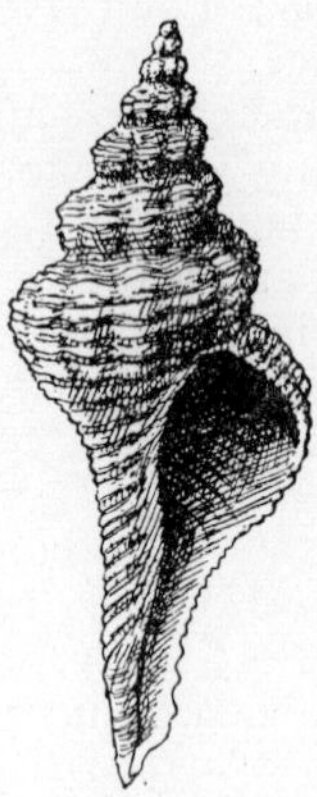

Figure 239

Figure 240

237. *Propefusus pyrulatus* $\times 1$
238. *Fusus novae-hollandiae* $\times \frac{2}{3}$
239. *Fusus australis* $\times 1$
240. *Microcolus dunkeri* $\times 3$

TYPE SPECIES *Murex colus* Linné, 1758.

This genus has been known as *Colus* Humphrey, 1797, but precedence must be given to the above name. Also Humphrey has been ruled as nonbinomial by the International Commission on Zoological Nomenclature.

New Holland Spindle Shell

Fusus novae-hollandiae Reeve, 1848

(Figure 238)

Shell white invested with a brown fibrous epidermis, solid, elongately fusiform. Spire less than half the length of shell. Whorls rounded, sculptured with strong concentric ridges and numerous fine longitudinal ribs, giving it somewhat of a latticed appearance. Interior white. Aperture oblong; outer lip sharp and operculum horny, brown. Size: 6 inches.

Localities: Snake Island; Port Albert; Lakes Entrance.

This is a very ornate shell and easily the longest Victorian Spindle. Dredged alive in Bass Strait on a sea-bed of sand and shell by the trawler *Endeavour,* one specimen was 8½ inches long. Distinctive features are its nodulose whorls and the extremely long anterior canal, rendering it easy to recognize. Found also in Tasmania and South Australia.

Southern Spindle Shell

Fusus australis Quoy & Gaimard, 1833

(Figure 239)

Shell light brown, solid, elongately fusiform. Spire about half the length of shell; spire-whorls slightly angular with small nodules at the angle; sculptured with irregular, strong, concentric ridges and numerous, fine longitudinal ribs presenting a somewhat latticed surface. Interior white. Aperture oblong; outer lip acute. Size: 4 inches.

Locality: Gippsland coast.

Rather a fine species, with a shorter spire, more angular whorls and finer sculpture than *novae-hollandiae.*

Microcolus Cotton & Godfrey, 1932

Spindle-shaped, small, turreted shells with thick axial ribs and obscure spiral lirae. Spire elevated, conical. Adult whorls numerous, rather convex, angulated, concave above. Canal very short, slightly curved.

TYPE SPECIES *Fusus dunkeri* Jonas, 1844.

Microcolus dunkeri (Jonas, 1844)

Dunker's Spindle Shell

(Figure 240)

Shell small, strong, fusiform, cream often exhibiting a narrow colour band just above the suture which in fresh examples is clearly visible through the aperture. Spire approximately half the length of shell. Whorls ventricose and somewhat angular. Sculpture consists of numerous irregularly spaced encircling ridges crossed by less frequent robust radial costae; earlier whorls finely, radiately ribbed. Aperture white, polished, oblong; outer lip sharp; anterior canal curved. Size: ¾ inch.

Localities: Sandringham; Western Port; off Lakes Entrance.

This shell, originally described under the genus *Fusus* is our smallest Victorian Spindle and by no means common on beaches. The colour band is not a constant feature and in water-worn examples only faintly visible. Recorded also from South Australia.

Family OLIVIDAE

Olives

Shells more or less cylindrical in shape, the enlarged body-whorl usually conceals all but the apex of the earlier whorls; usually polished and brightly coloured. The outer lip simple with an anterior notch; operculum horny, small, absent in the typical Olives.

Zemira H. & A. Adams, 1858

Shell ovate, oblong, moderately umbilicated. Spire acuminate. Whorls more or less convex, suture channelled. Aperture oval with a tooth on the outer lip; inner lip spreading, often covering the umbilicus in the adult.

TYPE SPECIES *Eburna australis* Sowerby, 1841.

Southern Zemira

Zemira australis (Sowerby, 1841)

(Figure 241)

Shell small, ovate, shining. Spire rather short. Brown, the body-whorl showing three rows of darker spots one just below the excavated sutures. Anteriorly about ten concentric grooves appear; the remainder of the shell to the apex similarly sculptured but with the grooves much narrower; the two areas divided by a rather deep furrow extending to the sharp outer lip and terminating in a tooth-like projection at about the anterior third; the whole surface crossed by faint irregular growth-lines. Interior white, polished. Aperture oblong-ovate; anterior canal short; operculum horny. Size: 1 inch.

Localities: Off East Moncoeur Island, Bass Strait (38 fathoms); Western Port.

The sole representative of the genus in Victoria. Occasionally washed up on the shores at the last-named locality and dredged alive between Phillip and French Islands (8 fathoms). It is a shapely little shell obtained on a sea-bed of sand, easily identified by the excavated sutures and the prominent tooth-like projection on the outer lip. Occurs in Tasmania and New South Wales.

Belloliva Peile, 1922

Shell somewhat slender, smooth, shining, with a moderately high spire and thickened suture. Mouth small, above acutely angled, less constricted below; operculum present, horny.

TYPE SPECIES *Olivella brazieri* Angas, 1877.

Wheat-Olives or Olivellas

Belloliva triticea (Duclos, 1835)
(Figure 242)

Shell small, fairly solid, oliviform. Spire produced, acute, a little less than half the length of shell. Whorls flatly rounded with two interrupted bands of reddish brown on the spire-whorls and three on the body-whorl; sutures canaliculate. Interior white. Aperture narrow; outer lip strong and straight; base of columella curved inwards and marked by two prominent plaits. Size: ½ inch.

Localities: Western Port; Port Albert; off Lakes Entrance.

Frequently washed up on beaches. In living condition the species of this genus bury themselves in sand. Occurring also in New South Wales and Tasmania.

This species often occurs in the literature as *Olivella pardalis* A. Adams & Angas, 1863, but the above name is the older and must be used. Some authors have divided *Belloliva* into a second genus *Gemmoliva* Iredale, 1924, in which this species is placed.

leucozona (A. Adams & Angas, 1863); the White-Zoned Wheat-Olive, from Western Port, Lorne and beyond Victoria in New South Wales and Tasmania, somewhat resembles the above but is reticulated with reddish brown zigzag lines and a concentric bluish white band at the anterior part of the body-whorl. These features easily distinguish the two species.

brazieri Angas, 1877; off Lakes Entrance (20 fathoms).

Cupidoliva Iredale, 1924 (*Olivella nympha* A. Adams & Angas, 1863).

nympha (Adams & Angas, 1863); White Wheat-Olive, from Port Phillip, Western Port and Portland, is the commonest Olivella in Victoria, smaller, about ½ inch, and easily separated by its slender form, semi-pellucid white colour and a definite opaque white, narrow, concentric band immediately below the sutures. Occurs also in Tasmania and New South Wales.

Alocospira Cossman, 1899

Shell fusiform oblong or ovate, thick, with the sutures calloused; base broadly emarginate. Spire coniform, covered with a thin callus and ornamented with some encircling spiral lirae; columella constricted and twisted, lower edge of shell with some small groove-like folds, one of which extends as a small tooth on the outer anterior mouth-edge.

Type Species *Ancilla papillata* Tate, 1889.

Iredale, 1924, pointed out the need for the separation of the southern Australian forms from delicate thin-shelled species of the restricted genus *Ancilla*. He showed that the fossil and recent species were closely related and that the former had already been divided into a number of sections by Fischer, Cossman and others. These genera were available for the recent species. He proposed the use of *Baryspira* Fischer, 1883, for the Australasian species with the Australian species separated subgenerically as *Alocospira*.

Cossman when he erected *Alocospira* gave it equal status with *Baryspira* and we, having considered the evidence, feel that these divisions constitute two natural groups. *Baryspira* is separated from *Alocospira* by a thicker

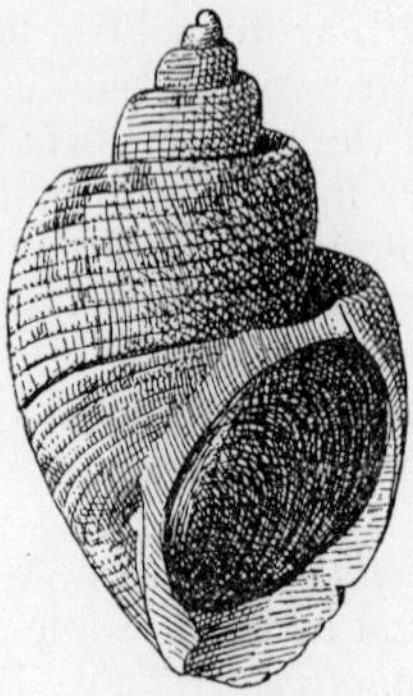

Figure 241

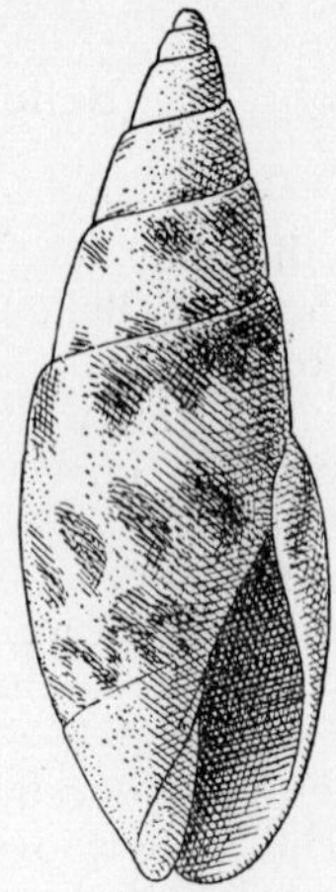

Figure 242

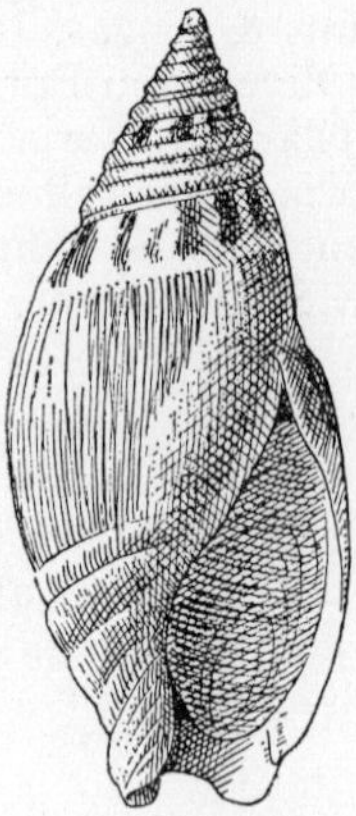

Figure 243

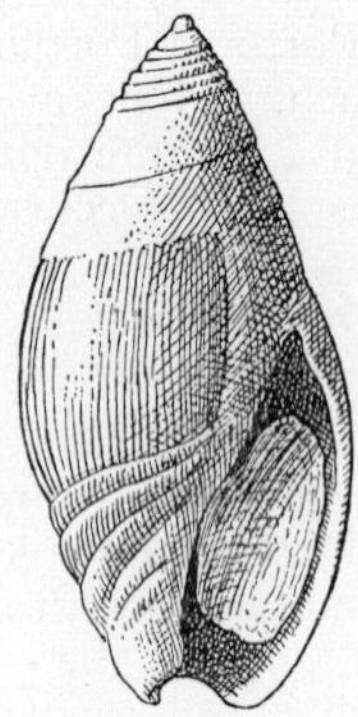

Figure 244

241. *Zemira australis* × 2
242. *Belloliva triticea* × 5
243. *Alocospira marginata* × 1⅓
244. *Alocospira fusiformis* × 2

shell, heavily calloused spire, either without or with indistinct, spiral lirae, and a flared mouth. Most New Zealand forms fall into *Baryspira* and the living Australian forms into *Alocospira*.

Margined Ancilla

Alocospira marginata (Lamarck, 1810)
(Figure 243)

Shell yellowish orange, ovately fusiform. Spire acuminated. Whorls smooth, polished, broadly calloused, margined at the upper part, callosity stained with orange brown and covering the sutures; earlier whorls strongly spirally ridged. Interior polished. Aperture long, terminating in a rather broad anterior canal; outer lip thin, acute; columella slightly twisted; operculum horny. Size. 1½ inches.

Localities: Mornington; Dromana; Point Nepean; Point Lonsdale; Western Port.

This shining, orange tinted form, neatly margined at the sutures with a prominent callus band, is a striking shell. With wave action, discoloration takes place, and being of rather light substance, the outer lip in shore specimens is seldom perfect. Inhabiting submerged sandbanks, living examples may be obtained in the neighbourhood of Port Phillip Heads and in Western Port. Occurs also in South Australia, Tasmania and New South Wales.

marginata tasmanica (Tenison Woods, 1876); a slightly smaller white variety of this species which occurs occasionally in Tasmania and Victoria.

dyspetes (Iredale, 1939); an allied species of smaller dimensions with a very short spire. It was described from southern New South Wales and the National Museum of Victoria has a specimen from Tasmania, so its occurrence in Victoria is likely.

Fusiform Ancilla

Alocospira fusiformis (Petterd, 1885)
(Figure 244)

Shell solid, highly polished, light pink with a distinct creamy white sutural band of callus. Spire acuminate, less than half the length of shell. Body-whorl showing fine microscopic, longitudinal striae. Interior light pink, similar to the exterior. Aperture elongately ovate; outer lip thin; anterior canal wide; columella twisted; operculum horny. Size: 1 inch.

Localities: Apollo Bay; Otway coast; off Portsea; Western Port; off Lakes Entrance.

A highly polished species distinguished from all Victorian members of the genus by its pinkish coloration even in faded shore water-worn examples. Occus also in Tasmania, South Australia and New South Wales.

This species shows considerable variation in breadth and callus, particularly in older shells. The broader more heavily calloused shell has been mistakenly identified as the following species.

gaza (Iredale, 1924); off Lakes Entrance.

petterdi (Tate, 1893); Port Fairy.

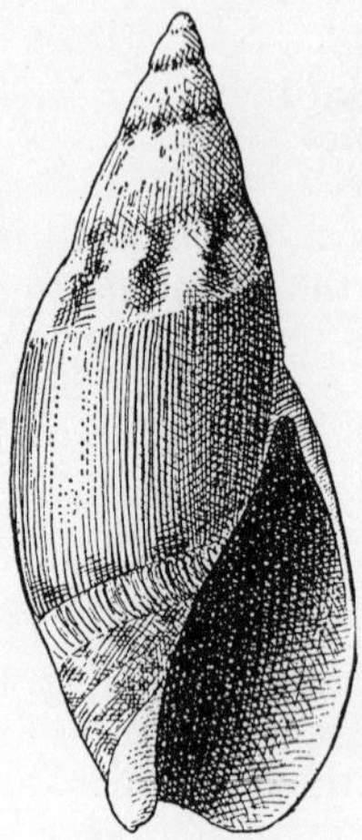

Figure 245

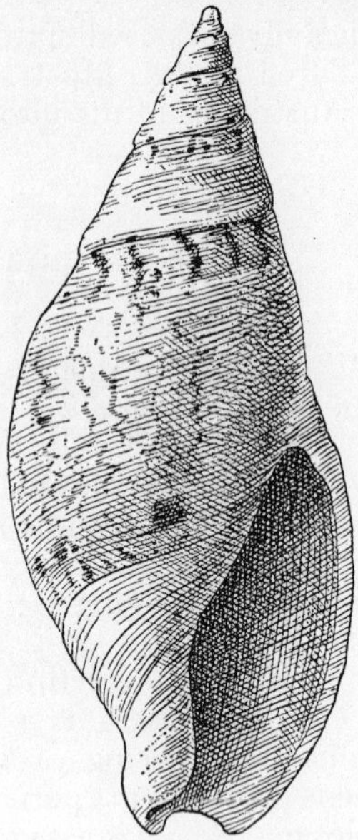

Figure 246

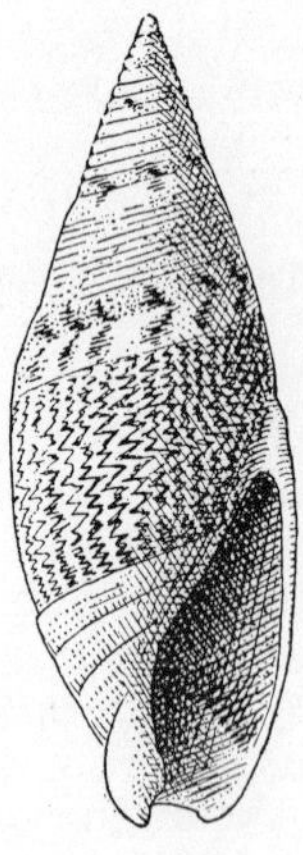

Figure 247

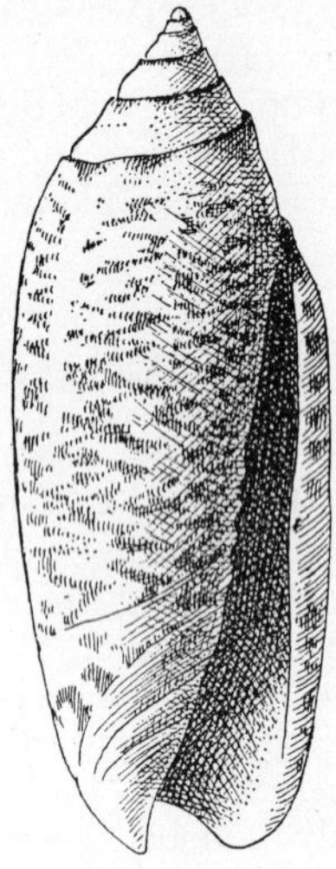

Figure 248

245. *Alocospira oblonga* × 2⅔
246. *Alocospira monilifera* × 3
247. *Alocospira edithae* × 3
248. *Oliva australis* × 2⅔

Oblong Ancilla

Alocospira oblonga (Sowerby, 1830)

(Figure 245)

Shell pinkish fawn, highly polished; banded at the sutures with a whitish, reddish flamed callosity, the body-whorl finely marked with zigzag lines; fusiformly oblong. Spire acuminated, much less than half the length of shell. Interior pinkish, similarly polished. Aperture oblong; columella a little twisted; anterior canal well open; operculum horny. Size: 1 inch.

Localities: Off East Moncoeur Island (38 fathoms); Western Port (8 fathoms) between Phillip and French Islands.

Occasionally cast up on the shore devoid of polish through wave action, but when alive is an extremely handsome species, the red-flamed sutural band serving as a useful recognition mark. Occurs also in Tasmania and South Australia.

Lineated Ancilla

Alocospira monilifera (Reeve, 1864)

(Figure 246)

Shell thin, whitish, shining, delicately lineated with orange brown, sutural band white and irregularly flamed with reddish brown; ovately fusiform. Spire acuminated, about half the length of shell. Body-whorl finely, microscopically striated. Interior white. Aperture elongately ovate; outer lip thin; anterior canal wide; columella twisted. Size: 1 inch.

Locality: Portland.

A beautifully polished shell, not of frequent occurrence in Victoria. It differs from its nearest ally *oblonga* Sowerby, by its lighter colour and broader outline. Occurs in South Australia and Western Australia.

Edith's Ancilla

Alocospira edithae (Pritchard & Gatliff, 1898)

(Figure 247)

Shell creamy white, with an interrupted brown band immediately above the upper basal groove; sometimes the coloration is continuous; the markings between the bands on the body-whorl are spirally zigzag, interrupted, and very light brown. Acuminately, elongately fusiform; fairly solid; whorls seven including two nuclear; from the dorsal aspect the body-whorl constitutes rather more than half the entire length, and is finely, longitudinally, and spirally striate. Columellar callosity starting from a little below the centre of the aperture, and extending across the body and penultimate whorls; well defined, but not conspicuously prominent, it has a peculiarly finely corroded appearance. There are about five distinct, slightly punctured, encircling grooves on the penultimate and spiral-whorls. The uppermost basal groove commences at the columella callosity, a little below the centre of the body-whorl, and terminates in a very slight projection of the outer lip. Aperture narrow-lanceolate; outer lip acute; columella twisted; operculum horny. Size: ¾ inch.

Localities: Rosebud; Sorrento; Western Port.

The above is from the authors' description and will serve to readily identify the shell. It is more narrowly acuminated, and the spire proportionately more lengthy than in any other Victorian species. In its living condition, located on submerged sandbanks. Recorded also from South Australia.

Habe, 1959, proposed a new genus *Austrancilla* for this species on the grounds that it differed from Japanese members of the genus *Baryspira* in having a more slender and maculate shell and in the anteriorly placed nucleus of the operculum. The anteriorly placed nucleus of the operculum is shared by other species of *Alocospira*, and as shell shape alone is not a generic character, the Habe name is not admitted.

Oliva Bruguière, 1789

Shell cylindrical. Spire short conical; suture channelled. Mouth long, narrow, widely notched below; columella obliquely plicate, grooved or striate in front, calloused behind; outer lip single; no operculum.

TYPE SPECIES *Voluta porphyria* Linné, 1758.

Southern Olive

Oliva australis Duclos, 1835

(Figure 248)

Shell whitish, ornamented with violet or purplish zigzag longitudinal lines; oblong, cylindrical, thick, smooth, polished. Spire short, about one-fifth the length of shell, with sutures distinct and somewhat excavated; subsutural edges lightly, irregularly splashed with a darker coloration. Interior white, sometimes with a thickened outer lip, slightly margined with reddish brown. Aperture elongated; columella fairly straight bearing numerous oblique plaits; anterior canal very short. Size: 1 inch.

Localities: Port Fairy; Portland.

This is the only representative of the genus in Victoria and not often obtained. In South Australia and Western Australia, however, it is of more frequent occurrence and shows much variation in coloration, the zigzag markings being brownish. Olives possess, as a rule, a smooth polish; they are chiefly tropical in distribution, displaying rich colours and patterns of extreme brilliance. They live on sand patches and are generally located by following the tracks they leave by burrowing just below the surface.

Family MITRIDAE

Mitres

Thick, solid, spindle-shaped shells with a sharply pointed spire. The mouth is small with a simple outer lip, anterior notch, and several plaits on the columella; the epidermis when present is very thin and the colouring can be seen quite clearly beneath it; operculum when present is small and horny. Southern species are much less brilliantly coloured than those of warmer seas.

Austromitra Finlay, 1926

Shell small, spindle-shaped, turreted; sculptured with axial ribs and spiral grooves on the base. Spire conical, short; protoconch of one and a half to two whorls, smooth. Whorls shouldered. Mouth long, narrow, with a short recurved anterior canal; columella with three or four oblique plaits.

TYPE SPECIES *Columbella rubiginosa* Hutton, 1885.

Thiele places this genus as a synonym of *Peculator* Iredale, 1924; in this we do not concur as *Austromitra* differs in the constriction of the anterior end of the body-whorl, the oblique plaits, and the spiral grooving of the base.

Tasmanian Mitre
Austromitra tasmanica (Tenison Woods, 1874)
(Figure 249)

Shell ovate, attenuate at both ends; brown, with two or three yellowish white transverse bands. Spire subelevate, acuminate. Whorls seven, flattened, thickly ribbed lengthwise with small obsolete ribs, and very finely girdled with transverse lirae. Aperture narrowed; columella quadriplicate. The fine ribs are most prominent on the upper whorls. Size: ½ inch.

Locality: Western Port, dredged in 8 fathoms on a sandy sea-bed.

It is by no means common and rarely seen on beaches. The above is the author's original description and, with the figure, no difficulty should be experienced in identifying the species. Occurs in Tasmania.

Several small species occur in Victorian waters and are likely to be cast up on the shores.

cinnamonea (A. Adams, 1854); a uniformly cinnamon coloured species with the spire strongly longitudinally ribbed, the ribs vanishing on the body-whorl. Size: ⅓ inch from Western Port.

scalariformis (Tenison Woods, 1875); pale yellow with a white band at about the centre of whorls which are strongly ribbed; columella triplicate. Size: ½ inch (coast generally).

analogica (Reeve, 1845); a brown, shining species with six smooth whorls; interior dark and lirate; columella triplicate. Size: ½ inch, Portland.

analogica tincta (A. Adams, 1854); an oblong-ovate shining form, dark brown with two white bands on the body-whorl; interior dark with the white bands plainly visible; columella triplicate. Size: ⅓ inch (Western Port).

acromialis (Hedley, 1914); off Ninety Mile Beach (40 fathoms).
bellapicta (Verco, 1909); Western Port.
bucklandi Gabriel, 1961; off Lakes Entrance (20 fathoms).
bucklandi bassiana Gabriel, 1961; east of Lakes Entrance (15 fathoms).
legrandi (Tenison Woods, 1875); Western Port; Port Phillip.
pumilio (May, 1915); off Ninety Mile Beach (40 fathoms).
retrocurvata (Verco, 1909); off Ninety Mile Beach (40 fathoms).
schomburgki (Angas, 1878); Port Phillip.
stadialis (Hedley, 1911); Bass Strait.
tatei (Angas, 1878); Western Port; Port Phillip.

Peculator Iredale, 1924 (*Peculator verconis* Iredale, 1924).
porphyria (Verco, 1897); off Wilson's Promontory.

Mitroidea Pease, 1865 (*Mitra ancillides,* Swainson, 1835).
pellucida Tate, 1886; San Remo, Western Port.

Cancilla Swainson, 1840

Shell spindle-shaped, slender. Sculptured with elevated spiral ridges and longitudinal growth-lines. Outer lip thin; columella with three to five plaits.
TYPE SPECIES *Voluta filosa* Born, 1780.

Strange's Mitre

Cancilla strangei (Angas, 1867)

(Figure 250)

Shell creamy with light brown bands, one on the spire whorls and two on the body-whorl; small, ovately fusiform. Spire about half the length of shell. Whorls seven to eight slightly convex, concentrically closely ridged, the ridges of body-whorl alternately larger and smaller. The whole surface crossed by fine longitudinal striae giving the shell a definite cancellated appearance. Interior creamy, polished. Aperture elongated; outer lip thin; columella pinkish; three to four plaited. Size: 1 inch.

Localities: Lorne; Western Port.

A charming shell easily recognized by its cancellated sculpture. Dead shells are of rare occurrence on beaches where they are much worn through wave action. Living examples have been dredged in 8 fathoms between Phillip and French Islands. For some time known under the name of *Mitra franciscana* Tenison Woods, 1876, a synonym. Beyond this state it occurs in Tasmania, New South Wales and Queensland.

Eumitra Tate, 1889

Medium-sized, dull, solid, elongate shells with a tapering spire; smooth or with a fine pattern of concentric and longitudinal striae. Mouth not flared; outer lip thickened and smooth within; columella usually with four plaits.

TYPE SPECIES *Mitra alokiza* Tenison Woods, 1879.

Iredale, 1929, erected *Vicimitra* (*Vicimitra prosphora* Iredale, 1929) for this southern Australian group. This name has been in general use until Ludbrook, 1958, pointed out that Tate had already proposed a prior name and she designated the above species as type.

Southern Mitre

Eumitra australis (Swainson, 1822)

(Figure 251)

Shell rather narrow, brown, encircled throughout with a single, light coloured, broad band; smooth, polished. Spire elevated, apex acute. Aperture narrow, less than one-quarter the length of shell; columella brown, bearing four distinct white plaits. Size: 1½ inches.

Localities: Sorrento; Point Nepean; Queenscliff; Western Port.

In the present form the encircling band is a constant feature, providing a useful diagnostic character. In the adult stage there is a remarkable expansion of the lip. The species is also found in Tasmania and South Australian waters.

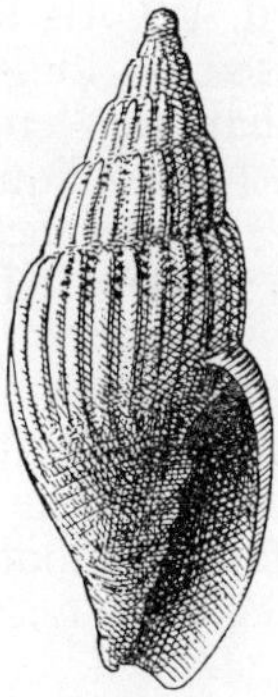

Figure 249

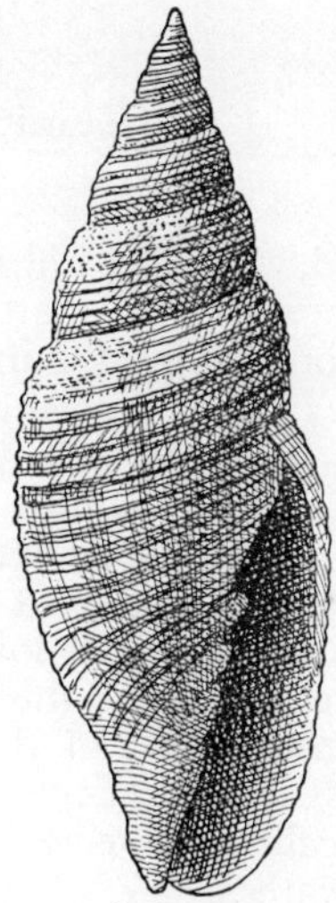

Figure 250

Figure 251

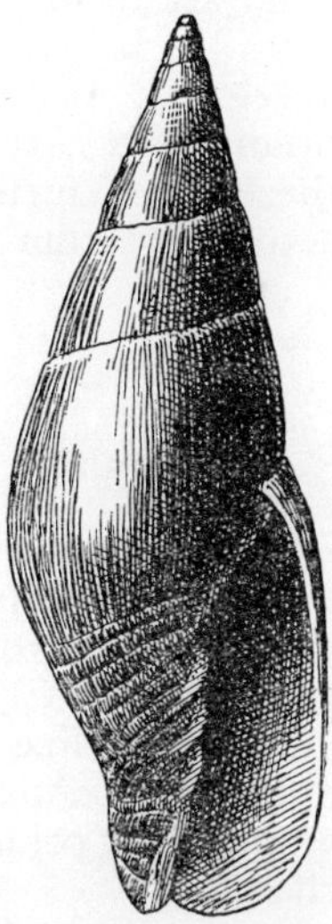

Figure 252

249. *Austromitra tasmanica* × 4
250. *Cancilla strangei* × 3
251. *Eumitra australis* × 1½
252. *Eumitra badia* × 2

Dark Brown Mitre

Eumitra badia (Reeve, 1845)
(Figure 252)

Shell solid, dark brown, polished, when living covered by a thin, olive brown epidermis; acuminately ovate. Spire elevated. Whorls flatly convex. Sculpture consisting of faint spiral grooves, crossed by longitudinal, extremely fine striae. Aperture narrow, about one-half the length of shell, smooth and light coloured within; columella with four oblique plaits, the anterior or lowest of which is somewhat obscure. Size: 1½ inches.

Localities: Frankston; Sorrento; Point Nepean; Queenscliff; Western Port.

Found under stones at low water; usually a trifle smaller than *australis* and easily separated by the uniform brown and absence of the encircling band so characteristic of that species. The animal is opaque white, and albino shells are occasionally obtained at Port Fairy and other western Victorian localities. The species, with its white variety, also occurs in South Australian waters.

There has been considerable confusion over the name for this shell and authors have disagreed on the status of Reeve's two species *Mitra badia* and *rhodia*. We have again examined the specimens from New South Wales and Victoria and are inclined to the view that they are conspecific, though Sydney shells tend to be narrower and the fourth anterior columella plait is not so well developed. As *badia* has page priority, we have used this name for the Victorian shells. According to Hedley, 1913, the locality of *rhodia* is Sydney; therefore if further examination of animals should prove the Sydney shells to differ from the Victorian shells, this name would apply.

Glabra Mitre

Eumitra glabra (Swainson, 1821)
(Figure 253)

Shell large, solid, elongately fusiform, flesh or fawn. Aperture a little more than one-third the length of shell. Spire with about nine flatly convex whorls, the whole surface sculptured with narrow, concentric, punctate grooves and numerous, fine, longitudinal striae, plainly seen with the aid of lens. Interior whitish. Aperture elongate; outer lip strong; anterior canal wide, slightly upturned; columella white, bearing four to five oblique plaits. Size: 3-4 inches.

Localities: Otway coast; Port Fairy; Portland; Western Port.

A very solid, fawn species, in life invested with a sordid brown epidermis. It is the largest of the Victorian Mitres and with the above description and figure no difficulty should be experienced in identifying it. Frequently washed up on beaches, and was dredged alive in 8 fathoms between Phillip and French Islands, a 'giant' measuring over 4 inches. Occurs in Tasmania, New South Wales and South Australia.

Iredale, 1936, pointed out that *Mitra carbonaria* Swainson, 1822, is a synonym of the above species.

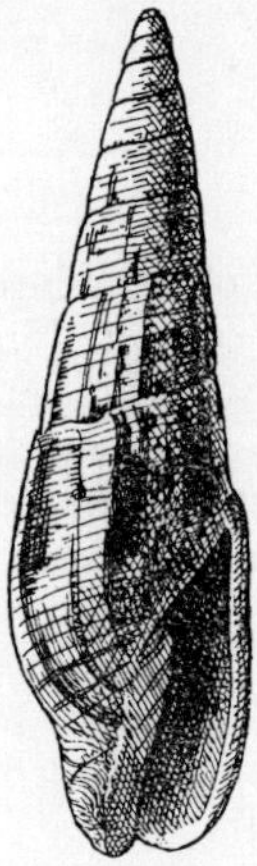

Figure 253

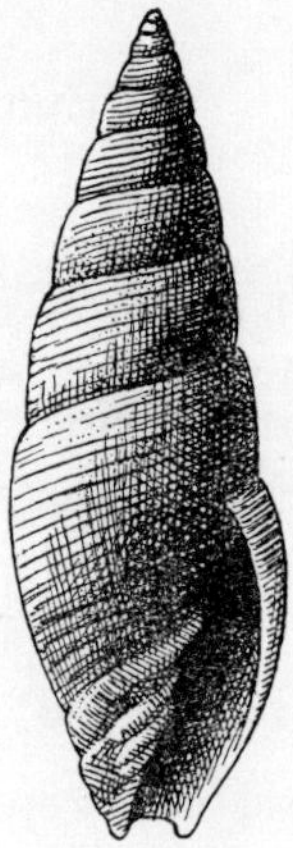

Figure 254

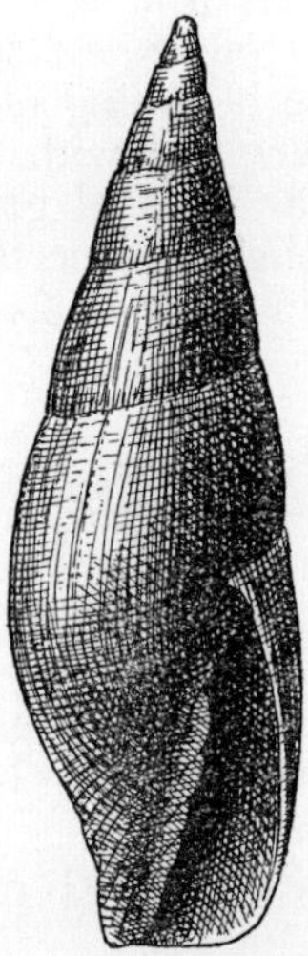

Figure 255

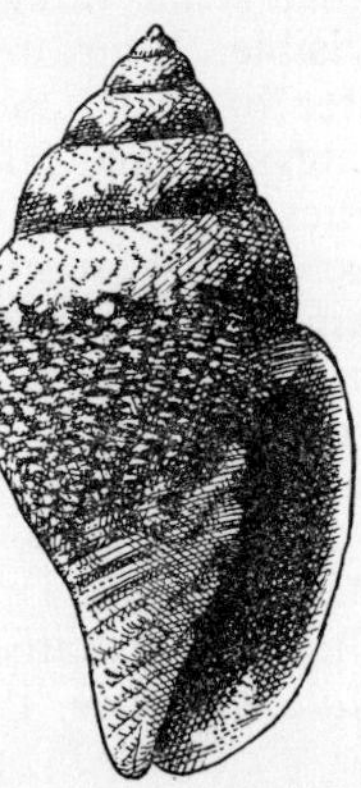

Figure 256

253. *Eumitra glabra* × 1
254. *Eumitra vincentiana* × 2
255. *Eumitra melaniana* × 1
256. *Proximitra pica* × 3

St Vincent's Mitre

Eumitra vincentiana (Verco, 1896)
(Figure 254)

Shell salmon, solid, elongately fusiform. Spire of eight whorls, slightly more than half the length of shell with the first three or four longitudinally ribbed, this character becoming obsolete in the adult whorls. Sculptured on the body-whorl with close longitudinal microscopic striae and numerous irregular, concentric, flatly rounded ribs, least evident at the central area. Interior light salmon. Aperture elongate; outer lip acute; columella salmon, quadruplicate. Size: 1 inch.

Locality: Western Port, dredged in 8 fathoms.

A strong unicoloured species somewhat similar in shape to *australis* (Swainson) but differing in the absence of the light coloured broad band so characteristic of that species. Occurs in South Australia.

Black Mitre

Eumitra melaniana (Lamarck, 1811)
(Figure 255)

Shell large, uniformly dark brown, elongately fusiform. Spire slightly more than half the length of shell; spire-whorls flatly convex and sculptured with narrow, concentric, punctate grooves, a feature becoming obsolete in the adult stage. The whole surface is further ornamented with very fine longitudinal striae, and in addition to this, irregular lines of growth are plainly visible. Interior generally purple, highly polished. Aperture elongate; outer lip thin; columella pinkish brown, bearing four to five oblique plaits; anterior canal fairly wide. Size: 2½ inches.

Localities: Point Roadknight; Western Port.

The second largest species of Mitridae in Victoria and easily identified by its size and uniformly dark brown colour. It somewhat approaches *glabra* Swainson, but is immediately separated by its relatively longer spire and internal darker coloration. Occurs also in New South Wales and South Australia.

There has been some confusion over the identity of this species as both Küster and Tryon mistook it for *Mitra carbonaria* Swainson, 1822, and later authors perpetuated the mistake.

prosphora (Iredale, 1929); off Lakes Entrance (20 fathoms).

persik (Verco, 1908); Portland; Port Phillip.

rosettae (Angas, 1865); Anderson's Inlet; Point Roadknight; Airey's Inlet; Apollo Bay.

Proximitra Finlay, 1927

Biconical, with a low spire, usually axially ribbed and spirally lirate; protoconch small. Adult whorls somewhat shouldered. Mouth oblique, narrow; columella with four plaits.

TYPE SPECIES *Vexillum rutidolomum* Suter, 1917.

Magpie Mitre

Proximitra pica (Reeve, 1845)
(Figure 256)

Shell small ovately fusiform, somewhat inflated, olive with an undulating white band just below the sutures. Body-whorl delicately patterned with two interrupted bands of lightly coloured triangular splashes. Spire short, much less than half the length of shell, earlier whorls longitudinally ribbed. Near the sutures the whorls are slightly shouldered with three or four concentric ridges appearing on this sloping area. Further concentric ridges on the anterior extremity number about ten and are stronger. Interior purplish brown, becoming white at the thin outer lip; columella white, quadruplicate; anterior canal fairly wide. Size: ½ inch.

Localities: Western Port; Otway coast; Lorne; Port Fairy.

This is a very delicately mottled species and with its consistent dark interior should be easy to identify. Found also in the waters of Tasmania and South Australia.

Family HARPIDAE

Harp Shells

Shell usually large with an expanded body-whorl, longitudinally ribbed, smooth and brilliantly coloured. Mouth expanded with a reflected outer lip, notched anteriorly; columella smooth; no operculum.

Austroharpa Finlay, 1931

Shell small. Inflated spire short; protoconch consists of two smooth whorls followed by two round shouldered adult whorls, and a large body-whorl. Sculptured with longitudinal and spiral ribs. Outer lip thickened.

TYPE SPECIES *Harpa pulligera* Tate, 1889.

Iredale, 1931, erected *Palamharpa* with *Palamharpa exquisita* Iredale, 1931, for the recent species of this group. We concur with Ludbrook in placing it as a synonym of the above genus.

Exquisite Harp

Austroharpa exquisita (Iredale, 1931)
(Figure 257)

Shell small, harp-shaped. Spire short, bearing slightly shouldered whorls and a well-defined two-whorled protoconch. Surface of shell ornamented with somewhat flattened spiral ribs, about thirteen to fourteen on the body-whorl, crossed by irregular, finer ridges with six to eight threads between. Covered with a very fine epidermis. Outer lip thickened; interior shining. Size: 1 inch.

Locality: Trawled off Lake Tyers (15 fathoms).

The sole living representative of a genus hitherto unrecorded in Victoria. Originally described from New South Wales.

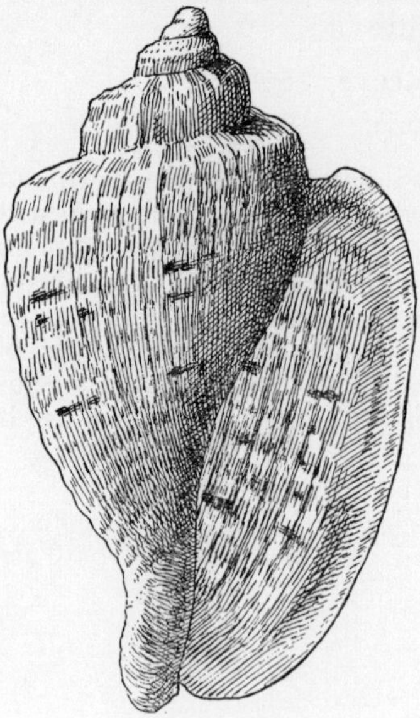

Figure 257

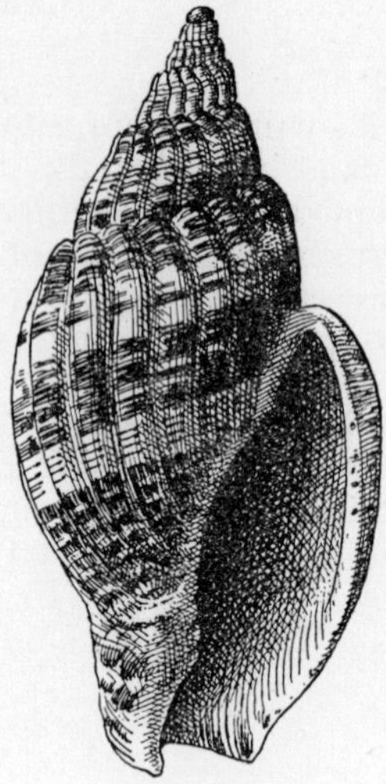

Figure 258

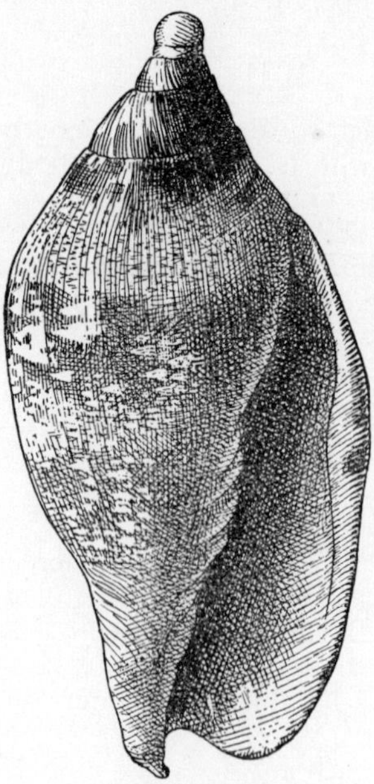

Figure 259

257. *Austroharpa exquisita* × 3
258. *Lyria mitraeformis* × 1⅓
259. *Ericusa papillosa* × ⅔

Family VOLUTIDAE

Volutes

Shell usually large spindle-shaped. Aperture oval with a simple outer lip which may become thickened and reflexed in old shells. It has an anterior notch and the columella is obliquely plaited; the operculum is horny but is absent in most of the larger species. The animals, which often have similar colouring to the shell, are carnivorous, feeding on a variety of animals which live in a similar habitat.

Lyria Gray, 1822

Shell small for the family, ovoid, thick, with strong axial ribs. Spire conical, elevated. Mouth simple with a thickened, slightly reflected outer lip; columella with three horizontal plaits and several weak wrinkles above; operculum present.

TYPE SPECIES *Voluta nucleus* Lamarck, 1822.

Mitre-shaped Lyria

Lyria mitraeformis (Lamarck, 1804)
(Figure 258)

Shell cream, with irregular, fine, brown, spiral lines and painted with scattered blotches of purple and brown; fusiformly ovate. Spire rather acuminated with a small papillary apex. Whorls longitudinally lyrately ribbed. Interior whitish. Aperture ovate; lip thickened, varicose. Columella with three prominent plaits at the anterior end and numerous thread-like plaits above. Size: 1½ inches.

Localities: Point Nepean; Point Lonsdale; Western Port.

The genus is composed of very elegant mitriform shells. The longitudinal ribbing is fairly constant, but much variation is noted in the colouring, some examples being more or less banded with various spots. When washed up on beaches they are very dull and the real beauty of this form is not revealed until fresh examples are examined. In living condition these shells are found on sandbanks. Occurs in South Australia and Tasmania; from the former locality the shell appears to show better development, both in dimensions and coloration.

Ericusa H. & A. Adams, 1858

Large, oval, smooth shells with an elevated spire; protoconch mamillate, oblique. Whorls convex, slightly shouldered. Mouth elongate, wide with a thickened, slightly sinuous outer lip; columella three to five plaits.

TYPE SPECIES *Voluta fulgetrum* Sowerby, 1825.

Iredale, 1929, erected the genus *Mesericusa* for *Voluta sowerbyi* Kiener, 1839, but we consider the differences shown between these two shells are specific rather than generic, and we prefer to consider *Mesericusa* as a synonym of the above genus.

Papillose Volute

Ericusa papillosa (Swainson, 1821)

(Figure 259)

Shell large, yellowish brown with darker brown markings or splashes on the whole surface, sometimes in interrupted rows. Spire short with a thick, yellowish, prominent papillose apex of two volutions. Interior pinkish, highly polished. Aperture elongately ovate; outer lip thick; columella with four to five folds; anterior canal fairly wide, slightly upturned. Size: 5 inches.

Localities: Portland; Western Port; Lakes Entrance.

This is a very solid and highly colured species and from the above description and figure is doubtless familiar to collectors. At the second-named locality taken alive in 8 fathoms. Beyond the state it is recorded from Tasmania and New South Wales.

papillosa kenyoniana (Brazier, 1907); from Bass Strait, Tasmania and New South Wales; the surface is sculptured with longitudinal folds in varying degree, an insufficient feature to be of specific value.

Spindle-shaped Volute

Ericusa sowerbyi (Kiener, 1839)

(Figure 260)

Shell flesh-coloured, with chestnut, irregular, angular or wavy markings; somewhat fusiform or spindle-shaped, thick, smooth. Spire acuminate, with a small papillary apex. Surface of shell, under the lens, minutely decussately striated. Interior pinkish. Aperture large, terminating anteriorly in a deep notch; lip slightly reflected; columella with three or four oblique plaits or folds. Size: 8 inches.

Localities: Point Nepean; Point Lonsdale; Lakes Entrance.

The present species in living condition, with its ever-varying colour pattern, is one of our most beautiful shells. In the deeper waters of Bass Strait it is a frequent species, and in the trawls off Bateman's Bay, New South Wales, magnificent examples are taken from time to time, specimens ten inches long being not uncommon. It is also found in Tasmania. These creatures are carnivorous and are found inhabiting the sand.

An albino variety of this shell has been called *porcellana* Jackson, 1954. Iredale, 1929, separated the New South Wales shells, which he said were larger, as *Mesericusa sowerbyi perspectiva*. In this we cannot concur as apparently large and small adult shells occur together in all localities. Sexual dimorphism may be the explanation but we have not as yet had the opportunity to confirm this.

Mamillana Crosse, 1871

Shell large, ovate, smooth. Protoconch mamillate, large, oblique. Mouth large; outer lip flared and reflected; columella with three plaits.

TYPE SPECIES *Voluta mamilla* Gray, 1844.

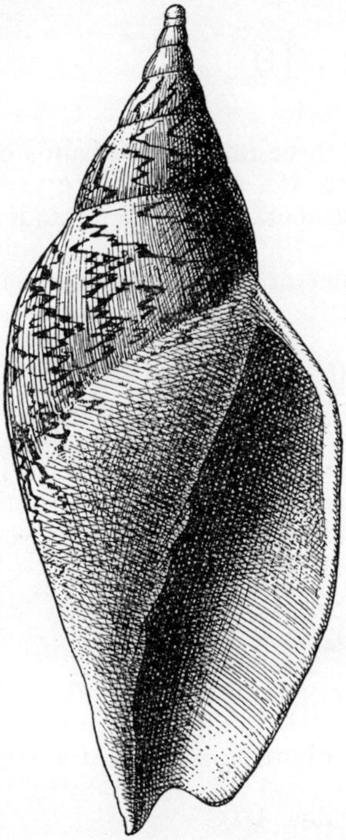

Figure 260

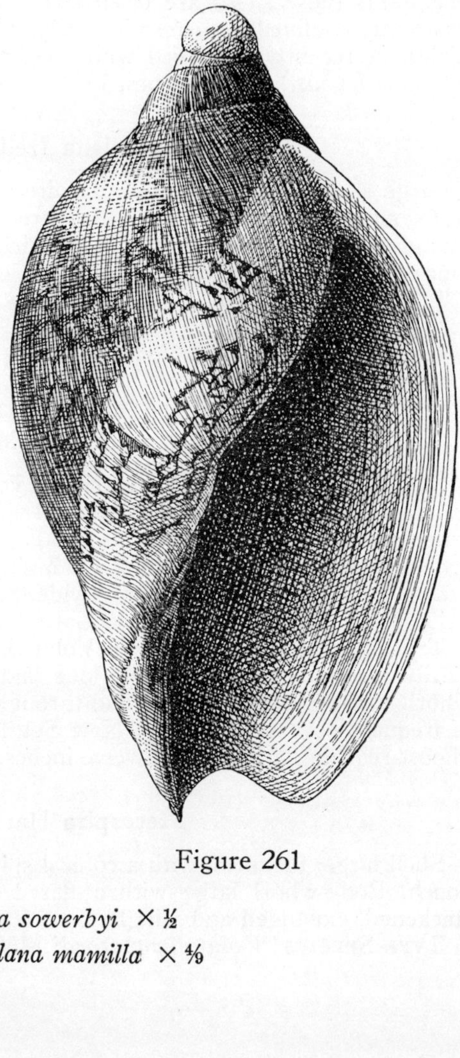

Figure 261

260. *Ericusa sowerbyi* × ½
261. *Mamillana mamilla* × 4/5

False Baler Shell

Mamillana mamilla (Gray, 1844)

(Figure 261)

Shell large, brownish, body-whorl with two or three interrupted bands of darker brown wavy or zigzag markings; oblong-ovate, globose. Spire very short displaying a conspicuous mamillate, obtuse apex. Body-whorl under lens showing a fine cancellate sculpture. Interior pinkish yellow, highly polished and frequently salmon tinted on the inner and outer lip which is slightly reflexed. Aperture very large, elongately ovate; anterior canal wide; columella three plaited. Size: 9 inches.

Localities: Cape Schanck; Gippsland coast; off Gabo Island.

This is an extremely handsome shell and the largest in Victoria, not infrequently attaining a length of twelve inches. With Hermit crabs as inhabitants these shells are often trapped in crayfish pots; the species is apparently confined to eastern Victoria, Tasmania and southern New South Wales. A recently described white variety of this species has been given the name *leucostoma* Mayblom, 1951.

Cymbiolena Iredale, 1929

Large shells with short conical spire and a very regularly wound small protoconch. Body-whorl large, shouldered, with an expanded oblique mouth extending nearly its length. Outer lip thickened; columella sinuous and with four delicate plaits on the anterior third.

TYPE SPECIES *Voluta magnifica* Shaw & Nodder, 1808.

Magnificent Volute

Cymbiolena magnifica (Shaw & Nodder, 1808)

(Figure 262)

Shell large, oblong-ovate, ventricose, yellowish or light brown with three concentric, darker broad bands marked with chestnut brown wavy or zigzag reticulations. Spire short with a small papillary apex. Body-whorl extraordinarily large. Interior pinkish, highly polished, bearing four distinct orange folds; anterior canal very wide. Size: 9 inches.

Locality: Trawled south from Gabo Island.

True to name this Magnificent Volute is by no means common in Victoria. Easily recognized by its rich colour and wavy reticulations on the body-whorl. Washed up on shores, and through trawling activities, it is a species of frequent occurrence on the New South Wales coast, specimens at times almost reaching a length of twelve inches.

Pterospira Harris, 1897

Shell large, fusiform, with a conical spire terminating in a bulbous protoconch. Body-whorl large with a flared elongate mouth, the outer lip is thickened, expanded and winglike.

TYPE SPECIES *Voluta hannafordi* McCoy, 1874.

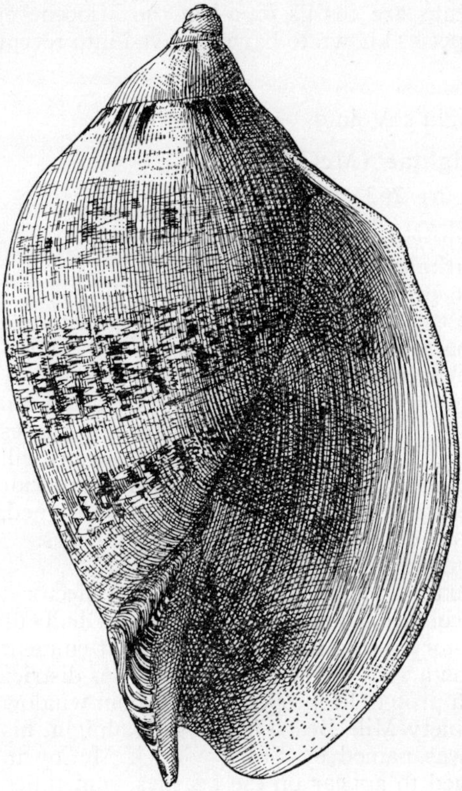

Figure 262

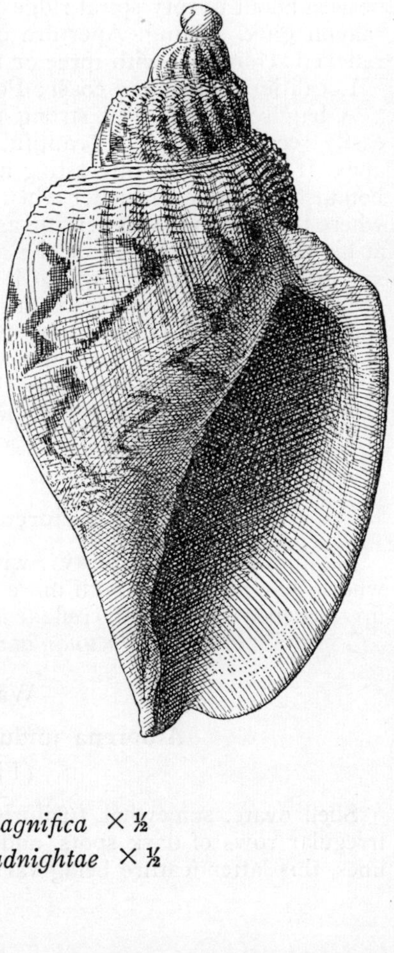

Figure 263

262. *Cymbiolena magnifica* × ½
263. *Pterospira roadnightae* × ½

Most of the members of this genus are fossils found in the Miocene of south-eastern Australia. The sole species known to have survived into recent times being that described below.

Roadnight's Volute

Pterospira roadnightae (McCoy, 1881)
(Figure 263)

Shell broad, fusiform, light brown with irregular, acutely angular, longitudinal, sparse, reddish brown markings. Spire conical about quarter the length of shell and terminating in a prominent bulbous apex. Spire-whorls longitudinally sculptured with about twenty well-developed obtusely flattened ribs, a feature becoming obsolete on the body-whorl where the only longitudinal sculpture appears as irregular lines of growth. Whorls are slightly shouldered towards the sutures, and on the sloping area there are about ten concentric ridges with parallel striae between them, the intervening spaces about twice as wide as the ridges. At the anterior end of the body-whorl will be seen about twenty spiral ridges similar to those near the sutures. Interior salmon pink, shining. Aperture oblong-ovate; outer lip strong, thickened, reflexed; columella with three or four oblique plaits. Size: 6 to 8 inches.

Localities: Gippsland coast; Portland; Cape Schanck.

A handsome volute of strong texture, much sought after by collectors; easily recognized by its beautiful colouring and very pronounced bulbous apex. It possesses an interesting history, being first noticed by that eminent botanist Baron von Mueller when on a visit to the Gippsland Lakes district where the specimen was serving as a prop to keep open the bedroom window at his hotel. It was found on the Ninety Mile Beach by Mrs Roadnight, his landlord's mother, after whom it was named by Sir Frederick McCoy in 1881. Occasional specimens continued to appear on the beaches, sometimes trapped in crayfish pots with hermit crabs in possession. One example in the collection of C.J.G. was found attached to the fluke of an anchor when hauled up at the last-named locality. Examples were taken in deep water east of Bass Strait by the *Endeavour*. The species has a wide distribution from Tasmania through Bass Strait to the Great Australian Bight and south-western Australia.

Amorena Iredale, 1929

Medium-sized shells, heavy, with a short conical spire and a large body-whorl. The elongate mouth more than twice the length of the spire; outer lip thickened and slightly reflexed, columella with four thick plaits.

TYPE SPECIES *Voluta undulata* Lamarck, 1804.

Wavy Volute

Amorena undulata (Lamarck, 1804)
(Figure 264)

Shell ovate, somewhat fusiform, smooth, shining; cream, marked with irregular rows of dark spots, and longitudinal, undulating chestnut brown lines, this latter feature being variable, as some specimens show the zigzag lines more closely set than others. Spire conical, apex blunt. Aperture elongated; outer lip thickened; columella bearing four distinct folds. Size: 3 inches.

Localities: Frankston; Queenscliff; Sorrento; Portsea.

In addition to the above localities, it is a frequent species at many places on the Victorian coastline and extends its range to New South Wales, Tasmania, South Australia and Western Australia. Dredged examples from Port Phillip and Western Port have provided specimens of a much darker colour; others may be practically colourless. When obtained alive it is a highly polished species and the characteristic undulating lines make it an easily recognizable form.

A colour variety with clear more distant lines on a light ground and with one or more broad smoky encircling bands has been given the name *angasi* Sowerby, 1864.

sclateri (Cox, 1869); a large white shell of similar form from King Island, Bass Strait, was once thought to be only an albino form of *undulata* but Ludbrook, 1953, in her recent work on the genus separates it as a species.

kingi (Cox, 1871); a pale pink squat form of the same shell.

Zebramoria Iredale, 1929

Shell medium sized, solid with a broad conical spire and large body-whorl. Mouth extends almost the length of the body-whorl; outer lip thickened and reflected; columella with four or five stout primary plaits.

TYPE SPECIES *Voluta zebra* Leach, 1814.

Zebra Volute

Zebramoria zebra (Leach, 1814)

(Figure 265)

Shell white or cream with numerous slightly undulating longitudinal streaks of brown; oblong-ovate, solid, smooth. Spire very short, conical and with a papillose apex. Interior whitish. Aperture much elongated; outer lip very thick; columella quadruplicate.

Locality: Gippsland coast.

A strongly textured shell somewhat resembling *Amorena undulata* Lamarck. It shows much variation in the number and width of markings, and is quite a common species in New South Wales and Queensland.

Microvoluta Angas, 1877 (*Microvoluta australis* Angas, 1877).

australis Angas, 1877; off Lakes Entrance (20 fathoms).

royana Iredale, 1924; 15 miles south-east of Lakes Entrance (25 fathoms).

Family CANCELLARIIDAE

Cancellarias

Shell oval, small, solid, usually sculptured with longitudinal and encircling ribs to give cancellations. Aperture elongate with a short anterior canal; columella strongly plicate and the outer lip ribbed within; no operculum.

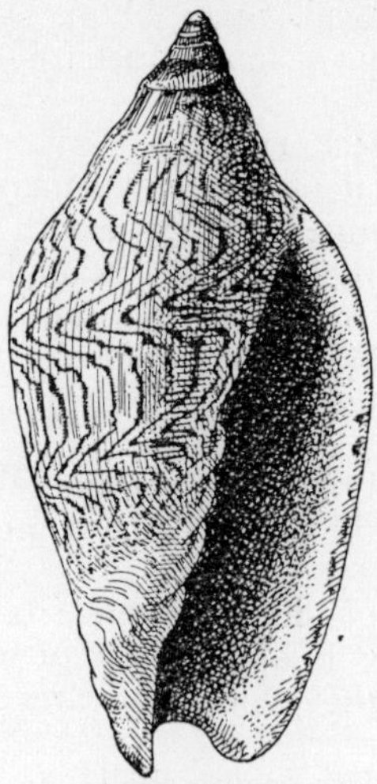

Figure 264

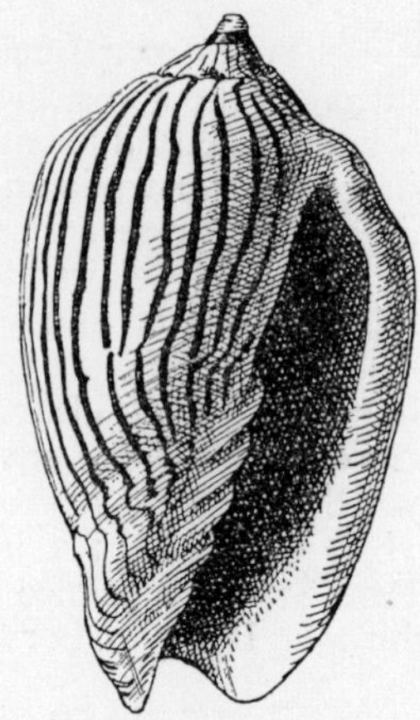

Figure 265

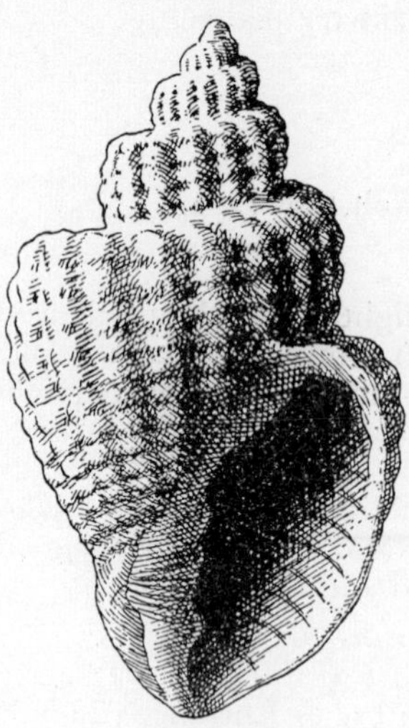

Figure 266

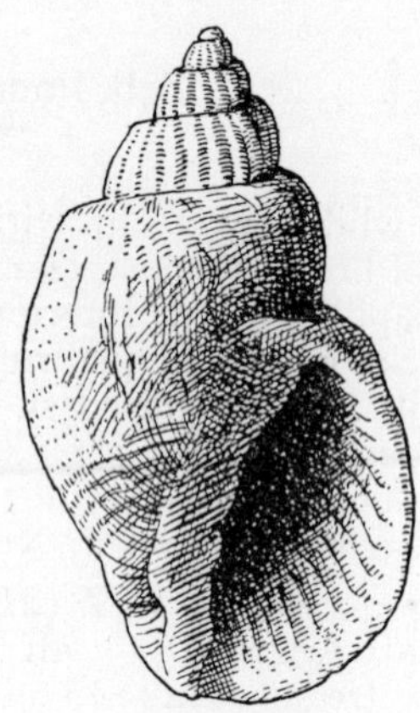

Figure 267

264. *Amorena undulata* × ⅔
265. *Zebramoria zebra* × 1⅓
266. *Sydaphera granosa* × 2
267. *Sydaphera lactea* × 1⅓

Cancellarias are vegetable feeders living from low tide down to approximately 40 fathoms.

Sydaphera Iredale, 1929

Medium-sized shells, imperforate, with the whorls strongly shouldered just below the suture. Sculpture is rounded rather than sharp. Mouth oval with a strong often thickened outer lip.

TYPE SPECIES *Sydaphera renovata* Iredale, 1929.

Grained Cancellaria

Sydaphera granosa (Sowerby, 1832)
(Figure 266)

Shell fawn or light brown, sparsely blotched with dark brown or red round the angle of the whorls; ovate. Spire acuminately turreted. Whorls flattened round the upper portion. Sculpture consisting of longitudinal ribs and numerous spiral ridges, this crossing of sculpture giving the shell a decussated or grained appearance. Umbilicus small. Aperture oblong, terminating in a short, anterior canal, lirate within and darker in colour; columella bearing three distinct plaits. Size: 1 inch.

Localities: Point Nepean; Point Lonsdale; Western Port.

The Cancellarias are shells of most elegant structure; the present species, with its characteristic granular character, is no exception. Inhabiting sandbanks at low water, these creatures may be obtained alive, and in addition to the Victorian coastline are found in Tasmania.

Smooth Cancellaria

Sydaphera lactea (Deshayes, 1832)
(Figure 267)

Shell ovate, solid, fawn with an interrupted brown band close to the sutures. Spire acuminate; less than half the length of shell, the earlier whorls of which are radiately ribbed, a feature almost absent on the body-whorl. Whorls slightly angled close to the sutures with the angles faintly noduled. Further ornamented with flat concentric ridges. Interior cream. Aperture ovate, with a slightly crenulated outer lip, lirate within; columella fairly straight with three plaits thereon; anterior canal very short, wide. Size: 1 inch.

Localities: San Remo; Lorne.

This species is not so bold in sculpture as *undulata* Sowerby, but is shorter, and with the body-whorl comparatively smooth. Occurs also in Tasmania. This shell was known for many years as *Cancellaria laevigata* Sowerby, 1832.

Grooved Cancellaria

Sydaphera purpuriformis (Küster, 1841)
(Figure 268)

Shell elongately ovate, cream with very light brown, encircling bands on the body-whorl and a third interrupted band of reddish brown just below the sutures. Spire acuminate, less than half the length of shell. Whorls rounded, slightly shouldered and fairly regularly, concentrically grooved throughout, the whole surface microscopically radiately striated. Interior white and if held to the light, bands of colour are visible from within. Aperture narrowly ovate; outer lip sharp, slightly crenulate, lirate within; columella a little curved; triplicate. Size: 1 inch.

Localities: Flinders; Shoreham; Western Port (dredged in 8 fathoms).

A delicate shell which with its narrow form should not be confused with any Victorian Cancellaria. *Cancellaria maccoyi* Pritchard & Gatliff, 1899, is synonym of this species. It occurs also in Tasmania and South Australia.

purpuriformis anxifer (Iredale, 1925); a more elongate and weakly longitudinally ribbed form from off Eden, New South Wales; may occur in eastern Victoria.

Waved Cancellaria

Sydaphera undulata (Sowerby, 1848)

(Figure 269)

Shell solid, fawn or light brown with an interrupted reddish brown band just below the sutures, erect, sharply turreted. Spire acuminate, about half the length of shell. The whole surface sculptured with somewhat distant and obliquely waved ribs, slightly noduled at the shoulder, and further ornamented with concentric grooves. Interior cream. Aperture ovate; outer lip sharp, slightly crenulated from the outside sculpture, lirate within; columella fairly straight, bearing three plaits; anterior canal short, wide. Size: 2 inches.

Localities: Anderson's Inlet; San Remo; Lorne; Port Fairy; Portland.

The largest Cancellaria in Victoria. A beautiful, solid, strongly sculptured species and quite easily identified by its longitudinal ribs and spiral grooves. Occurring also in Tasmania and South Australia.

Nevia Jousseaume, 1887

Cancellaria with a short, turreted spire and deeply excavated sutures and a small umbilicus.

TYPE SPECIES *Cancellaria spirata* Lamarck, 1822.

Excavated Cancellaria

Nevia spirata (Lamarck, 1843)

(Figure 270)

Shell oblong-ovate, cream or fawn with an interrupted band of reddish brown close to the sutures. Spire turreted, less than half the length of shell. Whorls deeply excavated at the sutures, somewhat keeled at the angles; earlier whorls radiately ribbed, becoming almost obsolete on the body-whorl. The whole surface irregularly, concentrically grooved. Interior pinkish white. Aperture ovate; outer lip crenulate, lirate within; columella slightly curved, bearing three plaits; umbilicus very small. Size: 1 inch.

Localities: Western Port; Airey's Inlet; Lorne; Portland.

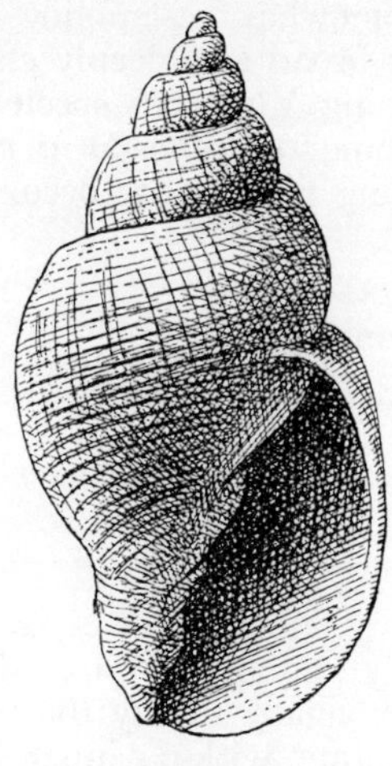

Figure 268

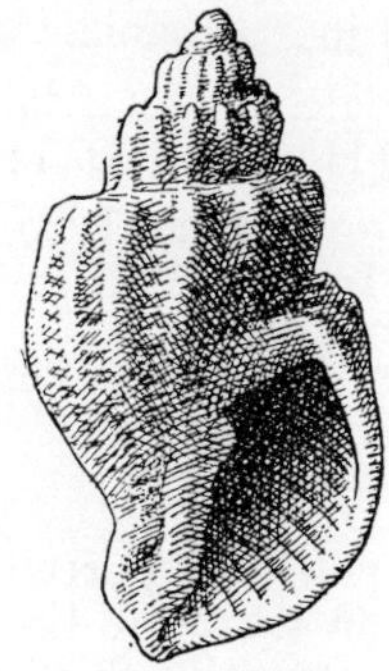

Figure 269

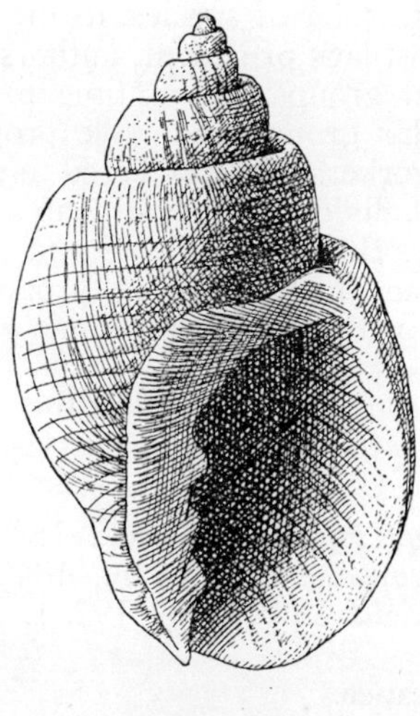

Figure 270

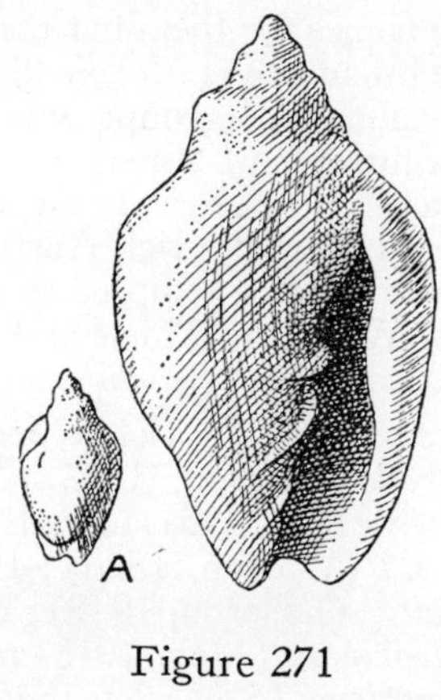

Figure 271

268. *Sydaphera purpuriformis* × 2⅔
269. *Sydaphera undulata* × 1
270. *Nevia spirata* × 2
271. *Austroginella muscaria* × 3
   A Dorsal view × 1

A very characteristic shell, in contour somewhat resembling *Zemira australis* Sowerby, and easily recognized by its broad and deeply excavated sutural channel, a feature absent in the remaining Victorian species of the family. For many years known under the name of *Cancellaria excavata* Sowerby, 1848, the above prior name precluding its use. Its occurrence is also noted in Tasmania.

Pepta Iredale, 1907 (*Admete stricta* Hedley, 1907).
stricta (Hedley, 1907); off Wilson's Promontory.

Microsveltia Iredale, 1925 (*Microsveltia recessa* Iredale, 1925).
recessa Iredale, 1925; off Cape Everard (65 fathoms).

Family MARGINELLIDAE

Margin Shells

Small polished porcellanous shells usually smooth. Spire short or immersed in the large body-whorl. Aperture elongate, nearly the length of the shell; outer lip thickened and may be dentate within; inner lip with several plaits on the columella; no operculum. Marginellas are generally located on sandbanks at low water where they swiftly glide along, and when observed in full activity are objects of great interest. Many of the minute forms are obtained in deeper waters. The animal covers the major portion of the highly polished shell with the mantle.

Though this family contains quite a large number of species, its members are all very similar in shape, and as they almost lack ornament, authors have had difficulty in dividing them into natural groups. From time to time authors have attempted to separate particular groups and have proposed generic names for them but the majority of workers have preferred to retain the one broad genus *Marginella*. Cotton, 1949, divided the Australian species into a number of groups which he called A, B, C, etc. Thiele and Wenz listed a number of genera in Marginellidae and Laseron, 1957, proposed a new classification for the Australian species with a number of new genera to accommodate the rich Australian fauna.

We have now attempted to reconcile Laseron's proposals with the earlier work of Wenz and Thiele and put forward the following tentative classification of Victorian species.

Cryptospira Hinds, 1844 (*Marginella quinqueplicata* Lamarck, 1822) = *Sinuginella* Laseron, 1957 (*Marginella inconspicua* Sowerby, 1846) = *Kogomea* Habe, 1951 (*Erato novemprovincialis* Yokohama, 1928).
agapeta (Watson, 1886); Portsea.
binivitta (Laseron, 1948); off Lakes Entrance.
caducocincta (May, 1915); off Wilson's Promontory.
inconspicua (Sowerby, 1846); off Gabo Island.
pipire (Laseron, 1948); off Lakes Entrance.
pulchella (Kiener, 1830); Portland.
pygmaeoides (Singleton, 1937); Western Port; Port Phillip; Lorne. The Volute-shaped Margin Shell, an ovate, ivory white, strong shell with outer lip varicosely reflected; somewhat volute-shaped, a feature sufficient to distinguish it from all the above species. Size: ⅛ inch.
schoutanica (May, 1912); off Ninety Mile Beach (40 fathoms).

stilla (Hedley, 1903) ; off Ninety Mile Beach (40 fathoms).
strangei (Angas, 1877) ; Port Fairy.
subbulbosa (Tate, 1878) ; Portsea.

Cloisia Gray, 1857 (*Marginella sarda* Kiener, 1834) = *Oviginella* Laseron, 1957 (*Marginella ovulum* Sowerby, 1846).

Wenz states that *Cloisia s.s.* has the inner edge of the outer lip toothed, whereas *ovulum* is smooth, otherwise they show no marked differences.

flindersi (Pritchard & Gatliff, 1898) ; Flinders.
ovulum (Sowerby, 1846) ; off Lakes Entrance.
petterdi (Beddome, 1883) ; Western Port. This shell was misidentified as *Marginella pisum* Reeve, 1865.
whani (Pritchard & Gatliff, 1900) ; Portland; Carrum.

Microginella Laseron, 1957 (*Marginella anxia* Hedley, 1909).
alternans (Pritchard & Gatliff, 1898) ; Western Port.
cratericula (Tate & May, 1900) ; no definite locality recorded.
cymbalum (Tate, 1878) ; Port Phillip; Western Port; Portland. A pure white, smooth polished species typifying a group with concealed spires of which there are many in Victoria. Interior white, with a narrow, curved aperture; outer lip thickened, columella with numerous plications, a feature separating it from all the above forms. Size: ¼ inch. Also occurs in Tasmania and South Australia.
halli (Pritchard & Gatliff, 1898) ; Shoreham.
nympha (Brazier, 1894) ; no definite locality recorded.
problematica (Gatliff & Gabriel, 1916) ; Western Port.
pumilio (Tate & May, 1901) ; Western Port; Brighton.

Triginella Laseron, 1957 (*Marginella malina* Hedley, 1915).
malinoides Gabriel, 1961; off Cape Everard (65 fathoms).

Euliginella Laseron, 1957 (*Marginella angasi* Crosse, 1870).
angasi Crosse, 1870; Port Phillip; Western Port; off Wilson's Promontory; off Ninety Mile Beach.
angasi simsoni (Tate & May, 1900) ; Western Port.
connectans (May, 1910) ; Western Port.
freycineti (May, 1915) ; off Rhyll, Western Port.
shorehami (Pritchard & Gatliff, 1898) ; Shoreham.
subauriculata (May, 1915) ; Western Port; off Wilson's Promontory.

Mesoginella Laseron, 1957 (*Marginella turbinata* Sowerby, 1846).
turbinata Sowerby, 1846; Turbinate Margin shell; Western Port; Port Phillip; Lorne; is a yellowish white or cream, turbinately ovate species with strong ribs at the angle on the upper part of body-whorl. Size: ½ inch.
turbinata pusilla (Laseron, 1948) ; off Lakes Entrance.
pattersoni (Cotton, 1944) ; off Lakes Entrance.

Deviginella Laseron, 1957 (*Marginella brachia* Watson, 1885).
victoriae (Gatliff & Gabriel, 1908) ; Western Port.

Volvarina Hinds, 1844 (*Marginella avena* Valenciennes, 1834) = *Haloginella* Laseron, 1957 (*Marginella mustellina* Angas, 1871).
mustellina (Angas, 1871) ; Western Port.
stanislas (Tenison Woods, 1876) ; off Wilson's Promontory.
tridentata (Tate, 1878) ; Port Fairy.

Pillarginella Gabriel, 1961 (*Marginella columnaria* Hedley & May, 1908).
columnaria (Hedley & May, 1908); off Lakes Entrance.

Glabella Swainson, 1840 (*Voluta jaba* Linné, 1758) = (*Plicaginella* Laseron, 1957 (*Marginella formicula* Lamarck, 1822).
formicula (Lamarck, 1822); Portland.

Austroginella Laseron, 1957

Medium to large, massive shells, strombiform in shape, with long spires, wide at the shoulders. Aperture relatively short, shoulders nearly square, sinuate within, outer margin nearly straight with a heavy varix, smooth within, the anterior sinus strong and deep. Inner margin with four heavy plications, the posterior pair transverse, occupying half or more of the aperture.

TYPE SPECIES *Marginella muscaria* Lamarck, 1822.

Fly-like Margin Shell

Austroginella muscaria (Lamarck, 1822)

(Figure 271)

Shell creamy white or flesh, solid, fusiformly ovate. Spire raised, about quarter the length of shell. Whorls gibbous, slightly angled near the suture. Interior cream tinted. Aperture elongated; outer lip with a thick, strong marginal varix; columella triplicate. Size: ½ inch.

Localities: Gippsland coast; Mallacoota.

This is our largest Marginella and cannot be confused with any Victorian member of the genus.

johnstoni (Petterd, 1884); Johnston's Margin Shell, found at Port Phillip; Western Port; off Lakes Entrance, and also New South Wales, Tasmania and South Australia. Pale yellowish white, polished, not so strong in texture but could almost be regarded as *muscaria* in miniature, and in the collection of C.J.G. appears a sinistral form of this interesting species. Size: ⅓ inch.

tasmanica (Tenison Woods, 1875); Tasmanian Margin Shell, white and shining, fusiformly oblong, spire acute, outer lip thickened, differing from the above three species in being narrower and with a longer spire. Size: ⅓ inch. From the above localities; also recorded in Tasmania, South Australia, and New South Wales.

vercoi (May, 1911); off Lakes Entrance.

Protoginella Laseron, 1957 (*Marginella laevigata* Brazier, 1876).
gatliffi (May, 1910); off Ninety Mile Beach (40 fathoms).
geminata (Hedley, 1912); Western Port; off Lakes Entrance. This shell was incorrectly identified as *laevigata,* a species which has not as yet been taken in Victorian waters.

Longinella Laseron, 1957 (*Marginella maugeana* Hedley, 1915).
everardensis Gabriel, 1961; off Cape Everard (65 fathoms).
gabrieli (May, 1911); off Wilson's Promontory.
kemblensis (Hedley, 1903); off Lakes Entrance.

Volvarinella Habe, 1951 (*Volvarinella makiyamai* Habe, 1951).
allporti (Tenison Woods, 1875); definite locality not recorded.
difficilis Gabriel, 1961; off Cape Everard (65 fathoms).
mayi (Tate, 1900); off Lakes Entrance.

Family TURRIDAE
Turrids

Shell turreted, subfusiform. Aperture with the fore part channelled, straight, and often much produced; posterior of outer lip detached from the body-whorl, forming a sinus, or with the margin fissured near the last whorl; operculum horny.

This family contains a large number of mostly small species and no attempt will be made to describe the many forms here. A number of authors have attempted to monograph all or part of the Australian species, and those wishing to study them further are referred to Hedley, 1922, Powell, 1942, Powell, 1944, Cotton, 1947A, Cotton, 1947B, Laseron, 1954B.

Subfamily TURRINAE

Epidirona Iredale, 1931

Shell large for the family, acuminate spire; protoconch of two smooth turbinate whorls; sinus broad and shallow; sculptured with a bead-row along the fasciole, and frequently another along the suture, separated by a broad furrow.

TYPE SPECIES *Epidirona hedleyi* Iredale, 1931.

Quoy's Turrid
Epidirona quoyi (Reeve, 1843)
(Figure 272)

Shell solid, elongately fusiform, brownish with numerous dots of similar colour but darker just below the sutures. Spire about half the length of shell. Whorls angulated at the upper portion and slightly canaliculate, granular at the angle and just below the sutures. The whole surface sculptured with numerous, irregular, concentric lirae and fine longitudinal striae producing a cancellated appearance. Interior brown, white towards the margin. Aperture elongately ovate; outer lip sharp with a broad and deep sinus; anterior canal short and wide. Size: 1 inch.

Localities: Warrnambool; Apollo Bay; Western Port.

The largest Turrid in Victoria, sometimes slightly exceeding the above dimensions. Washed up on the shores, more often decollated than not; also obtained in the waters of Tasmania and South Australia.

candida Laseron, 1954; off Cape Everard (70-80 fathoms).
gabensis (Hedley, 1922); off Gabo Island (80 fathoms).
molleri Laseron, 1954; off Lakes Entrance.
philipineri (Tenison Woods, 1876); Western Port.

Fenestrosyrinx Finlay, 1926 (*Turris nexilis bicarinatus* Suter, 1915).
mayi (Verco, 1909); 30 miles south of Cape Nelson (300 fathoms).

Benthofascis Iredale, 1936 (*Bathytoma biconica* Hedley, 1903).
sarcinula (Hedley, 1905); off Lakes Entrance (30 fathoms).

Epidirella Iredale, 1931 (*Hemipleurotoma tasmanica* May, 1911).
tasmanica (May, 1911); off Lakes Entrance (20 fathoms). Laseron erected *Austrogemmula* for this species, apparently having missed Iredale's previous work.

Vexitomina Powell, 1942 (*Drillia metcalfei* Angas, 1867).
aemula (Angas, 1877); a misidentification of the next species.
coxi (Angas, 1867); Portland; Western Port.
garrardi Laseron, 1954; off Lakes Entrance (20 fathoms).
radulaeformis (Weinkauff, 1876; Bass Strait. This species is not known to Australian workers.
suavis (Smith, 1888); Western Port.

Paracuneus Laseron, 1954 (*Mangelia immaculata* Tenison Woods, 1875).
immaculatus (Tenison Woods, 1875); off Phillip Island; Western Port.
spadix (Watson, 1886); off Gabo Island (80 fathoms).
spadix tumulus Laseron, 1954; off Lakes Entrance (20 fathoms).

Subfamily CLAVINAE

Austrodrillia Hedley, 1918 (*Pleurotoma angasi* Crosse, 1863).
angasi (Crosse, 1863); San Remo.
beraudiana (Crosse, 1863); Flinders; San Remo; off Phillip Island; Port Phillip; Kilcunda. An olive brown species about ¾ inch long, acuminate, strongly longitudinally ribbed and finely concentrically striate with whorls somewhat nodosely angled.
saxea (Sowerby, 1896); off Wilson's Promontory.

Splendrillia Hedley, 1922 (*Drillia woodsi* Beddome, 1885).
eburnea (Hedley, 1922); off Gabo Island (80 fathoms).
lygdina (Hedley, 1922); off Gabo Island (150-200 fathoms).
nenia (Hedley, 1903); off Ninety Mile Beach (40 fathoms).
woodsi (Beddome, 1882); Gippsland coast.

Antimelatoma Powell, 1942 (*Drillia maorum* Smith, 1877).
harpularia (Desmoulins, 1942); Portland.
subviridus (May, 1910); off Gabo Island (80 fathoms).

Maoritomella Powell, 1942 (*Pleurotoma albula* Hutton, 1873).
foliacea Laseron, 1954; off Cape Everard (70-80 fathoms).

Subfamily BORSONIINAE

Mitrithara Hedley, 1922 (*Columbella alba* Petterd, 1879).
alba (Petterd, 1879); Western Port. A small species, about ¼ inch, from Flinders Beach, Western Port, easily recognized by its robustly fusiform shape with whorls densely concentrically lirate and fine longitudinal striae.
axiscalpta (Verco, 1909); 6 miles south of Lakes Entrance (on seaweed).
bassiana Gabriel, 1956; 25 miles south-east of Lakes Entrance (30 fathoms).

incerta (Pritchard & Gatliff, 1902); off Rhyll, Western Port.
macphersonae Gabriel, 1956; 25 miles south-east of Lakes Entrance.
proles Hedley, 1922; off Gabo Island (80 fathoms).

Scrinium Hedley, 1918 (*Mitromorpha brazieri* Smith, 1891).
furtivum Hedley, 1922; off Gabo Island (80 fathoms).
gatliffi (Verco, 1909); Inverloch (8 fathoms); transferred to *Aesopus.*

Austrocarina Laseron, 1954 (*Leucosyrinx recta* Hedley, 1903).
recta (Hedley, 1903); off Ninety Mile Beach (40 fathoms).

Teleochilus Harris, 1897 (*Daphnella gracillima* Tenison Woods, 1876).
royanus Iredale, 1924; off Gabo Island (80 fathoms); Bass Strait.

Subfamily MANGELIINAE

Filodrillia Hedley, 1922 (*Drillia tricarinata* Tenison Woods, 1878).
columnaria Hedley, 1922; off Wilson's Promontory.
lacteola (Verco, 1909); off Ninety Mile Beach (40 fathoms).
mucronata Hedley, 1922; off Lakes Entrance (25 fathoms).
ordinata Laseron, 1954; off Cape Everard (70-80 fathoms).
steira Hedley, 1922; off Gabo Island (80 fathoms).
tricarinata (Tenison Woods, 1878); off Wilson's Promontory.
vitrea Laseron, 1954; off Cape Everard (70-80 fathoms).

Austropusilla Laseron, 1954 (*Mangelia hilum* Hedley, 1908).
hilum Hedley, 1908; off Wilson's Promontory.

Etrema Hedley, 1918 (*Mangelia aliciae* Melvill & Standen, 1895).
bicolor (Angas, 1871); Western Port.
crassina (Angas, 1880); a synonym of the following species.
denseplicata (Dunker, 1871); Portland; Port Phillip; off Phillip Island, Western Port. A small, ornate, acuminate, fawn, occasionally white-banded, species occurring frequently on the shores of Port Phillip and Western Port; about ½ inch long and easily identified by its dense longitudinal ribs and fine concentric striae.
kymatoessa (Watson, 1886); off East Moncoeur Island.
levicosta Laseron, 1954; off Lakes Entrance (20 fathoms).
nassoides (Reeve, 1845); Western Port.

Guraleus Hedley, 1917 (*Mangelia picta* A. Adams & Angas, 1863).
alucinans (Sowerby, 1896); this is a synonym of *vincentinus.*
brazieri (Angas, 1871); Port Albert.
cuspis (Sowerby, 1896); Port Phillip; Western Port.
delicatulus (Tenison Woods, 1876); off Phillip Island, Western Port.
flaccidus (Pritchard & Gatliff, 1899); San Remo.
incrustus (Tenison Woods, 1876); Port Phillip; Western Port.
meredithiae (Tenison Woods, 1875); common along whole coast.
pictus (Adams & Angas, 1863); a species widely distributed on the Victorian coastline. Extremely variable in colour, normally white with brown bands, very wide and centrally situated on the body-whorl; some examples almost white, with narrow bands of brown just below the sutures, others pure white; sculptured with prominent longitudinal ribs between which will be seen fine concentric striae visible only with the aid of lens.
vincentinus (Crosse & Fischer, 1865); Western Port; Port Phillip; Portland.

Mitraguraleus Laseron, 1954 (*Bela mitralis* A. Adams & Angas, 1863).
australis (A. Adams & Angas, 1863); a variety of the following species confined to South Australia and Western Australia.
mitralis (Adams & Angas, 1863); common along whole coast. An elongately fusiform white shell, body-whorl marked with encircling spots or lines; whorls rounded, somewhat angular near the sutures and sculptured with longitudinal ribs and numerous concentric striae; size: ½ inch.

Euguraleus Cotton, 1947 (*Euguraleus anisus* Cotton, 1947).
lallemantianus (Crosse & Fischer, 1865); Port Phillip.
tasmanicus (Tenison Woods, 1875); Port Fairy.

Paraguraleus Powell, 1944 (*Paraguraleus balcombensis* Powell, 1944).
emina (Hedley, 1905); off Lakes Entrance (20 fathoms).
kingensis (Petterd, 1879); off Phillip Island, Western Port.

Turella Laseron, 1954 (*Clathurella tenuilirata* Angas, 1871).
granulosissimus (Tenison Woods, 1878); off Phillip Island, Western Port.
letourneuxiana (Crosse & Fischer, 1865); a misidentification of *G. brazieri*.
morologus (Hedley, 1922); Western Port.

Marita Hedley, 1922 (*Cithara compta* A. Adams & Angas, 1863).
bella (Adams & Angas, 1863); Port Phillip; Western Port; Kilcunda.
compta (Adams & Angas, 1863); Flinders; San Remo; Sorrento. A small cream shell about ½ inch, narrowly fusiform, longitudinally plicate, the whole surface finely concentrically grooved; aperture narrowly ovate, with a rather strong outer lip.
fallaciosa (Sowerby, 1896); off Wilson's Promontory.
insculpta (A. Adams & Angas, 1863); Port Albert.
schoutenensis (May, 1910); Bass Strait.
varix (Tenison Woods, 1876); is a synonym of *M. compta*.

Heterocithara Hedley, 1922 (*Clathurella bilineata* Angas, 1871).
bilineata (Angas, 1871); off Rhyll, Western Port; Port Albert.

Apispiralia Laseron, 1954 (*Clathurella albocincta* Angas, 1871).
albocincta (Angas, 1871); Port Albert.

Paramontana Laseron, 1954 (*Clathurella modesta* Angas, 1877).
modesta (Angas, 1877); Port Phillip; Western Port.
tincta (Reeve, 1846); Port Phillip; Western Port; Cape Otway coast. C.J.G. examined the types of this species and *Clathurella rufozonata* Angas, 1877, in the British Museum and considers them identical.
trachys (Tenison Woods, 1877); Brighton.

Macteola Hedley, 1918 (*Purpura* [*Cronia*] *anomala* Angas, 1877).
anomala (Angas, 1877); Portland; Port Phillip; Western Port.

Subfamily DAPHNELLINAE

Daphnella Hinds, 1844 (*Pleurotoma lymneiformis* Kiener, 1840). Laseron, 1954, erected *Paradaphne*, with *Daphnella botanica* Hedley, 1918, as type, for the Australian species on the grounds that it was unlikely that North Atlantic and southern Australian species would have close relationship.

However, Laseron was mistaken in the type locality of Kiener's shell, which was Indian Ocean.
botanica Hedley, 1918; off Rhyll, Western Port.

Asperdaphne Hedley, 1922 (*Daphnella versivestita* Hedley, 1912).
bastowi (Gatliff & Gabriel, 1908); Western Port.
bitorquata (Sowerby, 1896); San Remo.
desalesii (Tenison Woods, 1876); Sorrento.
perplexa (Verco, 1909); exact locality not recorded.
sculptilis (Angas, 1871); between Phillip and French Islands, Western Port.
tasmanica (Tenison Woods, 1876); off Rhyll, Western Port.
vestalis (Hedley, 1903); exact locality not recorded.
walcotae (Sowerby, 1893); Portland.

Exomilus Hedley, 1918 (*Mangelia lutraria* Hedley, 1907).
telescopialis (Verco, 1896); Portsea.

Nepotilla Hedley, 1918 (*Daphnella bathytoma* Verco, 1909).
excavata (Gatliff, 1906); Portsea; Ocean Beach, Point Nepean.
lamellosa (Sowerby, 1896); off Wilson's Promontory.
microscopica (May, 1915); off Wilson's Promontory.
mimica (Sowerby, 1896); Western Port.
triseriata (Verco, 1909); off Ninety Mile Beach (40 fathoms).

Aspertilla Powell, 1944 (*Drillia legrandi* Beddome, 1882).
legrandi (Beddome, 1882); Portland; San Remo.

Gatliffena Iredale, 1929 (*Donovania fenestrata* Tate & May, 1900).
fenestrata (Tate & May, 1900); Flinders.

Family CONIDAE

Cone Shells

Cone-shaped shells of many whorls, the last whorl enveloping the earlier ones. Aperture long, and narrow; outer lip thin, free or notched at the hind part near the suture.

Floraconus Iredale, 1930

Medium-sized cone shells with a short spire and inflated body-whorl; colour pattern of irregular blotches of rose, brown or orange on a cream ground and covered by a thin periostracum.
TYPE SPECIES *Conus anemone* Lamarck, 1810.

Anemone Cone Shell

Floraconus anemone (Lamarck, 1810)

(Figure 273)

Shell conical, generally with a short spire. Whorls numerous; covered with a thin epidermis. Surface spirally striated, grooved near the base. Aperture narrow, lengthened; outer lip thin; interior purplish; operculum small, pointed, horny. Size: 2 inches.
Localities: In rock-pools, Western Port; Port Phillip.

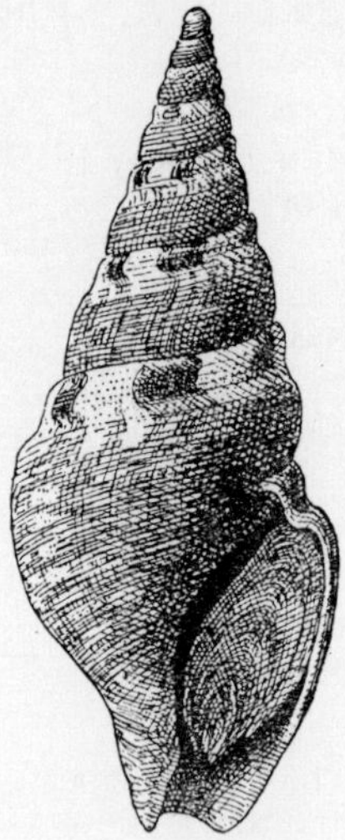

Figure 272

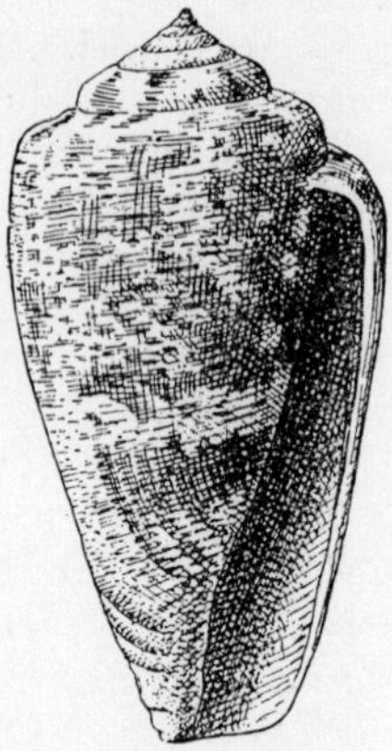

Figure 273

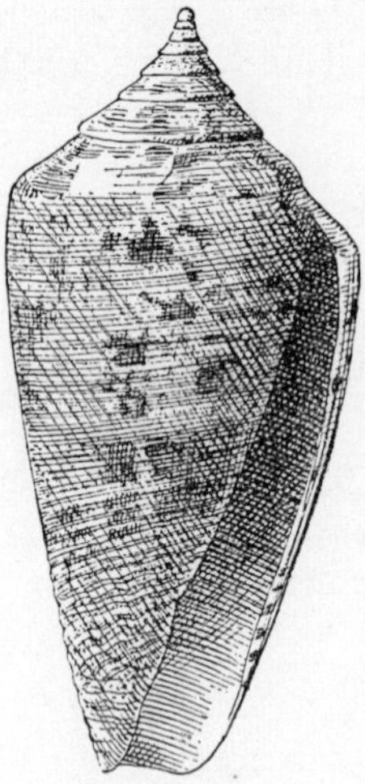

Figure 274

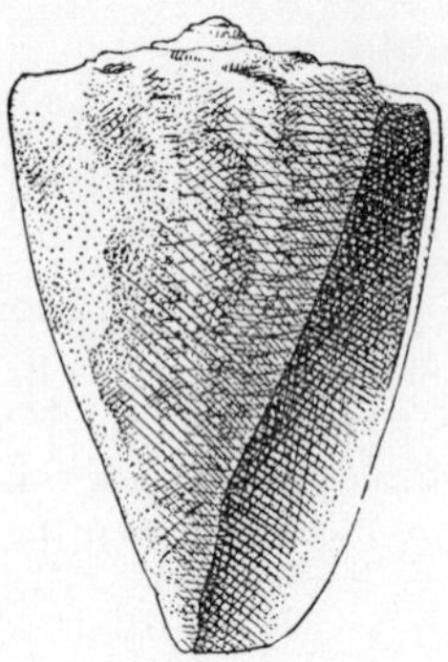

Figure 275

272. *Epidirona quoyi* × 2
273. *Floraconus anemone* × 1
274. *Floraconus segravei* × 2
275. *Parviconus rutilus* × 3

Consistency in shape is not apparent as the length of the spire varies considerably; it may be slightly elevated; occasionally it is produced to one-third of the entire length of the shell, and in some specimens a distinct scalariform character is evident. The variety of marking and the many delicate tints may include cinnamon, rose, brown, chestnut and orange. A pure white variation is frequently obtained near the Heads, particularly at Ocean Beach, Portsea; it has been given the name *singletoni* Cotton, 1945.

Segrave's Cone

Floraconus segravei (Gatliff, 1890)

(Figure 274)

Shell with spire somewhat concavely elevated, coarsely concentrically striated and slightly grooved, maculated with orange brown; base closely striately grooved, striae slightly waved. Shoulder angle well defined. Orange brown, with irregularly sized, tentoriform, white reticulations, encircled with many dotted lines of rather darker colour, narrow white band on the shoulder, and another broader one below the middle of the body-whorl. Interior roseate. Length, 1¼ inches.

Localities: Shoreham; dredged between Phillip and French Islands, Western Port (8 fathoms); off Portsea, Port Phillip.

A rare species occasionally washed up on the shores of Western Port. In living condition possesses a light brown epidermis, the removal of which discloses a delicately patterned shell. This feature with the smoother surface and pinkish interior will serve as a means to separate it from the commoner littoral form *anemone*. Recorded also from South Australia.

Parviconus Cotton & Godfrey, 1932

Shell small, thin, subtriangular, subinflated, body-whorl smooth, shining; all adult whorls coronate.

TYPE SPECIES *Conus rutilus* Menke, 1843.

Fiery Cone

Parviconus rutilus (Menke, 1843)

(Figure 275)

Shell small, thin, smooth, shining, somewhat pyriform-turbinate but showing much variation in the length of spire which is coronated. Whorls turreted, radiately lirate. Colour very variable may be uniformly reddish, violet, brown, pink, yellow or with chestnut splashes and spots of similar or darker colour. Interior more frequently pinkish; aperture wide, outer lip rather sharp. Size: ½ inch.

Localities: Western Port, frequently cast up on the beaches; coast generally.

A handsome cone, the smallest of the genus in Victoria, easily recognized by its peculiar pyriform shape and coronate spire. For many years known under the name of *Conus smithi* Angas, 1877. It has a wide distribution from New South Wales, Tasmania, through Bass Strait to South Australia and Western Australia.

Family TEREBRIDAE

Auger Shells

Slender, elongate, many-whorled shells with a small mouth. The outer lip is simple and there is a short anterior canal and no plaits on the columella, but it may be twisted at the notch. Found living in sand.

Nototerebra Cotton, 1947

Shell turreted, solid, whorls flat, increasing in size rapidly, sutures impressed. Protoconch of one and a half whorls, slightly swollen, smooth and round.

TYPE SPECIES *Terebra albida* Gray, 1834.

White Auger

Nototerebra albida (Gray, 1834)

(Figure 276)

Shell white, smooth, solid, long, pointed; whorls flattened, showing numerous, oblique growth-lines, last whorl rounded. Aperture small, ovate; columella short, twisted; operculum horny. Size 1½ inches.

Localities: Portsea; Point Nepean; Point Lonsdale; Western Port.

This white species is easy to recognize. From the first-named locality specimens have been obtained exhibiting small dark brown spots on the shoulder of the whorl, numbering about twelve on the penultimate whorl. The species also occurs in South Australia and Tasmania.

Acuminia Dall, 1908

Adult whorls slender, smooth not nodulous, summit narrowly shouldered; sutural groove absent; columella with one fold.

TYPE SPECIES *Terebra lanceolata* Linné, 1758.

Brazier's Auger

Acuminia brazieri (Angas, 1871)

(Figure 277)

Shell small, subulate, creamy with longitudinal streaks of reddish brown on the flatly convex whorls. The body-whorl possessing an encircling band just below the periphery, and between this and the anterior extremity a concentric brown band, sometimes interrupted. Sculptured throughout with feeble, longitudinal ribs. Interior whitish; outer lip acute; columella somewhat twisted. Size: 1 inch.

Localities: Western Port; off Lakes Entrance.

Occasionally washed up on the beaches. With its comparative smoothness, longitudinal streaks of reddish brown and flatly convex whorls cannot be confused with any Victorian species. Distribution New South Wales, Tasmania and South Australia.

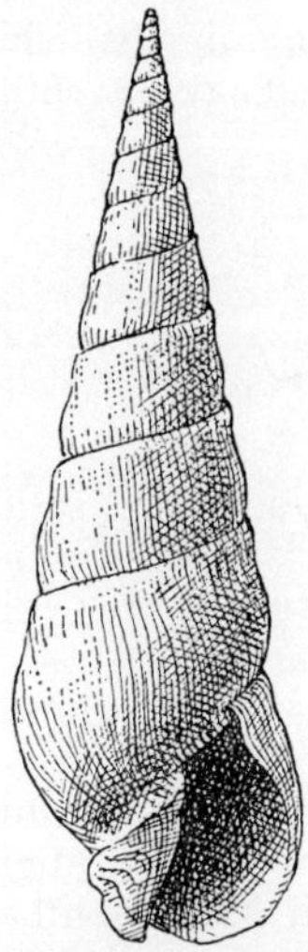

Figure 276

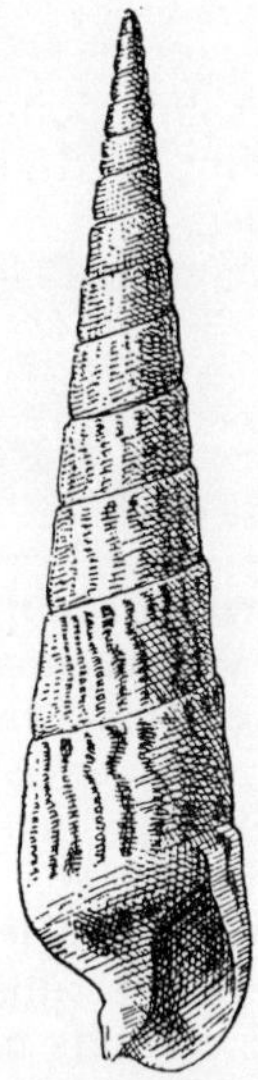

Figure 277

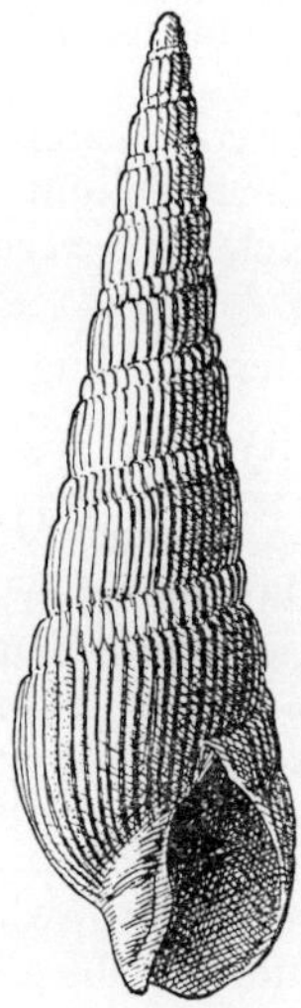

Figure 279

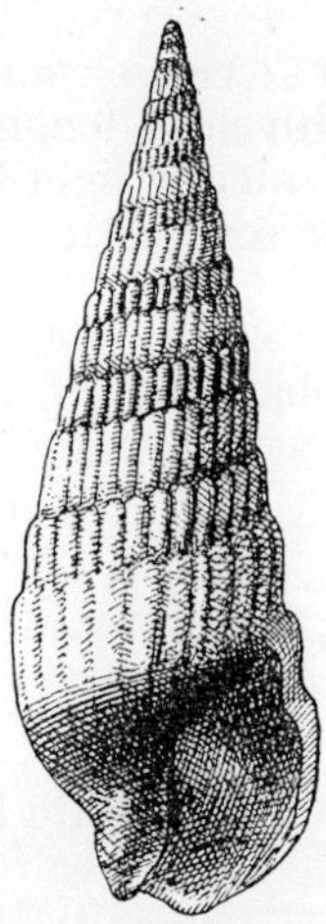

Figure 278

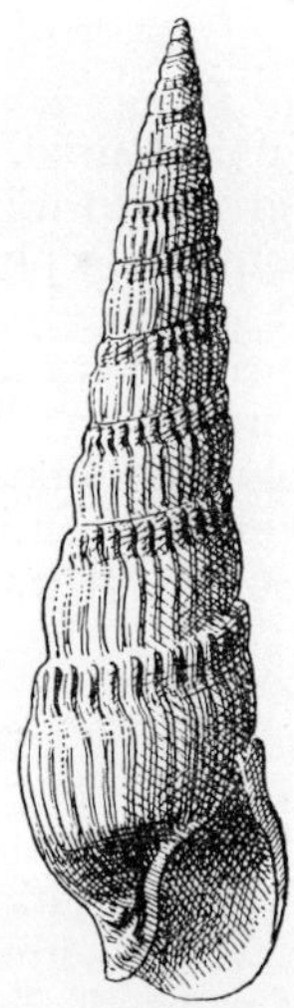

Figure 280

276. *Nototerebra albida* × 1½
277. *Acuminia brazieri* × 2⅔
278. *Pervicacia ustulata* × 2
279. *Pervicacia kieneri* × 2
280. *Pervicacia bicolor* × 3

Pervicacia Iredale, 1924

Shell solid, pointed, whorls convex, having close axial ribs which are deeply impressed at the upper portion; fold on columella represented by a basal twist.

TYPE SPECIES *Terebra ustulata* Deshayes, 1857.

Scorched Auger

Pervicacia ustulata (Deshayes, 1857)

(Figure 278)

Shell white, stained with brown at the base and apex, subulate, solid, long, pointed. Whorls numerous, convex, ornamented with close, longitudinal ribs, which are deeply impressed at the upper portion. Aperture ovate; columella slightly twisted; operculum horny. Size: 1 inch.

Localities: Portsea; Point Nepean; Point Lonsdale; Western Port; Portland.

A variable species, as regards coloration; besides the normal white and brown, some examples are uniformly light brown while a specimen from Portland exhibits a distinct violet. It is recorded from New South Wales and Tasmania.

Kiener's Auger

Pervicacia kieneri (Deshayes, 1859)

(Figure 279)

Shell unicoloured, brown, subulate, whorls numerous; convex and with a spiral groove forming a posterior band. Sculptured with close, longitudinal ribs which are deeply impressed at the upper portion. Interior light brown. Aperture elongately ovate; outer lip acute; columella somewhat twisted; anterior canal very short, wide. Size: 1 inch.

Localities: Port Phillip; Western Port.

A unicoloured, finely ornamented shell, similar in character to *ustulata* (Deshayes) but with the longitudinal ribs about twice as numerous. Found also in South Australia.

Two-toned Auger

Pervicacia bicolor (Angas, 1867)

(Figure 280)

Shell whitish with the anterior half of the body-whorl chocolate brown, subulate. Whorls nine to ten, slightly rounded with a spiral groove forming a posterior band. Sculpture consisting of fine longitudinal ribs. Interior white with the chocolate brown plainly visible within. Aperture elongately oval; outer lip acute; columella somewhat twisted. Size: ¾ inch.

Localities: Portsea; Western Port; Port Albert.

A small narrow species with the two-toned feature providing a useful recognition mark, occurring also in New South Wales, Tasmania through Bass Strait to South Australia.

fictilis (Hinds, 1844); a rarer form from Port Fairy and dredged in Western Port and found also in New South Wales and South Australia approaches the above species, but may be distinguished by its stronger sculpture.

assecla Iredale, 1924; off Lakes Entrance (20 fathoms).

Euterebra Cotton & Godfrey, 1932 (*Terebra inconspicua* Pritchard & Gatliff, 1902).

inconspicua (Pritchard & Gatliff, 1902); dredged in Western Port and recorded from Tasmania and South Australia. Size: ½ inch. Nine flatly convex whorls, sculptured by numerous, feeble, radial riblets and concentric microscopic striae visible only under lens. Colour cream; quite distinct from the above two species.

Parviterebra Pilsbry, 1904 (*Parviterebra paucivolvis* Pilsbry, 1904).
brazieri (Angas, 1875); Western Port.

Order CEPHALASPIDEA

Family ACTEONIDAE
Pupa Shells

Shell solid, ovate with a large body-whorl. Spire conical, many whorled; spiral sculpture. Mouth long, narrow, rounded anteriorly; outer lip sharp; columella with one or two strong folds; operculum horny, lamellate.

Acteon Montfort, 1810 (*Voluta tornatilis* Linné, 1758).
fructuosus Iredale, 1936; Cape Everard.
retusus Verco, 1907; off Lakes Entrance.
subroseus Iredale, 1936; Montague Island (60-70 fathoms).

Pupa Roding, 1798 (*Bulla solidula* Linné, 1758).
nivea (Angas, 1871); off Lakes Entrance (20 fathoms).
tragulata Iredale, 1936; off Lakes Entrance.

Pugnus Hedley, 1896 (*Pugnus parvus* Hedley, 1896).
parvus Hedley, 1896; Bass Strait.

Family RINGICULIDAE
Helmet Bubbles

Shells small, rounded, as a rule conical in shape, with a small spire; smooth or spirally striate. Mouth large, with an anterior notch; outer lip thickened and reflexed; columella with two strong plaits; operculum absent.

Ringiculadda Iredale, 1936 (*Ringicula semisculpta* Hedley, 1911).
australis (Hinds, 1844); Western Port.
grandinosa (Hinds, 1844); Port Albert.

Family BULLARIIDAE
Bubble Shells

Globose, oval or oblong-cylindrical, the body-whorl enveloping the remainder of the whorls. Spire sunken, vertex umbilicated. Mouth as long as the shell, rising above the vertex; columella concave, calloused.

Bullaria Rafinesque, 1815

Description as for the family.
TYPE SPECIES *Bulla ampulla* Linné, 1758.
Iredale, 1929, suggested that there was some confusion regarding the correct generic name for these shells and separated the Australian species as *Quibulla* with *Bulla botanica* Hedley, 1918, as type species. This is such a uniform group and the number of species are so few that we do not consider this is warranted.

Botany Bay Bubble Shell

Bullaria botanica (Hedley, 1918)
(Figure 281)

Shell mottled, thin, smooth, oval. Spire sunken. Aperture long, extending the entire length of body-whorl, gradually narrowing towards the spire; outer lip thin. Size: 1½ inches.
Localities: Extremely abundant in Corio Bay; Portarlington; Werribee; Altona; Rosebud.
The species lives concealed in sandy mud and under seaweed. The animal is bulky, partially investing, but not entirely covering, the shell, and is capable of complete retraction into the shell. Occasionally in cleaning shells of this species, collectors find three small, hard plates about ¼ inch long; these are gizzard plates and aid the animal in crushing and digesting its food. Recorded from New South Wales, Tasmania, South Australia and Western Australia, it is a common shell in a favourable habitat. *Bulla australis* Gray, 1825, is a synonym and it has also mistakenly been called *Bullaria tenuissima* Sowerby, 1868, a small very thin-shelled species from Swan River, Western Australia.

Family ATYIDAE
Little Bubble Shells

Shell white, brittle, body-whorl enveloping remainder of shell. Spire hidden, involute. Columella concave ending in an anterior fold.

Haminoea Turton, 1830

Thin, fragile, horny, oval or globose, transversely striate and usually covered with a thin epidermis. Columella concave, thin, with a slight fold at the junction with the body-whorl.
TYPE SPECIES *Bulla hydatis* Linné, 1758.

Short Bubble Shell

Haminoea brevis (Quoy & Gaimard, 1833)
(Figure 282)

Shell small, fragile, white, without colour markings, invested with a thin, brownish epidermis; cylindrical vertex slightly impressed. Surface of shell striated at the anterior and with about fifteen to twenty concentric lines seen only with the aid of lens, the remainder being quite smooth; irregular growth-lines are plainly visible. Interior white, shining. Aperture wide throughout its length, a little narrower at the posterior extremity; outer lip, simple, acute. Size: ½ inch.

Localities: Port Albert; Western Port; Port Phillip.

This Bubble Shell is rather common in Victoria, but owing to its fragile nature is washed up in more or less fragmentary condition. Easily recognized by its pure white, cylindrical form and the faint concentric lines at the anterior end. The animal is white; its stomach possesses three oblong deeply toothed plates. Living in muddy localities, this species is exceedingly rapacious, preying on the animals of bivalves. Recorded also from New South Wales, Tasmania through Bass Strait to South Australia.

Thin Bubble Shell

Haminoea tenera (A. Adams, 1850)

(Figure 283)

Shell small, very thin, fragile, cream or light brown, ovate, vertex very slightly impressed. Surface of shell showing numerous, fine, microscopic, longitudinal striae and a few minute, concentric lines at the anterior extremity, with lines of growth plainly visible under the lens. Interior shining. Aperture wide throughout but narrower at the posterior end; outer lip sharp. Size: ⅜ inch.

Localities: Port Phillip; Western Port.

A species of common occurrence at the above localities. Similar in habit to *brevis* living on a sandy mud sea-bed, but easily distinguished from that species by its more globose form. Found also in Tasmania, New South Wales and South Australia.

cuticulifera Smith, 1872; Sorrento; Portsea.

Family RETUSIDAE

Elongate Canoe Shells

Shell rolled or pyriform, with a raised or sunken spire.

Retusa Brown, 1827 (*Bulla truncatula* Bruguière, 1792).
amphizosta Watson, 1886; Portsea.
apiculata Tate, 1879; Sorrento; Portsea.
eumicra Crosse & Fischer, 1865; Sorrento; Shoreham.

Cylichnina Monterosato, 1884 (*umbilicata* Montagu, 1804).
atkinsoni Tenison Woods, 1875; a synonym of the next species.
pygmaea A. Adams, 1854; a much smaller species obtained at Portsea and Western Port. About ¼ inch, white, subcylindrical and slightly contracted in the middle. It is known as the Little Peanut Shell.
sculpta Gatliff & Gabriel, 1913; off Wilson's Promontory.

Volvulella Newton, 1891 (*Bulla rostrata* A. Adams, 1850).
rostrata (A. Adams, 1850); San Remo.

Figure 281

Figure 282

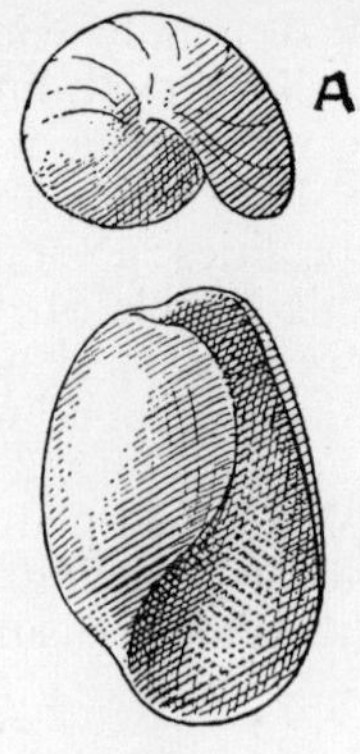

Figure 283

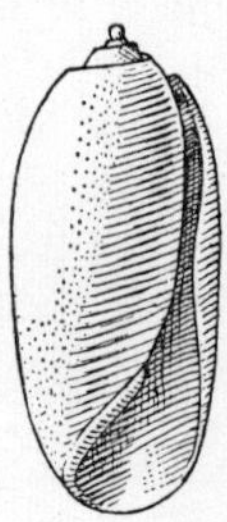

Figure 284

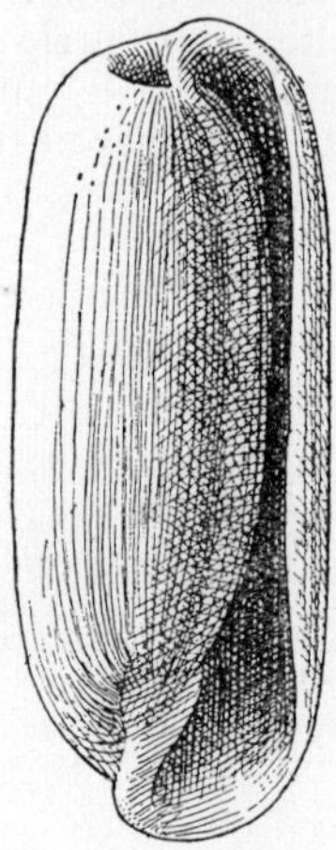

Figure 285

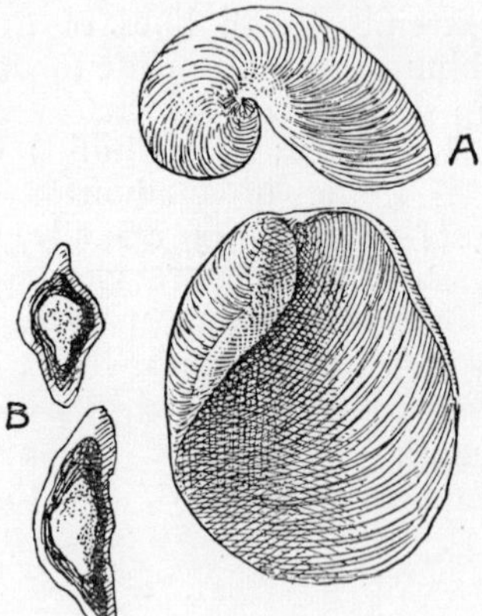

Figure 286

281. *Bullaria botanica* $\times \frac{2}{3}$
282. *Haminoea brevis* $\times 2\frac{2}{3}$
283. *Haminoea tenera* $\times 3$
A Posterior elevation
284. *Acteocina fusiformis* $\times 4$
285. *Cylichna arachis* $\times 2$
286. *Philine angasi* $\times 1$
A Posterior elevation B Gizzard plates

Family SCAPHANDRIDAE

Canoe Shells

Shell unicoloured, usually white, smooth or with spiral sculpture, which is often formed by rows of small punctures or hollows; cylindrical or egg shaped. Spire may be either elevated or sunken. The columella may or may not have a fold; outer lip simple. The animal is carnivorous, 'bulldozing' through the sand or mud in search of its prey.

Acteocina Gray, 1847

Shell cylindrical or fusiform, covered with an epidermis. Spire distinct, produced, apex mamillated; suture canaliculated. Aperture narrow; columella with a single plait at the anterior.

TYPE SPECIES *Bulla voluta* Quoy & Gaimard, 1833.

Shells now located in this genus were originally placed in *Tornatina* A. Adams, 1850, a synonym.

Spindle-shaped Tornatina

Acteocina fusiformis (A. Adams, 1854)

(Figure 284)

Shell small, white, thin, smooth, cylindrically fusiform, polished. Spire exsert. Whorls five, separated by deeply channelled sutures. Interior white. Aperture a little more than three-quarters the length of shell, long, narrow, much wider anteriorly; outer lip acute; columella slightly plicated, plait callous. Size: ¼ inch.

Localities: Sorrento; Balnarring; Flinders; Port Albert.

This quaint little shell represents a very interesting genus and should be easily recognized by its characteristic exsert spire, with canaliculate or channelled sutures. It is an inhabitant of deeper water, with restricted powers of movement, the foot of the animal being extremely short and narrow. Beyond Victoria it occurs in New South Wales, Tasmania, and South Australia. Synonyms are *Tornatina brenchleyi* Angas, 1877; *T. apicina* Gould, 1859; *T. avenarius* Watson, 1886.

hofmani (Angas, 1877; Hofman's Tornatina is a close Victorian ally from Western Port, of similar dimensions but readily separated by its more shouldered body-whorl and by being slightly contracted in the middle of whorl.

Cylichna Loven, 1846

Shell subcylindrical. Spire sunken. Mouth narrow behind, widening anteriorly, columella with a single fold.

TYPE SPECIES *Bulla cylindracea* Pennant, 1777.

Peanut Shell

Cylichna arachis (Quoy & Gaimard, 1833)

(Figure 285)

Shell small, fairly solid, white, invested with a brownish or cinnamon epidermis. Cylindrical spire perforated. Surface of shell sculptured with extremely fine, microscopic, concentric striae and moderately close, irregular lines of growth. Interior white. Aperture fairly wide throughout its length, much narrower at the posterior end; outer lip simple, margin minutely crenulated. Size: ¾ inch.

Localities: Port Phillip; Western Port.

A species easily recognized by its long cylindrical shape and fine microscopic, concentric striae. At Shoreham these shells are frequently cast up on the beach; living examples appearing in dredgings between Phillip and French Islands. Found also in New South Wales, Tasmania and South Australia.

protumida Hedley, 1903; off Wilson's Promontory.
tenuis Hedley, 1905; Ocean Beach, Point Nepean.

Damoniella Iredale, 1918 (*Bulla cranchii* Fleming, 1828).
exigua (A. Adams, 1854); off Rhyll, Western Port; Sorrento.

Family PHILINIDAE

Philines

Shell capable of containing but a small part of the body, entirely internal, covered by the reflexed and united mantle; whitish, fragile, open from in front or below, consisting of two or one whorl.

Philine Ascanius, 1772

Shell ovate or squarish, thin and fragile, smooth, spirally striate or punctate, or latticed, translucent, pale. Consists of a few loosely convoluted whorls, which are entirely open from below. Spire sunken. Aperture very large. The shell is completely hidden in the mantle of the large, usually pale coloured animal.

TYPE SPECIES *Philine aperta* Linné, 1766.

Angas' Philine or Bubble Shell

Philine angasi (Crosse & Fischer, 1865)

(Figure 286)

Shell white, smooth, shining, thin, fragile; ovately subquadrate. Destitute of distinct spire. Longitudinally striate. Aperture large, wide, smooth within. Size: 1½ inches.

Localities: Port Melbourne; St Kilda; Rickett's Point; Point Cook; Western Port.

Inhabiting mud-banks, these creatures are predacious, living on small bivalves. The prey is seized and swallowed entire, the shells being crushed by the attacker's gizzard plates. It is a very common species and is found in South Australia, New South Wales and Tasmania.

columnaria Hedley & May, 1908; Western Port.

Family AGLAIIDAE

Tailed Slugs

Shell internal, body oblong, with a head shield and a posterior shield or mantle separated by a transverse furrow. Posterior of mantle produced into two lobes. Foot wide with the sides expanded into fleshy lobes which stand erect on each side of the body. Shell small with a minute spire.

Aglaia Rénier, 1804 (*Aglaia depicta* Rénier, 1804).
queritor Burn, 1958; Portarlington; Portsea; Torquay.
taronga Allan, 1933; Swan Bay.

Order ANASPIDEA

Family APLYSIIDAE

Sea Hares

(Figure 287)

Rather large, soft-bodied animals with four large ear-like tentacles and a high back formed by the large, free mantle lobes. The internal shell is thin and plate-like.

These soft-bodied molluscs, commonly known as Sea Hares, have a nearly world-wide distribution in tropical and temperate seas, usually inhabiting shallow water. They are vegetable feeders and often frequent rock-pools, living amongst the algae that grows in such places. Many of them expel a large amount of either a violet or a milky, odorous fluid when irritated.

Aplysia Linné, 1766

Shell convex, thin, horny, shield-like, with a membraneous epidermis which usually extends beyond the calcareous layer.

TYPE SPECIES *Laplysia depilans* Linné, 1766.

Small Sea Hare

Aplysia parvula Mörch, 1863

(Figure 288)

Shell internal, thin, brown, horny, fragile, clypeiform, covered by a very thin membranous epidermis extending a little beyond the calcareous layer, apex slightly incurved. Interior light brown with a pearly lustre; margin slender. Size: ¾ inch.

Localities: Anderson's Inlet; Western Port.

No difficulty should be experienced in separating it from *sydneyensis* Sowerby. The shell is much higher, smaller, not so thin and generally of a darker colour.

This Sea Hare is the smallest species known from Australian waters and the brown or greenish yellow animal often with the mantle lobe outlined with a thin, dark line will be easily recognized. It has a wide distribution in the Indo-Pacific and is known from every state in Australia. Previously it has been recorded as *concava* Sowerby, 1822, and *norfolkensis* Sowerby, 1869.

Sydney Sea Hare

Aplysia sydneyensis Sowerby, 1869

(Figure 289)

Shell internal, very thin, brown, smooth, fragile, moderately convex, shield-shaped, placed on the back of the animal with convex side uppermost. Apex slightly incurved. Size: 1 inch.

Localities: Sorrento; Queenscliff; San Remo; Mallacoota; Portland.

Frequent in pools under stones or low water during spawning season (about January). The animal moves actively by means of 'swimming lobes', and when molested exudes a rich purple fluid in such quantities that the water is coloured for some yards around it. Possessing a certain similarity to a hare crouching, the animal has been called by the ancients *Lepus marinus* or Sea Hare. Like many of the Sea Hares, this species has a wide distribution and is recorded from every state in Australia. It has previously been recorded in the literature under a number of names including *tasmanica* Tenison Woods, 1876; and *hyalina* Sowerby, 1869.

dactylomela Rang, 1928; Cowes, Phillip Island.

juliana Quoy & Gaimard, 1832; Portland; Flinders.

nigra d'Orbigny, 1835; Portland.

sowerbyi Pilsbry, 1896; *tigrina* Angas, 1867 (non Rang, 1828); Western Port; Portland.

Family AKERATIDAE

Paper Bubble Shells

Shell somewhat thin and elastic, involute, ovate. Spire not raised or just slightly. Mouth edge open more or less broadly detached so that a slit or groove is formed.

Cylindrobulla Fischer, 1856 (*Cylindrobulla beauii* Fischer, 1856).

fischeri A. Adams & Angas, 1864; Port Phillip Heads.

Akera Müller, 1776 (*Akera bullata* Müller, 1776).

soluta (Gmelin, 1791); San Remo.

tasmanica Beddome, 1882; Shoreham; Portsea.

Order THECOSOMATA

Family SPIRATELLIDAE

Sea Butterflies

Pelagic animals with fins attached to the sides of the mouth, and united ventrally by an operculigerous lobe. Shell minute, spiral, sinistral, calcareous. Operculum few-whorled, vitreous.

This and the next family belong to a group of the mollusca sometimes called the Pteropoda and some writers such as Tryon have even gone so far as to make them a separate class. Modern systematists have reduced them to an order in the subclass Opisthobranchiata.

They are all small molluscs living gregariously in the open ocean and so are only occasionally carried ashore by storms or currents. They are commonly known as Sea Butterflies or whale food, the first because of their delicate appearance and the constant movement of the large swimming lobes; and second because they are a substantial part of the plankton of tropical and temperate sea.

Spiratella Blainville, 1817 (*Clio helcina* Philps, 1774).

bulimoides (d'Orbigny, 1836); eastern Victoria.

inflata (d'Orbigny, 1836); eastern Victoria.

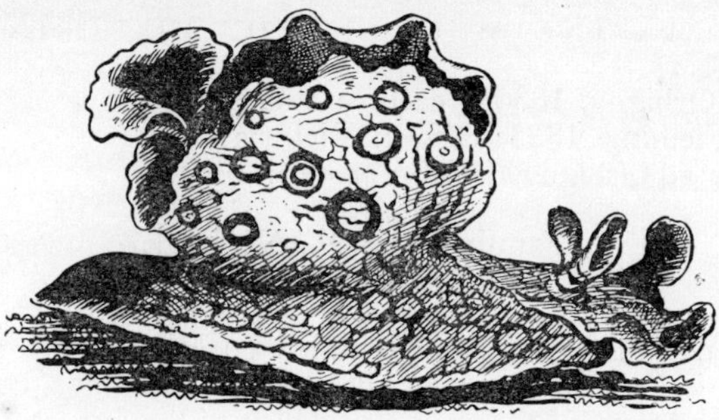

Figure 287

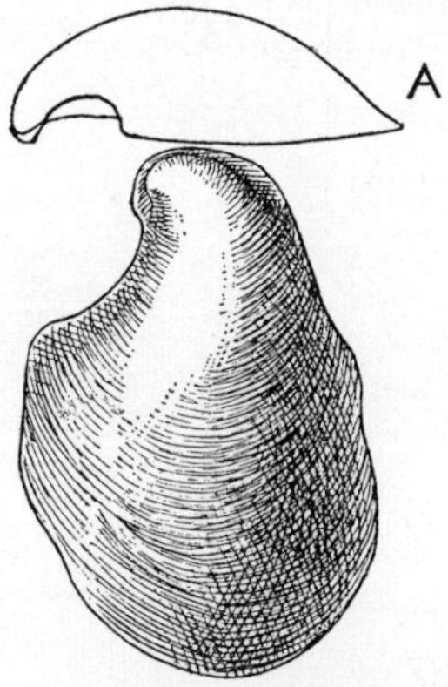

Figure 288

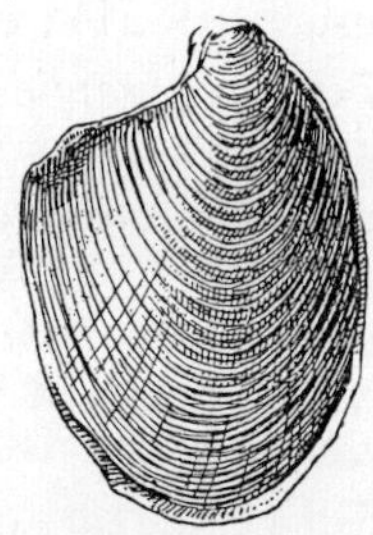

Figure 289

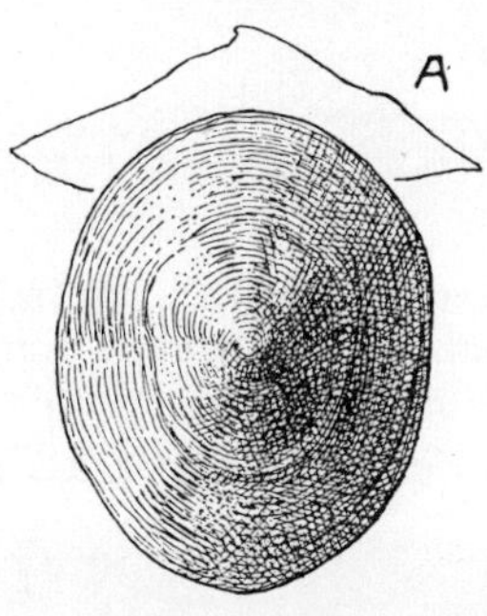

Figure 290

Figure 291

287. Aplysiidae
288. *Aplysia parvula* × 2⅔
    A Lateral elevation
289. *Aplysia sydneyensis* × 1⅓
290. *Tylodina corticalis* × 1⅓
    A Lateral elevation
291. Pleurobranchidae

lesueuri (d'Orbigny, 1836) ; eastern Victoria.
retroversa Fleming, 1823; eastern Victoria.
trochiformis (d'Orbigny, 1836) ; eastern Victoria.

Family CAVOLINIIDAE
Shelled Sea Butterflies

Pelagic animals with the foot expanded on each side to form two large fin-like projections for swimming. These are attached by a columella muscle running from the apex of the shell to the base of the fins. Body enclosed in a mantle. Shell thin, translucent; straight or curved, never spiral, globular or needle-shaped, symmetrical. Operculum absent.

Cavolina Abildgaard, 1791 (*Anomia tridentata* Forskal, 1773).
gibbosa Rang, 1836; eastern Victoria.
longirostris Lesueur, 1821; eastern Victoria.

Diacria Gray, 1847 (*Hyalaea trispinosa* Lesueur, 1821).
trispinosa (Lesueur, 1821) ; eastern Victoria.

Clio Linné, 1767 (*Clio pyramidata* Linné, 1767).
pyramidata Linné, 1767; off Melbourne.

Creseis Rang 1828 (*Creseis virgula* Rang, 1828).
virgula Rang, 1828; eastern Victoria.

Styliola Lesueur, 1827 (*Styliola recta* Lesueur, 1827).
recta Lesueur, 1827; eastern Victoria.
subula (Quoy & Gaimard, 1827) ; off Melbourne.

Herse Gistel, 1848 (*Cuvieria columella* Rang, 1827).
minor McCoy, 1885; Portland.

Order SACOGLOSSA

Family CALIPHYLLIDAE
Cyerces

Animal without shell or mantle, more or less depressed, with many rows of laminated dorsal appendages. Head with frontal lobes and cleft rhinophores. Anterior foot somewhat broad.

Branchophyllum Pruvot-Fol, 1947 (*Proctonotus orientalis* Kelaart, 1859).
pallens (Burn, 1958) ; Queenscliff; Flinders.

Family OXYNOIDAE
Oxynoids

Shell, spiral, very thin, subflexible with an epidermis. Animal slender, with mantle reflected into side-lobes partly covering the shell and united behind.

Oxynoe Rafinesque, 1819 (*Oxynoe olivacea* Rafinesque, 1819).
viridis Pease, 1861; Torquay; Ocean Beach, Portsea.

Lobiger Krohn, 1847 (*Lobiger philippi* Krohn, 1847).
wilsoni Tate, 1889; south channel, Port Phillip Heads (7-16 fathoms).

Family ELYSIIDAE

Elysids

Animal slug-like, without shell, distinct mantle or breathing organ. Head with more or less large rhinophores and behind them small eyes.

Elysia Risso, 1818 (*Laplysia viridis* Montagu, 1803).
furvacauda Burn, 1958; Torquay.

Burn, 1960 A, B, has shown that *Edenttellina typica* Gatliff & Gabriel should be placed in this order but its exact relationships have not yet been determined, see p. 322.

Order NOTASPIDEA

Family UMBRACULIDAE

Umbrella Shells

Shell patelliform, depressed, sinistral, more or less conically elevated; the apex to the left and somewhat behind the centre, curved backwards. Animal much larger than the shell which forms a plate on its back. Head small and not projecting very far beyond the mantle whose thin edge is fringed with numerous flat processes. The family is widely distributed but contains few species.

Tylodina Rafinesque, 1814

Shell cap-shaped, somewhat deep, apex with two small whorls to the left and posterior usually eroded. Shell edge not calcified.
TYPE SPECIES *Tylodina punctulata* Rafinesque, 1814.

Small Umbrella Shell

Tylodina corticalis (Tate, 1889)

(Figure 290)

Shell primrose yellow, thin, outline orbicular and showing little variation in contour, margin acute; moderately elevated; apex fairly prominent, almost central, slightly incurved. Surface more or less covered with a tough weed-like epidermis, removable in one piece; fine concentric lines of growth are visible, with a few obscure, broad, radial ridges. Interior showing a central, callous, slightly coloured disc, surrounded by an irregular muscular impression. Size: 1 inch.

Locality: dredged off Portsea; Mornington; Flinders.

The present form was described from a specimen dredged in Port Phillip over 60 years ago. Since then very few examples have appeared in Victoria, but one in excellent condition was obtained adhering to a rock under the pier at the second locality, and it is suggested further specimens may be collected in this area. Their detection certainly presents difficulty since the shell is more or less covered by the weed-like epidermis with very little of the shell on view. The animal is of a golden yellow colour, changing to deep port-wine on preservation, and possesses a broad muscular foot, by which it adheres to rocks. It is recorded also from South Australia.

Umbraculum Schumacher, 1817 (*Patella sinicum* Gmelin, 1791).
sinicum (Gmelin, 1791); off Marlo.

Family PLEUROBRANCHIDAE

Side-Gilled Slugs

(Figure 291)

Slug-like molluscs in which the feathery gill-plume arises above the middle of the right side and extends backwards; the dorsal shield is fleshy, and may be stiffened by a delicate shell or spicules. Animals are often brightly coloured and inhabit shallow waters of tropical and temperate seas.

Berthella Blainville, 1825 (*Bulla plumula* Montagu, 1803).

medietas Burn, 1961; Western Port; Port Phillip. Body smooth, elongate, flattened above, rounded at each end, mantle covering the foot at the side but not behind. Colour clear orange with a row of white spots down each side of the dorsal shield. Veil wide, arcuate, terminating in two rounded points, tentacles very long. Shell small, ear-shaped and thin. Size: 2 inches.

Localities: Western Port; Point Lonsdale; Point Addis.

This species which will be easily recognized by its shape and colour is recorded also from New South Wales, Tasmania and South Australia.

Berthellina Gardiner, 1936 (*Berthellina engeli* Gardiner, 1936).

punctata (Quoy & Gaimard, 1832); Western Port.

Berthellinops Burn, 1959 (*Berthellinops serenitas* Burn, 1959).

serenitas Burn, 1961; Flinders.

Susania Gray, 1857 (*Pleurobranchus tuberculatus* Meckel, 1808).

hilli Hedley, 1894; Western Port. A large, round, dark plum-coloured species, the fleshy mantle covered with warty protuberances, notched in front to expose the short rhinophores. It was originally described from Port Jackson, New South Wales, and has been taken by both of us in Western Port, Victoria.

Pleurobranchaea Meckel, 1813 (*Pleurobranchidium meckelii* Blainville, 1826).

maculatus Quoy & Gaimard, 1832; Sorrento; Western Port; and coast generally, with a larger, rounder body (3 inches) than the above species, cream with light brown markings and a pale foot and with a shell. It is found throughout south-eastern Australia.

Order NUDIBRANCHIATA

Naked Sea Slugs

All the members of this order are shell-less, soft bodied molluscs, slug-like and usually brightly coloured. Because they lack a characteristic shell and their soft bodies are difficult to preserve and even when preserved lose their colour, Nudibranchs are the least well known of the Victorian molluscs. A number of species are known to occur in our waters and the more intensive recent collecting is continually bringing new records and in some instances new species to light.

Fortunately, this neglected group has at last found a champion in Robert Burn, an enthusiastic worker who has published a number of papers in the last few years, and to whom we are grateful for supplying the following list of Victorian species. We have discussed only the two most common species in detail.

Family POLYCERIDAE

Tambja Burn, 1961 (*Nembrotha? verconis* Basedow & Hadley, 1905).
verconis (Basedow & Hedley, 1905); Port Phillip.

Polycera Cuvier, 1816 (*Doris quadrilineata* Müller, 1776).
janjukia (Burn, 1961); Torquay.
parvula (Burn, 1958); Torquay.

Family GYMNODORIDIDAE

Gymnodoris Stimpson, 1855 (*Gymnodoris maculata* Stimpson, 1855).
arnoldi (Burn, 1957); Torquay.

Paliolla Burn, 1958 (*Polycera cooki* Angas, 1864).
cooki (Angas, 1864); Breamlea; Torquay; Airey's Inlet.

Family TRIOPHIDAE

Kaloplacamus Bergh, 1884 (*Euplocamus croceus* Philippi, 1836).
ramosus (Cantraine, 1835); Torquay.

Limacia Müller, 1781 (*Limacia claviger* Müller, 1781).
multidigitalis (Burn, 1957); Torquay.

Family ACANTHODORIDIDAE

Lamellidoris Alder & Hancock, 1855 (*Doris bilamellata* Linné, 1767).
maugeansis Burn, 1958; Torquay; Breamlea.

Family GONIODORIDIDAE

Goniodoris Forbes & Goodsir, 1839 (*Doris nodosa* Montagu, 1808).
meraculus Burn, 1958; Torquay.

Family DORIDIDAE

Body mostly with a more or less broad margin to the mantle, may be smooth, warty or matted. Gills with a permanent pouch into which they can be withdrawn. Rhinophores foliated and usually retractile. True jaw plates absent; however, the lips are often armed with plates or a ring of hard corpuscles.

Glossodoris Ehrenberg, 1831 (*Actinodoris xantholeuca* Ehrenberg, 1831).
alternata Burn, 1957; Portarlington; Torquay; Western Port.
haliclona Burn, 1957; Breamlea; Portarlington.
perplexa Burn, 1957; Torquay.
tasmaniensis (Bergh, 1905); Torquay.
victoriae Burn, 1957; Torquay; Portarlington.

Hallaxa Eliot, 1909 (*Halla decorata* Bergh, 1878).
indecora (Bergh, 1905); Torquay.

Rostanga Bergh, 1879 (*Doris coccinea* Alder & Hancock, 1848 = *Rostanga rufescens* Iredale & O'Donoghue, 1923).
arbuta (Angas, 1864); Torquay; Lorne; Breamlea.

Jorunna Bergh, 1876 (*Doris tomentosa* Cuvier, 1804).
hartleyi (Burn, 1958); Flinders; Torquay; Breamlea; Sutherland's Bay, Phillip Island.

Ceratosoma A. Adams & Reeve, 1848

Body elongate, smooth; tentacles clavate, non-retractible, without sheaths; forehead simple, rounded; gills plumose; back produced behind into a single elevated, conical appendage.

TYPE SPECIES *Ceratosoma cornigera* A. Adams & Reeve, 1848.

Short-tailed Sea Slug

Ceratosoma brevicaudata Abraham, 1876

(Figure 292)

Animal large, elongate, dorsally flat, rounded in front, sides nearly parallel and tapering to an obtuse point behind; sides elevated; mantle oval with an undulate margin, and ending posteriorly in a peculiar nipple-like protuberance. Colour shades of pink and buff with deeper colour at the edges of mantle and foot and speckled with circular spots of light violet purple and larger markings of violet purple surrounded by a ring of deeper colour. Gills twelve, reddish yellow retractile into the gill-pouch. Size: 4 inches.

Localities: Port Phillip; Western Port; off Lakes Entrance.

This beautiful and easily recognized species is recorded from Western Australia, South Australia and New South Wales.

Praegliscita Burn, 1957 (*Doris chrysoderma* Angas, 1864).
chrysoderma (Angas, 1864); Queenscliff; Torquay; Flinders.

Alloiodoris Bergh, 1904 (*Alloiodoris marmorata* Bergh, 1904).
nivosus Burn, 1958; Portarlington; Western Port; Port Phillip.

Archidoris Bergh, 1878 (*Doris tuberculata* Alder & Hancock, 1878).
pustulata (Abraham, 1877); along whole Victorian coast.
wellingtonensis (Abraham, 1877); Portland.

Austrodoris Odhner, 1926 (*Archidoris rubescens* Bergh, 1898).
peculiaris (Abraham, 1877); Queenscliff; Breamlea.

Trippa Bergh, 1877 (*Trippa ornata* Bergh, 1877).
albata Burn, 1961; Sutherland's Bay, Phillip Island; Western Port.

Thordisa Bergh, 1877 (*Thordisa maculigera* Bergh, 1877).
sabulosa Burn, 1957; Torquay; Queenscliff.

Halgerda Bergh, 1880 (*Halgerda formosa* Bergh, 1880).
graphica Basedow & Hedley, 1905; Torquay.

Aphelodoris Bergh, 1879 (*Aphelodoris antillensis* Bergh, 1879).
berghi Odhner, 1924; Torquay.
varia (Abraham, 1877); Victoria (no locality recorded).

Platydoris Bergh, 1877 (*Doris argo* Linné, 1767).
galbanus Burn, 1958; Sutherland's Bay, Phillip Island.

Dendrodoris Ehrenberg, 1831 (*Dendrodoris lugubris* Ehrenberg, 1831).
aurea (Quoy & Gaimard, 1832); a very common species along the whole Victorian coast. The general body colour is orange with or without small white patches on the usually smooth, soft mantle. The five branchiae are large and bushy.
albopurpura Burn, 1957; Western Port; Flinders.
denisoni (Angas, 1864); San Remo.

Figure 292

Figure 293

292. *Ceratosoma brevicaudata* × 1
293. *Doriopsilla carneola* × 1

gunnamatta Allan, 1932; Inverloch.
maugeana Burn, 1961; Flinders.
nigra (Stimpson, 1855); Portarlington; Port Phillip.
vadisi Burn, 1961; Portarlington; Torquay.

Doriopsilla Bergh, 1880 (*Doris areolata* Bergh, 1880).
carneola Angas, 1864; common on the whole Victorian coast (Figure 293). Distinguished from the last by the tough, usually granulose mantle, dull red dorsal and yellow ventral. The five branchiae are tripinnate.
staminea (Basedow & Hedley, 1905); Port Phillip; Flinders; Torquay.

Family MADRELLIDAE

Madrella Alder & Hancock, 1866 (*Madrella ferruginosa* Alder & Hancock, 1866).
sanguinea (Angas, 1864); Breamlea; Torquay.

Family ANTIOPELLIDAE

Janolus Bergh, 1884 (*Janolus australis* Bergh, 1884).
hyalina (Alder & Hancock, 1854); Torquay.

Proctonotus Alder & Hancock, 1844 (*Venilia mucronifera* Alder & Hancock, 1844).
affinis Burn, 1958; Torquay.

Family ARMINIDAE

Armina Rafinesque, 1814 (*Armina tigrina* Rafinesque, 1814).
cygnea Bergh, 1876; Port Phillip.

Family TRITONIDAE

Paratritonia Baba, 1949 (*Paratritonia lutea* Baba, 1949).
lutea Baba, 1949; Port Phillip.

Family SCYLLAEIDAE

Scyllaea Linné, 1758 (*Scyllaea pelagica* Linné, 1758).
pelagica Linné, 1758; Port Phillip.

Family TETHYIDAE

Melibe Rang, 1829 (*Melibe rosea* Rang, 1829).
australis (Angas, 1864); Torquay; Queenscliff.
maugeana Burn, 1960; Torquay.

Family DOTONIDAE

Doto Oken, 1815 (*Doris maculata* Montagu, 1808).
ostentus Burn, 1958; Torquay; Flinders.

Family CORYPHELLIDAE

Coryphellina O'Donoghue, 1929 (*Coryphellina rubrilineata* O'Donoghue, 1924).
poenicia (Burn, 1957); Torquay; Breamlea; Portarlington.
poenicia aurantia Burn, 1961; Portarlington.
rubrilineata (O'Donoghue, 1929); Torquay.

Family CUTHONIDAE

Cuthona Alder & Hancock, 1855 (*Eolis nana* Alder & Hancock, 1842).
bractea Burn, 1961; Torquay.

Catriona Winkworth, 1941 (*Eolis aurantia* Alder & Hancock, 1842).
viridiana Burn, 1961; Torquay.

Tergipes Cuvier, 1805 (*Limax tergipes* Forskal, 1775 = *Doris lacinulata* Gmelin, 1790).
pauculas Burn, 1961; Portarlington.

Family FACELINIDAE

Facelina Alder & Hancock, 1855 (*Eolida coronata* Forbes & Goodsir, 1839 = *Doris longicornis* Montagu, 1808).
hartleyi Burn, 1961; Flinders.
newcombi (Angas, 1864); Portarlington; Torquay.

Family FAVORINIDAE

Favorinus Gray, 1850 (*Eolis alba* Alder & Hancock, 1844) (non Van Hasselt, 1824 = *albidus* Iredale & O'Donoghue, 1923).
pannuceus Burn, 1961; Flinders.

Cratena Bergh, 1864 (*Doris coerulea* Montagu, 1808).
macphersonae Burn, 1961; Flinders.
serrata (Baba, 1949); Torquay.

Austraeolis Burn, 1961 (*Flabellina ornata* Angas, 1864).
fucia Burn, 1961; Queenscliff.
ornata (Angas, 1864); Torquay; Flinders.

Echinopsole Macnae, 1954 (*Echinopsole fulvus* Macnae, 1954).
brevicerata Burn, 1961; Torquay; Flinders.

Family AEOLIDIIDAE

Aeolidiella Bergh 1867 (*Eolida soemmeringii* Leuckart, 1828).
faustina Bergh, 1900; Torquay; Breamlea; Flinders; Sutherland's Bay, Phillip Island; Blanket Bay, Cape Otway.
macleayi (Angas, 1864); Torquay.

Order BASOMMATOPHORA

Family ELLOBIIDAE
Air-Breathing Ear Shells

Shell spiral, solid, usually thick, covered by an epidermis. Spire more or less elevated. Whorls sometimes flattened. Mouth elongate, contracted by columella teeth and often also bearing teeth within the lip. The inner walls of the whorls are usually absorbed, leaving a single cavity. These shells usually inhabit salt-marshes and mangrove areas in the vicinity of the sea. Most species are found in tropical regions, only a few small members of the family occur in Victoria.

Marinula King, 1831

Shell oblong-ovate, imperforate, solid, smooth. Spire mostly somewhat small, sharp. Mouth oval; inner lip thick with three deep plications, upper largest; outer lip simple.

TYPE SPECIES *Marinula pepita* King, 1831.

Delicate Air Breather

Marinula zanthostoma H. & A. Adams, 1854

(Figure 294)

Shell small, white, cream or pinkish, oblong-ovate, smooth, fairly solid. Spire approximately one-third length of shell. Whorls flatly rounded. Interior creamy. Aperture elongately ovate; outer lip sharp; columella slightly curved, bearing three conspicuous plaits, the posterior one of which is the largest. Size: ⅜ inch.

Localities: Frankston; Western Port; Back Beach, Phillip Island; Lorne.

A common species distinguished by its peculiar oblong-ovate shape and three-plaited columella. Occurs commonly in the Victorian literature under the name of the allied species *Marinula patula* Lowe, 1835.

meridionalis (Brazier, 1877); Meridion Air Breather, a horn coloured, more elongated form about ⅜ inch with spire about half the length of shell; whorls flatly convex, smooth but showing fine lines of growth; aperture elongately oval, with the inner lip generally displaying two prominent teeth.

Ophicardelus Beck, 1837

Shell elongate-ovate, smooth or spirally grooved. Spire elevated conical. Mouth oval-elongate; inner lip reflected with two strong plications; outer lip thin.

TYPE SPECIES *Auricula australis* Quoy & Gaimard, 1833.

Common Mangrove Air Breather

Ophicardelus ornatus (Ferussac, 1821)

(Figure 295)

Shell small, thin, elongately ovate, brown, invested with a darker brown epidermis. Spire about one-third the length of shell, the whorls of which are flatly convex. Body-whorl rounded, showing fine lines of growth and possessing two or three more or less light encircling bands. Interior brown, shining. Aperture elongately oval; outer lip sharp; inner lip with two prominent anterior plaits on the columella. Size: ½ inch.

Localities: Williamstown; Western Port; Lakes Entrance.

This amphibious, brown, elongately oval species is extremely common, located crawling about on mud-banks and among the tangled roots of mangroves. It is easily recognized by its peculiar oval shape and the two well-defined columella plaits. Often appearing in collections under *australis* (Quoy & Gaimard, 1833) and *cornea* (Swainson, 1856). The above prior name precluding their use. Recorded also from Tasmania and New South Wales.

Figure 294

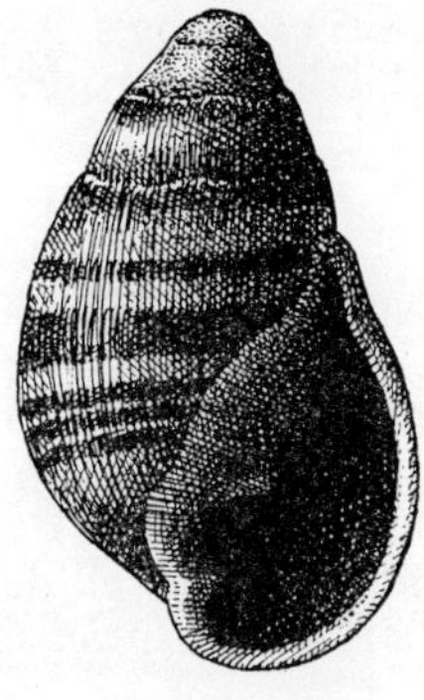

Figure 295

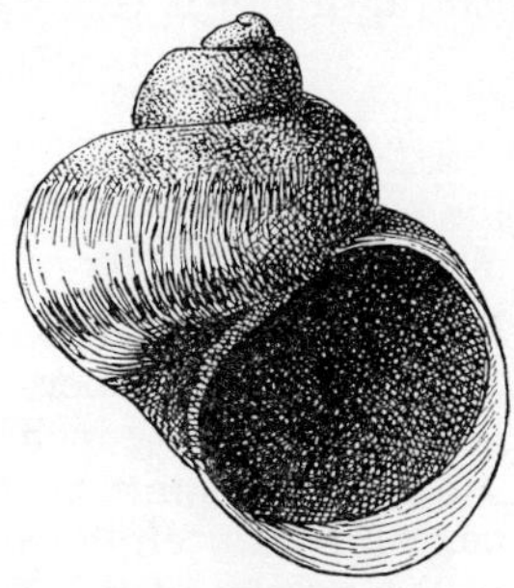

Figure 296

Figure 297

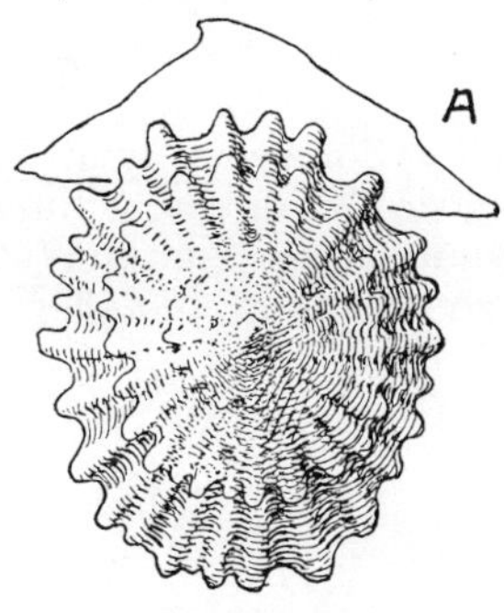

Figure 298

294. *Marinula zanthostoma* × 3
295. *Ophicardelus ornatus* × 3
296. *Salinator fragilis* × 3
297. *Salinator solida* × 3
298. *Gadinia conica* × 5⅓
A Lateral elevation

quoyi (H. & A. Adams, 1855); Quoy's Air Breather; occurs at Lakes Entrance and is easily separable by its broader form, more distant growth-ridges and the presence of an incised line below the sutures, the outer lip is thickened bearing a single tooth internally centrally situated.

sulcatus (H. & A. Adams, 1855); the Grooved Air Breather; also from Lakes Entrance, is another species of similar dimensions, but immediately distinguished by its strongly, concentrically, striated spire.

Leuconopsis Hutton, 1883 (*Leuconopsis obsoleta* Hutton, 1883).

pellucidus (Cooper, 1841); from Frankston; Portsea; and Stony Point. A diminutive (1/12 inch), white, imperforate, ovate shell with flattish whorls and apex not quite centrally situated; a well-developed central tooth on the inner lip is a useful recognition mark. *victoriae* Gatliff, 1905, is now regarded as a synonym of this species.

Family AMPHIBOLIDAE

Air Breathers

Shell subglobose, umbilicated. Spire short. Mouth sub-oval; outer lip sinuous; columella simple or calloused; operculum horny, subspiral. Inhabitants of salt marshes near the sea, they are able to seal the shell with the operculum and so live over dry periods buried in the mud.

Salinator Hedley, 1900

Shell of somewhat small size and thin-walled, roundly ovate. Body-whorl symmetrically convex, without sculpture. Mouth without a sinus. Nucleus of operculum larger than in *Amphibola*.

TYPE SPECIES *Ampullaria fragilis* Lamarck, 1822.

Fragile Air Breather

Salinator fragilis (Lamarck, 1822)

(Figure 296)

Shell small, thin, fragile, globose, umbilicated; colour light brown, with, normally, three slightly darker encircling bands of varying widths. Spire approximately one-third the length of shell. Whorls much rounded, with irregular fine lines of growth; sutures deeply impressed. Interior creamy, shining. Aperture ovate, through which the concentric colour bands are plainly visible; outer lip thin, entire; umbilicus fairly wide and deep; columella slightly oblique; operculum very thin, light horn colour. Size: ½ inch.

Localities: Stony Point; Port Phillip; Lakes Entrance.

This umbilicated species is readily recognized by its fragile globose form, and conspicuous brown colour bands encircling the whorls. In addition to these features, specimens frequently occur showing irregular, wavy radial markings, a colour form to which the name *quoyana* (Potiez & Michaud, 1838) has been given; however, so much variation exists, it is impossible to separate the two forms and the latter sinks as a synonym.

solida (von Martens, 1878); the Solid Air Breather (Figure 297), is more elongate than *fragilis*, possessing close longitudinal zigzag markings, and with an outstanding strong spiral rib on the inner area of the operculum, a feature absent in *fragilis*. Both species appear in Queensland, New South Wales, Tasmania, South Australia and Western Australia.

These Estuarine Sand Snails, as they are sometimes called, are gregarious in their habits, living in backwaters of beaches more or less under the influence of fresh water. At Altona and many other localities on the Victorian coastline *fragilis* Lamarck appears in countless numbers and Freame (1948) supplies some interesting observations: 'Altona provides excellent specimens of the species, which, as the tide recedes, will be seen burrowing through the sand. Tracking for a short distance, these small animals deposit their eggs, cementing them with grains of sand to form a small band or girdle. The eggs take from fourteen to sixteen days to hatch. On the return of the tide, the animals disappear in the sand; they crawl fairly quickly, and when contracted, enter their shell, which they close with their horny operculum.'

Family GADINIIDAE

Gadinias

Shell patelliform, depressedly conical, surface radially ribbed. Interior showing muscle impression, horseshoe-shaped and entirely or almost entirely open frontally; without an aperture for the siphon on the right side. Jaw and gills missing.

Gadinia Gray, 1824

Description as for the family.
TYPE SPECIES *Patella mammillaris* Linné, 1758.

Conical Gadinia or Siphon Shell
Gadinia conica Angas, 1867
(Figure 298)

Shell small, white or cream, depressedly conical, margin circular; apex subcentral, situated a little posteriorly. Surface ornamented with thirty to thirty-six strong, irregular radiating ribs, the concentric lines of growth being plainly visible. Interior white, shining; margin denticulate; muscular impression horseshoe-shaped, the siphonal groove very evident, anteriorly. Size: ½ inch.

Localities: Portsea; Western Port.

A curious genus represented in Victoria by the above species. In general outline and sculpture it approaches *Siphonaria* but is immediately separated by its flatness and in the position of the siphonal groove which is situated more anteriorly. In living condition these air-breathing molluscs are located on rocks between tide marks, often encrusted with Serpuloid worm, the animal possessing a very peculiar flat foot. Recorded also from New South Wales and Tasmania.

Family SIPHONARIIDAE

Siphon Shells

Shell patelliform, depressedly conical surface radially ribbed; muscle impression entire except for an aperture for the siphon on the anterior right side. Jaw and gills present.

Siphonaria Sowerby, 1924

Description as for the family.
TYPE SPECIES *Patella laciniosa* Linné, 1758.

Van Diemen's Land Siphon Shell

Siphonaria diemenensis Quoy & Gaimard, 1833

(Figure 299)

Shell brown, ovate, limpet-shaped; apex sharp, nearly central, inclining towards the posterior margin. The whole surface is sculptured with numerous, irregular, white, radiating ribs. Interior polished, the margin of which is conspicuously lineated with chestnut and white, and on the right side, passing from the apex to the margin, is a siphon or canal which is extended in a slight projection beyond the margin. Aperture wide, irregular. Size: 1 inch.

Localities: Sandringham; Frankston; Mornington; Rye; Portsea; Queenscliff; Lorne.

This is one of our commonest shells, somewhat gregarious in habit, found adhering to jetty piles, and rocks between tide marks. In shape it recalls the common limpets, but is immediately separated by the presence of the internal siphonal groove, a characteristic of this pulmoniferous or air breathing genus *Siphonaria*. There may be possible confusion with the three remaining species in Victorian waters but distinguishing features are evident and these will be noted under each heading. The ova appears in white irregularly curved masses, attached to rocks. The species extends its range to Western Australia, South Australia, Tasmania and New South Wales.

Blue Siphon Shell

Siphonaria tasmanica Tenison Woods, 1876

(Figure 300)

Shell ovate, conical, bluish, concentrically, irregularly zoned with light bands; apex brown, subcentral, slightly inclining to posterior end. The sculpture consists of forty to fifty flatly rounded radial ribs, the whole surface frequently becoming corroded. Interior rich purplish brown, polished, spatula much lighter in colour generally of bluish tint more or less entering the siphonal groove; entire margin crenulated. Size: ¾ inch.

Localities: Port Phillip; Western Port; Anderson's Inlet; Sorrento; Torquay; Lorne; Cape Otway; Portland.

With its less prominent ribs, constant blue exterior and purplish brown, polished interior, no difficulty should be experienced in identifying this shell. It is of common occurrence on rocks, frequently associated with *diemenensis* Quoy & Gaimard, and *baconi* Reeve. It occurs also in South Australia and Tasmania.

Corded Siphon Shell

Siphonaria funiculata Reeve, 1856

(Figure 301)

Shell conical, much elevated; apex somewhat sharp, subcentral inclining slightly to posterior end; colour brown. Ornamented with numerous irregular

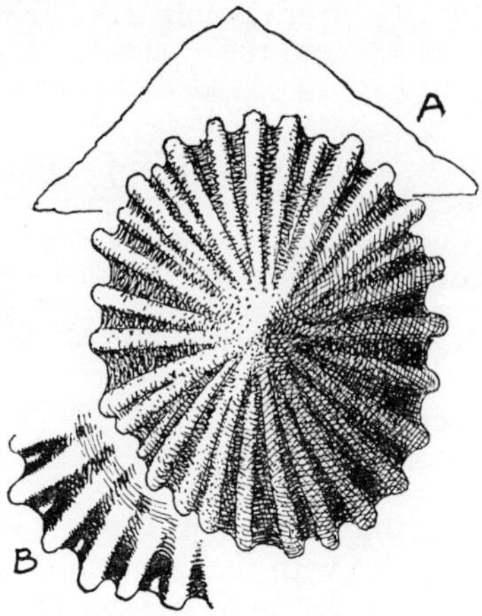

Figure 299

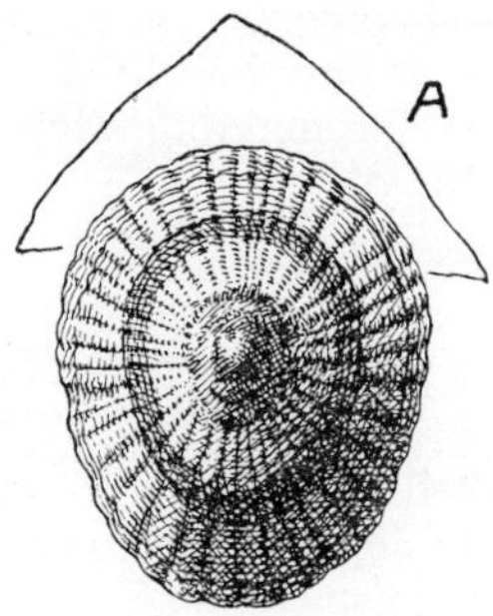

Figure 300

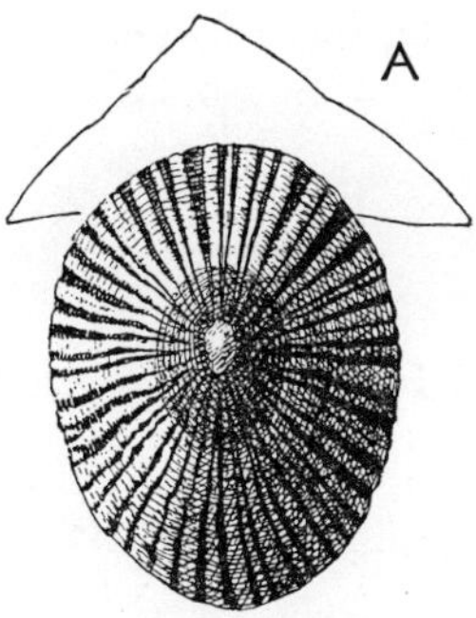

Figure 301

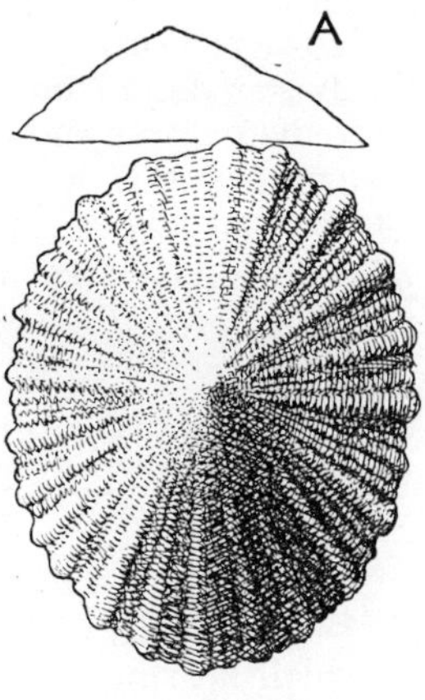

Figure 302

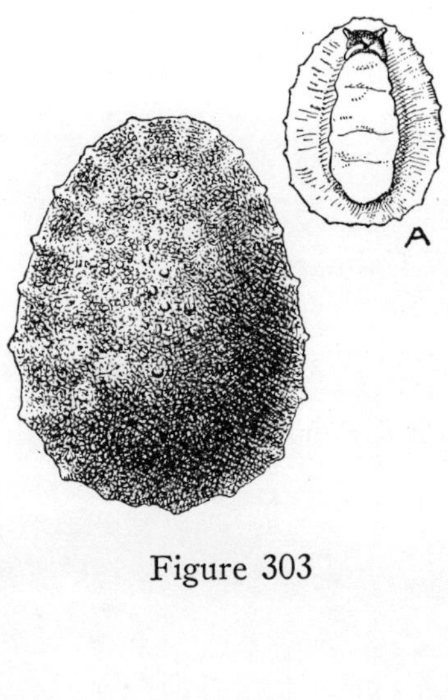

Figure 303

299. *Siphonaria diemenensis* $\times 2$
A Lateral elevation
B Underside of margin

300. *Siphonaria tasmanica* $\times 2$
A Lateral elevation

301. *Siphonaria funiculata* $\times 1\frac{1}{3}$
A Lateral elevation

302. *Siphonaria baconi* $\times 2$
A Lateral elevation

303. *Onchidella patelloides* $\times 2$
A Anterior view $\times 1$

radial ribs with interspaces of varying widths. Interior chocolate colour with the spatula much lighter, more or less continuing into the siphonal groove; margin crenulated. Size: 1 inch.

Localities: Port Phillip; Torquay; Lorne.

Compared with other representatives of the genus in Victoria it is nearest to *diemenensis* Quoy & Gaimard, from which it may be distinguished by its finer sculpture and rich chocolate coloured interior. Frequently in association with the remaining Victorian forms. Recorded also from Tasmania and New South Wales.

Bacon's Siphon Shell

Siphonaria baconi Reeve, 1856

(Figure 302)

Shell internally and externally cream, frequently showing faint brownish spots; depressedly conical, outer margin irregular with the siphonal groove very conspicuous. The sculpture consists of irregular radiating ridges with the interstices finely striated; apex subcentral. Interior polished, with the muscular impression darkened orange to brown. Size: ¾ inch.

Localities: Sorrento; Cape Schanck; Western Port; Anderson's Inlet; Airey's Inlet; Lorne.

Easily separated from the other Victorian representatives of the genus by its extreme flatness and constant light colour. Often found in company with *diemenensis* but is less common. It occurs also in South Australia and Western Australia.

Pugillaria Iredale, 1924 (*Siphonaria stowae* Verco, 1906).
stowae (Verco, 1906); Portsea.

Order STYLOMMATOPHORA

Family ONCHIDIIDAE

Air-Breathing Sea Slugs

Slug-like animals with a thick, leathery mantle entirely covering the back. Head small bearing short non-retractile tentacles with terminal eyes. Gills and shell absent.

Onchidella Gray, 1850

Onchidium with short tentacles, mantle smooth or granular but without tufts or radiating processes on the dorsal surface.

TYPE SPECIES *Onchidium celtica* Cuvier, 1817.

Ocean Beach Slug

Onchidella patelloides (Quoy & Gaimard, 1832)

(Figure 303)

Body oval, back elevated, mantle border more or less notched; head small, tentacles short and conical; mantle granulated between the scattered, large papillae. Colour yellowish brown or green with darker streaks and whitish marginal glands.

Localities: Port Phillip; Western Port; Lorne.

These small air-breathing slugs are found living near low-tide mark on most rock platforms in Victoria. It is also recorded from New South Wales, northern Tasmania and New Zealand.

Onchidina Semper, 1882 (*Onchidina australis* Semper, 1882).

australis Semper, 1882. This large elongate, dark slug lives in sand and mud, preferring the quiet waters and salt marsh conditions at the heads of bays and inlets. It occurs in New South Wales and reaches as far west as Wingan Inlet, Victoria.

Class SCAPHOPODA

Tusk Shells

THE TUSK SHELLS have an elongated animal which is enclosed in a conical, tapering shell open at both ends. The larger anterior opening is the entrance to the mouth and through it the foot can be projected. The small posterior opening serves for the entrance and exit of water (from which the animal absorbs oxygen through the mantle) and the excretion of waste and genital products. There is no definite head but the central nervous system and the mouth are concentrated on the concave or dorsal side of the shell at the base of the foot. The mouth is a long proboscis-like tube projecting into the cavity above the foot; attached at base of the proboscis is a group of cephalic filaments with the power to move in all directions. These filaments, which are replaced when lost or damaged, are ciliated and contractile with club-shaped extremities to aid in the capture of food, which consists mostly of Foraminifera and other small organisms. The mouth has a dorsal jaw and a radula which are used to break up the larger food particles ready for digestion.

The foot, from which the class takes its name, is wedge-shaped and very muscular and it can be well extended beyond the end of the shell for use as a digging organ.

The Tusk Shells are all marine, mostly inhabiting deep water, where they live embedded at an oblique angle in the mud or sand, with only the smaller end of the shell projecting.

The class is divided into two families based on shell, foot and radula characters.

Family DENTALIIDAE

Tusk Shells

Shell tapering uniformly from the broad anterior or mouth end to the narrow posterior apex; at least a portion of shell sculptured. Foot conical; central tooth of radula twice as wide as long.

Fissidentalium Cossman, 1888

Shell large and solid for the family, usually sculptured with fine longitudinal ribs. Apex usually with a long slit but may be entire or the slit replaced by a series of fissures.

TYPE SPECIES *Dentalium ergasticum* Fischer, 1882.

Polished Tusk Shell

Fissidentalium lubricatum (Sowerby, 1869)

(Figure 304)

Shell elongated, white, rather straight, narrow, thin, smooth, slowly increasing; apex entire. Size: 1 inch.

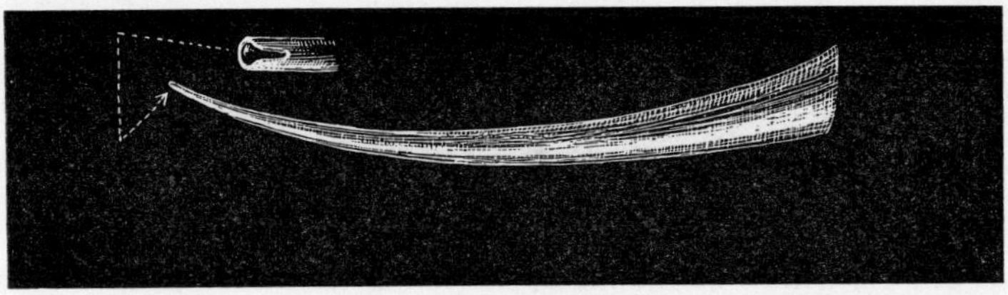

Figure 304

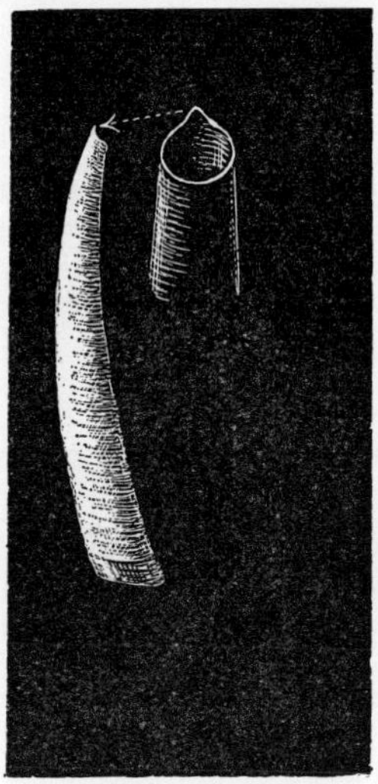

Figure 305

304. *Fissidentalium lubricatum* × 4
305. *Cadulus vincentianus* × 6⅔

Localities: Western Port; Ninety Mile Beach.

This is also recorded from Tasmania. Three other species are recorded from Victoria and collecting in the correct habitat would probably produce a number of additions to the list.

Compressidens Pilsbry & Sharp, 1897 (*Dentalium* [*Compressidens*] *pressum* Pilsbry & Sharp, 1897).

platyceras (Pilsbry & Sharp, 1898); from Port Albert. It is small (½ inch), glossy, white, with oblique encircling wrinkles but no longitudinal sculpture. It is common in New South Wales.

Dentalium Linné, 1758 (*Dentalium elephantium* Linné, 1758).

tasmaniensis Tenison Woods, 1877; off Ninety Mile Beach (40 fathoms).

Episiphon Pilsbry & Sharp, 1897 (*Dentalium sowerbyi* Goulding, 1834).

virgula (Hedley, 1903); off Ninety Mile Beach (40 fathoms).

Family SIPHONODENTALIIDAE
Tooth Shells

Shell inflated near the middle and as a rule with the mouth end contracted. Foot vermiform, but capable of expansion into a rosette-like disc. Central tooth of radula almost square.

Cadulus Philippi, 1844

Shells small, white, smooth, arcuate, swollen centrally or towards the anterior end. Aperture oblique and larger than in Dentaliidae, entire or cut by two or four slits.

TYPE SPECIES *Dentalium ovulum* Philippi, 1844.

St Vincent's Tooth Shell
Cadulus vincentianus Cotton & Godfrey, 1940
(Figure 305)

Thin, moderately arcuate, not much swollen medially, gradually tapering towards the rather large ends; tube somewhat flattened throughout but rather more so at the ends; glossy and smooth to the eye but with very fine crowded, transverse scratchings under the lens. The oval posterior end is irregular with a minute triangular spine projecting from the convex border of the shell. Size: ⅙ inch.

Localities: Western Port; Port Albert; Portland.

This species was originally confused with *acuminatus* Tate, 1887, a Miocene fossil, from which it differs in being more obese, less arcuate and in its transverse sculpture. It is also recorded from New South Wales, Tasmania, and South Australia.

gibbosus Verco, 1911 (⅜ inch); a smooth, polished, narrow shell greatly inflated at the anterior third, and with the dorsal surface obtusely angled at this point. Originally described from South Australia, it has been taken in 40 fathoms off the Ninety Mile Beach.

angustior Verco, 1911; Port Albert.

spretus Tate & May, 1900; off Ninety Mile Beach (40 fathoms).

Class PELECYPODA

Bivalves

THE BIVALVES are perhaps the most modified group of molluscs and externally are unlike the other classes (Figure 306). Two symmetrical mantle flaps enclose the whole body and secrete the paired shell valves which are hinged on the dorsal line. The valves can be tightly closed by retraction of the adductor muscles, typically situated at each end of the animal. When the adductor muscles relax, the shell is opened by the ligament, an elastic connecting strip between the two valves. The ligament is usually situated slightly behind the embryonic shell (umbo). The ligament may be external (dorsal to the hinge) or internal (ventral to the hinge). Along the hinge-line teeth that interlock in various ways may be developed to prevent fore and aft displacement of the valves.

The paired gills consist of two plate-like flaps hanging into the mantle cavity on each side of the foot. The cilia of the gills draw a powerful water current into the mantle cavity, and as the water passes between the gill filaments, food particles are strained back and then carried by other cilia towards the mouth where they are carefully graded for size by the ciliated labial palps. The current of water then passes out behind.

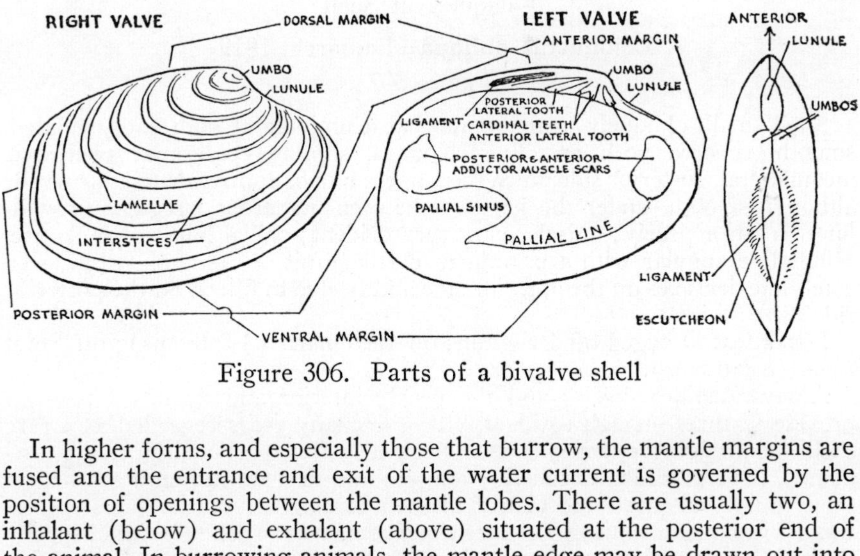

Figure 306. Parts of a bivalve shell

In higher forms, and especially those that burrow, the mantle margins are fused and the entrance and exit of the water current is governed by the position of openings between the mantle lobes. There are usually two, an inhalant (below) and exhalant (above) situated at the posterior end of the animal. In burrowing animals, the mantle edge may be drawn out into tubular extensible siphons that can reach the surface of the substratum. The line of attachment of the mantle to the shell shows as a scar line (pallial line) on the shell, and the indentation of the line left by the free lobe of the siphonal area is called the pallial sinus.

Most bivalves are sedentary, either anchored to the substratum or burrowing in it. Their means of locomotion is a wedge-shaped muscle foot which can be elongated and thrust forward in the substratum to haul the animal forward. At the base of the foot lies a gland from which many bivalves, for at least part of their life, can produce a bundle of tough threads (byssus) for temporary or permanent attachment.

Order TAXODONTA

Family NUCULIDAE

Nut Shells

Shell oval or trigonal, small; umbos turned towards the short anterior end; pearly within. Hinge composed of two long, lateral rows of transverse teeth separated at the umbos by a central ligament pit or chondrophore.

Leionucula Quenstedt, 1930

Shell small to medium sized, mostly somewhat slender, round or oval, always smooth, epidermis olivaceous; the chondrophore is under the hinge-line, inclined forward, and situated under the hindmost posterior tooth of the anterior row; the margin is uniform without radial ribs.

Type Species *Nucula albensis* d'Orbigny, 1845.

Thiele, 1935, considers *Ennucula* Iredale, 1931, a synonym of the above genus and in this we concur.

Oblique Nut Shell

Leionucula obliqua (Lamarck, 1819)

(Figure 307)

Shell small, obliquely ovate, somewhat acuminate at both ends, inflated, smooth, covered with an olive coloured, shining epidermis; equivalve, inequilateral, posterior side much the longer; umbos approximated, incurved; although smooth, under the lens will be seen numerous irregular growth-lines. Interior pearly; muscle scars very evident; pallial line simple, entire. Hinge-line angular with a prominent cartilage pit at the angle with seven comb-like denticles on the anterior side and twelve to fifteen on the posterior side. Size: ½ inch.

Localities: Dredged off Brighton; off Beaumaris (3 fathoms); off Point Cook (8 fathoms).

A very distinct species and the pearly interior and definite hinge-line provide features enough to identify it. For many years regarded as a rare shell but dredging operations at various localities, particularly at the last-named, have provided excellent examples and in quantity, on a sea-bed of sandy mud. Pritchard and Gatliff in their catalogue remark: 'This species was very abundant during Tertiary time, being a very common fossil species in the Victorian Miocene (or Kalimnan) beds.' Occurs also in Tasmanian and New South Wales waters.

beachportensis (Verco, 1907); dredged in 40 fathoms off the Ninety Mile Beach. It is distinguished by its smaller size (⅕ inch) by the faint sculpture of 'rounded concentric striae' and 'scanty microscopic radial striae'.

Pronucula Hedley, 1902 (*Pronucula decorosa* Hedley, 1902).

micans (Angas, 1878); a small (⅛ inch) smooth species from Rhyll, Western Port and Corio Bay.

hedleyi (Pritchard & Gatliff, 1904); polished, with distinct radial ribs, from Western Port.

Family LEDIDAE

Beaked Nut Shells

Shell not nacreous, oblique oval or more or less elongate, often with the posterior acuminate; shell margin more or less sinuous, toothed. Ligament with smaller or larger chondrophore between the teeth rows.

Scaeoleda Iredale, 1929

Shell oblique oval, rounded anteriorly and produced posteriorly into a short beak. Distinct concentric sculpture and a rim running from under the umbos to the posterior ventral margin.

TYPE SPECIES *Nucula crassa* Hinds, 1843.

Heavy Nut Shell
Scaeoleda crassa (Hinds, 1843)
(Figure 308)

Shell olive yellow, solid, shining, oblong, rounded anteriorly and produced posteriorly into a sort of beak, this beaked area being separated from the remainder of the shell by a distinct furrow running from the umbos to posterior-ventral margin; equivalve, almost equilateral; umbos approximate, incurved. Sculptured by numerous, fairly strong equally spaced concentric ridges. Interior pearly or bluish white, polished; muscle scars plainly visible; pallial line distinct. Hinge-line angulated with teeth divided into numerous V-shaped denticles on each side. Size: 1 inch.

Localities: Portland (8 fathoms); Port Albert; off Port Phillip (33 fathoms).

The present shining form is one easy to recognize with its characteristic rostrate or beaked posterior and ornament of strong concentric ridges. It is dredged on a sandy bottom. Recorded also in New South Wales and Tasmania.

dohrni (Hanley, 1861); a much smaller and narrow form from Bass Strait cable and New South Wales; it is whitish yellow, about ¼ inch in length and a more delicate shell than *crassa*.

hanleyi (Angas, 1873); off Lakes Entrance.

Poroleda Hutton, 1893

Shell elongate, sickle-shaped, beaked posteriorly. Hinge-line slightly curved with a series of longitudinal, imbricating teeth on each side.

TYPE SPECIES *Poroleda lanceolata* Hutton, 1893.

Beaked Nut Shell
Poroleda spathula Hedley, 1915
(Figure 309)

Shell yellowish green, very depressed, thin, fragile, elongated, equivalve, very inequilateral; anterior end rounded, posteriorly produced into an

Figure 307

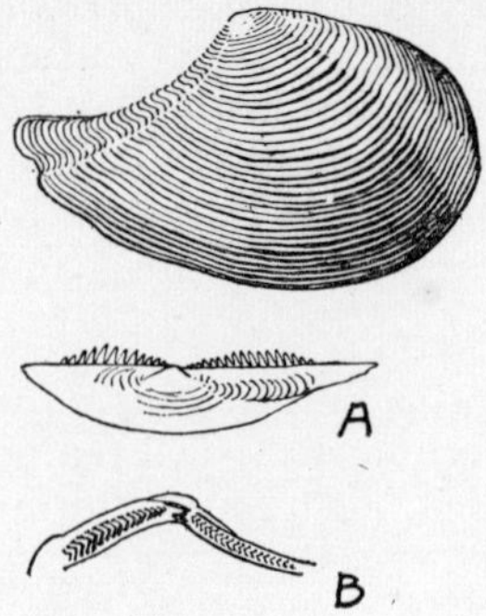

Figure 308

Figure 309

Figure 310

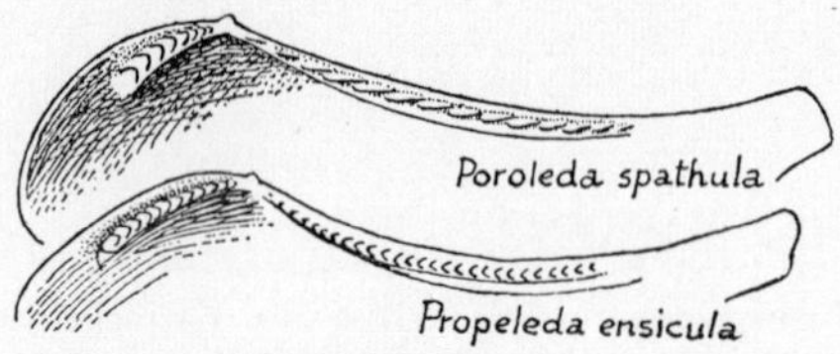

Figure 311

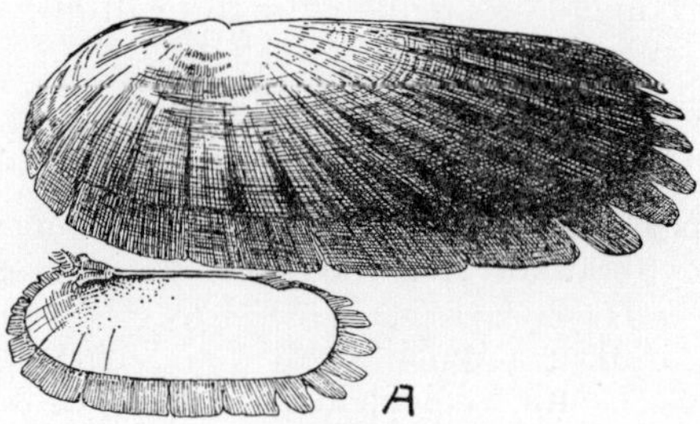

Figure 312

307. *Leionucula obliqua* × 1

308. *Scaeoleda crassa* × 1⅓
   A Dorsal view, showing hinge teeth
   B Interior lateral view of hinge

309. *Poroleda spathula* × 2

310. *Propeleda ensicula* × 2

311. Hinges of *Poroleda* and *Propeleda* × 4

312. *Solemya australis* × 1
   A Interior of valve × ½

elongated, narrow rostrum curving upwards; umbos situated at the anterior one-quarter of entire length of shell. On surface of valves concentric lines of growth are plainly visible. Interior shining. Hinge-line angular with about twelve closely set V-shaped teeth on the anterior side and fifteen to eighteen imbricating teeth on the posterior side of umbos (Figure 311). Size: ¾ inch.

Locality: Dredged in 40 fathoms off Ninety Mile Beach.

Another remarkable elongate Nut Shell somewhat resembeling *Propeleda ensicula* (Angas), but differing in the arrangement of teeth. It is flatter, more narrowly elongate and with the rostrum less acute. Its range of distribution is from New South Wales, Tasmania, through Bass Strait to South Australia.

Propeleda Iredale, 1924

Shell sickle-shaped with an elongate beak produced into a blunt end, the ligamental knob (chondrophore) sloping backwards; teeth thin, and arranged almost parallel to the edge.

TYPE SPECIES *Leda ensicula* Angas, 1871.

Sword Nut Shell

Propeleda ensicula (Angas, 1877)

(Figure 310)

Shell pale yellowish green, thin, fragile, depressed, ovately oblong, equivalve, very inequilateral, anteriorly rounded; posterior side attenuated into a long, narrow rostrum curving upwards; umbos incurved, small. Surface of valves showing fine, irregular growth-lines. Running from the umbos will be seen two, narrow, diverging ridges extending to the end of the rostrum; ventral margin convex. Interior polished. Hinge-line angular with numerous elongate, very acute, closely set teeth on each side of the umbos, especially on the posterior rostrate side (Figure 311). Size: ½ inch.

Locality: Dredged off Port Phillip, 33 fathoms on a sandy bottom.

A very frail and curiously shaped shell and no difficulty should arise in identifying the species. It was obtained originally off Port Jackson, New South Wales, on the tangles of the dredge, and occurs also in Tasmania.

Ledella Verrill & Bush, 1897 (*Ledella messenensis* Verrill & Bush, 1897).
miliacea (Hedley, 1902); off Ninety Mile Beach (40 fathoms).

Teretileda Iredale, 1929 (*Nuculana oculata* Iredale, 1925).
fortis (Hedley, 1907); off Ninety Mile Beach (40 fathoms).

Family SOLEMYIDAE

Date Shells

Shell elongate, rounded at each end, equivalve, inequilateral, gaping, thin, covered with a thick epidermis that extends beyond the shell margin as a fringe. Ligament inserted in a small oblique process. Hinge toothless.

Solemya Lamarck, 1818

Shell elongate, cylindrical, gaping at each end; epidermis dark brown; umbos posterior.

TYPE SPECIES *Solemya australis* Lamarck, 1818.

Iredale, 1931, erected *Solemyarina* for the Australian forms but as the type of *Solemya* is *australis* there is no valid reason for the change and his genus must be considered a synonym.

Southern Date Shell

Solemya australis Lamarck, 1818

(Figure 312)

Shell thin, fragile, elongated, gaping at both ends, anterior and posterior extremities obtuse, dorsal margin straight, ventral margin thin and sharp; equivalve inequilateral. Valves covered with a rich, chocolate brown epidermis, lighter at the umbonal region and extending some distance beyond the margins and developing into a sort of fringe. Surface of valves sculptured by broad flat ribs radiating from the umbos to margins, interstices shallow; concentric growth-lines are also plainly visible. A peculiar feature which is fairly constant is the presence of a slit in the 'fringe' at the termination of each rib. Interior greyish white; pallial line indistinct. Hinge without teeth, but with a callosity partly surrounding the posterior muscle scar, which, like the anterior scar, is well impressed, the latter being slightly larger. Size: 2 inches.

Localities: Shoreham; Anderson's Inlet; Otway coast.

This elongated bivalve is truly a remarkable shell, usually it is located burrowing in mud in about one or two fathoms of water. It is on record that the animals of this genus, which, in species, is sparsely represented throughout the world, have been seen leaping and swimming about in the water for some time without touching the bottom, the operation being performed by suddenly drawing in the foot at the same time that water is expelled from the posterior opening by the closing of the valves. Its range is extended to Tasmania, South Australia and Western Australia.

Family ARCIDAE

Ark Shells

Shells elongate or rounded, heavy, equivalve, with the posterior side produced. Sculptured with radiating ribs, usually with a thick, persistent periostracum. Ligament exterior. Hinge of two series of similar teeth meeting below the umbos.

Anadara Gray, 1847

Shell heavy, subcordiform or subquadrangular, equivalve, strongly, radially ribbed, with a thick periostracum. Hinge-line nearly straight with numerous teeth increasing in size regularly towards each end.

Type Species *Arca antiquata* Linné, 1758.

Mud Ark

Anadara trapezia (Deshayes, 1840)

(Figure 313)

Shell thick, somewhat boat-shaped, equivalve, very oblique, valves bearing about twenty-five, prominent, radiating, nodulose ribs wider apart at the posterior end and covered with a smooth, brown epidermis. Ligament

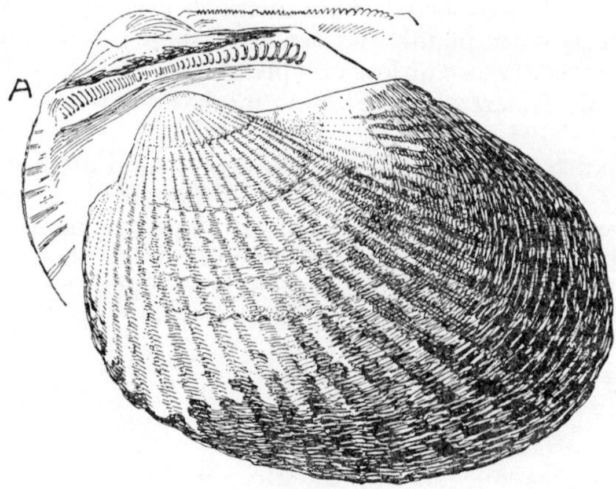

Figure 313

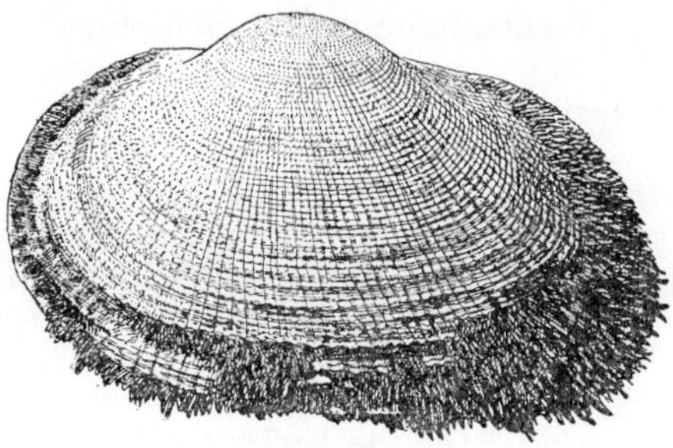

Figure 314

Figure 315

313. *Anadara trapezia* × 1
A Interior of valve, showing hinge
314. *Barbatia pistachia* × 1
315. *Barbatia squamosa* × 1

external, ligamental area broad with distinct cartilage grooves. Interior white, with muscle scars plainly discernible. Hinge a long unbroken line of somewhat oblong teeth more or less oblique at the outer ends. Size: 2½ inches.

Localities: Port Albert; Port Phillip; mud banks near Churchill Island, Western Port.

To early colonists it was known to be living near Port Melbourne, but is now apparently extinct in Hobson's Bay, although a few specimens have been taken recently in other parts of Port Phillip. Frequently referred to as Cockle or Mud Cockle but definitely with no affinity to shells associated with that name; it is a common shell on mud flats in New South Wales. For its size, it is an extremely heavy shell and the animal, although possessing a large foot, is slow in its movements and usually sunken in the mud for more than half its length, often with tufts of weed attached to the shell. The existence of great quantities at the aboriginal camps or kitchen middens indicates the importance of this mollusc as an article of food to the blacks. The genus possesses one species in Victoria but is widely represented, chiefly in tropical seas. It is recorded from Queensland, New South Wales, South Australia (subfossil), Tasmania (dead valves only), and south Western Australia.

Barbatia Gray, 1842

Shell oblong, oval or subquadrangular. Sculptured with fine ribs and covered with a coarse, hairy, persistent periostracum. Ligamental surface narrow. Hinge-line straight or slightly curved; teeth numerous, centrals smallest.

TYPE SPECIES *Arca barbata* Linné, 1758.

There has been some confusion regarding the correct usage of some of the genera consigned to Arcidae but the ruling of the International Commission on Zoological Nomenclature in 1944 that the type species of *Arca* Linné, 1758, should be *Arca noae* Linné, 1758, has allowed later workers to place other generic names. Thus *Barbatia* is no longer a synonym of *Arca* but has been reinstated for the *barbata* series of the family.

Hairy Ark

Barbatia pistachia (Lamarck, 1819)

(Figure 314)

Shell creamy white, concentrically stained with interrupted colour bands of brown or purple brown, covered with a thin epidermis of similar colour, and with bristles at the posterior end; equivalve, inequilateral, oblong, sides rounded; ventral margin slightly sinuate, valves gaping. Umbos prominent, directed towards the anterior extremity, separated by a grooved area upon which the ligament is spread. The shell is sculptured by radial riblets and fine concentric grooves, giving it a somewhat decussated appearance. Interior uniformly light brown. Hinge straight, consisting of numerous, regular teeth extending on each side of the umbo in each valve. Size: 2 inches.

Localities: St Kilda; Mordialloc; Frankston; Portsea; Geelong; Queenscliff; also in Western Port.

A common shell in Port Phillip, showing much variation in colour and contour. Ark is a term applied to numerous species, so called from their boat-like shape. Though sometimes free, they often anchor themselves by means of a strong byssus to rocks and stones, or may be located in holes, crevices and cavities of other shells. The present species is not confined to the shore, being frequently obtained in dredgings at many depths, and one of us (C.J.G.) recalls an occasion when a small stone was dredged in Western Port with numerous living examples in all stages of growth firmly attached. It is also found in Tasmania, New South Wales and Queensland.

Scaly Ark

Barbatia squamosa (Lamarck, 1819)

(Figure 315)

Shell yellowish white or light brown, more or less covered with a light brown epidermis; transversely oblong, inflated, equivalve, inequilateral; posterior side much the longer; anterior end rounded, short; posterior end larger and near the dorsal margin somewhat angled; ventral margin crenulated, flatly convex, and slightly sinuate, with valves gaping; umbos prominent, anteriorly incurved and separated by a grooved area upon which the ligament is spread; from the umbos to the posterior-ventral margin runs a strong keel defining the posterior area. Sculpture consists of fairly strong concentric ribs crossed by numerous, close radial ribs which are separated by fairly deep furrows giving the surface a cancellated appearance much resembling the tiles of a roof. Interior white; muscle scars fairly evident, about the same size and moderately rounded. Hinge-line slightly curved, narrow in the centre and widening at both ends, teeth regularly disposed. Size: 1 inch.

Localities: Port Phillip; Western Port; Otway coast.

A very common species on the whole coastline, located and living gregariously under stones from low water to several fathoms. Consistency in shape is not apparent, and much erosion often takes place. Easily recognized by its tile-like surface and the characteristic grooved area between the umbos. Found also in Tasmania, South Australia, Western Australia and New Zealand.

This species has been called *laminata* Angas, 1865, but the above name has priority.

Family GLYCYMERIDAE

Dog Cockles

Shell suborbicular, equivalve, nearly equilateral, smooth or radially ribbed, periostracum velvety; umbos central. Hinge with a semicircular row of transverse teeth; interior margin crenulate.

Glycymeris Da Costa, 1778

Shell smooth or with faint radial ribs, otherwise as described for the family.

TYPE SPECIES *Arca glycymeris* Linné, 1758.

Iredale, 1931, erected *Veletuceta* for the weakly ribbed Australian species with *Pectunculus flammeus* Reeve, 1843, as type. However, recent authors do not consider this is justified and in this we concur.

Flame Dog Cockle

Glycymeris flammeus (Reeve, 1843)

(Figure 316)

Shell solid, orbicular, ventricose, white or cream coloured, and in varying degrees stained with longitudinal flames or angular streaks of chestnut brown; equivalve, almost equilateral; umbos prominent, ligamental area deep with strong diverging grooves. Surface of valves smooth but faint concentric growth-lines are discernible under lens; covered with a dark brown epidermis rarely seen except on examples in living condition. Ventral margin well rounded, dentate within. Interior white stained with brown or a bluish tint; muscle scars very evident, pallial line entire. Hinge semicircular with about ten prominent teeth on each side of the umbos. Size: 2 inches.

Localities: Lakes Entrance, Port Albert.

This is the most highly coloured of the Victorian Dog Cockles, extremely variable, the splashes assuming all sorts of shapes and specimens with definite wavy or zigzag markings are not uncommon; a good percentage are almost albinos. Located on sandy beaches. Found also in Tasmania and New South Wales.

Tucetilla Iredale, 1939

Shell moderate size, a little globose, almost equilateral, equivalve; margin strongly denticulate. Surface sculptured with distant beaded ribs. Hinge shallowly arched, teeth few.

TYPE SPECIES *Glycymeris capricornea* Hedley, 1906.

Striated Dog Cockle

Tucetilla striatularis (Lamarck, 1819)

(Figure 317)

Shell solid, orbicularly ovate, normally light brownish white, variegated at the beaks with a few brown zigzag markings; equivalve almost equilateral; ventral margin well rounded, dentate within; umbos prominent; ligamental area deep, with strong diverging grooves. Sculpture consisting of flattish radial ribs which with the interstices are further ornamented with fine microscopic striae; under lens the concentric growth-lines are plainly visible, the whole surface invested with brown, soft, velvety epidermis. Interior white, stained with bluish brown; muscle scars well excavated; pallial line entire. Hinge semicircular with strong teeth. Size: 1½ inches.

Localities: Portland; Port Fairy; Port Phillip; Shoreham; San Remo; Lakes Entrance; Oberon Bay.

A very strong shell with a close ally in *radians* (Lamarck) from which it may be distinguished by its finer radiating microscopic striae. The umbonal variegation is not a constant feature as the valves are more or less splashed over the whole surface. Although of strong texture, specimens are frequently washed up perforated by the radula of some univalve mollusc. The species is gregarious, located on sandy banks, with a range of distribution from Tasmania through Bass Strait to South Australia and south Western Australia.

mayi (Cotton, 1947); off Wilson's Promontory, previously recorded as *Glycymeris tenuicostata* Reeve, 1843.

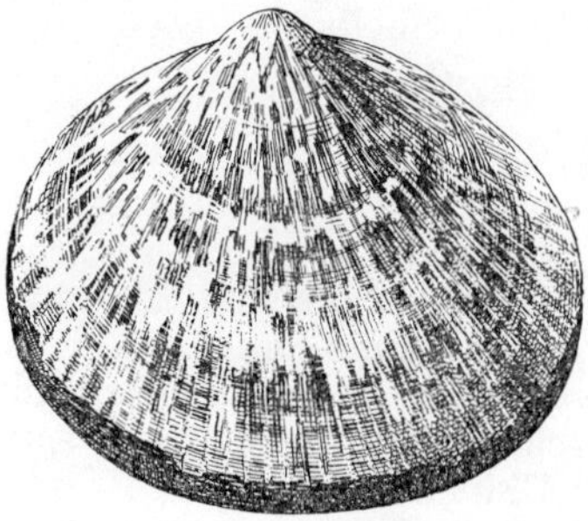

Figure 316

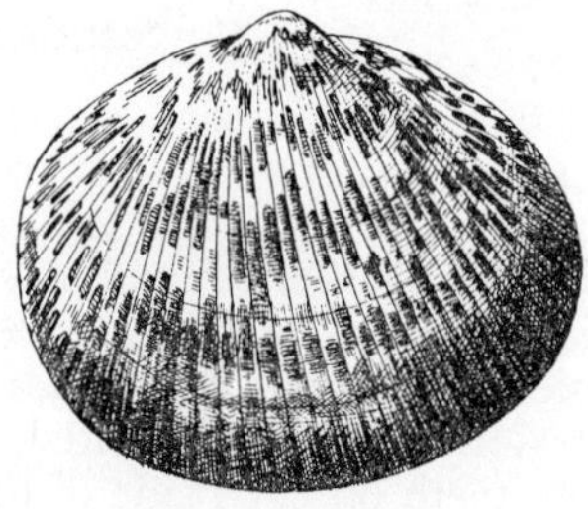

Figure 317

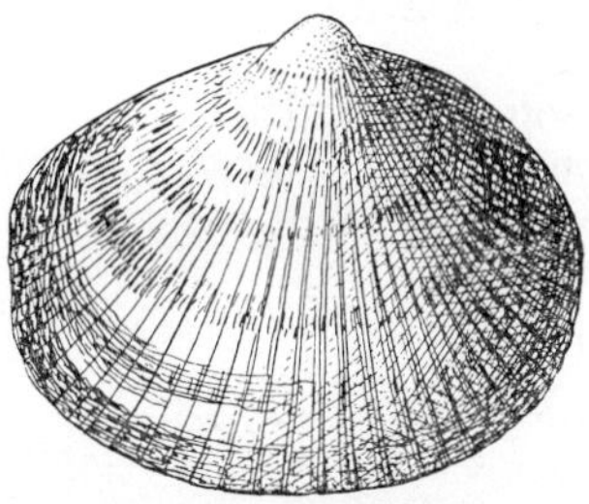

Figure 318

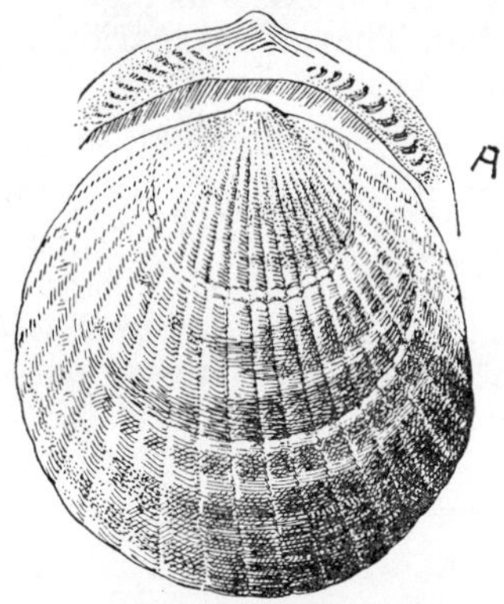

Figure 319

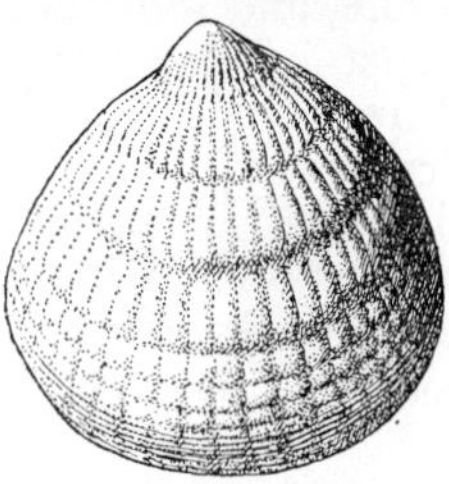

Figure 320

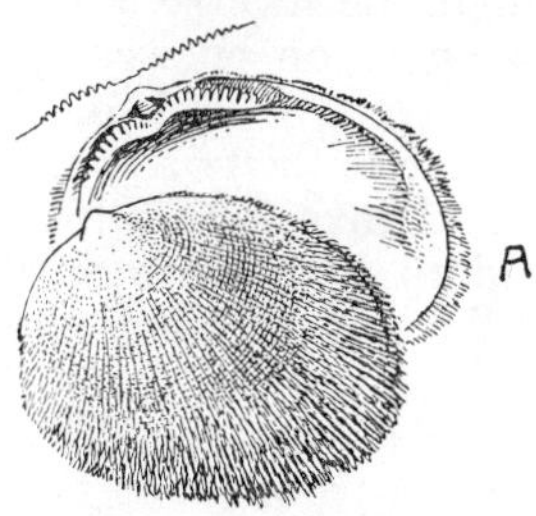

Figure 321

316. *Glycymeris flammeus* × 1
317. *Tucetilla striatularis* × 1
318. *Tucetilla radians* × 1⅓
319. *Tucetona flabellatus* × 1
    A Interior of valve, showing hinge
320. *Tucetona sordidus* × 1
321. *Limopsis tenisoni* × 1
    A Interior of valve, showing hinge

Radiating Dog Cockle or Comb Shell

Tucetilla radians (Lamarck, 1919)

(Figure 318)

Shell cream or yellowish brown, somewhat stained, in a more or less zoned manner, with reddish brown, covered with a brown velvety epidermis; solid, orbicularly ovate, equivalve, subequilateral. Sculpture consisting of regular, radiating ribs, which are longitudinally striated and sometimes rather obsolete. Ligament external, its area with diverging grooves. Interior white, ventral margin denticulated. Hinge semicircular, with a series of small, strong teeth on each side of the umbos. Size: 1½ inches.

Localities: Portsea; Point Nepean; also Western Port.

Variation exists as regards coloration, some examples being considerably darker. They are gregarious, and are found on sandy banks. Recorded also from Western Australia, South Australia and Tasmania.

Tucetona Iredale, 1931

Shells with strong ribs. Few hinge teeth set on a deeply curved hinge plate, and strongly denticulate internal margin.

TYPE SPECIES *Glycymeris flabellatus* Tenison Woods, 1877.

Thiele, 1935, considers *Tucetona* to be a synonym of *Pectunculus* Lamarck, 1799, but as the latter is preoccupied by Da Costa, 1778, it is necessary to use Iredale's name.

Fan-like Dog Cockle

Tucetona flabellatus (Tenison Woods, 1877)

(Figure 319)

Shell solid, ventricose, orbicularly ovate, in the juvenile slightly longer than high, the reverse in the adult stage; creamy white, and sometimes with brown spots or encircling bands; equivalve, almost equilateral; ventral margin strongly arched, dentate within. Sculpture consisting of about twenty-five to thirty strong, regular radiating ribs crossed by fine concentric striae and invested with a thin, light brown epidermis. Ligamental area fairly deep with diverging grooves. Interior white stained with brown; pallial line entire. Hinge semicircular with a series of teeth on each side of the umbos; muscle scars very distinct, much excavated and consistently coloured dark brown. Size: 1½ inches.

Localities: Portland; Lakes Entrance.

This is an ornate species inhabiting sandy coasts but seldom washed up on the beaches with conjoined valves. It is easily separated from the Victorian representatives of the genus by its strong radiating ribs and concentric striae, and like the other Dog Cockles from Victoria is frequently seen with the peculiar shell of *Myochama* adhering. It is recorded from Tasmania and South Australia.

Sordid Dog Cockle

Tucetona sordidus (Tate, 1891)

(Figure 320)

Shell solid, triangularly ovate, convex; colour sordid white with a few small reddish brown spots near the umbos and blotches or flames of similar colour crossing the hinge-line; equivalve, almost equilateral. Valves sculptured with about twenty flatly rounded radiating ribs and further ornamented with extremely fine, closely set concentric striae. A striking feature is the rest periods in the growth of the shell to the number of about four to six, producing concentric steps from ½ to 1 millimetre in depth, becoming less evident towards the rounded ventral margin. Ligamental area narrow, with several diverging grooves. Interior white, more or less tinged with chestnut; muscle scars well defined; pallial line entire; ventral margin strongly dentate within. Hinge semicircular with about twelve prominent teeth on each side of the umbo. Size: 1½ inches.

Locality: 40 fathoms off Wilson's Promontory.

Its dirty white surface, very triangular form, and peculiarly stepped surface provide features making it easily distinguishable from any Victorian Dog Cockle. It is not of frequent occurrence and unknown to the writers from any other Victorian locality. Dredgings at depths of 9 to 22 fathoms in various parts of South Australia have yielded numerous specimens in excellent condition, and in 100 fathoms seven miles each of Cape Pillar, Tasmania, worn and separate valves were taken.

Family LIMOPSIDAE
False Dog Cockles

Shell orbicular, convex, slightly oblique. Ligamental area with a triangular cartilage pit in the centre. Hinge with two curved series of transverse teeth.

Limopsis Sasso, 1827

Shell oval, slightly oblique posteriorly, periostracum hairy; umbos central, prominent. Hinge plate thick, broad, arched with two short series of teeth.

TYPE SPECIES *Arca aurita* Brocchi, 1814.

Iredale, 1931, erected the genus *Versipella,* with *V. soboles* Iredale, 1931, as type, for the *tenisoni* series; however, we consider the differences are only specific and do not warrant generic separation.

Tenison's False Dog Cockle
Limopsis tenisoni Tenison Woods, 1877
(Figure 321)

Shell orbicular, at times somewhat oblique, thick, tumid, dirty white or cream; equivalve, inequilateral; umbos small. Ligamental area narrow. Sculptured with about thirty-six rounded, subgranulose, radiating ribs, occasionally with fine riblets between, crossed by regular, fine, close concentric striae, the whole surface giving a cancellated appearance, markedly near the umbos; valves covered with a pilose epidermis extending far beyond the margins. Interior white, radiately striated; muscle scars well defined, pallial line entire. Hinge semicircular with about ten to twelve teeth on each side of the umbos. Size: 1 inch.

Localities: Inverloch; Flinders; San Remo; Warrnambool; Airey's Inlet; Portland; 38 fathoms off East Moncoeur Island.

Its peculiar shape and cancellated surface are features sufficient to identify the species. On our shores it is not of frequent occurrence but in South Australian waters it has appeared in abundance through the dredging operations, at all depths from 10 to 150 fathoms. At the last-named Victorian locality it was obtained on a sea-bed of sand and shell, and is recorded also from Tasmania and New South Wales as well as South Australia.

Cyrillona Iredale, 1929 (*Cyrilla dalli* Hedley, 1902).
dalli (Hedley, 1902); off Ninety Mile Beach (40 fathoms).

Lissarca Smith, 1877 (*Lissarca rubrofusca* Smith, 1877).
picta (Hedley, 1899); Western Port.
rhomboidalis Verco, 1907; dredged between Phillip and French Islands, Western Port.
rubricata (Tate, 1886); Torquay; Port Phillip; Western Port.

Philobrya Carpenter, 1872 (*Philobrya setosa* Carpenter, 1872).
fimbriata Tate, 1898; Port Phillip Heads (7-9 fathoms).
pectinata Hedley, 1902; Ninety Mile Beach (40 fathoms).

Notomytilus Hedley, 1916 (*Philippiella rubra* Hedley, 1904).
rubra (Hedley, 1904); Western Port (7 fathoms); Portsea; Torquay.

Micromytilus Cotton, 1931 (*Myrina crenatuliferus* Tate, 1892).
crenatuliferus (Tate, 1892); Barwon Heads; Flinders; Western Port.

Order ANISOMYARIA

Family MYTILIDAE

Mussels

Shell elongate, equivalve, inequilateral; with a periostracum usually thick and closely adhering. Ligament usually external, deep seated. Anterior adductor muscle impression small and narrow, posterior large and obscure. Animal generally attached by a byssus.

Solamen Iredale, 1924

Shell elongate globose, translucent, equivalve, inequilateral, slightly restricted dorsally and expanded ventrally. Sculpture of radiating ribs crossed by concentric growth-lines which may give a cancellated appearance; margin finely denticulate, the denticles corresponding to the ends of the radial ribs. In some shells these denticles are stronger on each side of the hinge area.

TYPE SPECIES *Solamen rex* Iredale, 1924.

Iredale, 1929, erected *Exosiperna* with *Arcoperna scapha* Verco, 1908, as type, for the small, more solid shells with a cancellated sculpture and later authors have stressed that species assigned to *Solamen* lack the crenulated margin of *Exosiperna*. This we do not find to be correct; in *Solamen s.s.* the crenulation of the margin is very fine but can be clearly seen with a glass of ten diameters magnification.

The species recorded here in *Solamen* were originally placed in an Eocene genus, *Arcoperna* Conrad, 1865, but later authors agree with Iredale that they do not belong there.

Boat Mussel

Solamen recens (Tate, 1897)

(Figure 322)

Shell thin, translucent, vitreous; obliquely oval, much inflated; rounded at the extremities, much more dilated behind than in front; umbos very large, obliquely inclined forwards, cordiform and submarginal. Posterior hinge-line arched, six millimetres long, with a well-defined subinternal, ligamental groove; anterior hinge-line nearly straight, about one-third the length of the posterior part. Exterior surface ornamented with concentric folds and intervening striae, and by radial costules; the intercrossing of the two sets of threads producing a fine reticulate ornament. The radial costules are sub-acute, serrated by the concentric striae, and are a little wider than the intervening sulci; in the medial part of the ventral margin there are five costules to each millimetre of width; the costules sharply serrate the margin. The adductor impressions are indicated by greater translucency of the shell; the posterior one is large and rounded, the anterior one is smaller, but conspicuous, narrow-oblong, and situated close under the anterior margin of the hinge-line. Size: ¾ inch.

Locality: Ninety Mile Beach (40 fathoms).

A rare shell. The above is the author's original description and should serve to identify the species. The fragile nature, inflated form in addition to the fine, microscopic, reticulate, ornament and vitreous lustre of the interior are outstanding features. Found also in South Australia and taken alive off Point Esperance, Tasmania.

Little Boat Mussel

Solamen scapha (Verco, 1908)

(Figure 323)

Shell small, solid, white, oval, inflated; umbos prominent, directed slightly to the anterior; equivalve, inequilateral, ventral margin rounded, finely crenulate within. The author describes the sculpture as follows: 'The whole dorsum is engraved with crowded, slightly curved radial lines, latticed by concentric lines nearly as valid; some of the latter at distant intervals in the early stages of growth are much deeper, so as to form slightly imbricating ledges.' Interior white; muscle scars plainly seen, somewhat elongated, close to margin, the posterior slightly larger. Hinge-line post-dorsally provided with many, closely set, low vertical teeth on its inner surface; anterior-dorsal border with a few denticulations and one tooth. Size: ⅕ inch.

Localities: Off Wilson's Promontory; Bass Strait cable.

A rare species easily recognized. Recorded also from South Australia and Tasmania.

Dacrydium Torell, 1859 (*Mytilus vitrea* Holböll in Möller, 1842).

radians Suter, 1908; off Rhyll, Western Port (6 fathoms). This shell was described by Gatliff & Gabriel, 1912, as a new species *Modiolaria rhyllensis* but they later, 1916, accepted it as a synonym of Suter's New Zealand species. Iredale, 1936, erected the subgenus *Quendreda* with *Dacrydium fabale* Hedley, 1904, as type and the present species belongs with it.

Figure 322

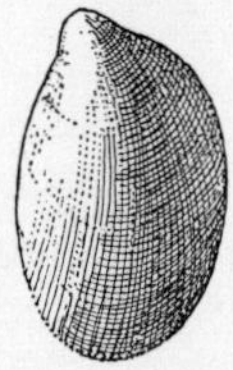

Figure 323

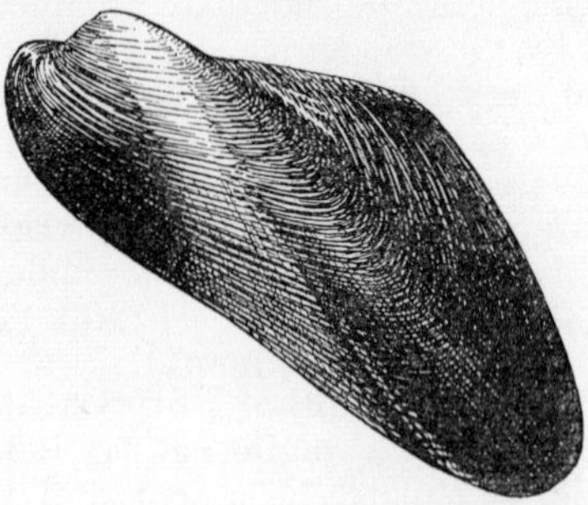

Figure 324

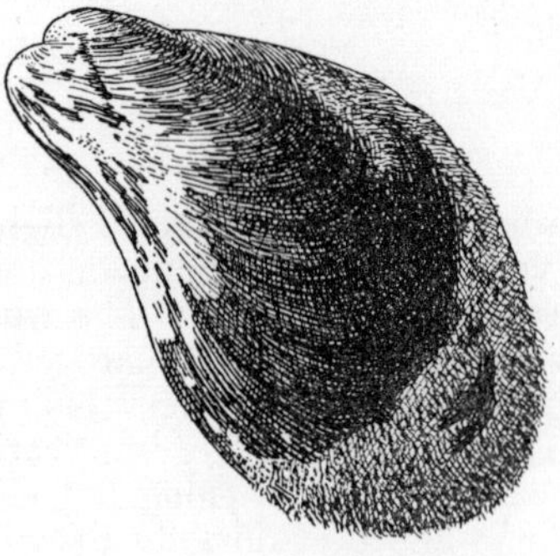

Figure 325

Figure 326

322. *Solamen recens* × 1
323. *Solamen scapha* × 5
324. *Modiolus albicostus* × ⅔
325. *Modiolus cottoni* × ⅔
326. *Modiolus inconstans* × 1

Modiolus Lamarck, 1799

Shell smooth, elongate, with a slightly curved ligamental margin and curved dorsal margin; umbos placed slightly behind the rounded anterior margin which is bent outwards; the shell is narrow anteriorly and expanded posteriorly. The periostracum is usually thick, and may bear hair-like processes.

TYPE SPECIES *Mytilus modiolus* Linné, 1758.

Modiolus are usually inhabitants of moderately deep water and have the habit of burrowing or nestling in a mass of byssal fibres in which broken shells and stones, etc., are interwoven.

White Ribbed Horse Mussel

Modiolus albicostus (Lamarck, 1819)

(Figure 324)

Shell brown, with a distinct whitish ray; somewhat triangularly oblong, inflated, with a broad, prominent, diagonal elevation, equivalve, inequilateral. Surface of valves smooth, concentrically striated; anterior end very short, narrow, slightly gaping to admit the passage of a byssus; posterior side elongated. Ligament long, partly external. Interior somewhat pearly. Size: 4 inches.

Localities: Portsea; Point Lonsdale.

Beyond Victoria, the species is recorded from Western Australia, South Australia, New South Wales and Tasmania. Iredale, 1924, rejected *albicostus* for this species on the grounds of its invalidity and proposed *delinificus* to replace it. However, recent authors have not been able to confirm Iredale's opinion and as there is no doubt to which species reference is made we prefer to retain the above well-known name.

Cotton's Bearded Horse Mussel

Modiolus cottoni Laseron, 1956

(Figure 325)

Shell large, thin, very oblique; very narrow at the anterior end, expanded posteriorly; dorsal margin nearly straight, angled with the posterior margin which is regularly curved; ventral margin sulcate just below the umbo, anterior margin small and auriculate; umbos not large, and near anterior margin. Colour pale yellow brown. Periostracum hirsute, the hairs near the posterior margin being long, thin, flattened and not forked. Interior pure white; adductor muscle scars hardly impressed, the anterior small and close to the umbo, the posterior larger, rather wide, close to the posterior extremity. Hinge-line long and nearly straight, edentulous, the ligament internal, thin and weak, set in a very narrow submarginal groove. Size: 3 inches.

Localities: Mornington; Portsea; Point Nepean; Queenscliff; Western Port.

Quite common and easily recognized. The peculiar shape and beard-like fringes on the posterior side are very characteristic. This species occurs also in southern New South Wales, Tasmania, South Australia, and Western

Australia. The Australian shells were originally placed in *australis* Gray, 1827, an indeterminate species based on an inadequate description made on one unlocalized water-worn valve.

In 1923, Hedley grouped all the Australasian shells of this type under the name of the New Zealand species *areolatus* Gould, 1850. Laseron, 1956, showed that the Australian shells differ from that species, and also that there is more than one species represented in the series. He named the smaller, stouter, northern New South Wales shell *Modiolus peronianus.*

Variable Horse Mussel

Modiolus inconstans (Dunker, 1850)

(Figure 326)

Shell oblong-ovate, equivalve, very inequilateral. Surface of valves smooth, with growth-lines plainly evident, the whole invested with a shining, brown epidermis. Ligament linear, almost internal and placed in a narrow canal. Interior of a bluish tint; muscle scars very unequal. Size: 1 inch.

Localities: Hobson's Bay; Williamstown; Sandringham; Corio Bay.

A species fairly consistent in shape but variable in colour from light brown to almost black. Extremely common, located in mud in shallow water. *semivestita* Tate, 1887, is a synonym of this species which also occurs in South Australia and Western Australia.

Little Black Horse Mussel

Modiolus pulex (Lamarck, 1819)

(Figure 327)

Shell bluish black, smooth, inflated, covered by a blackish, shining periostracum; from the umbo to the posterior portion of the basal margin runs a rather strong rounded keel. Sculptured with well-defined, concentric growth-lines. Interior brownish blue with slight iridescence; muscle scars long and narrow. Size: 1 inch.

Localities: Port Phillip; Western Port; Wilson's Promontory; Cape Otway; Port Fairy; Lakes Entrance.

It is an extremely common species living in sheets on rocks in areas where sand and rocks meet and on jetty piles above mid-tide level; the tiny bivalve *Kellia australis* Lamarck is often entangled in the byssal threads. Occurs also in Tasmania, South Australia, Western Australia and New South Wales. *M. confusus* Angas, 1871, is a synonym.

Victorian Horse Mussel

Modiolus victoriae (Pritchard & Gatliff, 1903)

(Figure 328)

Shell elongate-oblong, tumid, with broad umbonal region, a broad and well-marked anterior, and a posterior rather remarkable for its uniformity of height; ventral margin slightly sinuated; umbos tumid, only slightly separated, and strongly incurved towards the anterior; lunule ill-defined,

Figure 327

Figure 328

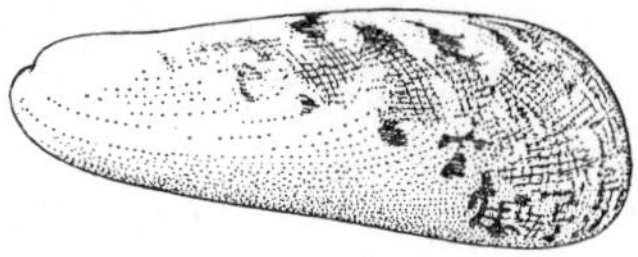

Figure 329

Figure 330

Figure 331

327. *Modiolus pulex* × ⅔
328. *Modiolus victoriae* × 1
  A Anterior elevation
329. *Amygdalum beddomei* × 1
330. *Hormomya erosa* × 1
331. *Brachidontes rostratus* × 1⅓

scarcely excavated; anterior margin regularly convex from the umbos to the ventral margin; colour rich dark brown, with the convex umbo-ventral keel of a lighter shade. The surface is marked by irregular lines of growth and fine parallel striae, with faint radial striae on the anterior region, tending to become obsolete posteriorly. Interior of a bluish tint. Size: 1½ inches.

Locality: Dredged alive in 6-8 fathoms off Rhyll and Tortoise Head, Western Port.

Very little variation in contour is evident; the colour is constant and compared with other Victorian species of the genus, *victoriae* may be easily separated by its regular tumidity and consistent uniform height.

It is also recorded from New South Wales, being dredged in shallow water in Twofold Bay.

lineus (Hedley, 1906); Ninety Mile Beach (40 fathoms).

vexillum Reeve, 1857; Port Melbourne; Gippsland Lakes.

Fluviolanatus Iredale, 1924 (*Modiolarca subtorta* Dunker, 1857).

amarus Laseron, 1956; Gippsland Lakes; Betka River; Port Fairy.

Amygdalum Megerle von Muehlfeld, 1811

Shell very inequilateral, equivalve, elongate, cylindrical, thin, smooth, translucent; umbos terminal; ligament weak, no hinge teeth.

TYPE SPECIES *Amygdalum dendriticum* Megerle von Muehlfeld, 1811.

There has been some confusion regarding the author of this genus and the above form and spelling is taken from the British Museum Catalogue.

Beddome's Mussel

Amygdalum beddomei Iredale, 1924

(Figure 329)

Shell oblong-cylindrical, smooth, shining, translucent white, covered by a very smooth, pale yellow epidermis; posterior slope with purplish brown divaricating lines forming a more or less network pattern. Interior white. Size: 2 inches.

Localities: Dredged alive in 6-8 fathoms off Rhyll and Tortoise Head, Western Port; Lakes Entrance.

One of our prettiest shells and easy to recognize. It is extremely fragile and requires careful handling. The animals form nests for themselves with the copious threads of the byssus. Recorded also from Tasmania, New South Wales, south Queensland and Western Australia. The form from the last locality shows slight variation and ultimately may be separated.

Hormomya Mörch, 1853

Mussels with anterior umbos, with several teeth (four or five in the type) of which the anterior ones are the stronger, followed by a few crenulations along the anterior part of the ventral margin, posterior ventral margin and posterior and dorsal margins crenulate. Sculpture unilateral bifurcating lines or ribs, below the strong keel four or five separate lines may arise from a single rib. Mantle has small contractile papillae.

TYPE SPECIES *Mytilus exustus* Linné, 1758.

Rough Beaked Mussel

Hormomya erosa (Lamarck, 1819)
(Figure 330)

Shell brown, dark, horn coloured or opal green, when periostracum is removed by erosion whitish, streaked with blue or purple; elongate, narrow anteriorly, expanded posteriorly; beaked, equivalve inequilateral, keeled. Radially coarsely ribbed, sculpture much finer below the keel. Several hinge teeth, the more anterior ones large but rapidly tapering ventrally to crenulations. Size: 2 inches.

Localities: San Remo; Portland.

A common species in sheltered waters, found on rocks between tides. The South Australian conchologists Cotton and Godfrey have noted that the shells are sometimes embedded by their posterior ends in a firm spongy green seaweed deforming the posterior half of the shell by compression in a remarkable manner. Recorded from Tasmania, South Australia and Western Australia.

Brachidontes Swainson, 1840

Mussels with simple bifurcating sculpture, umbos not quite terminal. Hinge with a few prominent teeth (three in type species) set on a plate immediately below the umbos, followed by a series of crenulations on the anterior margin which continue the hinge at a different angle. The posterior mantle margin furnished with branched tentacles.

TYPE SPECIES *Modiola sulcata* Lamarck, 1819 = *Arca modiolus* Linné, 1767.

For many years the species listed here as *Hormomya* and *Brachidontes* were both grouped in the latter genus. Laseron, 1956c, reviewed the New South Wales Mytilidae and pointed out the differences in the hinge-line of the two genera. He suggested that *Hormomya erosa* was close to *Brachidontes* but *B. rostratus* required a new genus and erected *Austromytilus* for it. Unfortunately Laseron depended on Swainson's incomplete and inexact generic description which does not correspond to the type *B. modiolus* which is as defined above. Therefore *B. rostratus* must be retained in *Brachidontes* and *Austromytilus* Laseron becomes a synonym. A detailed review of these genera was made by Soot-Ryen, 1955.

The spelling adopted for this genus is that originally used by Swainson. Later authors considered it incorrect and amended it to *Brachyodontes,* but modern usage is to revert to the original unless it is an obvious misspelling or typographic error.

Beaked Mussel

Brachidontes rostratus (Dunker, 1857)
(Figure 331)

Shell purplish black, rather elongated, beaked, equivalve, inequilateral; radially finely striated; anterior and posterior extremities rounded; gaping near the anterior to admit passage of a byssus. Ligament long, deep-seated, partly internal. Interior purplish or bluish black. The number of teeth on the hinge plate is variable, two or three in the left valve and one or two in the right; these are followed at an oblique angle on the anterior margin by a series of small denticles tapering posteriorly. Size: 1½ inches.

Localities: Portsea; Point Nepean; Queenscliff; Western Port; Cape Otway; Cape Bridgewater.

Attached by the byssus to exposed rocks, this species is extremely abundant and gregarious at midtide level on ocean rock platforms. With finely striated ornament and the umbos produced into a sharp beak, little difficulty should be experienced in its identification. It is mostly black, only showing purple when worn. Recorded also from South Australia, Tasmania and southern New South Wales. The small bivalve *Kellia australis* is always associated with this species living in the shelter of the byssus threads.

Trichomya Ihering, 1900

Mussels with simple bifurcating sculpture and fibrous-hirsute periostracum; umbos terminal. Hinge with two or three small knob-like teeth in right valve not interlocking but fitting into an irregular pit, followed by a tapering series of crenulations that extend almost to the centre of the ventral margin.

Type Species *Mytilus hirsutus* Lamarck, 1819.

Hairy Mussel

Trichomya hirsutus (Lamarck, 1819)

(Figure 332)

Shell dark brown, strongly curved, surface of valves with very fine riblets and covered with long, barbed hairs. Under lens the presence of several barbs in each hair is revealed—a consistent character. Posterior-dorso-ventral margin crenulated. Interior of a bluish tint, iridescent. Hinge as described for the genus. Size: 2½ inches.

Localities: Cape Otway; Kilcunda; Mallacoota.

Not uncommon on the beaches. Found living in masses on rocks and sea-bottom at about 6 to 8 fathoms. It occurs in Queensland, New South Wales, Tasmania and South Australia.

Lanistina Gray, 1847

Shell divided into three areas, anterior and posterior areas sculptured with radial ribs, central area smooth; valves rounded, inflated; periostracum persistent, shining. Interior translucent; muscle scars faint, anterior adductor in front of umbos, posterior retractor has a strong, elongated branch fastened along the dorsal margin and a single slender branch above the posterior adductor.

Type Species *Mytilus discors* Lamarck, 1819 (non Linné) = *Modiolaria impacta* Hermann, 1782.

The members of this genus differ from *Musculus* Röding, 1798, in their more inflated form and the form of the posterior retractor muscle. The differences have been recognized for many years but confusion has arisen over the generic names applied. Swainson, 1840, erected *Lanistes* with *M. discors* Lamarck as type; but this was preoccupied. Gray, 1843, used

Modiolarca for *M. impacta,* but this had previously been used by him for a quite distinct shell in 1840. Then in 1847 he erected *Lanistina* to replace *Lanistes* Swainson. Soot-Ryen, 1955, outlined this history using *Modiolarca,* but did not realize that it had been used by Gray previously to 1843 for another shell. A second source of confusion has been the type species; Iredale, 1915, designated *Mytilus discors* Linné, 1767, an American shell, as type of *Musculus* and not all later authors have realized that *Mytilus discors* Lamarck, 1819, is not this species. *Lanistina* must therefore be used for this group and is not a synonym of *Musculus.*

Three-area Mussel

Lanistina impacta (Hermann, 1782)
(Figure 333)

Shell thin, fragile, tumid, brown, sometimes olive tinted; valves divided into three areas with the anterior and posterior ends strongly ribbed, the intermediate area usually devoid of radial sculpture; this last feature is not constant as the shell is sometimes more or less radially ribbed throughout, the whole surface showing distinct concentric lines of growth. Epidermis coriaceous, olive tinted and slightly shining. Umbos incurved, situated near the anterior end. Ligament short, mostly internal. Interior greyish blue, iridescent through which the radial sculpture is plainly seen; muscle scars large, pallial line continuous and simple. Margins of valves smooth at the intermediate area but crenulated anteriorly and posteriorly. Size: 1 inch.

Localities: Western Port; dredged off Wilson's Promontory.

These mussels are gregarious, living in crevices of rocks at low-tide level to 25 fathoms, sometimes embedded in sponge-like byssal masses. The animal possesses a long, tapering foot enabling it to crawl rather rapidly by extending it anteriorly and fixing the point, movement taking place by dragging itself forward. This species has been frequently called *Musculus cumingianus* Reeve, 1857, but having examined a series we are of the opinion that the Australian and New Zealand shells are conspecific, and that the South Australian shell called *M. nanum* Cotton & Godfrey, 1938 (non Dunker, 1856) also falls into synonymy. If this is so the species has a range from southern Western Australia to southern Queensland, Tasmania and New Zealand.

ulmus (Iredale, 1936); a smaller, variably coloured species with finer sculpture found living gregariously in association with Tunicates. It occurs in most Victorian bays and was dredged by the *Challenger* in 6-15 fathoms in Port Jackson, New South Wales. Mistakenly identified as *M. cuneatus* Gould, 1861, a South African species.

paulucciae (Crosse, 1863); a small thin form of emerald green turning to yellow, shining. Occurs also in Western Australia, South Australia, Tasmania and New South Wales.

Gregariella Monterosato, 1884

Shell divided into three areas, anterior and posterior areas radially ribbed, central area smooth; covered with a periostracum which is hirsute on the posterior half. Hinge with rudimentary, oblique teeth and the dorsal margin crenulate.

Figure 332

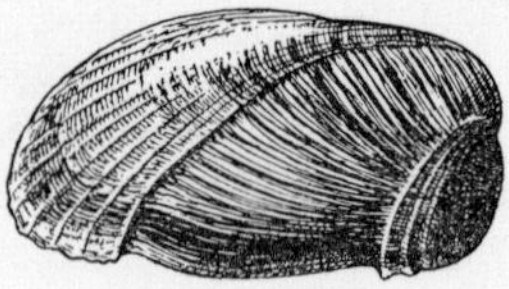

Figure 333

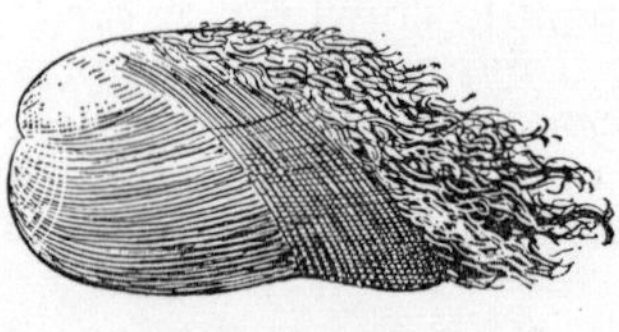

Figure 334

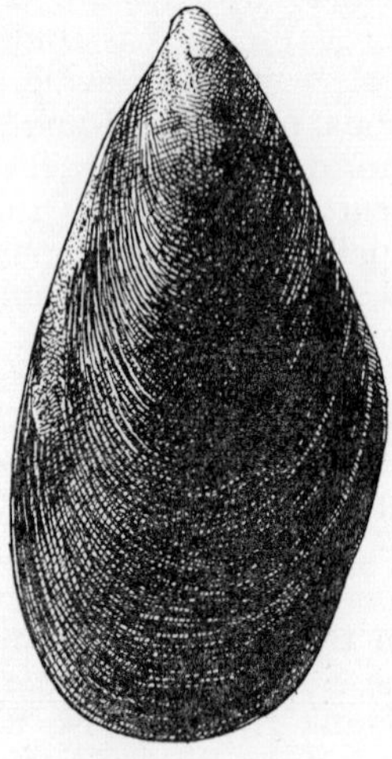

Figure 335

332. *Trichomya hirsutus* × ⅔
333. *Lanistina impacta* × 1⅓
334. *Gregariella barbatus* × 4
335. *Mytilus planulatus* × ⅔

Type Species *Modiolus sulcatus* Risso, 1826 (non Lamarck, 1805, 1819) = *Modiola opifex* Gray, 1825.

Iredale, 1924, erected the genus *Trichomusculus* for the Australian hirsute forms of *Musculus* with *Lithodomus barbatus* Reeve, 1858, as type, but Soot-Ryen, 1955, considers it a synonym of the above genus and in this we concur.

Hairy Three-area Mussel

Gregariella barbatus (Reeve, 1858)
(Figure 334)

Shell small, horn coloured or yellow, fragile, oblong ovate, tumid, hirsute posteriorly. Running from the umbo to the posterior end of the basal margin will be seen a broadly rounded angle. Posterior and anterior ends sculptured with fine radial riblets. The generic character of 'three areas' on the valves is plainly evident with the median section quite smooth excepting for lines of growth. Epidermis shining, produced posteriorly in rather long bristles. Interior white, showing slight iridescence through which the radial sculpture is plainly discernible; muscle scars slightly impressed. Size: ½ inch.

Localities: Frankston; Ocean Beach, Portsea; dredged Western Port (8 fathoms); off Cape Howe and Wilson's Promontory.

The species is located embedded on a sandy mud bottom. Also found in Tasmania, New South Wales, South Australia and Western Australia.

Mytilus Linné, 1758

Shell wedge-shaped, inequilateral, equivalve, umbos terminal, acute; periostracum thick, smooth. Hinge with few, small, teeth on the short thickened anterior margin; anterior adductor muscle scar small, anterior retractor elongate and posterior to the umbos, large posterior adductor and retractor scars continuous, resilial ridge pitted.

Type Species *Mytilus edulis* Linné, 1758.

Mussel

Mytilus planulatus (Lamarck, 1819)
(Figure 335)

Shell purplish black, in juvenile state invariably brown; somewhat wedge-shaped, moderately inflated; equivalve, very inequilateral; umbos acute, terminal and anterior; posterior extremity broad, rounded, near the anterior, slightly gaping to admit the passage of a byssus. Ligament long, deep-seated, partly internal. Interior bluish white. Size: 5 inches.

Localities: Port Melbourne; Brighton; Mornington; Williamstown; Western Port.

This species is extremely common in Port Phillip where they are found in countless numbers on exposed rocks and jetty piles. The mode of attachment, characteristic of the genus, is by a byssus. The animal is much sought after for food, being eaten raw or boiled, and is regarded by many as one of the most savoury of shell-fish; as a bait for fishing it appears to have few superiors. For some time this mussel has been attacked by the large starfish, *Coscinasterias calamaria,* a common species in all parts of Port Phillip.

Common in Western Australia, South Australia, New South Wales and Tasmania.

The Pea-Crab *Pinnotheres* is frequently found within the shells of this common mussel; it is a curious and interesting little creature, the female of which is commensal in bivalve and, more rarely, in univalve molluscs.

Family VULSELLIDAE

Sponge Finger Oysters

Oyster-like shells, inequivalve. Ligament usually with several cartilagenous furrows, rarely with only one; gill lamellae as a rule smooth, byssal threads simple.

Foramelina Hedley, 1914

Large oyster-like shells, in which the byssus, instead of escaping by a lateral notch, penetrates the right valve by a median subumbonal orifice and groove. A suture extends from the orifice posteriorly to the edge of the hinge. Ligamental furrows narrow and crowded. Adductor muscle single, concentrated and reduced.

TYPE SPECIES *Foramelina exempla* Hedley, 1914.

Perforated Oyster
Foramelina exempla Hedley, 1914
(Figure 336)

Shell equivalve, equilateral, large, flat, discoidal, margin irregular in outline. Valves thick, composed of brittle, imbricating lamellae, which both include, and are overlaid by, a thin membranous epidermis, where eroded of a silvery sheen. Interior dark bronze. Perforation in the right valve about 10 millimetres in diameter, median and subumbonal, internally with a raised margin. Byssus a dense bundle of threads about ½ inch long. Pallial margin entire, about ½ inch within the ventral margin. Size: 5 inches.

Locality: South from Gabo Island (100-250 fathoms).

The above is from the author's original description. An extraordinary bivalve, *exempla* alone representing the genus; one living specimen and a separate right valve obtained by the trawler *Endeavour*, providing the material for description. It is quite possible valves may be cast upon the shore in eastern Victoria and with their peculiar outline they should be easily recognized.

Vulsella Röding, 1798

Shell oblong, striated, subequivalve, with an inner pearly and outer fibrous layer; umbos straight, earless.

TYPE SPECIES *Vulsella major* Röding, 1798 = *Mya vulsella* Linné, 1758 (by tautonymy).

Members of this genus are usually found embedded in sponge.

Sponge Finger
Vulsella spongiarum Lamarck, 1819
(Figure 337)

Shell horn coloured or brown, generally paler at the umbos and sides of valves, elongated, irregular, inequivalve, inequilateral, very variable in

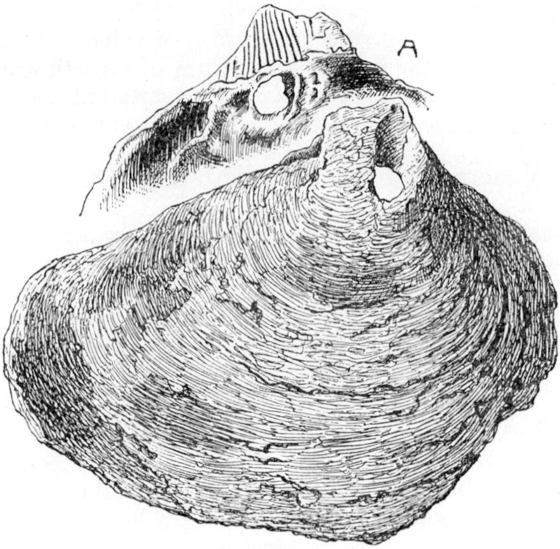

Figure 336

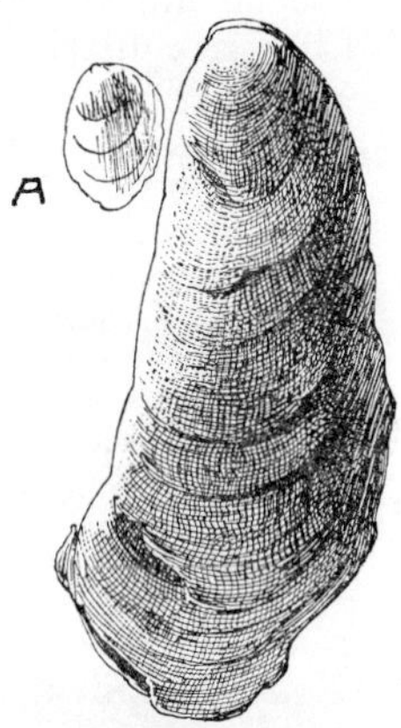

Figure 337

336. *Foramelina exempla* × ⅔
A Interior of valve, showing hinge

337. *Vulsella spongiarum* × 1
A Juvenile shell

shape. Sculptured with fine, concentric striae which are crossed by numerous closely set radiating riblets giving the surface of valve a reticulated appearance, particularly down the central area; lines of growth are more evident towards the ventral margin, producing a somewhat squamose surface. Interior pearly. Hinge furnished with a conical, obliquely arcuated pit for the reception of the ligament. Size: 2 inches.

Localities: Western Port; Port Phillip; Cape Otway; Warrnambool; Portland; Mallacoota.

This is a most curious bivalve. In life found embedded in living sponges at various depths. Valves, sometimes encrusted with species of Bryozoa, are frequently washed up on the beaches and occasionally clumps of sponges as large as one's hand appear on the shores containing one or two dozen specimens, with only their ventral margins visible. The shell is easily recognized by its irregular shape and characteristic sculpture. Its range of distribution is Tasmania, through Bass Strait to South Australia and Western Australia.

Malleus Lamarck, 1799 (*Malleus vulgaris* Lamarck, 1799).

albus Lamarck, 1819; recorded from the Gippsland coast by Pritchard & Gatliff, 1904, but not taken in recent years.

Family PTERIIDAE

Winged Pearl Oysters

Shell inequivalve, very oblique, resting on the smaller (left) valve, attached by a byssus; umbos anterior, eared, the posterior ear wing-like. Interior nacreous; posterior muscular impression large, subcentral; anterior small, within the umbo; pallial line irregularly dotted. Hinge-line straight; elongated.

Electroma Stolickza, 1871

Shell oblique, thin, smooth, inequivalve, left valve slightly flattened. Hinge-line short and posterior wing very short.

Type Species *Avicula smaragdina* Reeve, 1857.

Butterfly or Wing Shell

Electroma georgiana (Quoy & Gaimard, 1835)

(Figure 338)

Shell cream, with irregular, radiating rays of reddish brown, semi-transparent, smooth, thin, fragile, oblique; inequivalve, right valve more convex, a sinus in the left for the passage of a byssus, inequilateral. Ligament partly external. Interior pearly. Size: 1½ inches.

Localities: Brighton; Sandringham; Frankston; Dromana; Altona.

In more or less fragmentary condition, this is an exceedingly common shell cast up on the beaches at many parts of Port Phillip. In life, these appropriately named molluscs are attached by the byssus to seaweeds, and after rough weather the beaches are strewn with these shells, frequently arriving in bunches where perfect examples may be obtained. It is a very fragile species and should be handled with care. Widely distributed, being

Figure 338

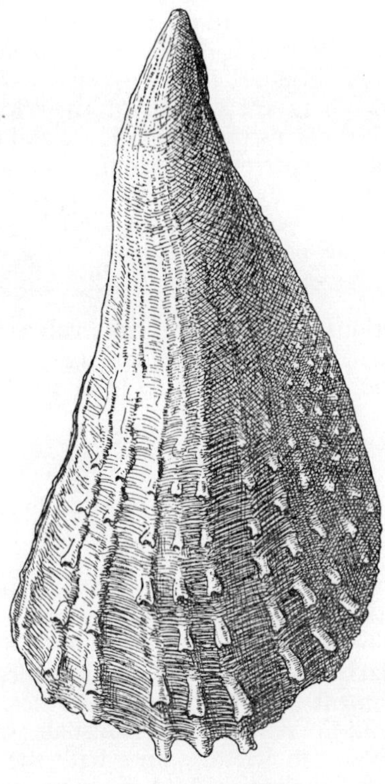

Figure 339

338. *Electroma georgiana* × 1
339. *Atrina tasmanica* × ⅔

recorded from Western Australia, South Australia, New South Wales and Tasmania. This species was known for many years under the name *Pteria papilionacaea* (Lamarck, 1819).

Pinctada Röding, 1798 (*Mytilus margaritifernus* Linné, 1758).

fucata (Gould, 1850) ; Mallacoota.

margaritifera (Linné, 1758) ; Barwon Heads. As recent work by Hynd, 1955, has shown that this species does not occur south of Moreton Bay, Queensland, it seems likely that the earlier record refers to the species above.

Family PINNIDAE

Fan Mussels

Shell equivalve, wedge-shaped, inequilateral, thin, fragile; umbos anterior, terminal; posterior truncated, gaping; ligamental groove linear, elongate. Hinge without teeth; anterior adductor scar apical, posterior subcentral, large, ill-defined.

Fan Mussels, the juveniles of which are very thin and brittle, are extremely interesting, living buried in the mud or sand with the sharp edges of the valves gaping and protruding from the surface. Pearls of an amber colour are sometimes found in the shells. Some species of this genus are used as articles of food, and others are valued for their long and silky byssus, which, mixed with silk, has been woven into gloves and other articles. Owing to their fragile nature, shore specimens seldom escape fracture. Numerically the family is not very strong in species, which, however, are widely distributed, many of them being remarkable for their large size.

Atrina Gray, 1847

Shell with apical portions of the valve entire; valves unisulcate or without the median carina. Internal nacreous layer entire.

TYPE SPECIES *Pinna nigra Chemnitz,* 1785 = *Pinna nigra* Dillwyn, 1817.

Fan Mussel

Atrina tasmanica (Tenison Woods, 1875)

(Figure 339)

Shell brown, ovately wedge-shaped, thin, fragile, equivalve, inequilateral, beaks terminal, margins rounded, gaping posteriorly. Surface rather obsoletely, longitudinally ribbed, ribs sometimes subnodose, and towards the margin slightly, irregularly scaly. Interior faintly iridescent, shining. Hinge straight, toothless. Ligament internal. Size: 12 inches.

Localities: Queenscliff; Sorrento; Point Lonsdale; Western Port.

This is the largest bivalve in Victoria and, with the univalve, *Mamillana mamilla* Gray, shares the distinction of being the largest mollusc of this state.

Some years ago numerous examples, protected by seaweeds, were washed ashore near Newhaven, Western Port; they were very large, one specimen attaining a length of fourteen inches. Off Port Phillip Heads, on a trawling expedition, some time ago, the government steamer *Lady Loch* obtained some fine specimens; and off East Moncoeur Island, Bass Strait, in 38 fathoms, the species was trawled by the *Challenger*. The author described the shell from Tasmanian material and it is also recorded from South Australia. This Fan Mussel, as with other members of the genus, is much more beautiful in the earlier stages, suffering considerably from abrasion of its sculpture with advancing age. The name of Razor Edge is also applied to Fan Mussels on account of their sharp edges.

Family PECTINIDAE

Scallops and Fan Shells

Shells inequivalve, inequilateral, eared; usually free, resting on the right valve. Internal ligament inserted in a cardinal pit under the umbos; adductor muscle single, subcentral.

Propeamussium Gregorio, 1883

Shell small, with interior radial strengthening ridges; right valve flatter with concentric sculpture, left valve with radial or netted sculpture.

TYPE SPECIES *Pecten* (*Propeamussium*) *ceciliae* Gregorio, 1884.

Iredale, 1929, considered that the Australian shells included here should be separated and erected *Ctenamusium* for them with *Amusium thetidis* Hedley, 1902, as type species. Iredale did not state how they differed and as we are unable to discern any generic differences between the faunas we prefer to include the Australian species in *Propeamussium*.

Grau, 1959, makes *Ctenamusium* a synonym of the subgenus *Parvamussium* Sacco, 1897 (*Pecten* [*Pleuronectes*] *duodecim-lamellatus* Bronn, 1832). This subgenus differs from *Propeamussium s.s.* in the presence of a byssal sinus, inequality of the auricles, and greater convexity.

Thetis Saucer Scallop

Propeamussium thetidis (Hedley, 1902)

(Figure 340)

Shell small, thin, white, translucent, compressed, equilateral; externally glossy and concentrically grooved by reverse imbricating sculpture, the internal lirae visible through the shell. The anterior auricle well developed. Hinge-line long, slightly concave, umbo projecting. Interior with ten slender raised radiating lirae, knobbed at the ventral margin and vanishing towards the umbo. Size: ¼ inch.

Locality: Ocean Beach, Point Nepean.

A characteristic frail little species and the only representative of the genus in Victoria. No difficulty should be experienced in its identification. Found also in New South Wales, South Australia and Western Australia.

Cyclopecten Verrill, 1897 (*Cyclopecten pustulosum* Verrill, 1897).

favus Hedley, 1902; Point Nepean; dredged between French and Phillip Islands, Western Port (8 fathoms).

Pecten Müller, 1776

Shells free-living, large, strongly inequivalved, equilateral, left valve flat or concave; ears equal-sized; byssal notch non-functional; sculptured with radiating costae.

TYPE SPECIES *Ostrea maxima* Linné, 1758.

In 1927 Finlay separated the Australasian species, erecting a new genus *Notovola* with *Pecten novaezelandiae* Reeve, 1852, as type species. Fleming, 1950, points out that this is unwarranted and misleading and in this we concur.

Scallop

Pecten alba Tate, 1886

(Figure 341)

Shell strong, orbicular, nearly equilateral, inequivalve; the upper or left valve flattened, slightly concave near the umbo, and shorter at the ventral margin; right valve very convex; auricles or ears nearly equal. Surface of shell rayed with about fifteen ribs, are flatly convex and not so wide as their interstices. Hinge-line straight; ligament internal. Size: 5 inches.

Localities: Port Phillip; Western Port; Bass Strait.

Detached valves are frequently obtained at any of our beaches, but the use of the dredge is necessary for living examples. Particularly fine specimens have been located off Laverton.

In the present species the prevailing colour of the left valve, which is always darker, is light brown, but examples appear in all shades of brown, yellow, orange, pink and flesh colour. Generally found in numbers, resting horizontally with the flat valve uppermost. The animals show great activity, moving through the water by flapping their valves together. When the valves gape, one is confronted with a series of beautiful eyes round the edge of the animal.

A similar shell known as *Pecten meridionalis* Tate, 1886, is fished extensively in southern Tasmania and commercially workable beds are continually being sought in other states.

Veprichlamys Iredale, 1929

Shells small, thin, oblique in shape; delicately sculptured.

TYPE SPECIES *Chlamys perillustris* Iredale, 1925.

Brilliant Fan Shell

Veprichlamys perillustris (Iredale, 1925)

(Figure 342)

Shell pale orange to orange brown; internally white, flattened, ears very unequal, thin, obliquely oval. Right valve apically smooth, then ornamented with about twenty slender radial ribs, the interstices between very broad; the ribs are a little wavy at first, later obliquely radiating with a gentle curve; lamellae arise and these are placed rather far apart, and, growing angulate and tallish, resemble thorns; the interstices between the ribs are minutely scratched longitudinally; the sculpture on the left valve is similar, but the prickles begin at an earlier stage, and are more closely packed posteriorly. Hinge-line straight, ligamental pit triangular, small, with scarcely perceptible ridge and corresponding groove very minutely serrated. The posterior auricle of the left valve is small and scarcely ribbed. The anterior auricle possesses half a dozen fine ribs bearing fine prickles. Size: 1 inch.

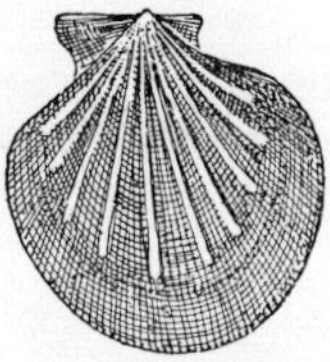

Figure 340

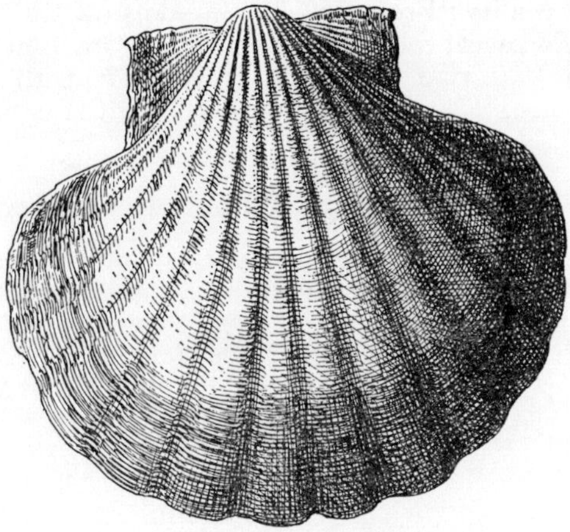

Figure 341

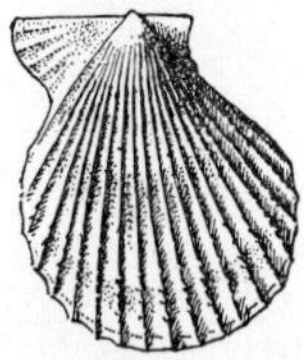

Figure 342

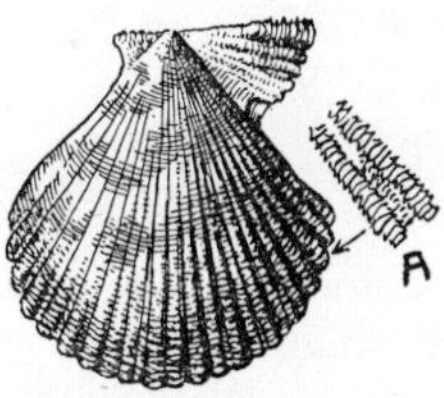

Figure 343

Figure 344

340. *Propeamussium thetidis* × 2
341. *Pecten alba* × ⅔
342. *Veprichlamys perillustris* × 1
343. *Camptonectes famigerator* × 1⅓
   A Sculpture (enlarged)
344. *Chlamys asperrimus* × ⅔

Locality: From 150-250 fathoms off Gabo Island.

To reveal the beauty of this deep sea Fan the use of the lens is necessary. No Victorian species approaches it with its peculiar prickly character. Appears also in the waters of South Australia, Tasmania and New South Wales.

Camptonectes Meek, 1864

Shell small, thin, equivalve; smooth or with weak radial sculpture; eared, posterior ear small; anterior of right valve; with a byssal sinus.

TYPE SPECIES *Pecten lens* Sowerby, 1843.

Iredale, 1929, erected *Talochlamys* for *Chlamys famigerator* Iredale, 1925, but pointed out its close affinity with *Camptonectes*. Thiele, 1934, considers the separation is not warranted and we agree.

Beautiful Fan Shell

Camptonectes famigerator (Iredale, 1925)

(Figure 343)

Shell small, fragile, somewhat compressed, colour extremely variable, frequently unicoloured, white, yellow, orange, pink or more often variegated with splashes of white or cream. Interior similarly coloured and shining. Equilateral, inequivalve; prodissoconch small, shining. Sculptured with about twenty-four ribs, unadorned at first, then developing lamellae in a discrepant manner on the left valve, more regularly on the right, which is a little more convex. On the left valve the lamellae occur on every third or fourth rib, the intervening ribs remaining smooth. At a little older stage the ribs on the right valve are regularly surmounted by smaller scaly lamellae, the broad interstices are at first concentrically lined but the sculpture becomes irregular and broken with age, and intercalating ribs spring up. Hinge-line straight, narrow, right valve; with long ridge bearing minute serration on each side of small ligament pit, with corresponding serrated groove in left valve. Size: ¾ inch.

Localities: Dredged off Portsea, Port Phillip; off Wilson's Promontory; off Bass Strait cable.

A graceful little Fan easily recognized by its rich colouring and peculiar concentric and radiating sculpture. Occurs also in New South Wales, Tasmania, and South Australia.

Chlamys Röding, 1798

Shell triangular-ovate, longer than broad, equivalve; sculptured with radiating ribs, eared, the posterior smaller; with a byssal notch and comb teeth on the anterior margin of the right valve at the byssal sinus.

TYPE SPECIES *Pecten islandicus* Müller, 1776.

Iredale, 1929, divided the south-eastern Australian species previously placed in *Chlamys* into a number of genera, including *Mimachlamys* with *Pecten asperrimus* Lamarck, 1819, as type species, and *Belchlamys* with *Chlamys atkinos* Petterd, 1885, as type species. Thiele, 1934, placed *Mimachlamys* in the synonymy of *Aequipecten* Fischer, 1887, type species of which is *Pecten opercularis* Linné, 1758. Ludbrook, 1953, thinks that the differences in the three species are specific rather than generic and therefore the Australian species placed in *Mimachlamys* should be retained in *Chlamys s.s. Chlamys atkinos* Petterd also falls within the range of generic variation and should be retained in *Chlamys.*

While we agree with Ludbrook that *islandicus* and *asperrimus* are congeneric and the latter should be retained in *Chlamys,* we disagree with the inclusion of *Pecten opercularis* which shows difference in basic shell-shape as well as in sculpture. Therefore we consider that if *Chlamys s.l.* is to be divided, this group warrants separation.

Fan Shell or Doughboy Scallop

Chlamys asperrimus (Lamarck, 1819)

(Figure 344)

Shell rather thin, equivalve, almost equilateral, ears unequal; ornamented with about twenty-five to thirty closely set angular ribs, which are finely, regularly and erectly scaled throughout. Size: 3 inches.

Localities: Mordialloc; Frankston; Sorrento; Queenscliff; eastern Bass Strait.

In the adult the primary ribs are flanked on each side by minor ribs. Much variation in colour is apparent. Most predominant is reddish brown, but specimens are seen in all shades of red, yellow, orange and purple, and it is not infrequent to observe specimens faintly but beautifully spotted. Shore specimens are invariably worn and generally devoid of colour. When dredged alive the shell is often covered with sponge, which protects the shell. Obviously, these produce the better preserved specimens, as will be seen on removal of the sponge. The colour is richer and the scale feature on the ribs is more pronounced. The movement of Fans is brought about in a peculiar manner. They flap through the water, movement being effected by the rapid opening and closing of the valves. Also recorded from New South Wales, Tasmania, South Australia and south Western Australia.

Shagreened Fan Shell

Chlamys atkinos (Petterd, 1886)

(Figure 345)

Shell strong, depressed, elongately orbicular, almost equivalve, extremely variable in colour, may be yellowish white, brown, reddish brown, pink and purplish with somewhat angular splashes of white sometimes resembling festooning. Upper valve ornamented with fifteen to twenty irregular, radiating, imbricated ribs with a further adornment of a peculiar shagreen character, discernible under lens, between the ribs on both valves. Ears, the anterior much larger; very unequal, radiately ribbed. Interior white or faintly tinted. Hinge area often much darker. Size: 1½ inches.

Localities: Lorne; Port Fairy; Sorrento; Shoreham; dredged 8 fathoms between Phillip and French Islands, Western Port.

Confusion with any Victorian Fan should not be experienced as its elongated, somewhat oblique shape and the characteristic shagreened surface of the valves should be sufficient to distinguish the species. It is also recorded from New South Wales, Tasmania, South Australia and Western Australia.

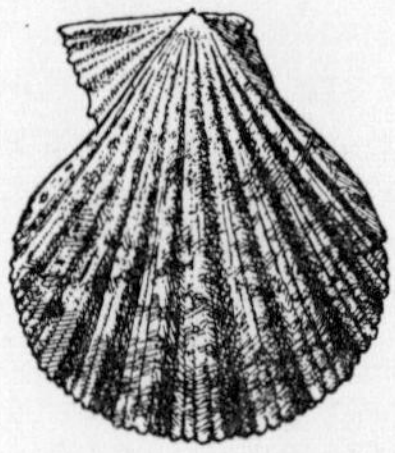

Figure 345

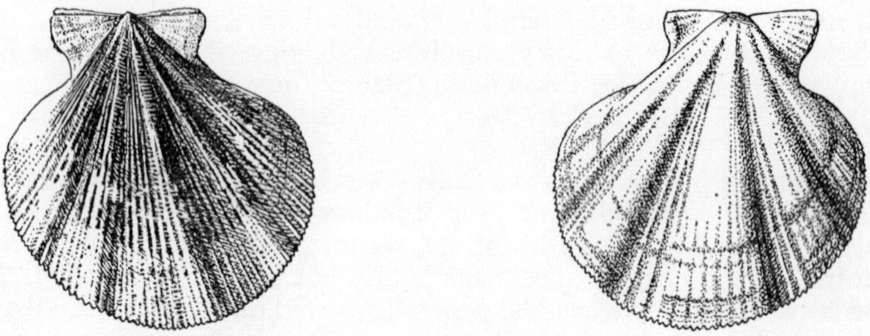

A B

Figure 346

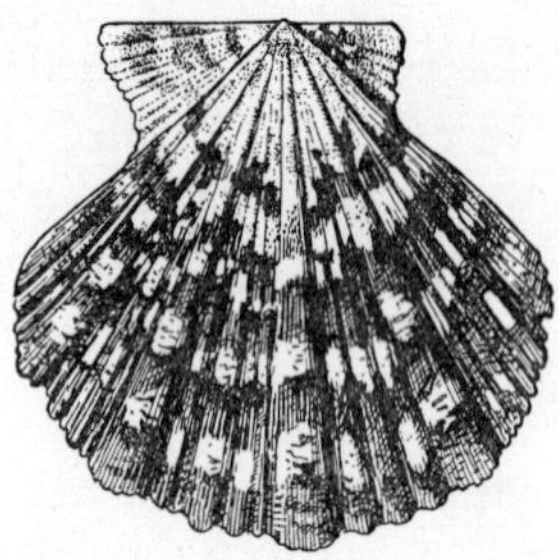

Figure 347

345. *Chlamys atkinos* × ⅔

346. *Mesopeplum caroli* × 1
A Left valve B Right valve

347. *Mesopeplum tasmanicum* × ⅔

Mesopeplum Iredale, 1929

Shell suborbicular, inequivalve, right valve more convex, ears small, subequal, byssal gap small, pectinidial teeth small in young shells, obsolete in adult; strongly radiately ribbed, interspaces wider than the ribs in the left valve.

TYPE SPECIES *Mesopeplum caroli* Iredale, 1929.

Cotton, 1930, erected *Notochlamys* for *C. anguineus* and *tasmanica* with the former as type species. However, we consider that the generic separation from *caroli* is not justified, the differences being specific rather than generic, and therefore would place *Notochlamys* in the synonymy of *Mesopeplum.*

Carol's Fan Shell

Mesopeplum caroli Iredale, 1929

(Figure 346)

Shell suborbicular, very solid, convex. Hinge-line short with the auricles comparatively small, well developed and bearing a few faint radials; colour variable, left or upper valve may be rose with white irregular splashes or blotches, or white with rich carmine over a greater portion of the shell and much darker on the ribs; lower valve white with a trace of rose pink at the umbo and sometimes yellowish in the interstices; almost equilateral, inequivalve, the right valve; deeper than the left, ventral margin well rounded. Sculpture of lower valve consisting of six prominent, broad, radially striated ribs with three or four radials in the narrow interstices; on the left or upper valve the ribs, also radially striated, are narrower with wider interstices which possess three to five radials. The shell is further ornamented with fine, microscopic threads appearing only in the interstices of the lower valve but over the whole surface of the upper valve. Interior white. Size: 1½ inches.

Localities: Ocean Beach, Point Nepean; Torquay.

Its greater convexity, the fine concentric ornament over the whole surface of valve, in adition to the rose or rich carmine colour provides adequate means to distinguish it from *C. anguineus*. It is seldom washed up on the beaches. Recorded also from New South Wales but appears to be a deeper water shell taken in depths down to 70 fathoms.

Tasmanian Fan Shell

Mesopeplum tasmanicum (A. Adams & Angas, 1863)

(Figure 347)

Shell triangularly orbicular, inequivalve, valves convex. Sculpture of valves dissimilar, right valve rays and interspaces about equal and marked with riblets, left valve four or five ribs with wide interspaces which bear three or four secondary riblets; colour of left valve variable and darker than right valve, general colour is red, purple or purple rose, maculated with white or cream. Ears very unequal. Size: 2 inches.

Localities: Lorne; Port Fairy; dredged between Phillip and French Islands, Western Port (8 fathoms); off Wilson's Promontory; Port Phillip; Port Albert.

An exceedingly attractive shell and easily recognized by its distinctive ribs and beautiful coloration. Its range of distribution is South Australia, Tasmania and dredged in Twofold Bay, New South Wales.

anguineus (Finlay, 1926); easily distinguished as the interspaces of the left valve lack the strong secondary ribs and the ventral margin of the right valve curve up to meet the flatter left valve. This species has the same distribution as *tasmanicum.*

Equichlamys Iredale, 1929

Shell large, inequivalve, valves convex, ears small, subequal; byssal gap small, the pectinate teeth becoming obsolete in adult shells; six to eight radial ribs, in the right valve the interspaces are narrow while the left valve has narrow ribs with wide interspaces. The whole shell has a shagreen texture.

TYPE SPECIES *Pecten bifrons* Lamarck, 1819.

Two-faced Fan Shell or Queen Scallop
Equichlamys bifrons (Lamarck, 1819)
(Figure 348)

Shell strong, lilac, triangularly orbicular, somewhat depressed, subequivalve, almost equilateral; left or upper valve slightly more convex and normally ornamented with seven to eight moderately narrow ribs and numerous linear ridges; the whole surface of the valves excepting ribs and ridges of a peculiar granulated shagreen character, the ribs being much darker than the remainder of the shell—almost a deep violet. Right valve slightly smaller and tightly clasped by the left. Auricles or ears nearly equal. Interior of a rich purple or magenta purple, much lighter in the umbonal region; ribs very conspicuous. Size: 4 inches.

Localities: San Remo; Portland; off Lakes Entrance.

One of our largest Fan Shells sometimes known as the Queen Scallop. Its peculiar ribbed sculpture, the shagreened surface, and purple colour of the interior should suffice to identify the species. It occurs in New South Wales, South Australia and Tasmania where it attains a diameter of 6 inches. In the last state it is one of the three principal edible Scallops and is obtained in sufficient quantity to be of commercial value.

Spondylus Linné, 1758

Shell irregular, attached by right valve, radiately ribbed, spiny or foliaceous, eared, inequivalve. Hinge with two strong cardinal teeth in each valve; a central ligamental pit; muscular impressions double; pallial line entire.

TYPE SPECIES *Spondylus gaederopus* Linné, 1758.

Thorny or Spiny Oyster
Spondylus tenellus (Reeve, 1856)
(Figure 349)

Shell scarlet rose, ovate, pectiniform, inequivalve, almost equilateral; ears not very prominent and rather unequal. Surface of shell is ornamented with

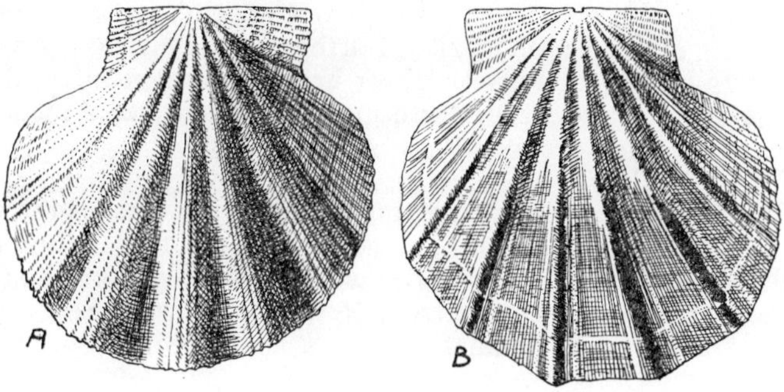

Figure 348

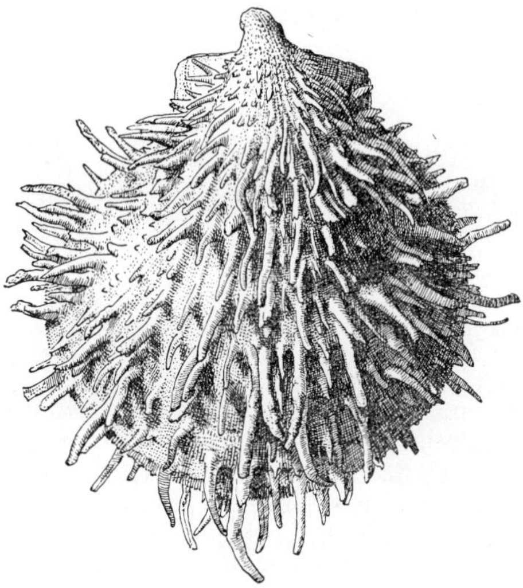

Figure 349

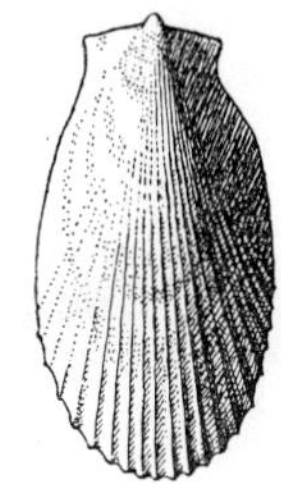

Figure 350

Figure 351

348. *Equichlamys bifrons* $\times \frac{2}{3}$
 A Right valve
 B Left valve

349. *Spondylus tenellus* $\times \frac{2}{3}$

350. *Limatula strangei* $\times 1$

351. *Lima nimbifer* $\times 1$

strong radial ribs with paler leafy and spinous appendages overlapping each other almost the entire length of ribs. Further ornamentation is seen in the crowded riblets of the interspaces. Dark brown ligament internal. Interior white. Hinge bearing two strong teeth in each valve. Size: 3 inches.

Localities: Portland; dredged Bass Strait.

Normally the shell is of a scarlet rose colour in varying degree, but from Bass Strait a light brown perfect living example appears in the collection of C.J.G. Thorny Oyster suggests a likeness to the oyster, but it is only superficially so as its affinity really lies with the scallop. Found attached to stones in deep water, it is recorded from New South Wales, South Australia, Western Australia and Tasmania.

Family LIMIDAE

File Shells

Shell eared, white, oblong, equivalve gaping at the sides. Hinge edentulous, with a central, triangular cartilage-pit.

Limatula Wood, 1839

Valves equilateral, radial ribs in central area becoming obsolete towards the anterior and posterior extremities.

TYPE SPECIES *Lima subauriculata* Montagu, 1808.

Strange's File Shell

Limatula strangei (Sowerby, 1872)

(Figure 350)

Shell white, strong, oblong-ovate, inflated, auricles almost equal, equivalve, equilateral. Central area of valves sculptured with about twenty strong, rather sharp ribs, and intervening spaces being much wider, anterior and posterior ends comparatively smooth; irregular concentric lines of growth are plainly seen. Interior white, shining. Hinge margin thick, fairly straight, edentulous; cartilage-pit central, large, projecting into the cavity. Size: 1 inch.

Localities: Rye; Portsea; off Wilson's Promontory; off East Moncoeur Island; Mallacoota; Western Port.

An ornate species easily recognized by its peculiar oval shape and characteristic sculpture. Dredged alive at the last-named locality (8 fathoms). Also occurs in Queensland, New South Wales, Tasmania through Bass Strait to South Australia and Western Australia.

Lima Cuvier, 1798

Shell equivalve, compressed, obliquely oval; anterior side straight, gaping; posterior rounded, usually close; umbos apart, eared, white, smooth, radially ribbed.

TYPE SPECIES *Lima alba* Cuvier, 1798 = *Ostrea lima* Linné, 1758.

Iredale, 1929, erected *Austrolima* for the Australian species with *Lima nimbifer* Iredale, 1924, as type but later authors such as Thiele, 1935, and Ludbrook, 1955, do not consider this is justified. Confusion has also arisen regarding the genus *Mantellum*; the type of *Mantellum* Röding, 1798, is

Ostrea lima Linné, 1758, and it is therefore a direct synonym of *Lima*. The type species of *Mantellum* Mörch, 1853, is *inflata* Chemnitz = *Ostrea inflata* Gmelin, 1798. Winckworth, 1930, designated *Ostrea inflata* Gmelin as type of *Limaria* Link 1807, thus allowing the latter to replace Mörch's preoccupied name. However, Iredale, 1939, pointed out that the type of *Limaria* by tautonomy is *Limaria vulgaris* Link, 1807, which is a synonym of *O. lima* Linné, 1758, and therefore *Limaria* is an absolute synonym of *Lima*. The group known as *Mantellum* was thus still without a name and Iredale, 1939, erected *Promantellum* for it with *Promantellum parafragile* Iredale, 1939, as type species.

Cloud File Shell

Lima nimbifer Iredale, 1924

(Figure 351)

Shell white, thick, compressed, obliquely subtrigonal; equivalve, inequilateral, gaping at the sides; ears small, unequal; anterior end straight; lunule excavated, showing fine radial striae, posteriorly convex, ventral margin rounded. Surface of valves sculptured with regular radiating, scaly ribs with intervening spaces of similar width but a little variable in this respect. Ligamental area elongate. Interior white, shining. Hinge-line straight, edentulous. Size: 1 inch.

Localities: Anderson's Inlet; Western Port; Mallacoota; Port Fairy.

This is a curious white bivalve and easy to recognize, the peculiar oblique shape and strong radial sculpture providing useful recognition marks. Like the Scallops, the animals swim freely, flapping through the water by means of the rapid opening and closing of the valves. Sometimes they spin a byssus and not infrequently weave into it fragments of shell grit, forming a nest in which they live gregariously. Occurs also in New South Wales, Tasmania and South Australia.

Promantellum Iredale, 1939 (*Promantellum parafragile* Iredale, 1939).
orientalis (A. Adams & Reeve, 1850); a fragile species easily separated from *nimbifer* by its smaller size and finer radial sculpture. Occurs at Portland and Ocean Beach, Point Nepean; also in New South Wales and South Australia.

Limea Bronn, 1831 (*Ostrea strigilata* Brocchi, 1814).
murrayi (Smith, 1891); off Ninety Mile Beach (40 fathoms).

Family ANOMIIDAE

Jingle Shells

Shell thin, inequivalve, nacreous; right valve flat with a hole or notch near the beaks for the passage of the byssal plug, left valve convex.

Monia Gray, 1849

Shell adherent, irregular, flattened, inequivalve; radially grooved; right or lower valve with two thick, divergent, elongated lamella, corresponding with two long pits in the upper or left valve; left valve with two muscle scars. Perforation of right valve, large; only slightly embracing the large, thin byssal plug.

TYPE SPECIES *Monia zelandica* Gray, 1849.

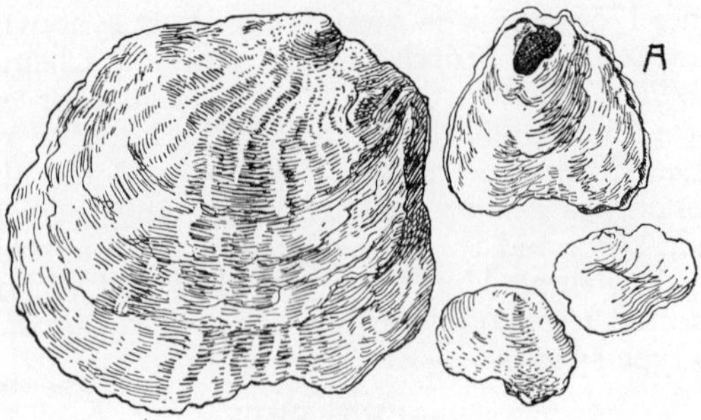

Figure 352

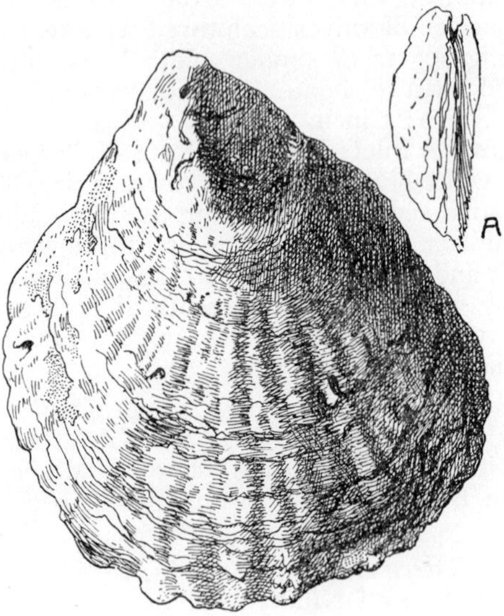

Figure 353

352. *Monia ione* × 1
A Lower or right valve, showing perforation

353. *Ostrea angasi* × ⅔
A Lateral elevation

Jingle Shell

Monia ione Gray, 1849

(Figure 352)

Shell yellowish golden, thin, suborbicular, irregular, left or upper valve fairly solid, convex; right or lower valve flattish, but conforming to the undulations, or grooves, of the object to which it is attached; byssal plug strong and somewhat calcified. Surface of upper valve rugosely radiately ribbed. Interior golden, shining; muscle scars one in each valve, central, and fairly rounded particularly in the lower valve. Size: 2 inches.

Localities: Portland; Balnarring.

Found attached to dead shells, stones, etc., and not confined to the shore line, being obtained down to 100 fathoms. Its range of distribution is New South Wales, Tasmania through Bass Strait to South Australia and Western Australia.

deliciosa Iredale, 1936; off Cape Everard (75 fathoms).

Family OSTREIDAE

Oysters

Shell inequivalve, slightly inequilateral, free or attached, resting on left valve; umbos central, straight; epidermis thin. Ligament internal. Adductor muscle impression single, behind the centre.

Ostrea Linné, 1758

Shell irregular, attached by left valve; upper valve flat or concave, lower convex, often plaited or foliaceous, with a prominent beak. Ligamental cavity triangular or elongated. Hinge toothless.

TYPE SPECIES *Ostrea edulis* Linné, 1758.

Common Mud Oyster

Ostrea angasi Sowerby, 1871

(Figure 353)

Shell large, rounded, cinereous or ash gray, generally darker towards ventral margin, compressed, inequivalve, irregular; lower valve adherent, thick, concave, radially ribbed, upper valve flattish, most frequently convex. Surface of valves ornamented with thin, smooth, imbricating, scales, somewhat fluted at the margins. Interior white and muscle scar almost central, elongated, slightly curved towards the umbo in both valves, young specimens show a few tooth-like processes in each valve. Size: 5 inches.

Localities: Port Phillip: Western Port; Port Albert.

Pritchard and Gatliff in their catalogue remark: 'Occurs commonly in the old beach deposits of Port Phillip, also in the Pleistocene beds of West Melbourne.' Also recorded from New South Wales, Tasmania, South Australia and Western Australia, it formed an important food of the aborigines. Known in South Australia as the Port Lincoln Oyster, it is fairly abundant and marketed in large quantities. Iredale, 1924, Cotton and Godfrey, 1938, and Cotton, 1957, referred this species to *Ostrea sinuata*

Lamarck, 1819, but Thomson, 1954, states that the type of Lamarck's species in the Muséum d'Histoire Naturelle, Paris, is not this shell but is a *Crassostrea.*

When young these oysters attach themselves to stones, and various species of shells. The size given above is average, by no means the largest, other examples exceeding these dimensions by far, one from Port Phillip in the collection of C.J.G. measuring 7 x 5 inches and another in the Victorian National Museum, 7½ x 7½ inches. Maturity is reached at about six years; from observations, the shell is about one inch in diameter at one year old, increasing in size at the rate of 1 inch per year for about five or six years. It is found at all depths from low tide to 90 fathoms.

glomerata Gould, 1850; has been recorded from Tellaburga Island but doubt exists as to the correct nomination of this form—very solid, subtrigonal in shape with surface of valves strongly and rugosely ribbed. Thomson, 1954, suggests that it is the deep-water form of the following species.

Crassostrea Sacco, 1897 (*Ostrea virginica* Gmelin, 1791).

commercialis (Iredale & Roughley, 1933); Sydney Rock Oyster, occurs at Ram's Head and Wingan Inlet, eastern Victoria, and is too well known to require description. It may easily be separated from *angasi* by its smaller size, crumpled valves and bluish coloration.

Iredale, 1936, proposed *Saxostrea* (*Ostrea commercialis* Iredale & Roughley, 1933) for the Australian species of rock oysters but Thompson, 1954, states the two genera are indistinguishable.

Order EULAMELLIBRANCHIATA

Family TRIGONIIDAE

Brooch Shells or Trigonia

Shell equivalve, trigonal, with the umbos directed posteriorly; ligament external. Interior nacreous. Hinge teeth few, diverging.

Neotrigonia Cossmann, 1912

Shell equivalve, trigonal, closely sculptured with radial ribs bearing scales, umbos directed posteriorly. Hinge teeth large, strongly striated, two in right, three in left valve.

TYPE SPECIES *Trigonia pectinata* Lamarck, 1819 = *Trigonia margaritacea* Lamarck, 1804.

Trigonia

Neotrigonia margaritacea (Lamarck, 1804)

(Figure 354)

Shell equivalve, inequilateral, surface externally ornamented with longitudinal, granulated ribs. Ligament external, thick. Pearly and iridescent within. Hinge composed of four oblong, diverging teeth in the left valve, receiving between their grooved sides two similar teeth in the right valve. Size: 1½ inches.

Localities: Near Point Nepean; dredged off Portsea; Western Port; off Lakes Entrance.

Valves are frequently collected on the shore of Western Port, and in living condition it may be obtained in all parts of the bay. The *Challenger* expedition recorded it from 38 fathoms off East Moncoeur Island, Bass Strait. Internally, it is of a pearly character tinged with varying degrees of silver, purple and golden brown. The strong nature of the shell, with its peculiar shape and brilliant coloration, has made it an article much sought after by jewellers. The animal shows great activity, being enabled to make considerable leaps by means of its angulated foot, and it is on record that a specimen, placed on the gunwale of a boat, leapt overboard, clearing a ledge of four inches. The removal of the animal should be undertaken with care, more particularly in the hinge area, as the fracture of the hinge teeth when the valves are pressed together may occur. To lessen this risk the closing of the valves should commence at the umbos. In addition to Victoria the species is recorded from New South Wales and Tasmania, and it is interesting to note that Australia furnishes the only living representatives of the genus.

gemma Iredale, 1924; 40 miles east of Lakes Entrance.

Family CRASSATELLIDAE

Crassatellas

Shell oblong, usually produced posteriorly, mostly concentrically striated or sulcated, covered with an epidermis. Hinge with few cardinal teeth and a cartilage pit in each valve; lateral teeth always small, elongated.

Talabrica Iredale, 1924

Shell small, transversely oval, compressed, equivalve subequilateral; interior of ventral margin finely crenulated; sculptured with regularly spaced concentric ridges.

TYPE SPECIES *Crassatella aurora* A. Adams & Angas, 1863.

Some authors consider that the Australian species referred to the above genus should be placed in *Crassatina* Weinkauff, 1881, the type species of which is *Venus contraria* Gmelin, 1798. However, because of the very distinctive sculpture of the latter species we have retained *Talabrica* but are inclined to think that the splitting off of the minute species under the genus *Salaputium* Iredale, 1924, with *Crassatella fulvida* Angas, 1871, as type, is unnecessary.

Rayed Crassatella

Talabrica aurora (A. Adams & Angas, 1863)

(Figure 355)

Shell fulvous brown, transversely oval, compressed, equivalve, subequilateral; ventral margin well rounded and finely crenulated on the inside. Sculpture consists of fairly regularly spaced concentric ribs which are scarcely seen at the dorsal margin. Two interrupted brownish red rays extend from the umbos to the ventral margin. Posterior and anterior margins rose tinted. Interior somewhat similar in colour. Size: ¾ inch.

Figure 354

Figure 355

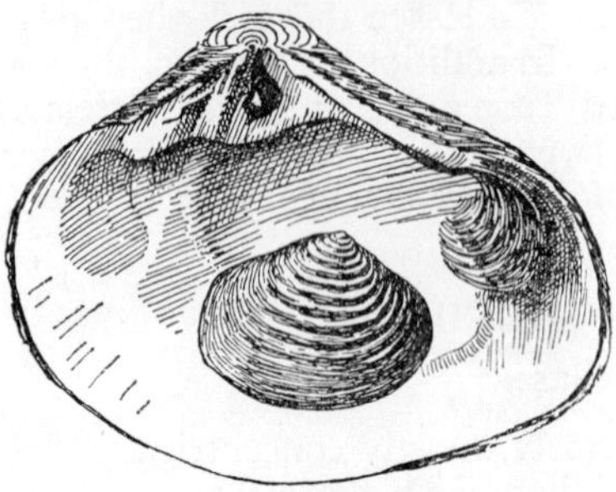

Figure 356

Figure 357

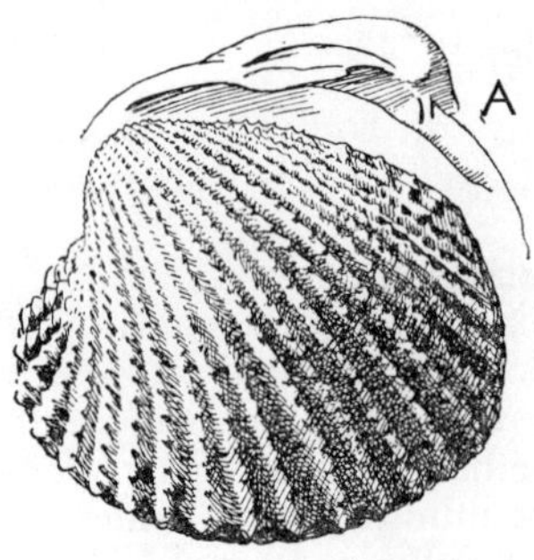

Figure 358

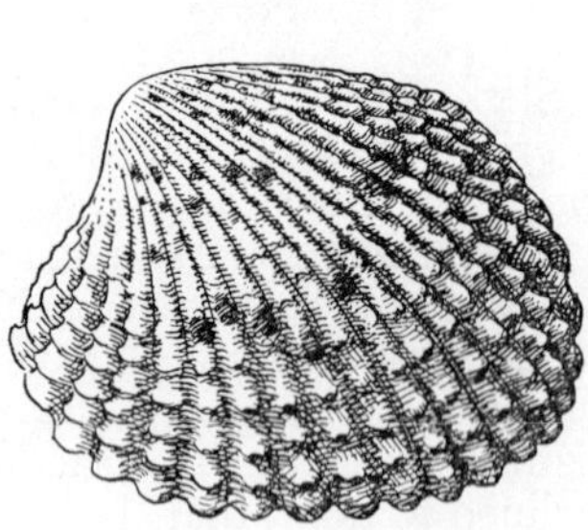

Figure 359

354. *Neotrigonia margaritacea* × 1
355. *Talabrica aurora* × 1⅓
356. *Eucrassatella kingicola* × ⅔
Interior of valve to show hinge,
with smaller shell,
showing external sculpture
357. *Venericardia bimaculata* × 1⅓
358. *Venericardia quoyi* × 1
A Hinge
359. *Venericardia rosulenta* × 2

Localities: 38 fathoms off East Moncoeur Island, Bass Strait; off Wilson's Promontory.

An extremely pretty shell, found on sand and shell, and easily recognized by its compressed, ovate form and the presence of two interrupted colour rays which are a constant feature. Recorded also from Banks Strait, Tasmania.

fulvida (Angas, 1871); off Lakes Entrance.

micra Verco, 1895; off Wilson's Promontory.

Eucrassatella Iredale, 1924

Shell large, thick, heavy, triangularly ovate, equivalve, produced posteriorly; left valve with two diverging cardinal teeth, right valve with three; ventral margin smooth.

TYPE SPECIES *Crassatella kingicola* Lamarck, 1804.

Two other names have been incorrectly applied to this genus, *Crassatella* Lamarck, 1799, which belongs to a different group of fossil molluscs, and *Crassatellites* Krueger, 1823, which is invalid.

King Island Crassatella

Eucrassatella kingicola (Lamarck, 1804)

(Figure 356)

Shell brown, solid, equivalve, inequilateral; umbos prominent. Valves in the early stages transversely furrowed, a feature which becomes obsolete in the adult, in which concentric growth-lines are plainly visible; the whole surface covered by a thick dark brown epidermis. Lunule very distinct. Ligament internal, inserted in a triangular pit in each valve. Interior white; pallial line simple, entire, almost parallel with the ventral margin; muscle scars rounded, deeply impressed and invariably coloured brown. Hinge broad, with very large teeth. Size: 4 inches.

Localities: Lakes Entrance; Port Albert; Anderson's Inlet; dredged 8 fathoms between Phillip and French Islands, Western Port.

A shell of very dense structure and with a powerful hinge. It is not uncommon for valves to be cast up, more particularly on the Gippsland beaches, sometimes with one or more examples of the curious bivalve *Myochama* firmly attached.

The South Australian member of the genus (*verconis* Iredale, 1936) has been used by natives as a hand axe, the strong nature of the valve with its sharp ventral edge rendering it most suitable as an implement. The species appears also in Tasmania and New South Wales.

Cuna Hedley, 1902 (*Cuna concentrica* Hedley, 1902).

comma Verco, 1908; off Ninety Mile Beach (40 fathoms).

concentrica Hedley, 1902; Torquay.

delta (Tate & May, 1900); off Rhyll, Western Port.

edentata Verco, 1908; off Ninety Mile Beach (40 fathoms).

planilirata Gatliff & Gabriel, 1911; between Phillip and French Islands, Western Port (5 fathoms).

Iredale, 1936, separates *C. delta* as *Volupicuna* because of its radial sculpture; in this we do not concur.

Saltocuna Iredale, 1936 (*Cuna particula* Hedley, 1902).
atkinsoni (Tenison Woods, 1876); off Ninety Mile Beach (40 fathoms).
particula (Hedley, 1902); off Ninety Mile Beach (40 fathoms).

Family CARDITIDAE

False Cockles

Shell equivalve, solid, usually with radial sculpture. Hinge with one or two cardinal teeth and sometimes one or two laterals.

Venericardia Lamarck, 1801

Shell rounded-trigonal, strongly radially ribbed, the ribs frequently beaded but beading may become obsolete with age; lunule small and deep, escutcheon linear. Internal margin crenate. Hinge with two transversely striated cardinal teeth in left and three in right valve, latter teeth small or absent.

TYPE SPECIES *Venericardia imbricata* Lamarck, 1801.

Splashed False Cockle
Venericardia bimaculata (Deshayes, 1852)
(Figure 357)

Shell small, triangularly oval moderately convex, cream, spotted with pale reddish colour upon the ribs; equivalve, inequilateral, posterior side much the longer; lunule brownish, slightly excavated, elongately heart-shaped; ventral margin flatly rounded, dentate within. Sculptured with about twenty rounded, radiating ribs as broad as the intervening furrows and carrying numerous small compressed nodules, the furrows showing distinct lines of growth, the whole surface more or less invested with a light brown epidermis. Ligament external. Interior white; muscle scars plainly seen, sometimes more or less coloured brown; pallial line faintly discernible. Hinge frequently tinted brown; right valve with a central tooth large, white and triangular. Size: ½ inch.

Localities: Wilson's Promontory; dredged between Phillip and French Islands, Western Port (8 fathoms); Port Phillip.

The genus is fairly well represented in Victoria and this is probably the most easily recognized species, a useful recognition mark being the pale reddish spots upon the radiating ribs and the presence of dark colour on muscle scars, although this is not a constant feature. A common shell frequently cast up on the beaches. After stormy weather specimens appear in quantity at St Kilda and South Melbourne. Also recorded from Tasmania and South Australia.

amabilis (Deshayes, 1852) (¾ inch); is separated from the above species by its more numerous ribs and broader lunule. It is found on a sea-grass bed of sand and shells in 4 fathoms off East Moncoeur Island, Bass Strait, and is also recorded from Lakes Entrance and Western Port in Victoria and from South Australia. A synonym of this species is *beddomei* Smith, 1885.

Quoy's False Cockle

Venericardia quoyi (Deshayes, 1852)

(Figure 358)

Shell white, solid, oblong-ovate, convex, equivalve, inequilateral, posterior side much the longer, anterior end very short; posterior end rounded, ventral margin moderately convex. Sculptured with about twenty-two rounded, radiating ribs bearing numerous tubercles, intervening spaces of similar width, the whole surface covered with a golden yellow epidermis. Lunule not very conspicuous. Ligament external. Interior whitish; muscle scars well excavated, broad; pallial line, distinct, dentate within. Hinge normal. Size: 1 inch.

Localities: Lorne; Flinders; off Lakes Entrance.

A very strong shell and the largest of the genus in Victoria, very close to *rosulenta* (Tate) but a much bolder species. Much variation exists and a large series may possibly unite the two species. Found also in South Australia.

Rosy False Cockle

Venericardia rosulenta (Tate, 1886)

(Figure 359)

Shell dirty white, with ribs pinkish and red spotted; oblong-ovate, moderately convex, equivalve, inequilateral, the posterior end much the longer; umbos oblique, incurved anteriorly; anterior side short, posterior end obliquely and roundly truncated; lunule inconspicuous, broadly heart-shaped. Valves sculptured with about twenty narrow sub-acute ribs bearing scale-like projections of varying strength, the interspaces wider and concentrically, distantly and irregularly striated. Epidermis very thin, golden yellow, minutely ciliate fringed. Interior white; flesh tinted at the umbonal region; muscle scars and pallial line easily seen. Hinge normal. Size: 1 inch.

Locality: From Bass Strait cable.

A species not of frequent occurrence on our shores, easily recognized by its narrow, more or less strongly scaled ribs. Its range of distribution is from New South Wales, Tasmania through Bass Strait to South Australia and Western Australia.

cavatica (Hedley, 1902); off Ninety Mile Beach (40 fathoms).
columnaria Hedley & May, 1908; Bass Strait.

Bathycardita Iredale, 1924

Shell oval, equivalve, inequilateral, the anterior end short; umbos small, erect. Sculpture with flattened ribs ornamented with short spines.

TYPE SPECIES *Venericardia raouli* Angas, 1872.

Raoul's False Cockle

Bathycardita raouli (Angas, 1872)

(Figure 360)

Shell white, tinged with salmon pink; solid, somewhat compressed, squarely ovate, equivalve, inequilateral, posterior side much the longer;

umbos not prominent; anterior end obtuse, posterior end arched, ventral margin rounded. Sculpture consisting of about thirteen fairly strong, radiating ribs with intervening spaces a trifle narrower; the ribs being ornamented with elevated spines, scaly posteriorly, and more numerous, nodulous, and obtuse anteriorly; concentric growth-lines plainly evident, the whole surface invested with a brownish epidermis. Interior whitish, with traces of salmon pink; ventral margin strongly dentate within. Size: 1½ inches.

Locality: off Lakes Entrance.

An ornate species with a resemblance to *V. Rosulenta* (Tate). Its salmon pink colour, squarely ovate form, and the characteristic arrangement of the rib scales are features enough to separate it. The original locality was southern Tasmania, dredged off Cape Raoul by Admiral Loring. It is a common species on the continental shelf of southern New South Wales.

Vimentum Iredale, 1925

Shell small, thin; sculptured with radial ribs bearing scales or knots. Hinge of right valve with one cardinal tooth, left valve two teeth.

TYPE SPECIES *Cardita dilecta* Smith, 1885.

Bald False Cockle
Vimentum calva (Tate, 1887)
(Figure 361)

Shell ovately subtrigonal, subequilateral, moderately depressed; rounded anteriorly, truncately rounded posteriorly. Umbos prominent, acute, slightly incurved; lunule depressed, broadly lanceolate, smooth, with well-defined margins. The umbonal area is smooth, with excentric distant sulcations, the rest of the surface of the valve radiately ribbed; ribs twenty covered with thick closely set granulations; the interspaces on the medial part of the surface are linear or almost obliterated, becoming wider towards the anterior and posterior margins. Size: ⅜ inch.

Localities: Dredged off Wilson's Promontory; from Bass Strait cable.

Comparison of Victorian shells from the first-named locality with authentic specimens of *Venericardia dilecta* variety *excelsior* Verco, 1908, proved their identity. Comparison with the type of the fossil, *Cardita calva* Tate, was made by Verco who accepted his *excelsior* as an absolute synonym, adding 'Tate's name is unfortunate, for the "Baldness" (calva) is probably accidental, and his type is barely bald at the extreme apex.'

Recorded also from New South Wales, Tasmania and South Australia.

delicatum (Verco, 1908); has a fan-like appearance; due to the narrow diverging ribs which bear fine scales. A small shell, ⅜ inch, it is dredged in 40 fathoms off Ninety Mile Beach, and is also recorded from South Australia and Western Australia.

dilectum (Smith, 1885); easily confused with the above species but has a more elongate shape and stronger sculpture.

jaffaensis Cotton & Godfrey, 1938 = *Carditella exulata* (of authors non Smith, 1885). Australian shells appear to differ from Smith's species in the fewer ribs and fewer and stouter scales. It is recorded from the Ocean Beach, Phillip Island.

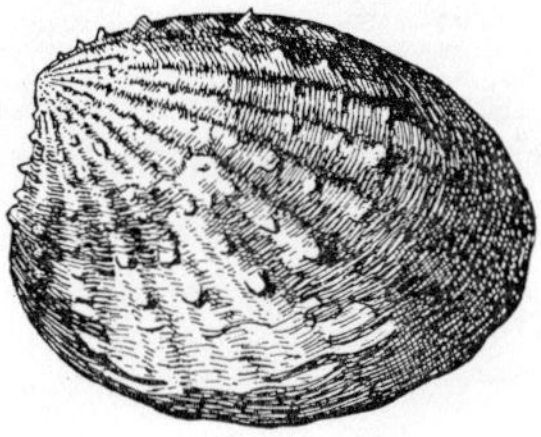

Figure 360

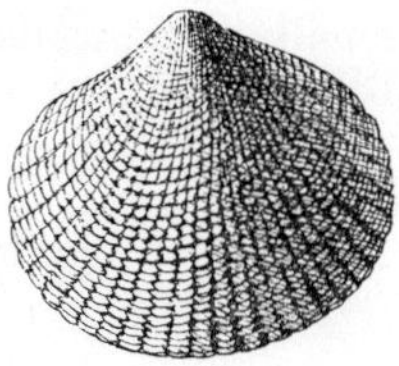

Figure 361

Figure 362

Figure 363

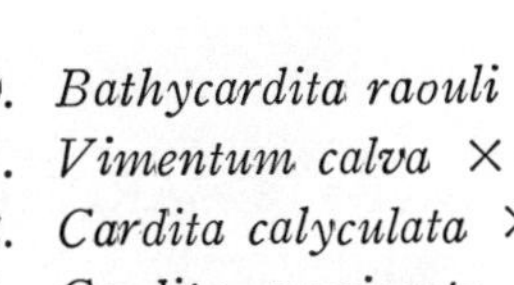

Figure 364

360. *Bathycardita raouli* × 1
361. *Vimentum calva* × 4
362. *Cardita calyculata* × 1⅓
363. *Cardita crassicosta* × 1
    A Hinge
364. *Edenttellina typica* × 6
    A Hinge
    B Coiled protoconch

Cardita Bruguière, 1792

Shell small, thick, radially ribbed, very inequilateral, with a slight ventral gape and byssus. Hinge oblique, right valve with two large parallel cardinals and a feeble anterior lateral, left valve with two cardinals and a small posterior lateral.

TYPE SPECIES *Cardita sulcata* Bruguière, 1792.

Bud-shaped Cardita

Cardita calyculata (Linné, 1758)

(Figure 362)

Shell solid, quadrately oval, white or cream and much darker between the ribs, inflated, inequivalve, very inequilateral, very short anteriorly; posterior end rounded; umbos almost at the end, byssiferous. Sculptured with about fifteen bold ribs of varying width and covered with fine imbricating scales; surface invested with a thin brownish epidermis. Interior white, invariably tinted brown; muscle scars well defined; pallial line plainly seen. Size: 1 inch.

Localities: Port Phillip; Western Port; Back Beach, Sorrento; Barwon Heads; Torquay; Lorne; 40 fathoms off East Moncoeur Island, Bass Strait.

A shell with peculiar features, somewhat similar in shape to *crassicosta* Lamarck but easily separated by its smaller size and the less prominent ribs. Found also in New South Wales and Tasmania.

Thick-ribbed Cardita

Cardita crassicosta Lamarck, 1819

(Figure 363)

Shell very solid, quadrately oval; salmon pink, and dark brown or bluish between the ribs; inflated, inequivalve, very inequilateral, extremely short anteriorly, posterior end rounded; umbos almost at the end. Sculpture consisting of about twelve ribs, four or five of much larger dimensions than the remainder and more or less imbricately scaled, the scales at times being fairly erect, intervening spaces narrower. Lunule roughly ovate, finely striated. Ligament external, moderately long. Interior generally white but sometimes with an orange tint; muscle scars well excavated, posterior larger than the anterior; pallial line clearly indicated. Size: 2 inches.

Locality: Ocean Beach, Sorrento.

This beautiful shell shows extreme variability in colour, shape and in the boldness of the ribs, but no difficulty should be experienced in identifying the species. Occurs also in Queensland, New South Wales, South Australia and Western Australia.

Family CONDYLOCARDIIDAE

Little Carditas

Shell small, ribbed or rarely smooth. Ligament partly or completely internal. Cardinal teeth weak, often with the lateral teeth hardly separated.

Members of this family are all small, deep-water forms and few collectors find examples. They are obtained by dredging and careful washing of fine sediments, or from shell sand. Specimens from the latter source are often badly eroded.

Carditellona Iredale, 1936 (*Carditella angasi* Smith, 1885).
angasi (Smith, 1885); Port Phillip Heads.
elegantula (Tate & May, 1901); Ocean Beach, Flinders.
vincentensis (Verco, 1908); off Ninety Mile Beach.

Condylocardia Bernard, 1896 (*Condylocardia sanctipauli* Bernard, 1896).
crassicostata Bernard, 1896; Frankston, Port Phillip; dredged Western Port = *Condylocardia adelaideana* Macpherson & Chapple, 1951.

Condylocuna Iredale, 1936 (*Condylocardia projecta* Hedley, 1902).
projecta (Hedley, 1902); off Rhyll, Western Port.
ovata (Hedley, 1905); dredged between Phillip and French Islands, Western Port.

Benthocardiella Powell, 1930 (*Benthocardiella pusilla* Powell, 1930).
chapmani (Gatliff & Gabriel, 1912); Portsea; Port Phillip; Ocean Beach, Point Nepean; Torquay.

Radiocondyla Iredale, 1936 (*Radiocondyla arizela* Iredale, 1936).
australis Bernard, 1896; Ocean Beach, Flinders.
pectinata Tate & May, 1900; off Rhyll, Western Port.
porrecta Hedley, 1906; off Wilson's Promontory.
subradiata Tate, 1888; Bass Strait.

Family CYAMIIDAE
Bean Shells

Shell small and thin, mostly hyaline, often with fine radial sculpture, more or less strongly arched. Hinge-line weak with inner, short cartilage; in the right valve a more or less pronounced bicuspid tooth, in the left two corresponding teeth.

Cyamiomactra Bernard, 1897 (*Cyamiomactra problematica* Bernard, 1897).
balaustina (Gould, 1861); Portsea, Port Phillip; Western Port.
communis Hedley, 1905; Port Fairy.
mactroides (Tate & May, 1900); Port Albert; off Rhyll, Western Port; Port Phillip.

Legrandina Tate & May, 1901 (*Legrandina bernardi* Tate & May, 1901).
bernardi Tate & May, 1901; Port Albert.

Family GAIMARDIIDAE
Gaimard Shells

Shell thin, smooth, mostly small, brown; umbos more or less towards the anterior end, often somewhat cuspidate. Ligament external or sunken. Hinge weak, mostly with a cardinal tooth on each side, but they may be completely absent; pallial line without a sinus.

Neogaimardia Odhner, 1924 (*Kellia rostellata* Tate, 1888).
rostellata (Tate, 1888); Portland; Torquay; Barwon Heads; Western Port.
tasmanica (Beddome, 1882); Portsea, Port Phillip; Portland.

Family JULIIDAE

Horned False Mussels

Shell small, oval; umbos more or less anterior. Ligament external, marginal; adductor muscle attached to a round scar placed towards the centre of the valve.

Edenttellina Gatliff & Gabriel, 1911

Shell small, thin, fragile, somewhat gaping behind; right valve bearing a horn on the umbo. Hinge with a single elongate cardinal tooth in each valve.

TYPE SPECIES *Edenttellina typica* Gatliff & Gabriel, 1911.

Horned False Mussel

Edenttellina typica Gatliff & Gabriel, 1911

(Figure 364)

Shall small, flattened, broadly ovate, very inequilateral, posterior side longer and broader; slightly convex; translucent, yellow. Slight irregular concentric sculpture. Posterior side somewhat swollen, dorsal margin arched. Exterior shining. The right valve only has an incurved projection of a rounded form proceeding from the umbo, it forms a coil, and overlaps the left valve. Ventral margin nearly straight. Interior vitreous, thin and fragile. Hinge-line short, with a single cardinal tooth in each valve. Size: 1/8 inch.

Localities: On beaches Portsea; Port Phillip; Ocean Beach, Point Nepean; Shoreham.

The original authors, having dealt with material not adult, erred in their selection of a generic name as the shell is by no means edentulous.

Verco, who obtained South Australian specimens, wrote in a personal communication, 'I obtained a perfect specimen (1/5 inch) with the two valves adherent, one had the horned umbo, the other not. My loose valves show a well marked cardinal tooth in each, and in the right valve is a long post dorsal groove to receive a rather long laminar tooth of the left valve, and a short anterior groove for the edge of the left valve. Your generic name is therefore somewhat unhappy.'

Kawaguti & Baba, 1959, published an account of a small green slug from Tamono, Japan, which they described as a two-valved Sacoglossan Gastropod and for which they erected a new genus and species *Tamanovalva limax*. The description of the shell suggests that it is closely related to the above species and it is possible that the Japanese species should be referred to *Edenttellina* and this genus transferred to Order Sacoglossa, Class Gastropoda. Then Burn, 1960, collected *E. typica* and has shown that it is also a sea-slug rightfully belonging to Class Gastropoda, Order Sacoglossa.

Keen, 1961 (*Proc. Calif. Acad. Sci.*, ser. 4, vol. 4 (2), pp. 47-66), places both species in *Berthelinia* Crosse, 1875 (*Berthelinia elegans* Crosse, 1875). See p. 251.

Family UNGULINIDAE

Globe Shells

Shell suborbicular, closed, sometimes a little irregular. Ligament marginal. Hinge with two bifid, divergent cardinal teeth, laterals obsolete or absent.

Diplodonta Brown, 1831

Shell suborbicular, smooth except for concentric growth-lines. Ligament double, rather long, submarginal. Muscle impression polished, anterior elongated. Hinge teeth two in each valve, of which the anterior left and posterior right are bifid.

TYPE SPECIES *Venus lupina* Brocchi, 1814.

The generic name *Diplodonta* was rejected by Finlay, 1926, and *Zemysia* substituted on the grounds of its similarity to *Diplodon*. However, this does not stand under the international rules of nomenclature, so *Diplodonta* is retained here.

Inflated Globe Shell

Diplodonta globularis (Lamarck, 1818)

(Figure 365)

Shell white, thin, very inflated particularly at the umbos, equivalve, inequilateral. Surface of shell showing fine, irregular, concentric growth-lines. Posterior side rounded, anterior and convex, ventral margin markedly rounded. Ligament external, long, narrow. Interior white; muscle scars very evident; pallial line distinct, entire. Hinge composed of two cardinal teeth, the posterior of the right, and anterior of left valve bifid in each valve. Size: 1 inch.

Localities: Dredged between Phillip and French Islands, Western Port; off Point Cook, Port Phillip (8 fathoms) on a sandy mud sea-bed.

A white, smooth shell and with its much inflated form cannot be confused with any Victorian Diplodont. Dredged specimens provide the best examples as the valves when washed ashore are invariably much worn. Its range of distribution is Tasmania through Bass Strait to South Australia and Western Australia.

Little Globe Shell

Diplodonta globulosa A. Adams, 1855

(Figure 366)

Shell small, whitish; almost circular, globose, equivalve, inequilateral, posterior side the longer; beaks somewhat pointed but not prominent; posterior and ventral margins well rounded. Sculpture consisting of fine, irregular, concentric striae, the whole surface covered with a thin light brown epidermis. Ligament external, short, anterior. Interior white; muscle scars plainly seen; pallial line simple. Hinge with the usual characteristic bifid cardinal teeth. Size: ½ inch.

Localities: Dredged Western Port; off Portsea; off Point Cook, Port Phillip (8 fathoms).

A white shell, but with no outstanding features, however its circular shape and the globose nature of the shell should materially help in the recognition of the species. At the last-named locality it was dredged on a sea-bed of sandy mud. Found also in New South Wales and Tasmania.

Figure 365

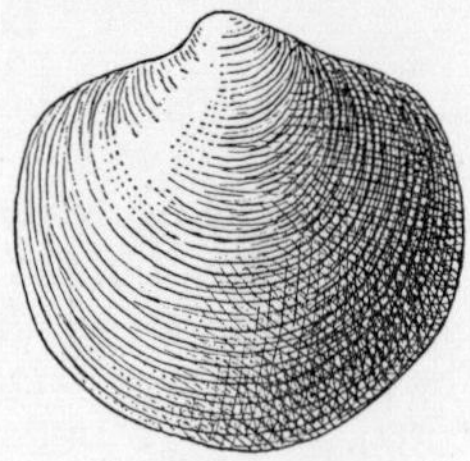

Figure 366

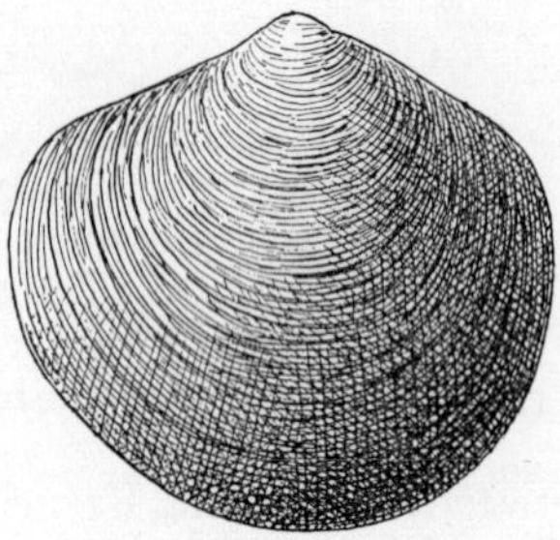

Figure 368

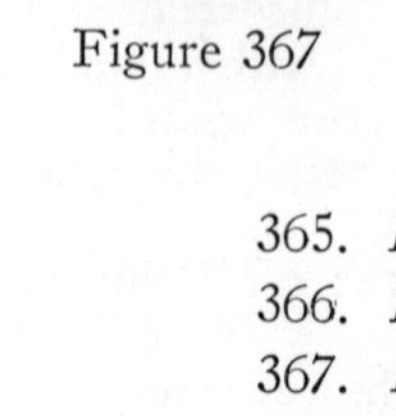

Figure 367

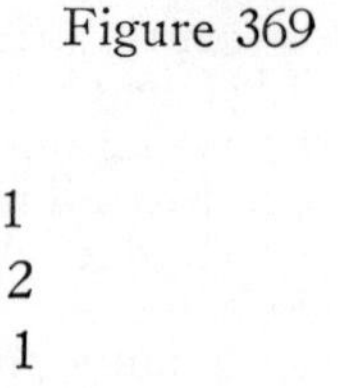

Figure 369

365. *Diplodonta globularis* × 1
366. *Diplodonta globulosa* × 2
367. *Diplodonta tasmanica* × 1
     A Hinge
368. *Diplodonta sublateralis* × 1⅓
369. *Numella adamsi* × 2

Tasmanian Globe Shell

Diplodonta tasmanica (Tenison Woods, 1876)

(Figure 367)

Shell white, solid, almost circular, moderately inflated, equivalve, inequilateral, anterior side the shorter; posterior and ventral margins circularly rounded; umbos acutely triangular. Valves sculptured with fine, irregular, concentric lines of growth, covered with a thin, light brown epidermis. Ligament external, short. Interior whitish; muscle scars well impressed; pallial line distinct, simple. Hinge with a strong, bifid, cardinal tooth in each valve. Size: 1 inch.

Locality: Lakes Entrance.

A white, solid, almost circular form distinguished from *globularis* (Lamarck) by its flatness and stronger hinge-line. Very worn specimens are washed up at the above-named locality but seldom with valves attached. The name *zelandica* (Gray, 1835) has often been applied to this shell but it has proved to be a distinct species. Recorded also from South Australia and Tasmania.

Depressed Globe Shell

Diplodonta sublateralis A. E. Smith, 1884

(Figure 368)

Shell white, thin, inflated, equivalve, inequilateral, posterior side the longer; umbos small, acute; posterior dorsal margin almost straight, ventral margin much rounded, posterior end less rounded than the ventral margin. Surface of valves sculptured with fine concentric lines of growth. Ligament external, long, narrow. Interior white; muscle scars plainly seen, the anterior one a little more elongated; pallial line distinct. Hinge-line very narrow. Size: 1 inch.

Locality: 8 fathoms off Point Cook, Port Phillip.

Not a common species and unknown to the writers from any other Victorian locality, where it was dredged from a sandy mud sea-bed. Its nearest Victorian ally is *globularis* Lamarck, from which it may be distinguished by its flatter form and more acute umbos. It was described by the author from Torres Strait. A fifth species *Diplodontă sphaericula* (Deshayes, 1854) was recorded by Pritchard and Gatliff, 1903, but they iater stated in an unpublished note that it was an incorrect identification and the name should be deleted.

Numella Iredale, 1924

Shell orbicularly elongate, equivalve. Lunule small. Sculpture of concentric lines crossed by fine, short, radial striae and covered with a thin, horny periostracum. Hinge plate large, two teeth in each valve, one of which is bifid.

TYPE SPECIES *Mysia adamsi* Angas, 1867.

Adams' Globe Shell

Numella adamsi (Angas, 1867)

(Figure 369)

Shell small, white, polished, orbicularly elongate, equivalve, nearly equilateral, anterior end the shorter. Lunule very small, ventral margin rounded. Sculpture consisting of fine irregular concentric striae, the whole surface in life invested with a very thin, horny epidermis. Interior white; muscle scars faintly discernible; pallial line distinct. Hinge with two teeth in each valve, one of which is bifid. Size: ½ inch.

Localities: Port Phillip; Western Port.

Reported also from New South Wales, Queensland and Tasmania. In worn condition, the concentric sculpture is detected only with the aid of lens. However, with its peculiar shape and the depressed nature of the shell, it should not be difficult to identify.

jacksonsiensis (Angas, 1867); from Western Port and Lakes Entrance, similarly sculptured but distinguished by its smaller size, oblique form and the consistent rose pink colour. It is also recorded from New South Wales.

Thyasira Lamarck, 1818 (*Tellina flexuosa* Montagu, 1803).

adelaideana Iredale, 1930; off Ninety Mile Beach (40 fathoms).

Family LUCINIDAE

Lucinas

Shell of moderate size, equivalve, rounded or oval, smooth, or with concentric radial or divaricate sculpture. Ligament subinternal, set in a deep groove. Hinge usually with two cardinal teeth in each valve, and anterior and posterior laterals in right valve with corresponding socket in left.

Myrtea Turton, 1822

Shell small, compressed, concentrically ribbed, often crossed by microscopical radial lines. Hinge with one cardinal tooth in right valve and two in left.

TYPE SPECIES *Venus spinifera* Montagu, 1803.

Iredale, 1924, erected *Notomyrtea* for the Australian representatives of this genus with *Lucina botanica* Hedley, 1917, as type species. However, Thiele does not consider the separation warranted, and we have not been able to find generic characters to separate them.

Botany Bay Lucina

Myrtea botanica (Hedley, 1917)

(Figure 370)

Shell white, covered by a light brown epidermis, flatly convex, equivalve, subequilateral; umbos small, acute; posterior and anterior ends rounded. Lunule narrow. Valves sculptured by irregular, concentric riblets. Interior white; muscle scars and pallial line plainly seen. Size: ½ inch.

Localities: Frankston; Western Port dredged (8 fathoms).

A flatly convex shell not uncommon on our shores. It is allied to *M. mayi* (Gatliff & Gabriel, 1911), but is without the radial riblets, a feature easily detected and which immediately separates the two species. Occurs also in Tasmania and New South Wales.

May's Lucina

Myrtea mayi (Gatliff & Gabriel, 1911)

(Figure 371)

Shell creamy white, flatly convex, subequilateral; umbos small, acute, slightly elevated. Lunule elongate and well defined. Escutcheon long and narrow, impressed. Sculpture of fine, sharp, concentric lamellae, with much wider interspaces, which become closer as they approach the ventral margin. Radial sculpture comprising irregular riblets, plainly seen under the lens, which traverse the concentric interspaces. Both valves present these features. On the right valve a shallow flexuosity is observed at the posterior-dorsal area. The concentric lamellae continue to the post-dorsal margin. Muscular impressions well defined, with the pallial line easily discernible. Hinge normal. Size: ⅜ inch.

Localities: Dredged off Point Cook, Port Phillip (8 fathoms); Lakes Entrance.

This little species was located on a sandy mud sea-bed, but only detached valves. It is allied to *botanica* (Hedley) but differs in the possession of radial riblets and a much sharper concentric sculpture. Occurs also in Tasmania and South Australia.

Bellucina Dall, 1901

Shell small, plump, equivalve, often equilateral. Sculptured with strong concentric ridges and less developed radial threads. Hinge with two cardinals and anterior and posterior laterals in each valve.

Type Species *Parvilucina eucosmia* Dall, 1901.

Densely Striated Lucina

Bellucina crassilirata (Tate, 1886)

(Figure 372)

Shell small, creamy white, orbicular, somewhat globose, solid, equivalve, almost equilateral. Sculpture consists of thick, irregular, concentric ridges which are a little broader than the intervening spaces, and numerous radial threads which somewhat crenulate the margins of the ridges without passing over them; anterior and ventral margins regularly rounded, posterior end showing a slight truncation. Ligament external, narrow. Interior white. Hinge strong; muscle scars well defined, the posterior one the broader; pallial line plainly seen, continuous; margins finely crenulate within. Size: ⅛ inch.

Locality: Western Port.

This small, rather ornate species is frequently seen on the shores of Western Port but seldom with the valves attached; dredging operations between Phillip and French Islands in 8 fathoms have provided C.J.G. with many excellent examples in living condition. Occurs in Tasmania and South Australia.

Divalucina Iredale, 1936

Shell suborbicular, convex, equivalve. Sculpture with concentric and divaricating ridges. Hinge with two cardinal teeth, and strong laterals.

TYPE SPECIES *Lucina cumingi* Adams & Angas, 1863.

V-marked Lucina

Divalucina cumingi (A. Adams & Angas, 1863)

(Figure 373)

Shell white with traces of a thin brownish epidermis at both ends of the valves, orbicular, equivalve, subequilateral, solid; beaks small, situated slightly to the anterior. Ventral margin well rounded, not serrated. Valves ornamented by a divaricating sculpture and distant concentric ridges with interstices considerably broader than the ridges and showing very fine lines of growth. Lunule narrow, well defined. Ligament external. Interior white; muscle scars well defined, the anterior more elongate; pallial line distinct. Hinge small, narrow. Size: 1 inch.

Localities: Port Phillip; Western Port; Ninety Mile Beach.

A handsome shell occasionally washed up on the beaches after storms, but in living condition buried in sand at varying depths. Easy to recognize by its rounded form and strong divaricate sculpture. Occurs in Tasmania, New South Wales, South Australia and Western Australia.

Wallucina Iredale, 1930

Globose, thin, equivalve, having a crumpled appearance. Hinge with one strong cardinal in the right valve and two small teeth in left, laterals obscure.

TYPE SPECIES *Lucina* (*Loripes?*) *jacksoniensis* Smith, 1885 = *Loripes assimilis* Angas, 1867.

Assimilated Lucina

Wallucina assimilis (Angas, 1867)

(Figure 374)

Shell dirty white or cream, fragile, somewhat triangularly orbicular, inflated; equivalve, almost equilateral. Lunule darker in colour, well defined, cordiform. Surface of valves sculptured throughout with concentric ridges and fine microscopic radial striae; ventral margin much rounded, finely crenulated. Interior white. Size: ½ inch.

Localities: Frankston; Point Nepean; Western Port; Wilson's Promontory.

A small fragile species with very little character, the sculpture being visible only with the aid of lens. A similar shell was described under the name *Lucina jacksoniensis* by Smith, 1885, and some authors have thought the two descriptions refer to separate species but we concur with Hedley, 1908, in regarding them as the same shell. He states: 'For good specimens show fine close concentric laminae but no trace, or hardly any, of radial lines. As wear of the shell proceeds, the lamellae go, and over the whole shell appear fine, close radial lines, which are not surface sculpture, but opaque substance within the shell.' We therefore retain the prior name *assimilis* Angas for the Victorian shell.

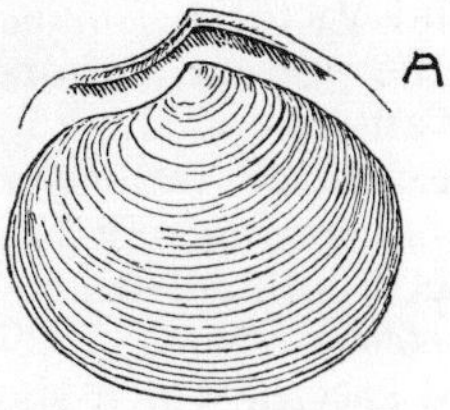

Figure 370

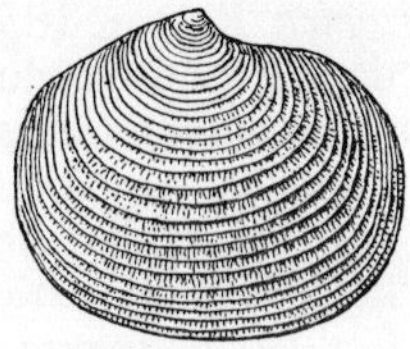

Figure 371

Figure 373

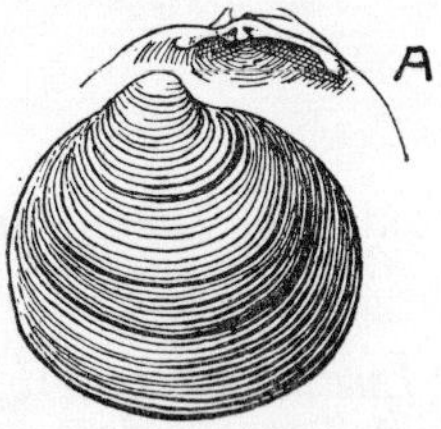

Figure 372

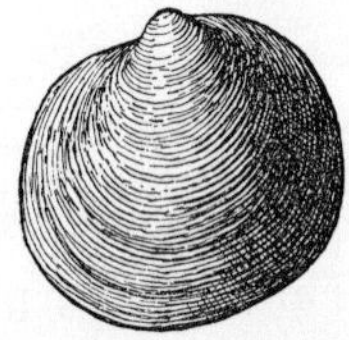

Figure 374

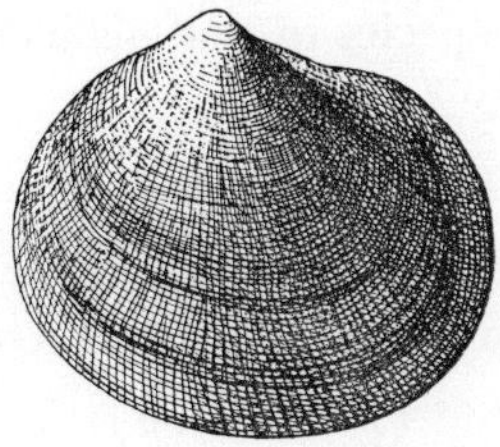

Figure 376

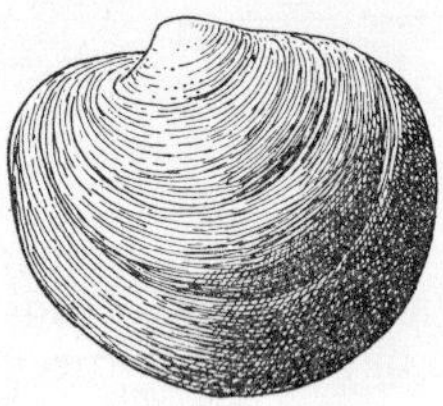

Figure 375

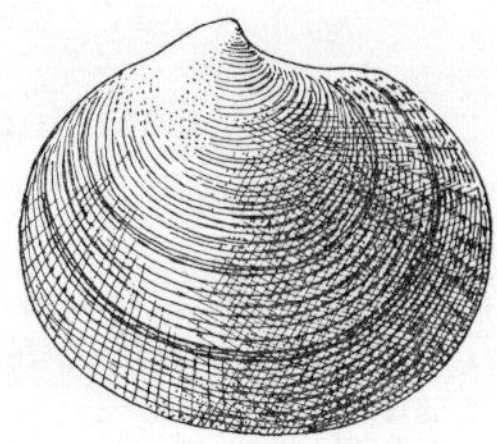

Figure 377

370. *Myrtea botanica* × 2
A Hinge
371. *Myrtea mayi* × 3
372. *Bellucina crassilirata* × 2
A Hinge
373. *Divalucina cumingi* × 1
374. *Wallucina assimilis* × 2
375. *Cavatidens perplexa* × 1
376. *Epicodakia minima* × 3
377. *Epicodakia tatei* × 3

A third species is also recorded in South Australia and Victoria, *Lucina icterica* Reeve, 1850, but independent examination by the present authors led us to the conclusion that the shell so-called was only an erosion stage in *Wallucina assimilis*. Ludbrook, 1955, states that she examined the types of *Lucina icterica* in the British Museum and that it is certainly not the South Australian shell so called. She goes on to say that the South Australian Pliocene shell is not a *Wallucina* but belongs in *Gonimyrtea* Marwick, 1929 (*Loripes concinna* Hutton, 1885) and is an undescribed species which she calls *G. salisburyensis*. She further states she has not been able to examine recent material and so decide if it is conspecific. South Australian Recent specimens in the collection of C.J.G. are definitely *Wallucina assimilis* and not Ludbrook's Pliocene species. The species as now constituted is recorded from Western Australia, South Australia, Victoria, New South Wales and Tasmania.

Cavatidens Iredale, 1930

Shell globose, equivalve, subequilateral umbos approximate. Ligament elongate, thin, internal. Hinge without teeth.

TYPE SPECIES *Cavatidens omissa* Iredale, 1930.

Thiele, 1934, places this genus in the synonymy of *Lucina* Bruguière, 1797, and quotes *Venus edentula* Linné, 1758, as typical, but R. Tucker Abbott, 1954, states that Anton designated a toothed shell *Venus pensylvanica* Linné, 1758, as type species of that genus in which case it is certainly not applicable to the following species.

Toothless Lucina

Cavatidens perplexa (Cotton & Godfrey, 1938)

(Figure 375)

Shell white, thin, globosely rounded, inflated, equivalve, inequilateral; lunule not very conspicuous; ventral margin, anterior and posterior ends regularly rounded; umbos approximate, inflated. Surface of shell sculptured with fine, irregular lines of growth. Interior white; muscle scars plainly seen, the posterior one broader, pallial line entire Hinge-line narrow, edentulous, but with a very elongated internal ligament. Size: 1 inch.

Locality: Port Fairy.

A somewhat frail shell readily recognized by its globose shape and narrow toothless hinge. Recorded also from South Australia.

This species was originally misidentified as *Thyasira globosa* Forskal, 1909.

Epicodakia Iredale, 1930

Shell small, equivalve, inequilateral. Sculptured with fine, concentric and radial striae. Ligament deeply sunken. Muscle scars prominent, anterior elongate. Hinge with one cardinal and two laterals in right valve and two cardinals and four laterals in left.

TYPE SPECIES *Epicodakia gunnamatta* Iredale, 1930.

Small Codakia

Epicodakia minima (Tenison Woods, 1875)
(Figure 376)

Shell white, small, convex, oblong-ovate, equivalve, inequilateral, anterior side the longer, ventral margin much rounded and crenulated. Valves ornamented with fine, regular, radial riblets and microscopic concentric striae. Interior white; muscle scars conspicuous; pallial line well defined and with a narrow triangular sinus. Size: 3/8 inch.

Localities: Point Nepean; Phillip Island.

A common species washed up on beaches. Its nearest ally is *tatei* (Angas) from which it is separated by its finer concentric and radial sculpture, features discernible only under a lens. Recorded also from Tasmania.

Iredale, 1930, renamed this species *consettiana* but we cannot find a valid reason for the change.

Tate's Codakia

Epicodakia tatei (Angas, 1878)
(Figure 377)

Shell white, convex, oblong-ovate, equivalve, inequilateral, anterior side the longer, ventral margin well rounded, posterior end wider than the anterior. Sculpture consists of numerous radial ribs and irregular concentric striae. Interior white; muscle scars and pallial line well impressed. Hinge normal. Size: 3/8 inch.

Locality: Western Port.

Quite a common species, cast up on the shore. In shape very similar to *minima* (Tenison Woods) but differs in both radial and concentric sculpture which are much coarser. Found also in Tasmania, South Australia and Western Australia.

perobliqua (Tate, 1892); Point Nepean.

Callucina Dall, 1901 (*Lucina radians* Conrad, 1841).

lacteola (Tate, 1897); Cotton & Godfrey, 1938, record this species from Victoria but do not state a locality.

Family ERYCINIDAE

Erycinids

Shell very small, thin, fragile, usually transparent, and sometimes gaping, rounded or transverse. Muscle scars well marked. Hinge narrow with one or two cardinal teeth, laterals when present elongate.

The family Erycinidae is a most interesting one, the foot of the animal being ligulate and furnished with a byssal groove. With the animal living, as is often the case, in the confinement of a rock crevice, in the cavity of a shell or at the roots of seaweeds, locomotion becomes restricted, but when free it is capable of considerable movement, crawling in all directions with the help of its lingulate or strap-shaped foot. The family is located in low water to many fathoms.

Kellia Turton, 1822

Shell minute, rounded, concentrically striated. Ligament long. Hinge with two cardinals in each valve.

TYPE SPECIES *Tellina rubra* Montagu, 1803.

There has been some confusion regarding the correct name of this genus, and *Lasaea* Brown, 1827, was in use for many years. However, Grant and Gale, 1921, showed that the type of both genera is the same and therefore the prior name *Kellia* must be used.

Southern Kellia

Kellia australis (Lamarck, 1818)

(Figure 378)

Shell small, yellowish white, stained with pink at the dorsal areas, more or less ventricose, triangularly oval, inequilateral; umbos prominent, incurved. Surface of valves concentrically striated, a character showing extreme variation in the thickness of striae which sometimes become rather coarse and have minute punctures. The whole shell covered by a yellowish brown epidermis. Interior white and shining through which the concentric sculpture is plainly seen. The internal coloration shows much variation as specimens appear in all shades of purple. Muscle scars distinct; pallial line entire. Hinge plate well developed and pink. Size: ⅛ inch.

Localities: Off Portsea; Queenscliff; Flinders; San Remo; Waratah Bay; Lakes Entrance.

An extremely common shell on the whole Victorian coastline. Attached by a byssus and gregariously living in the crevices of rocks between tide levels and among congregations of barnacles, mussels, galeolaria worm tubes, and roots of seaweeds and corallina. Its globular shape and pinkish tint at the dorsal area will serve as useful recognition marks. The valves are often distorted due to its gregarious habit. It occurs also in Tasmania, New South Wales, South Australia and Western Australia.

angasiana Tate, 1886; a small triangularly oval, dull, semitransparent shell ⅛ inch, originally found in shell and sand at Holdfast Bay, South Australia, and has been taken at San Remo, and may easily be distinguished by its smaller size and fine concentric threads.

Melliteryx Iredale, 1924 (*Erycina acupuncta* Hedley, 1902).

acupuncta (Hedley, 1902); Western Port.

helmsi (Hedley, 1915); Port Phillip; Port Albert; Lakes Entrance.

Bornia Philippi, 1836 (*Erycina corbuloides* Philippi, 1836).

Iredale, 1924, proposed *Borniola* to replace this genus but we consider it was not warranted.

lepida (Hedley, 1905); San Remo.

trigonale (Tate, 1879); Port Albert; Western Port; Port Phillip. Laseron, 1956, suggests this species should be included in his genus *Pileatona*. In this we do not concur.

Marikellia Iredale, 1936

Shell smooth, white, oval, inflated inequilateral. Ligament internal with a wide resilium posterior to cardinal tooth. Hinge with one conical cardinal

Figure 378

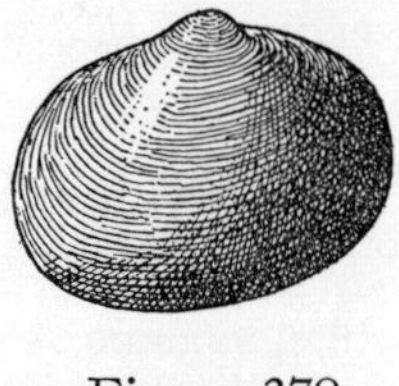

Figure 379

Figure 380

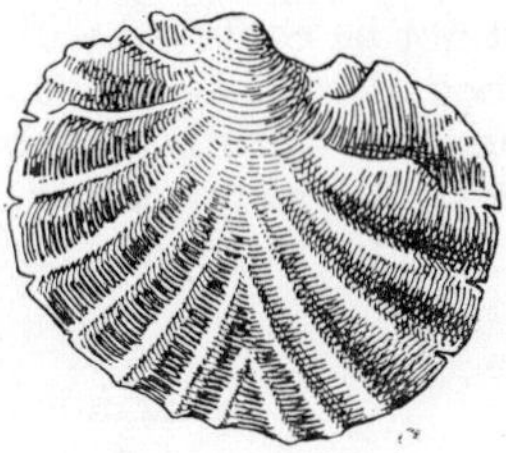

Figure 381

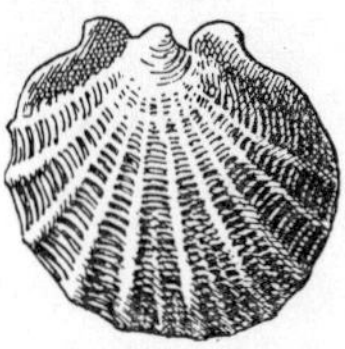

Figure 382

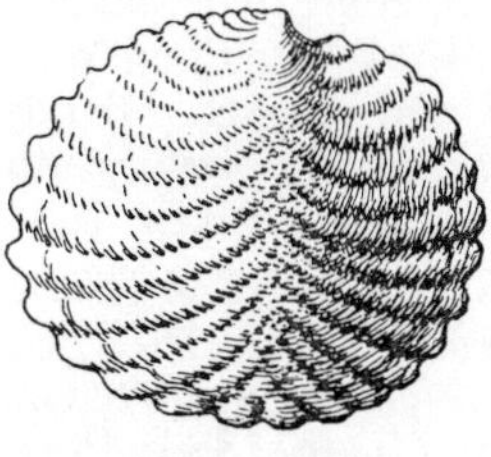

Figure 383

378. *Kellia australis* × 4
379. *Marikellia solida* × 3
380. *Marikellia rotunda* × 2
381. *Myllita deshayesi* × 3
382. *Myllita auriculata* × 4
383. *Myllita tasmanica* × 4

tooth and anterior and posterior lamelliform teeth in left valve, and two cardinals in the right valve.

TYPE SPECIES *Kellia solida* Angas, 1877.

Solid Kellia

Marikellia solida (Angas, 1877)

(Figure 379)

Shell white, opaque, shining, quadrately oval, equivalve, inequilateral, posterior side the longer; umbos small; anterior end obtusely rounded; posterior side arched, ventral margin flatly rounded. The shell is smooth, somewhat iridescent, and under lens shows fine, microscopic lines. Ligament internal. Interior white, shining, muscle scars and pallial line faintly discernible. Size: ¼ inch.

Locality: Dredged (8 fathoms) Western Port.

A solid, pure white, highly polished shell and with its more oblong shape should not be confused with any Victorian member of the genus. Found on a sea-bed of sand and shell, and occurs also in New South Wales, Tasmania and South Australia. Cotton & Godfrey, 1938, separated the South Australian shells as *vincentensis* but after examination of a series from the three states we find them inseparable.

Rotund Kellia

Marikellia rotunda (Deshayes, 1855)

(Figure 380)

Shell white, glossy, ovately trigonal, ventricose, equivalve, inequilateral, anterior end the shorter; umbos rather prominent, anteriorly curved; anterior bluntly rounded, posterior end very convex, ventral margin much rounded. Sculptured with fine, microscopic, concentric lines. Ligament internal. Interior white, shining, muscle scars well defined, the posterior one more elongated; pallial line distinct, entire. Hinge normal. Size: ½ inch.

Localities: Dredged off East Moncoeur Island, Bass Strait; between Phillip and French Islands (8 fathoms), Western Port. Occurs also in Queensland and New South Wales.

jacksoniana (Smith, 1884); off Phillip Island, Western Port.

Lepton Turton, 1822 (*Solen squamosum* Montagu, 1803).

australis Angas, 1878; Sorrento, Port Phillip.

frenchiensis Gatliff & Gabriel, 1916; off French Island, Western Port.

ovatum Tate, 1886; Portsea, Port Phillip.

Notolepton Finlay, 1926 (*Kellia antipodium* Filhol, 1880).

antipodium (Filhol, 1880); Port Phillip Heads.

novacambricum (Hedley, 1915); off Ninety Mile Beach; off Wilson's Promontory.

sanguineum (Hutton, 1884); Torquay; Point Nepean (Ocean Beach).

Myllita d'Orbigny & Recluz, 1850

Shell trigonal, sculpture concentric, radial or divaricate; ligament external. Hinge with one cardinal and anterior and posterior laterals in left valve, one cardinal and two anterior and posterior laterals in right valve.

TYPE SPECIES *Myllita deshayesi* d'Orbigny & Recluz, 1850.

Deshayes' Myllita

Myllita deshayesi d'Orbigny & Recluz, 1850
(Figure 381)

Shell white, solid, convex, subcircular, equivalve, equilateral; beaks small. Surface of valves divaricately sculptured with about sixteen to eighteen bold ribs. Ligament external, narrow. Interior white; muscle scars rounded, plainly seen; pallial line simple with a small triangular sinus. Size: ½ inch.

Localities: Warrnambool; Sorrento.

An ornate shell, located on sandy beaches. This is the largest of our Victorian species, easily separated from *tasmanica* (Tenison Woods) by its larger size and bolder sculpture, and from *auriculata* A. E. Smith in the absence of the characteristic ears. Occurs also in Tasmania, South Australia and Western Australia.

Eared Myllita

Myllita auriculata A. E. Smith, 1891
(Figure 382)

Shell white, small, solid, subcircular, equivalve, equilateral; umbos acute, auriculate; valves are adorned with about twelve radial, rather sharp, prominent costae, with an additional sculpture of fine concentric riblets. Ligament external. Interior white; muscle scars hard to detect; pallial line simple with a small triangular sinus. Hinge normal. Size: ⅛ inch.

Locality: Flinders, Western Port.

In solidity, colour, and dentition this very remarkable shell agrees exactly with the other two members of the genus in Victoria, but differs from them in having the superficial costae arranged in a radiating instead of a divaricating manner. The valves, when viewed inside with the umbo upward, resemble a bat, the outwardly recurved auricles representing the ears. Recorded also from Tasmania.

Tasmanian Myllita

Myllita tasmanica (Tenison Woods, 1875)
(Figure 383)

Shell white, suborbicular, convex, equivalve, equilateral, elegantly latticed with concentric sulci and ribs radiating in opposite ways; ribs angulate and curved, and slightly raised behind; umbos submedian, oblique and small. Interior white; muscle scars small, but plainly seen; pallial line simple, with a slight triangular sinus. Hinge normal. Size: ⅛ inch.

Localities: San Remo; Ocean Beach, Phillip Island.

A beautiful little shell somewhat approaching *deshayesi* d'Orbigny & Recluz, but is smaller and with more numerous divaricating ribs. Found also in Tasmania, South Australia and Western Australia.

Ephippodonta Tate, 188 (*Ephippodonta lunata* Tate, 1886).

lunata Tate, 1886; Balnarring. Mrs J. Plant collected a single live specimen in March, 1959, under a rock on the reef.

Family MONTACUTIDAE

Montacutas

Shell small, ovate, posterior end usually short. Ligament external. Hinge with an anterior, narrow lamina with a minute cardinal hook at the proximal end in the right valve, cardinal hook small or absent in left valve; resilium strong, posterior of cardinal.

Mysella Angas, 1877

Shell small, thin equivalve, inequilateral, quadrately cuneate, concentrically striated; resilium small, triangular, internal. Hinge of left valve with a single large, flattened cardinal posterior to the resilium, right valve with two elongate tooth-like projections which interlock above left cardinal.

TYPE SPECIES *Mysella anomala* Angas, 1877.

Irregular Mysella
Mysella anomala Angas, 1877
(Figure 384)

Shell small, white, thin, triangularly ovate, equivalve, inequilateral, posterior side the longer; ventral margin almost straight, posterior end rounded. Sculpture consisting of fine, regular, concentric ridges. Interior white, polished; muscle scars well defined, posterior one the longer; pallial line entire, distinct. Size: ⅕ inch.

Localities: Dredged off Point Cook, Port Phillip; off Mornington; and between Phillip and French Islands, Western Port (8 fathoms).

A flattish form with a fine, microscopic, concentric sculpture. At the first-named locality it was obtained on a sea-bed of sandy mud. Recorded also from New South Wales, Tasmania, through Bass Strait to South Australia.

donaciformis Angas, 1878; the Donax-like Mysella, found on all parts of the coast. It is a smaller shell (⅛ inch), possesses a similar concentric sculpture but is immediately separated by its more cuneate form. It is found also in Tasmania, New South Wales and South Australia.

dromanaensis (Gatliff & Gabriel, 1912); Dromana, Port Phillip; Western Port.

lactea (Hedley, 1902); Western Port.

Montacuta Turton, 1822 (*Ligula substriata* Montagu, 1808). *Coriareus* Hedley, 1907, is now placed in the synonymy of this genus.

nitens Gatliff & Gabriel, 1911; Flinders; San Remo.

semiradiata Tate, 1889; parasitic on the sea urchin *Echinocardium*; Port Phillip; Western Port; Port Fairy.

Benthoquetia Iredale, 1930 (*Turquetia integra* Hedley, 1907).

integra (Hedley, 1907); this species is recorded from New South Wales and South Australia so may be expected in Victoria.

Family CARDIIDAE

Cockles

Shell heart-shaped, equivalve, inequilateral. Sculpture of radiating ribs, posterior slope often with scale or nodulose ornament on the ribs. Hinge with two cardinals and two laterals in each valve; ligament external, short, prominent.

Regozara Iredale, 1936

Shell large, elongate oval, convex, sculptured with strong radial ribs, valves tightly closed, no lunule or escutcheon; teeth strong.

TYPE SPECIES *Regozara olivifer* Iredale, 1936.

Oblique Cockle

Regozara cygnora (Deshayes, 1854)

(Figure 385)

Shell creamy white, sometimes spotted with reddish brown; strong, ovate, somewhat oblique, equivalve, equilateral, cordiform; umbos prominent. Sculptured by forty to forty-five elevated, regularly spaced radiating ribs. Margin crenulated; anterior and ventral margin rounded. Lunule narrow. Ligament external with surrounding ribs yellowish or cream and pink interstices. Interior white, shining; muscle scars elongated. Size: 2 inches.

Localities: Carrum; Rye; Portsea; Lorne; and dredged on sandy bottom (8 fathoms) Western Port.

Odd valves appear occasionally on the beaches. In Victoria it could only be confused with *Fulvia tenuicostata* Lamarck, but is easily separated by its stronger texture and oblique form. Recorded also from Tasmania, South Australia and Western Australia.

Fulvia Gray, 1853

Shell transversely oblong, very inequilateral, posteriorly produced, radiately ribbed.

TYPE SPECIES *Cardium apertum* Bruguière, 1792.

Thin-ribbed Cockle

Fulvia tenuicostata (Lamarck, 1819)

(Figure 386)

Shell creamy white, fragile, equivalve, cordiform or heart-shape, transversely ovate; umbos prominent, tipped with rose purple. Sculpture consisting of about fifty fine ribs radiating from the beaks; margins crenulated. Ligament external. Size: 2 inches.

Localities: Brighton; Point Cook; Western Port.

This Thin-ribbed Cockle is also recorded from Queensland, South Australia and Western Australia. It is gregarious in habit and may be dredged in sand and mud in considerable numbers at various parts of Port Phillip. Cockles inhabit all depths, from very few feet to some of the deepest seabeds. In other countries they are regarded as valuable articles of food and some are noted for their excellence of flavour.

Hedley, 1917, rejected Lamarck's name and replaced it with *Cardium racketti* Donovan, 1825, on the ground that Delessert's figure of Lamarck's type represented a different shell. However, Ludbrook, 1953, has examined the holotype of *tenuicostata* and states that it is the southern Australian shell.

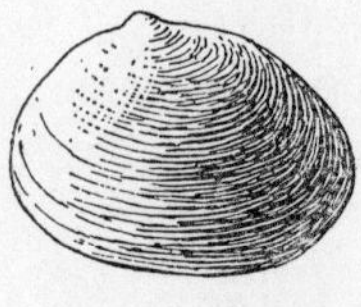

Figure 384

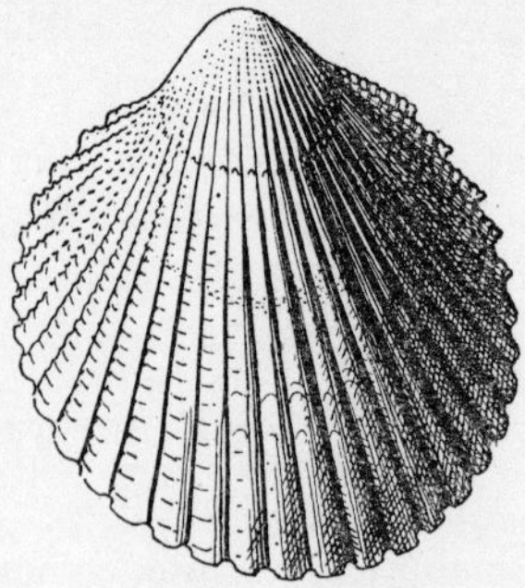

Figure 385

Figure 386

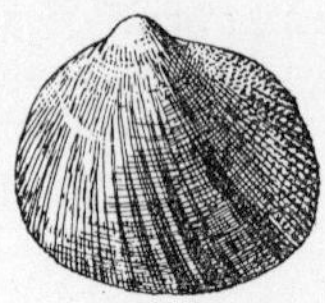

Figure 387

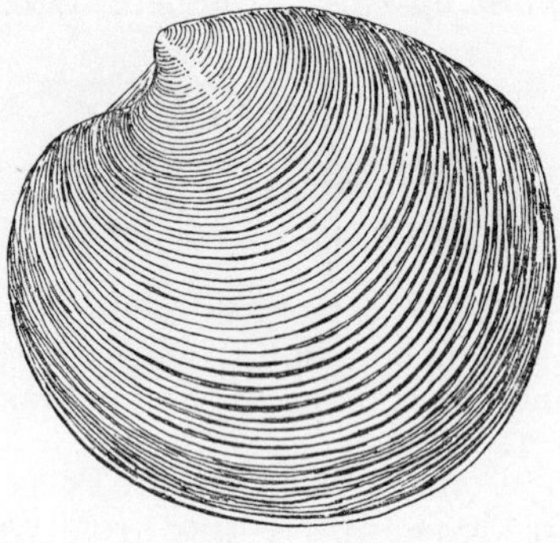

Figure 388

384. *Mysella anomala* × 4
385. *Regozara cygnorum* × 1
386. *Fulvia tenuicostatum* × 1
387. *Pratulum thetidis* × 1⅓
388. *Dosinia grata* × 1

Pratulum Iredale, 1924

Shell small, thin, equivalve, almost equilateral, finely radially ribbed. Muscle scars faint.

TYPE SPECIES *Cardium thetidis* Hedley, 1902.

Thetis Cockle

Pratulum thetidis (Hedley, 1902)

(Figure 387)

Shell orbicular, thin, equivalve, almost equilateral; radiately and closely striated, striae raised and covered posteriorly with minute scales; a few of the striae white but the majority, in rays, are of a bright orange red coloration; in some instances shells are of a uniform canary yellow with the exception of a narrow streak of pink or light brown on the dorsal margin. Ventral margin crenulated. Ligament external. Interior white with faint traces of pink at the anterior and posterior ends, and two yellowish rays diverging from the umbonal region in each valve; muscle scars faintly discernible. Size: ½ inch.

Localities: Portsea, Port Phillip; Western Port; off East Moncoeur Island; off Wilson's Promontory; off Ninety Mile Beach.

This Cockle, located on a sandy mud bottom, is readily recognized by its fine regular sculpture and rays of orange red. Occasionally washed up on the beaches but obtained in greater quantity during dredging operations along the Victorian coastline. Occurs also in New South Wales, Tasmania and South Australia. This shell has been confused for many years with the allied New Zealand shell *Cardium pulchellum* Gray, 1844.

Family VENERIDAE

Venus Shells

Shell regular, closed, suborbicular or oblong, equivalve, inequilateral. Ligament external. Hinge usually with three cardinals in each valve and a single obsolete lateral in one valve.

This very large family is usually divided into three subfamilies, which some recent authors even accept as of full family rank. They are:

Dosiniinae—Shell suborbicular. Hinge with three cardinals in left and three or four in right valve, left valve with an anterior lateral fitting into a groove in the right valve, and sometimes a posterior right lateral.

Meretricinae—Shell oblong. Hinge with three cardinals in each valve and an anterior left lateral.

Venerinae—Shell oblong. Hinge usually with three cardinals in each valve and no lateral tooth.

Subfamily DOSINIINAE

Dosinia Scopoli, 1777

Shell circular, sculpture of fine concentric ridges. Lunule deeply sunk; escutcheon with raised edges in left valve. Pallial sinus fairly long, ascending. Hinge with right posterior cardinal narrow, median cardinal triangular, strongly bevelled.

TYPE SPECIES *Venus exoleta* Linné, 1758.

Delicate Dosinia

Dosinia grata Deshayes, 1823

(Figure 388)

Shell rounded, inequilateral, light brown, darker towards the umbos; umbos directed anteriorly. Lunule elongately cordiform and slightly lighter in colour. Surface of shell sculptured with about one hundred and thirty thin, erect lamellae, the interstices of which are concentrically striate. Interior creamy white and with the muscle scars well pronounced. Size: 1½ inches.

Locality: Dredged between Phillip and French Islands, Western Port.

Few specimens exist in Victorian collections and the shell has been confused with *Phacosoma circinaria* (Deshayes), but is immediately separated by its sharp, erect concentric lamellae. Found also in Tasmania.

Phacosoma Jukes-Brown, 1912

Shell circular, sculpture of fine, concentric ridges. Lunule deeply sunken; escutcheon without raised edges in left valve. Pallial sinus fairly long, ascending. Right posterior cardinal narrow, right median cardinal triangular, strongly bevelled.

TYPE SPECIES *Artemis japonica* Reeve, 1850.

Blue-tinged Dosinia

Phacosoma coerulea (Reeve, 1850)

(Figure 389)

Shell large, greyish white; the umbos tinged with pink, orange or sometimes blue; thick, heavy, equivalve, inequilateral, moderately convex. Sculptured with regular, fine, concentric riblets. Lunule cordate, striated. Ligament brown, elongated, external. Interior white. Hinge strong and broad. Size: 2½ inches.

Localities: Point Nepean; Queenscliff; Western Port.

This is the largest species of the genus in Victoria, and with its stronger texture and finer concentric sculpture, no difficulty will be experienced in distinguishing it from the other members of the genus. Generally found inhabiting sandy ground. Common on the beaches of eastern Victoria and occurs also in New South Wales, Queensland and Tasmania.

Circular Dosinia

Phacosoma circinaria (Deshayes, 1853)

(Figure 390)

Shell light brown, darker towards the umbos; equivalve, inequilateral, fairly convex; closely concentrically striated. Lunule cordate, striated. Ligament narrow, external. Interior white. Size: 1½ inches.

Localities: South Melbourne; Mordialloc; Frankston; Dromana; Portsea; Queenscliff; Western Port; Lakes Entrance.

The present form may be differentiated from *coerulea* by its smaller size and stronger sculpture. Found on sandy ground, and certainly more frequently in Port Phillip than that species. It is also recorded from South Australia and New South Wales.

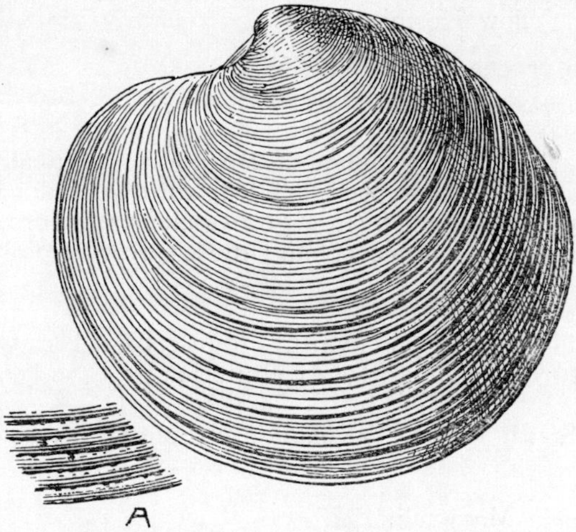

Figure 389

Figure 391

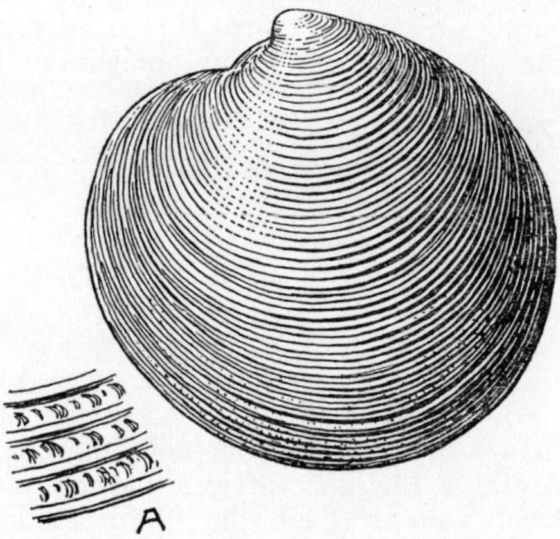

Figure 390

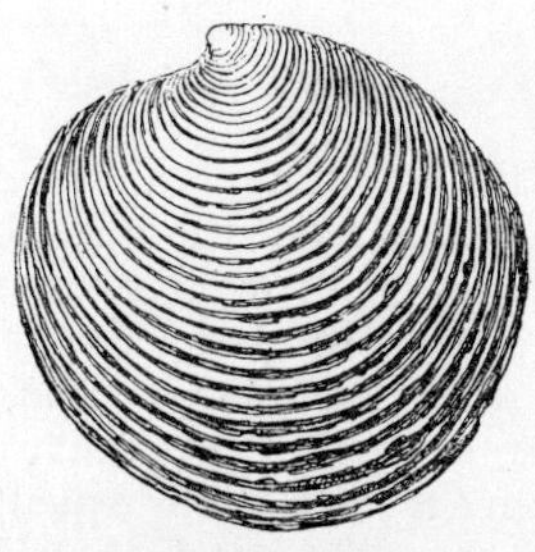

Figure 392

389. *Phacosoma coerulea* × 1
A Sculpture (enlarged)
390. *Phacosoma circinaria* × 1
A Sculpture (enlarged)
391. *Phacosoma crocea* × 1
392. *Kereia victoriae* × 1

Yellow Dosinia

Phacosoma crocea (Deshayes, 1853)

(Figure 391)

Shell fragile, creamy white, shining, subequilateral, surface concentrically lamellated; umbos small, smooth. Lunule well defined, cordate, reddish brown. Ligament fairly strong, partly hidden in a groove along the dorsal margin. Internally apricot or yellow orange colour; muscle scars plainly discernible. Size: 1½ inches.

Localities: San Remo; Lakes Entrance.

This species has been considered as a synonym of *circinaria* but because of its consistently finer sculpture and pink or orange interior we have kept them separate.

Also recorded from New South Wales, Tasmania and South Australia.

Kereia Marwick, 1927

Shell inflated, lunule shallow, bounded by deeply incised line. Sculpture with distant, raised, sharp, concentric lamellae. Hinge with strongly divergent teeth; right valve; with short, lamellar, anterior cardinal; triangular grooved median, and very broad, grooved, posterior cardinal; anterior pit long, rugose on lower side with a well-defined lateral above, weaker below; left valve with long, triangular, anterior cardinal; broad grooved median, and high lamellar posterior cardinal; anterior lateral long, moderately high, rugose.

TYPE SPECIES *Dosinia greyi* Zittel, 1864.

Victorian Dosinia

Kereia victoriae (Gatliff & Gabriel, 1914)

(Figure 392)

Shell cream, rather brittle, inequilateral, the anterior side shorter; umbos fairly prominent, slightly incurved and situated about one-third of the length from the anterior. The surface of shell has four subequally spaced rays of brown, arrow-shaped markings. The concentric sculpture resolves itself into about sixty-five excessively thin erect lamellae, the interstices of which are microscopically striate. The interstices are further sculptured by fine obsolescent radial riblets which are not seen to ascend the lamellae, and are a little less frequent than the minute striae. Lunular area of a brownish tinge, elongately cordiform, medially elevated and ornamented with imbricating lamellae. The ligament is sunken, narrow and long. Interior white. Hinge area normal. Size: 1½ inches.

Localities: Western Port; dredged off Portsea.

No difficulty should be experienced in identifying this species, the fewer and very erect lamellae separating it from any other Victorian member of the genus. Consistency of coloration is not apparent as some specimens are without markings whatsoever. Recorded also from Tasmania and South Australia.

Subfamily MERETRICINAE

Notocallista Iredale, 1924

Shell strong, smooth. Sculpture of concentric grooves and ridges which may be obsolete. Pallial sinus generally wide, ascending and pointed in front. Hinge of left valve with a strong, anterior lateral and three cardinal teeth, the two anterior united to form an inverted V; right valve with two anterior laterals separated by a pit and three separate cardinals, the posterior superficially grooved.

TYPE SPECIES *Cytherea kingii* Gray, 1827.

There has been some confusion in the placing of the shells of this genus; originally placed by most authors in *Macrocallista* Meek, 1876; they were later transferred to *Paradione* Dall, 1909. Iredale erected the above genus and it has been accepted by later authors, some of whom have again divided it into subgenera (Marwick, 1938; Ludbrook, 1955) on the length of the ligament.

King's Notocallista

Notocallista kingii (Gray, 1827)

(Figure 393)

Shell strong, pinkish or cream, variously rayed with reddish brown; ovate, equivalve, inequilateral, concentrically, irregularly striated throughout. Lunule long, defined. Ligament external, rather long. Anterior extremity rather short, rounded; posterior rounded. Interior white. Size: 2 inches.

Localities: St Kilda; Brighton; Frankston; Dromana; Western Port.

A compressed, solid species, remarkable for its elegance of form and brilliancy of colour. The ray-like markings which, as regards width, show extreme variability, are most striking and serve as a useful recognition mark. These rays are often exceedingly broad; some are linear, and on other occasions interrupted, a feature which characterizes the subspecies *lamarcki* (Gray, 1838). Examples frequently appear with large angular spots on the dorsal margin and it is not uncommon to procure pinkish specimens almost devoid of markings. These molluscs live in sand and may be obtained by dredging in four to five fathoms off Brighton. The *Challenger* records it from off East Moncoeur Island, Bass Strait, and it occurs in South Australia, New South Wales and Tasmania.

Tasmanian Notocallista

Notocallista diemenensis (Hanley, 1844)

(Figure 394)

Shell shining, oblong-ovate, extremely variable in colour, generally light brown, with several, more or less wide, radiating bands of darker colour, which in some cases are but faint lines. In addition to this, specimens are frequently seen with distinct V-shaped markings on the greater part of the shell. Equivalve, very inequilateral; somewhat acuminate posteriorly, which side is the longer. Sculpture resolving itself into a series of fairly regular concentric sulci. Lunule concave, longitudinally striated and a little darker than the surrounding areas. Ligament external, elongated. Interior white, polished, with a purplish brown blotch under the umbos and descending posteriorly; muscle scars and pallial line plainly seen; pallial sinus somewhat triangular with the end obtusely rounded and almost reaching the centre. Hinge normal. Size: 1 inch.

Localities: Port Albert; San Remo; Western Port.

This smooth, shining shell is easily recognized by its sulcate sculpture. As regards colour pattern, consistency is not always apparent as almost white examples occur frequently. The internal purplish brown blotch is a useful recognition mark but this feature is sometimes absent. The species is not uncommon in Victoria. Its range of distribution is extended to South Australia and Tasmania. This species has been confused with *Cytherea planatella* Lamarck, 1818, a north Western Australian species.

Broken-banded Notocallista

Notocallista disrupta (Sowerby, 1853)

(Figure 395)

Shell smooth, shining, oblong-ovate, moderately convex; colour creamy, irregularly stained and blotched all over with a pinkish brown, blotches more or less radiating and interrupted; equivalve, inequilateral, rather more acuminate posteriorly then anteriorly, posterior side the longer; umbos fairly prominent, usually with the apices pink-tinted. Valves finely concentrically striated and showing rather coarse growth-lines. Lunule narrow, excavated, longitudinally very finely striated. Ligament external, elongated. Interior white, young specimens having a purple blotch at the umbonal region; muscle scars and pallial line plainly visible; pallial sinus wide and deep, somewhat triangular, obliquely truncated at the end and almost reaching the centre of the valves. Hinge teeth fine. Size: 1½ inches.

Localities: Port Phillip Heads; Port Albert.

This is a fine species, of rare occurrence in Victoria and may be confused with *Notocallista diemenensis* (Hanley) which, however, is less oval in shape. It occurs also in New South Wales and southern Queensland.

regularis (Smith, 1885); a tropical species and mistakenly identified from Victoria.

Sunemeroe Iredale, 1930

Shell round, solid, posterior side longer than the anterior, left posterior cardinal tooth long and thick, right cusp grooved.

TYPE SPECIES *Sunetta adelinae* Angas, 1857.

Broad Hinged Venerid

Sunemeroe aliciae (A. Adams & Angas, 1863)

(Figure 396)

Shell shining, purplish with wavy or wedge-shaped lines and darker rays of similar colour and varying widths; solid, thick, ovately trigonal, compressed, equivalve, subequilateral; ventral margin well rounded, crenulated; posterior end more rounded than the anterior. Valves sculptured with

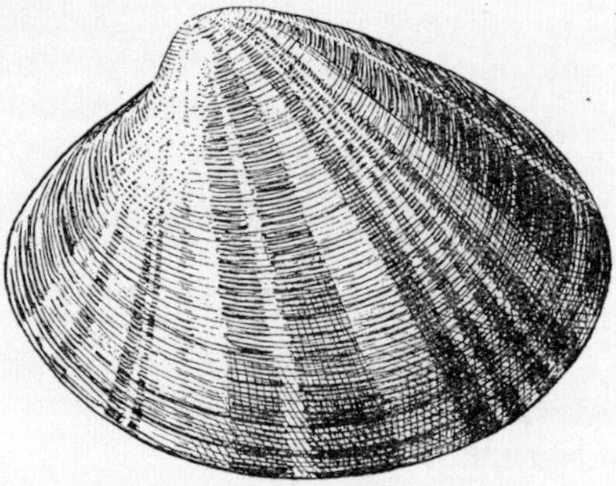

Figure 393

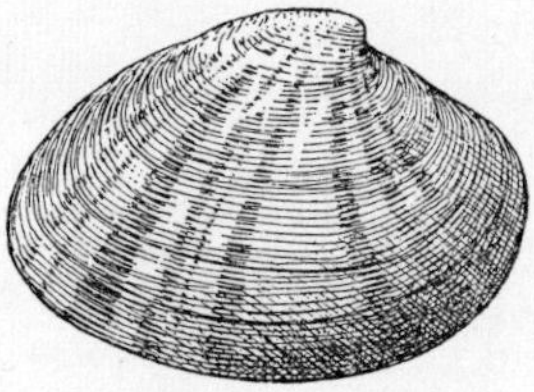

Figure 394

Figure 395

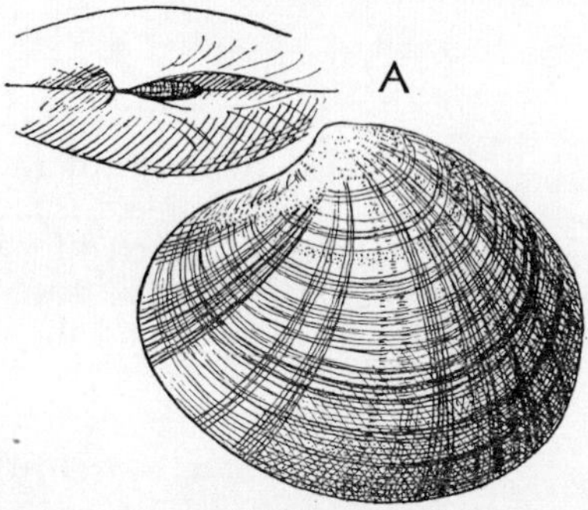

Figure 396

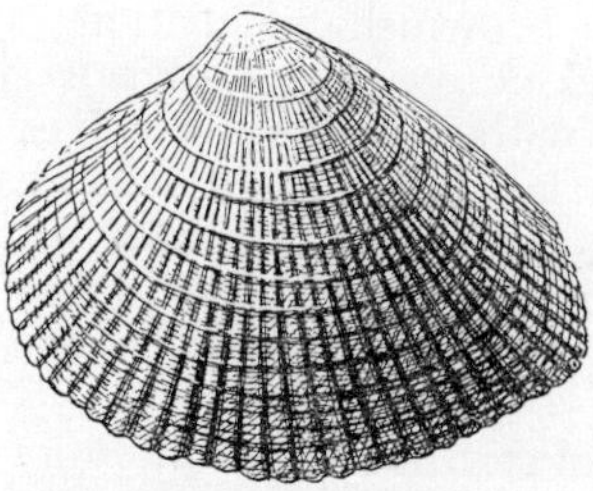

Figure 397

Figure 398

393. *Notocallista kingii* $\times 1$
394. *Notocallista diemenensis* $\times 1\frac{1}{3}$
395. *Notocallista disrupta* $\times 1$
396. *Sunemeroe aliciae* $\times 1\frac{1}{3}$
     A Dorsal view
397. *Chioneryx cardioides* $\times 3$
398. *Tawera gallinula* $\times 1$

regular, fine radial striae which are discernible only under lens. Concentric lines of growth are plainly visible. Lunule lanceolate. Ligament external, situated in a deep escutcheon. Interior whitish or pale yellow, flesh coloured in the umbonal region; muscle scars deeply impressed; pallial line very evident, pallial sinus oval, posteriorly situated. Hinge broad and flat with rather strong teeth. Size: 1 inch.

Locality: Portland; Apollo Bay; Western Port; Lakes Entrance; Port Albert; Mallacoota.

The genus is represented in Victoria by one species only. A charming shell recognized by its peculiar colour pattern; rays exhibit great variation in width and colour density, specimens of a rich, deep purple appearing frequently. It is by no means an uncommon shell on the Victorian coastline. Occurs also in Tasmania and South Australia.

Subfamily VENERINAE

Chioneryx Iredale, 1924

Shell small, oval, with small radial ribs and faint concentric lamellae. Shell edge crenulate. Ligament somewhat sunken. The two anterior cardinal teeth of the right valve directed forward, the middle very long, the posterior fissured; in the left valve the anterior tooth is long, the edge parallel, the middle one thick, fissured and the posterior small.

TYPE SPECIES *Erycina cardioides* Lamarck, 1818.

Heart Venerid

Chioneryx cardioides (Lamarck, 1818)

(Figure 397)

Shell small, triangularly oval, somewhat compressed; colour may be uniformly cream or yellow but frequently brownish maculations appear; equivalve, inequilateral, slightly acuminated posteriorly, posterior side the longer, ventral margin rounded, crenulated within. Surface of valves reticulated with radiating riblets and concentric striae. Lunule well defined, darker in colour than the remainder of the shell, longitudinally striated. Ligament external, small, narrow. Interior white, polished; muscle scars and pallial line very evident; pallial sinus small, somewhat triangular. Hinge small. Size: ½ inch.

Localities: St Kilda; Sandringham; Frankston; dredged off Point Cook; Western Port; off Wilson's Promontory.

A beautifully sculptured species, easily recognized by its characteristic latticed surface. Found in a sea-bed of sandy mud and after stormy weather examples are often washed up on beaches. Previously it was known under the name of *Venus striatissima* Sowerby, 1890, the above prior name precluding its use. Recorded also from New South Wales, Tasmania, South Australia, and Western Australia.

There has been some doubt amongst authors as to the correct name to apply to this shell. Sowerby, 1890, placed it in the genus *Venus* and as there was already a species *Venus cardioides* Lamarck, 1818, he proposed the new name *striatissima*. Iredale, 1924, when erecting the new genus

Chioneryx for it, said it should now revert to Lamarck's name. With this Ludbrook, 1955, does not agree and she uses *striatissima* on the grounds that *cardioides* is a suppressed homonym and cannot be used again. We cannot agree with this as according to International Rules of Zoological Nomenclature (Article 35) *Erycina cardioides* Lamarck has page priority to *Venus cardioides* Lamarck and therefore the latter is the name to be suppressed.

Tawera Marwick, 1957

Shell oval, valve margin crenulate. Sculpture of bevelled, smooth, concentric ridges, sometimes with radials corresponding to the marginal crenulations. Hinge teeth widely divergent, left valve with long, high, posterior cardinal; moderate, bevelled, fairly deeply but unequally divided median; and long, entire, triangular widely diverging anterior one, outer side of which tends to run forward along hinge margin. Right valve with broadly grooved posterior cardinal; median triangular bevelled, rather widely grooved; anterior entire, parallel to lunular margin.

Type Species *Venus spissa* Deshayes, 1835.

Feathered Venerid

Tawera gallinula (Lamarck, 1818)

(Figure 398)

Shell white or light brown, with zigzag colour markings, and occasionally with four distinct, widely separated, dark brown rays; transversely ovate, equivalve, inequilateral; umbos slightly anterior. Anterior end a little shorter and narrowly convex; posterior narrowly rounded. Sculpture consisting of numerous, prominent, concentric, slightly reflected ribs, and radial striation. Lunule ovate, striated. Ligament external, narrow. Interior violet; margins finely crenulated. Size: 1½ inches.

Localities: Frankston; Dromana; Sorrento; Portsea; Western Port.

This ornate shell, although showing remarkable latitude of colouring, is easy to recognize, and should not be confused with any Victorian species excepting, perhaps, *lagopus* (Lamarck). In general outline they are similar, but in the latter, the rays are interrupted, the interior is uniformly white, and in fresh examples there is an absence of lustre, a feature so characteristic of *gallinula*. The species is recorded from Western Australia, South Australia, New South Wales and Tasmania.

Banded Venerid

Tawera lagopus (Lamarck, 1818)

(Figure 399)

Shell transversely ovate, moderately convex, creamy, with four to six dark brown blotches or maculations arranged in three widely separated radial rows; equivalve, inequilateral, anterior end shorter, anterior-dorsal margin slightly convex, posterior end more acuminate than the anterior; ventral margin rounded, finely crenulated. Surface of valves sculptured with rather erect, regular concentric ridges, crossed by numerous radial striae. Lunule very distinct, ovate, dark brown and longitudinally striated. Ligament external, elongated. Interior white, shining; muscle scars and pallial line well defined; pallial sinus broad, rounded at the end, almost reaching the centre of valve. Hinge normal. Size: 1 inch.

Localities: Portsea, Port Phillip; San Remo; Western Port.

A species similar to *gallinula* Lamarck but distinguished by its erect, concentric ridges and in the absence of colour internally. The external coloration is not confined to the interrupted brown bands as in the spaces will be seen small splashes of brown and irregular zigzag markings in varying degrees. The species is located on sandy patches at low water with odd valves frequently cast up on the beaches. Its range of distribution is New South Wales, Tasmania, South Australia and Western Australia.

Delicate Venerid

Tawera spissa (Deshayes, 1835)

(Figure 400)

Shell ovate, subtrigonal, moderately convex; colour creamy white with splashes of light brown and zigzag lines of similar colour; equivalve, inequilateral, posterior side the longer; extremity subtruncate; anterior end narrowly convex, ventral margin flatly rounded, finely crenulate within. Sculptured with numerous, fairly regular concentric ribs, and at the posterior end two ribs merging into a stronger one. Lunule very narrow, slightly darker. Ligament external, prominent, deeply set. Interior whitish, stained with purple; muscle scars well defined, the anterior oval, posterior much rounder; pallial line distinct, pallial sinus somewhat triangular. Hinge normal. Size: ¾ inch.

Localities: 40 fathoms off Wilson's Promontory; from Bass Strait cable.

Its peculiar colour pattern appears to be a fairly constant feature and should serve to identify the species. Occurs also in South Australia and New Zealand.

Placamen Iredale, 1925

Shell small, thick, trigonal. Sculptured with strong, fringe-edged, concentric lamellae. Ligament external, elongated. Lunule heart-shaped, deeply impressed. Hinge with left median cardinal not divided. Pallial sinus blunt, ascending.

TYPE SPECIES *Venus placida* Philippi, 1844.

Placid Venerid

Placamen placida (Philippi, 1844)

(Figure 401)

Shell white or cream, sometimes with three broad rays of fawn or pink. Solid, trigonal, cordate, somewhat inflated, equivalve, inequilateral, posterior side the longer; post-dorsal margin conspicuously arched, excavated; ventral margin much rounded, finely crenulated within. Sculpture consists of twenty to twenty-five smooth, blunt, concentric, inwardly curved lamellae. Lunule heart-shaped, deeply impressed, finely longitudinally striated. Liga-

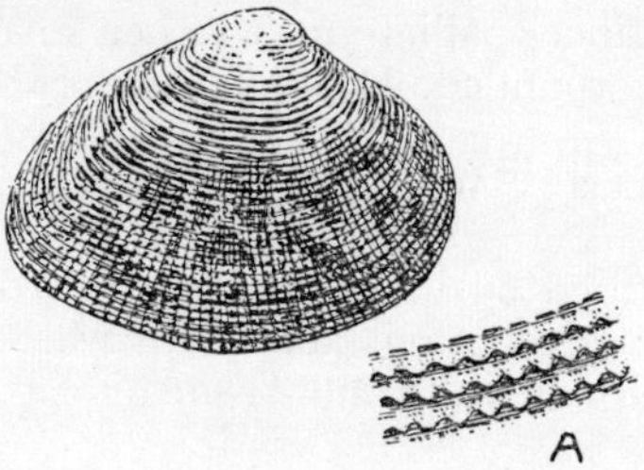

Figure 399

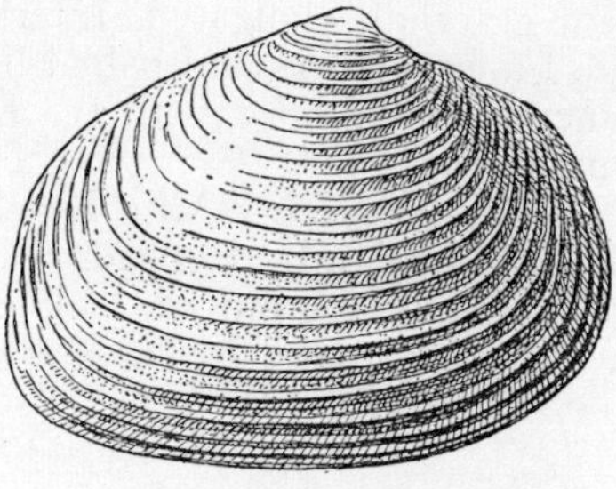

Figure 400

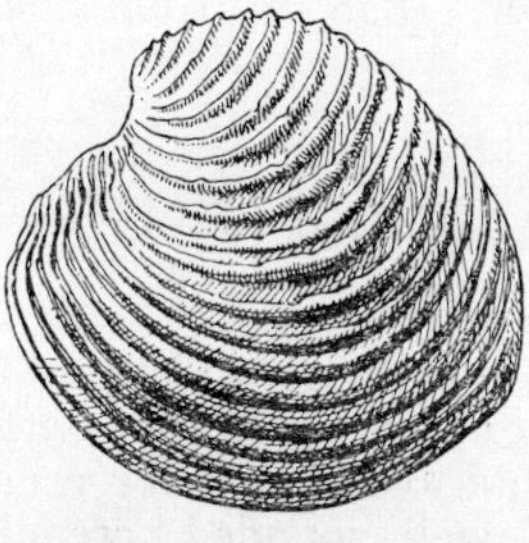

Figure 401

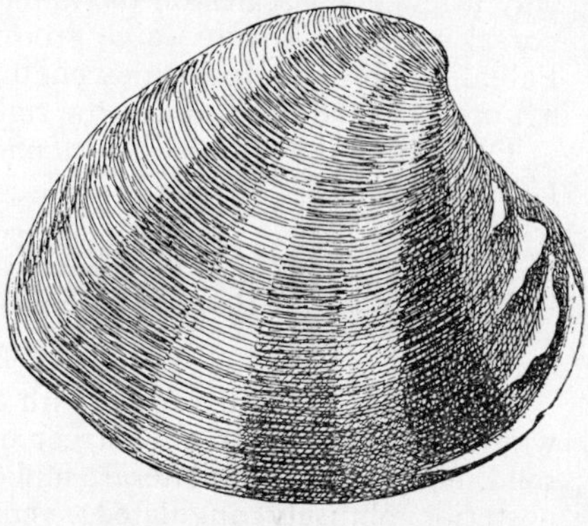

Figure 402

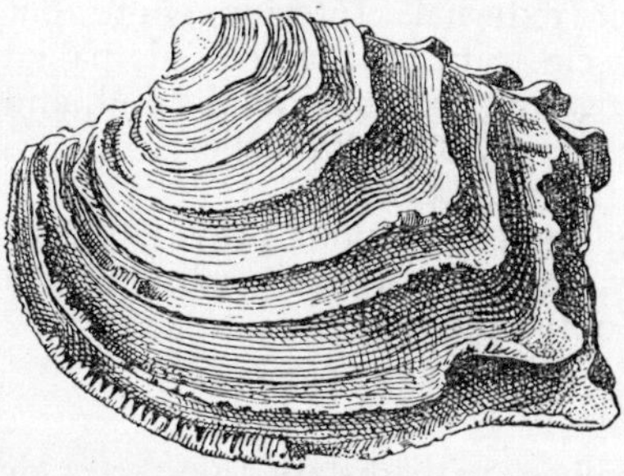

Figure 403

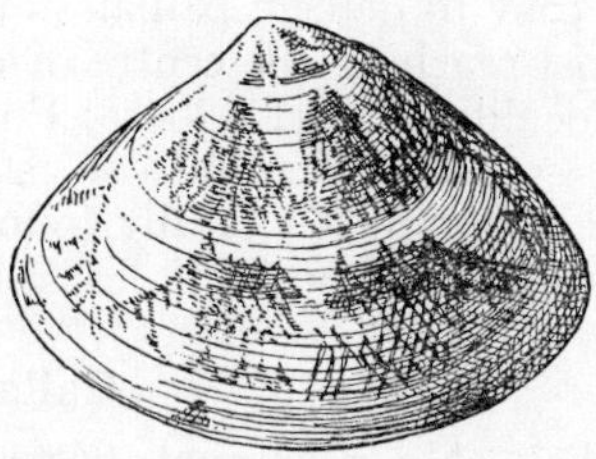

Figure 404

399. *Tawera lagopus* × 1
A Sculpture (enlarged)
400. *Tawera spissa* × 2
401. *Placamen placida* × 1⅓
402. *Bassina pachyphylla* × 1
403. *Callanaitis disjecta* × 1
404. *Gomphina undulosa* × 1⅓

ment external, elongated. Interior white, tinted with light purple at both ends; muscle scars and pallial line very distinct; pallial sinus rather small, somewhat triangular with the end obtusely rounded. Hinge strong. Size: 1 inch.

Localities: Ninety Mile Beach; Port Albert; San Remo; Portsea, Port Phillip; Portland (dredged).

A strongly sculptured, ornate shell identified by its numerous inwardly curved concentric lamellae. Located on a sandy bottom and often washed up on the beaches. Recorded also from South Australia and Tasmania.

Bassina Jukes-Brown, 1914

Shell subtrigonal, thick, convex, brownish; concentrically striated with only a few erect scales on the anterior slope; escutcheon not defined; both dorsal margins of right valve grooved, and those of the left bevelled to fit. Pallial sinus linguiform, descending. Hinge with large unequally divided left median cardinal and narrow right cardinals.

TYPE SPECIES *Venus paucilamellata* Dunker, 1858 = *Venus pachyphylla* Jonas, 1839.

Faintly Frilled Venerid

Bassina pachyphylla (Jonas, 1839)

(Figure 402)

Shell fawn or light brown with three broad, radiating bands of varying widths and of similar but darker colour; triangularly oval, convex, thick, solid, equivalve, inequilateral; umbos curved anteriorly. Anterior rounded, posterior obtusely angulated; ventral margin well rounded, crenulated. Surface of valves shows concentric lines of growth, which are rather coarse, and at the anterior end will be seen three to five erect lamellae. Lunule narrow, well defined, striated. Ligament external. Interior white but occasionally yellow tinted at the ends; muscle scars well impressed; pallial line easy to detect; pallial sinus deep, triangularly rounded at the end, and almost reaching the centre of valve. Hinge strong. Size: 2½ inches.

Localities: Port Albert; Portland; Port Fairy; Lorne.

Its solid nature and consistent radiating bands of colour provide means of recognition. Generally found in shallow water. Its range of distribution is from New South Wales, Tasmania to South Australia.

Callanaitis Iredale, 1917

Shell oblong or oval, flattened at the umbos, with distant thin, erect or recurved concentric lamellae. Lunule small, lanceolate. Escutcheon only defined in left valve. Umbos smooth. Margins feebly crenulate; pallial sinus moderately deep. Median teeth very bifid.

TYPE SPECIES *Venus yatei* Gray, 1835.

Marwick, 1927, Powell, 1946, and Ludbrook, 1953, are all of the opinion that this genus is inseparable from *Bassina*. Marwick states that 'the differences between *V. yatei* and *V. pachyphylla* are of degree, not of kind. While in the main in agreement we have retained *Callanaitis* for *disjecta* because of the extreme development of the lamellae and generally more delicate structure.

Frilled Venerid

Callanaitis disjecta (Perry, 1811)
(Figure 403)

Shell cream, thin, subglobose, equivalve, inequilateral. Sculpture consisting of distant, thin, recurved, concentric, lamellae, which are pink tinted when fresh, and striated on the outer side, the interstices showing fine growth-lines. In addition to the concentric sculpture there are two posterior ridges descending from the umbos to the ventral margin. Umbos situated anteriorly about one-fourth of the length of shell. Lunule distinct, lanceolate. Ligament strong, elongate. Size: 2½ inches.

Localities: Port Melbourne; South Melbourne; Frankston; Western Port.

This is one of our most attractive shells and the peculiar shape and characteristic lamelliform sculpture will serve to distinguish it. Perfect examples are obtained by dredging on sandy mud-banks. The species is also recorded from New South Wales and Tasmania, and the genus is apparently restricted to Australia and New Zealand.

Gomphina Mörch, 1853

Shell medium size, trigonal, anterior end narrowly rounded, posterior end produced, obliquely truncated to form a well-marked posterior area. Lunule long, very narrow, shallowly impressed. Sculpture of anterior area of weak, bevelled, concentric grooves, posterior area with weak radials crossed by strong growth-ridges. Pallial sinus moderate, oval, horizontal. Valve margin is smooth. Hinge of left valve with three rather narrow, straight divergent cardinals, median grooved; right valve; with straight narrow, grooved posterior cardinal, broadly triangular, and a grooved median and lamellar-like anterior one.

TYPE SPECIES *Venus donacina* Chemnitz, 1782 = *Donax veneriformis* Lamarck, 1818.

Waved Venerid

Gomphina undulosa (Lamarck, 1818)
(Figure 404)

Shell white, more or less marked near the umbonal region with reddish brown, triangular, equivalve, subequilateral. Surface concentrically, finely striated. Lunule elongated, not distinctly marked. Ligament external. Interior white. Size: 1 inch.

Localities: Portsea; Point Lonsdale; dredged off Balnarring, Western Port.

A solid shell, of an unusual triangular form. It may be pure white or marked with reddish brown, and sometimes with more or less undulating and angular reddish brown lines; the last-named feature is more evident in Western Australian examples, where it is found, in addition to South Australia and Tasmania. These molluscs live in sand or sandy mud.

Gouldiopa Iredale, 1924 (*Gouldia australis* Angas, 1865).

australis (Angas, 1865); off Wilson's Promontory; off Lakes Entrance (30 fathoms).

Eumarcia Iredale, 1924

Shell triangularly oval. Lunule not impressed, bounded by weak line; surface smooth, polished. Pallial sinus wide, not very long, ascending. Valve margins smooth. Left valve three cardinals, posterior narrow, median and anterior triangular deeply grooved; right valve with moderate triangular, grooved, posterior cardinal, very narrow, grooved median and entire anterior one.

TYPE SPECIES *Venus fumigata* Sowerby, 1853.

Smoked Venerid

Eumarcia fumigata (Sowerby, 1853)

(Figure 405)

Shell cream or light fawn, with two or three light brown rays; equivalve, inequilateral, ovately triangular; anterior end rounded; posterior somewhat acuminate or pointed. Surface of valves smooth and shining, with faint indication of grooves, more noticeable at the extremities. Ligament external. Interior white, with traces of pink or yellow. Size: 1½ inches.

Localities: South Melbourne; St Kilda; Mordialloc; Dromana.

Although exceedingly common, this species, in living condition, is one of our most variously painted bivalves, the colour of which fades considerably on exposure. The rays in varying degrees of colour pattern, darkest near the umbos, may be two, three or rarely four in number. After stormy weather, the beaches are strewn with this shell, in company with *Katelysia rhytiphora* and several other bivalves. It is found in shallow water living in sand and sandy mud, and with the aid of the dredge may be obtained in quantity quite close to shore. It also occurs in New South Wales, South Australia and Tasmania.

Katelysia Römer, 1857

Shell obliquely oval, compressed or convex, anterior side very short. Sculpture of strong, concentric ridges which may be corrugated by radial ribs. Pallial sinus small, obtuse or rounded. Hinge with three cardinals, nearly straight, divergent and more or less rugose.

TYPE SPECIES *Venus scalarina* Lamarck, 1818.

Stepped Venerid

Katelysia scalarina (Lamarck, 1818)

(Figure 406)

Shell white or creamy white, generally grey on the posterior slope, whole surface freely dotted with light brown or grey, obliquely wedge-shaped, flatly compressed. Sculptured throughout with fairly strong concentric ridges. Anterior extremity very short, posterior elongately produced. Ligament external, narrow. Size: 1½ inches.

Localities: Frankston; Dromana; Sorrento; Portsea; Western Port.

Along with *rhytiphora* Lamy this is a very common species. It is somewhat similar, but its smoothness, more oblique shape, and the absence of the

Figure 405

Figure 406

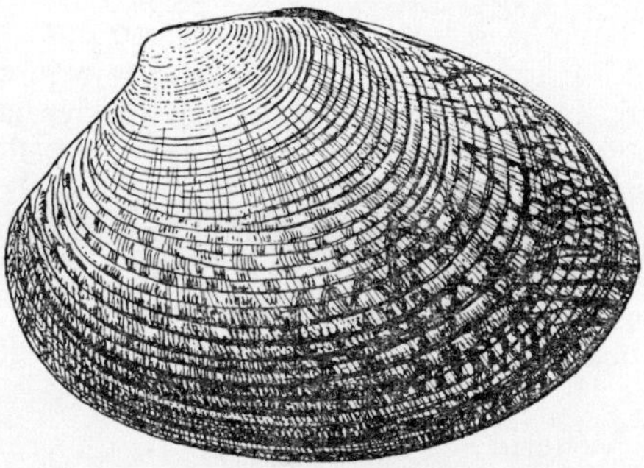

Figure 407

Figure 408

Figure 409

Figure 410

405. *Eumarcia fumigata* × 1
406. *Katelysia scalarina* × 1
407. *Katelysia rhytiphora* × 1⅓
408. *Katelysia peronii* × 1
409. *Venerupis crenata* × 1
410. *Venerupis iridescens* × 2

radiating striae distinguish this species. It is also recorded from South Australia, New South Wales and Tasmania.

Ridged Venerid

Katelysia rhytiphora Lamy, 1937

(Figure 407)

Shell light brown, with darker brown, wavy, or zigzag markings; equivalve, inequilateral, ovately wedge-shaped. Ornamented with fairly regular, concentric ridges, which are crossed by crowded striae. Ligament external, narrow. Interior yellowish, with blotches of purple at each extremity. Size: 2 inches.

Localities: Port Melbourne; South Melbourne; Frankston; Sorrento; Western Port.

This may be regarded as one of our commonest shells, occurring anywhere on our sandy beaches. Few localities have been noted above. It is so distributed around Port Phillip that it would be safe to name every locality where sand or sandy mud exists. As regards form and coloration, the species is subject to considerable variation. Some specimens are decidely more rounded. South Australian examples appear to be stronger shells.

There has been confusion over the name of this species but Lamy, 1939, showed that the two names most commonly used were not available. *Venus corrugata* Lamarck, 1818, was preoccupied and *Venus strigosa* Lamarck, 1818, was not this shell, the mistake being Sowerby's who figured this shell as *strigosa* Lamarck.

Peron's Venerid

Katelysia peronii Lamarck, 1818

(Figure 408)

Shell roundly oval, inflated; colour extremely variable, normally fawn or light brown with darker brown, wavy, or zigzag markings which are often absent, and some specimens are rich violet at the umbos and light blue towards the ventral margin; equivalve, inequilateral, posterior side the longer, somewhat acuminate posteriorly; ventral margin much rounded. Sculpture consisting of broad, flattened, irregular concentric ridges wider at the umbonal region. Lunule well defined, narrow, finely, longitudinally striated. Ligament external, elongated. Interior yellowish and dark rich violet, occasionally almost entirely the latter colour; muscle scars deep; pallial line well marked; pallial sinus wide and truncated at the end. Hinge with three prominent teeth. Size: 1½ inches.

Localities: Port Albert; Lakes Entrance.

A species located in sandy mud, in estuaries near half-tide mark. Found also in the waters of Tasmania, South Australia, Western Australia.

The genus is represented in Victoria by three species which both externally and internally show great variation in colour. However, the following which appear to be constant features will serve to differentiate the forms and no difficulty should be experienced in separating them.

rhytiphora; ovate, wedge-shaped form, anastomosing concentric ridges, crossed by radiating striae.

scalarina; obliquely wedge-shaped form, regular concentric ridges, radiating striae absent.

peronii; more inflated form, rounder, with flatter concentric ridges.

Venerupis Lamarck, 1818

Shell oblong. Sculptured with radial lines or riblets crossed by concentric ridges or striae. Lunule indefinite. Hinge plate very short, narrow, excavated and curtailed posteriorly; teeth short, close together and parallel, two in each valve bifid or grooved. Ventral margin smooth.

TYPE SPECIES *Venus perforans* Montagu, 1803.

Crenated Boring Venerid

Venerupis crenata Lamarck, 1818

(Figure 409)

Shell white with a tinge of fawn or brown during life; ovately oblong, tumid, inequilateral. Posterior-dorsal edge of right valve slightly overlapping the left. Sculpture consists of numerous irregularly spaced, concentric lamellae which become somewhat foliaceous or frill-like at the posterior region and are crossed by numerous, close, radial ribs, producing a reticulated or net-like surface to the shell. Lunule indistinct. Ligament long, partly immersed. Interior white with varying degrees of violet at the central and posterior areas; muscle scars and pallial line very pronounced; pallial sinus deep, elongated, triangular, posteriorly situated. Hinge narrow, tridentate in one valve and bidentate in the other. Size: 1½ inches.

Localities: Portarlington; Western Port; Anderson's Inlet.

This is a common species. Shells of this genus are found inhabiting the holes and crevices of rocks to which they are usually attached by the byssus; as a result, considerable abnormality and deformity is seen in the specimens. Occurs in New South Wales, Tasmania and South Australia.

Shining Boring Venerid

Venerupis iridescens Tate, 1889

(Figure 410)

Shell creamy white with a few splashes of reddish brown and covered with a thin, light brown epidermis, transversely ovate, inflated, inequivalve, inequilateral, posterior side the longer; anterior end short and narrow, posterior end truncated; ventral margin slightly curved at the posterior half. Sculptured with numerous concentric laminae and further adorned with four to five concentric striae in the interspaces, the whole crossed by fine, close, microscopic striae. Ligament external, narrow. Interior white; muscle scars well defined; pallial line plainly seen; pallial sinus very large, rounded and reaching the centre of shell. Hinge normal. Size: ¾ inch.

Locality: Port Albert.

An irregularly shaped shell easily recognized by its fine concentric ornament. Juvenile examples may possibly be confused with the Little Bean, *Pullastra fabagella* Deshayes, but distinction is at once evident in the irregular shape and stronger texture of the present species.

Stout Boring Venerid

Venerupis obesa Deshayes, 1853

(Figure 411)

Shell white, obese, inequilateral. Valves sculptured by about fifteen to twenty erect lamellae and numerous irregularly spaced ribs. Lunule not well defined; ligament external. Interior white, with a brownish tint at the posterior end; muscle scars and pallial line indistinct; pallial sinus posteriorly situated, fairly deep, triangular. Hinge small. Size: ½ inch.

Localities: Portsea, Port Phillip; San Remo.

This is a small obese species, somewhat allied to *exotica* Lamarck, from which it may be distinguished by its more numerous and more erect lamellae, its squatter form, and in the presence of interior coloration.

mitis Deshayes, 1853; Corio Bay.

Exotic Boring Venerid

Venerupis exotica Lamarck, 1818

(Figure 412)

Shell dirty white or cream, oblong, subglobose, inequilateral; posterior angle usually brown near the umbo, fading on approach to shell margin; posterior-dorsal edge of right valve slightly overlapping the left. Surface of shell sculptured with about ten to twelve distant concentric lamellae and numerous radial ribs. Lunule ill-defined; ligament external, partly sunken. Interior white; muscle scars well marked; pallial line distinct with a posterior, rather deep, triangular sinus. Hinge narrow, tridentate, teeth small. Size: 1½ inches.

Localities: Port Phillip; Western Port; Portland.

This is a distinct species, usually with posterior area brown, but this feature is sometimes almost absent. It is easily separated from *crenata* Lamarck by its flatness, more regularity in contour, fewer concentric lamellae, and fewer radial ribs. Recorded also from Tasmania and South Australia.

Claudiconcha Fischer, 1887 (*Venus monstrosa* Gmelin, 1791).

cumingi (Deshayes, 1853); an elongately oval form from Port Albert somewhat resembling *exotica* Lamarck but may be immediately separated by its violet posterior area internally and externally, and in the absence of the pronounced concentric lamellae so characteristic of that species.

Pullastra Sowerby, 1826

Shell with shallow radial or corrugated concentric sculpture; escutcheon not defined. Pallial sinus large and deep, sometimes touching the pallial line below; pedal scar separate.

Type Species *Venus pullastra* Montfort, 1810.

Milk Stone

Pullastra galactites (Lamarck, 1818)

(Figure 413)

Shell white, elongately oblong, equivalve, very inequilateral. Surface of shell decussated throughout. Anterior and posterior extremities rounded. Ligament external, brown. Interior white. Size: 2 inches.

Figure 411

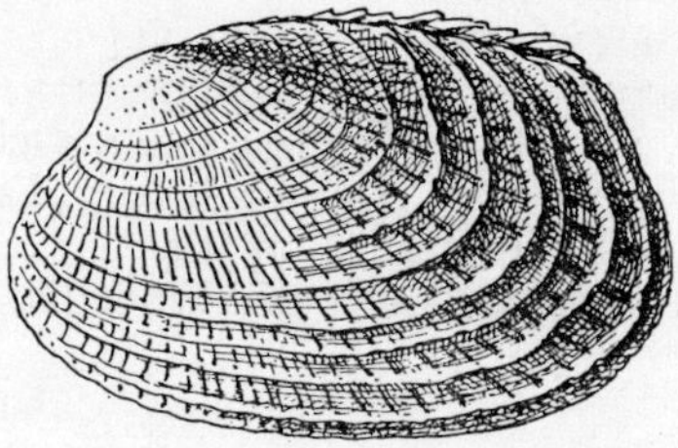

Figure 412

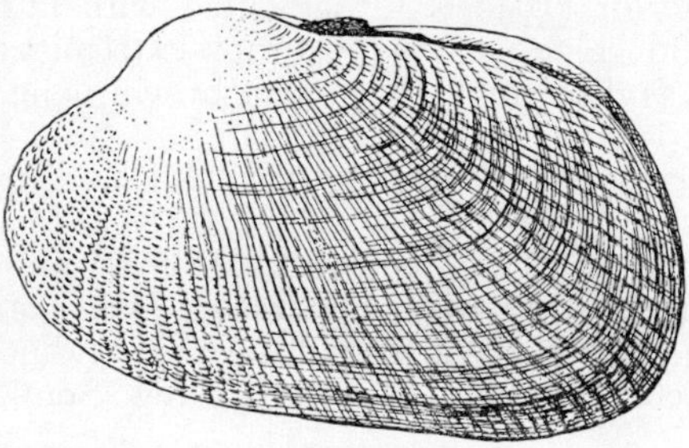

Figure 413

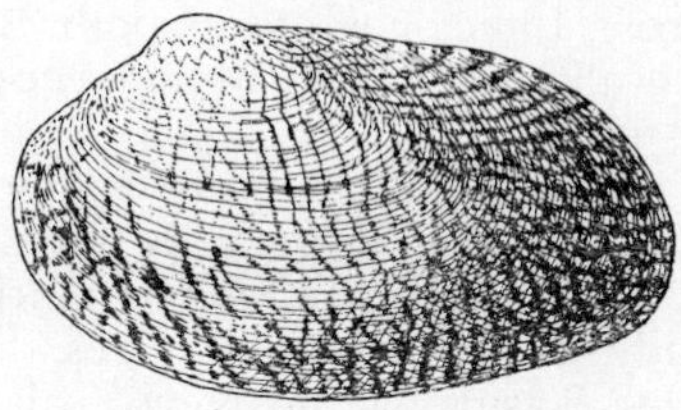

Figure 414

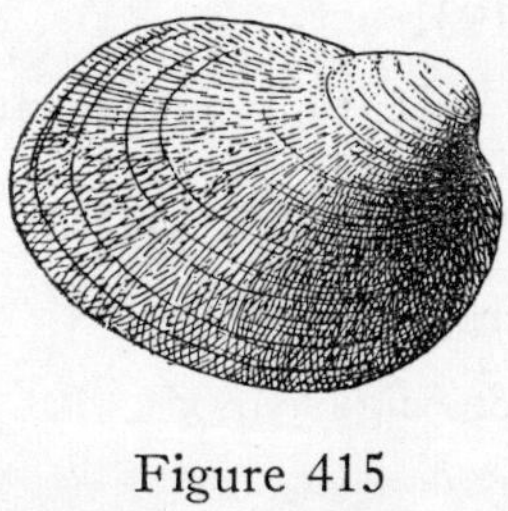

Figure 415

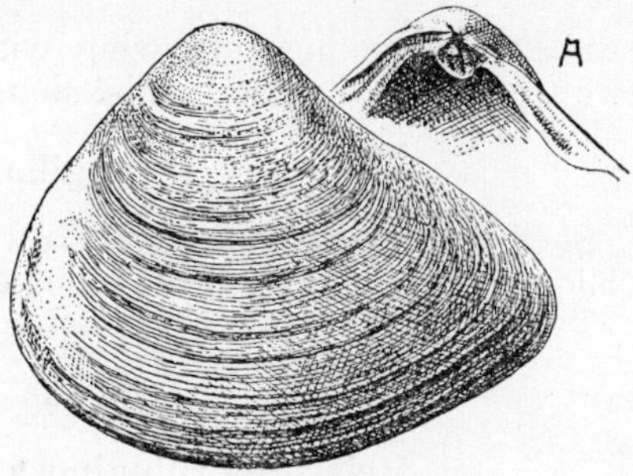

Figure 416

411. *Venerupis obesa* × 3
412. *Venerupis exotica* × 1⅓
413. *Pullastra galactites* × 1
414. *Pullastra flabagella* × 1⅓
415. *Velargilla rubiginosa* × 2
416. *Anapella cycladea* × 2
A Hinge

Localities: St Kilda; Brighton; Mordialloc; Sorrento; Western Port.

The vernacular name appears most appropriate, being derived from galactites—a precious stone of a white colour. The shell is always white, and with its striking shape and characteristic decussated sculpture, should prove a species easy to identify. Living in gravelly sand, it is fairly common in Port Phillip and on the Victorian coast generally. In addition to this it is recorded from South Australia, Western Australia and Tasmania.

Little Bean

Pullastra fabagella (Deshayes, 1853)
(Figure 414)

Shell white with zigzag markings; fragile, equivalve, inequilateral; anterior end narrowly rounded, posterior broader. Ornamented with fine concentric striae, which are fewer and bolder at the posterior extremity; under the lens a finer radial striation is discernible. Ligament brown, long, narrow. Interior white, smooth. Size: 1 inch.

Localities: South Melbourne; Brighton; Portarlington; Queenscliff; Western Port.

The Little Bean is quite common in Port Phillip, and with *galactites* may be found living in gravelly sand. The two are easily separated, *fabagella,* with its smaller size, colour markings and concentric sculpture, providing readily distinguishing features. It occurs in South Australia, New South Wales, Tasmania and New Zealand.

Family PETRICOLIDAE

Mud and Rock Borers

Shell oval, gaping, thin, white with a very thin epidermis. Hinge narrow, two teeth in each valve; pallial sinus deep.

Velargilla Iredale, 1931

Shell moderately long, reddish with brown rays; thin, with divaricating oblique irregular ribs; shell gaping at the anterior end.

TYPE SPECIES *Naranio rubiginosa* A. Adams & Angas, 1863.

Tinted Mud Borer

Velargilla rubiginosa (Adams & Angas, 1863)
(Figure 415)

Shell thin, fragile, cream, rubiginous in the umbonal region with several darker rays of similar colour; ventricose, equivalve, inequilateral; umbos small, acute; anterior end the shorter and much rounded, posterior somewhat oblique; ventral margin flatly convex. Radial sculpture consists of microscopic divaricating striae; concentric growth-lines plainly seen. Lunule distinct. Ligament in deep groove, short. Interior creamy with the coloration and rays easily seen through the shell. Posterior muscle scar rounded, slightly larger than the elongated anterior one; pallial line distinct, pallial sinus rounded and very deep. Hinge small. Size: ¾ inch.

Localities: Frankston; dredged off Portsea; Point Cook (8 fathoms).

Found burrowing in sandy mud; its peculiar oblique shape and characteristic divaricating sculpture provide the means of recognition. Worn examples are almost devoid of colour or radiating rays. Occurs also in Tasmania and New South Wales.

Naranio Gray, 1853 (*Naranio divaricata* Gray, 1853).

lucinalis (Lamarck, 1835); San Remo.

Family DONACILLIDAE
Wedge Shells

Shell thick, elongate, wedge-shaped, equivalve, slightly gaping. Hinge with one cardinal in each valve anterior to the central resilium, anterior and posterior external teeth well developed.

Anapella Dall, 1895

Shell triangularly ovate, ventricose, rounded anteriorly. Hinge with single cardinal in each valve, right furrowed, left prong-like, laterals smooth, subequal, compressed.

TYPE SPECIES *Mesodesma triquetra* Hanley, 1843.

Smooth-toothed Triangle

Anapella cycladea (Lamarck, 1818)
(Figure 416)

Shell dirty white, covered with a brownish epidermis, solid, ventricose, triangularly ovate, equivalve, inequilateral, posterior side the longer and angulated, anterior end rounded, ventral margin convex; umbos very prominent. From umbo to posterior-ventral margin runs a well-defined keel. Surface of valves sculptured with fine concentric striae and lines of growth. Ligament small, narrow. Interior white; muscle scars plainly seen, the posterior more rounded than the anterior and connected by a distinct pallial line without sinus. Hinge thick with a strong, smooth, lateral tooth on each side of the umbo. Size: 1 inch.

Localities: Port Fairy; Port Albert; Lakes Entrance.

This is a shell of somewhat similar shape to the Little Triangle, *Notospisula parva*. However, it is immediately separated by the absence of cross-grooved lateral teeth so characteristic of that species. Of common occurrence particularly in estuarine localities and mud flats. Occurs in New South Wales, Tasmania, and South Australia. Previously known as *A. triquetra* (Hanley, 1943), a name still used by some authors for the estuarine form.

Donacilla Blainville, 1819

Shell subtrigonal, depressed, anterior side produced. Hinge thick, cardinal tooth bifid; two anterior and two posterior laterals in right valve, one in left.

TYPE SPECIES *Mactra donacia* Lamarck, 1818.

Lamarck, 1818, proposed the genus *Amphidesma* and listed a number species only superficially related. Authors have since dispersed most of these into different genera but have failed to find a satisfactory resting place for the 'doniform' shells. *Amphidesma* has been used for them but the earliest designation of a type species, by Children, 1823, was *Amphidesma variegata* Lamarck, 1818, a species considered by subsequent workers to belong to *Semele*. *Amphidesma* thus becomes a subjective synonym of that genus.

Lamy, 1913, suggested the use of *Mesodesma* Deshayes, 1830, which has as type species *Mactra donacia* Lamarck, 1818, a 'doniform' shell and the second member of Lamarck's original assemblage. However, Dawson, 1959, has pointed out that *Mesodesma* is an absolute synonym of *Donacilla* Blainville, 1819, which must therefore be used.

Wedge Shell

Donacilla cuneata (Lamarck, 1818)

(Figure 417)

Shell white, shining; solid, triangularly ovate or wedge-shaped, equivalve, inequilateral; posterior side shorter, ventral margin curved, umbos small, acute, approximate. Surface of valves sculptured with irregular lines of growth. Interior white; muscle scars deep, the posterior one rounder than the anterior; pallial line well defined with a small, rounded sinus. Hinge very strong. Size: 1 inch.

Localities: San Remo; Balnarring; Port Fairy; Lakes Entrance.

A common white species located in sand between tide marks; it is heavier in texture than the other representatives of the genus in Victoria, and extremely variable in shape, at times almost equilateral; further variation is evident at the ventral margin with examples flatly convex, while others are much rounded. It ranges from Tasmania through Bass Strait to South Australia and Western Australia. *glabrella* Lamarck, 1818, is a synonym of this species.

Narrow Wedge Shell

Donacilla angusta Reeve, 1854

(Figure 418)

Shell white, smooth, shining, solid, covered with a thick horn-coloured epidermis; equivalve, inequaliteral. Anterior end truncated, ventral margins straight. Sculptured by faint irregular concentric striae. Interior bluish-white; muscle scars rather deep; pallial line well defined, pallial sinus somewhat rounded. Hinge white. Size: 1 inch.

Localities: San Remo; Airey's Inlet; Apollo Bay; Lorne.

Like the Shining Wedge Shell it is found living buried in the sand just below the surface. Valves are cast up on the shore in countless numbers, their white colour and strong texture rendering them useful for fancy boxes and ornaments. Common also in Tasmania, New South Wales, South Australia and Western Australia.

Eryx Wedge Shell

Donacilla erycinaea (Lamarck, 1818)

(Figure 419)

Shell cream with a light brown epidermis; solid, ovate-triangular, equivalve, inequilateral, anterior side the longer; ventral margin convex;

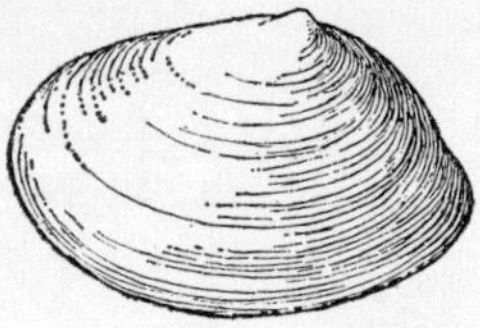

Figure 417

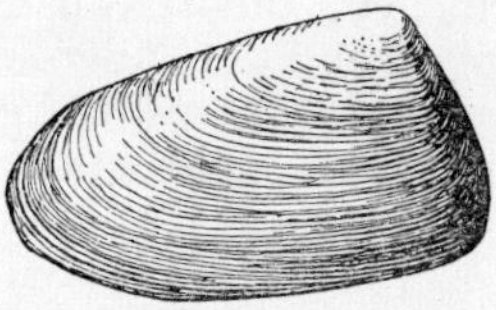

Figure 418

Figure 419

Figure 420

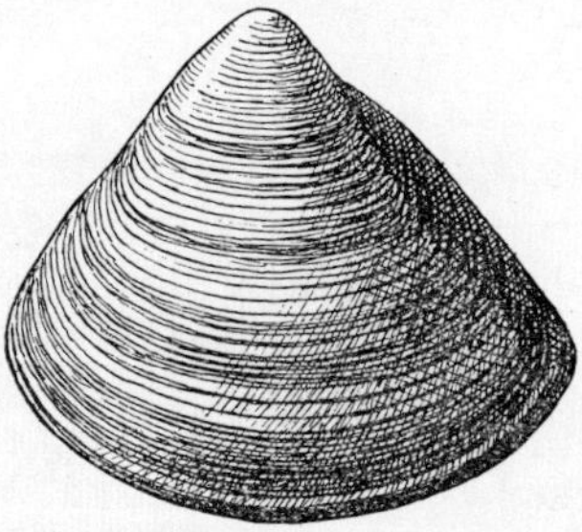

Figure 421

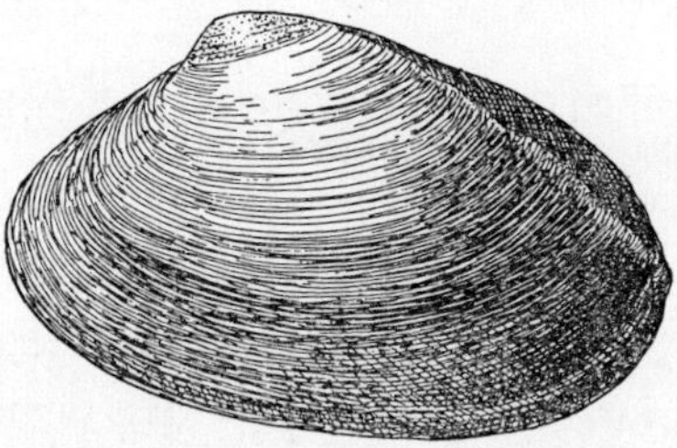

Figure 422

417. *Donacilla cuneata* × 1
418. *Donacilla angusta* × 1
419. *Donacilla erycinea* × 1
420. *Donacilla nitida* × 2
421. *Notospisula parva* × 2
422. *Electromactra antecedens* × 1

umbos small, acute, approximate. Surface of valves sculptured with irregular lines of growth and fine, microscopic striae. Interior bluish white, somewhat pearly towards the ventral margin; pallial line distinct with a small rounded sinus; muscle scars well sunken, anterior one more elongated. Hinge strong. Size: 1 inch.

Localities: Port Albert; Mentone.

At the latter locality taken alive on sandbank. Similar in shape to *cuneata* (Lamarck), but is more equilateral and invested with a much darker epidermis; added to these distinctive features is the bluish interior which will assist in identifying the species.

Shining Wedge Shell

Donacilla nitida (Deshayes, 1854)

(Figure 420)

Shell white, smooth, shining, covered with a thick yellowish, horny epidermis; solid, somewhat wedge-shaped, equivalve, very inequilateral. Sculpture consisting of feeble, irregular, concentric striae. Interior bluish white. Size: 1 inch.

Localities: South Melbourne; Elwood; Mordialloc; Mornington; Dromana.

Cast up on the beaches, Wedge Shells are extremely abundant, at times in such numbers as to render them of commercial value as shell grit. These molluscs live buried in sand at a small depth beneath the surface. For their size, these creatures show great activity, and are responsible for the numerous tracks observed on the sand at low water.

All along the coastline these shells are washed up, and are of common occurrence; in New South Wales and Tasmania.

Family MACTRIDAE

Trough Shells

Shell equivalve, trigonal. Hinge with two cardinal teeth in each valve, the posterior small compressed, often rudimentary; the anterior triangular, more or less deeply notched and situated in front of the pit for the resilium.

Notospisula Iredale, 1930

Shell trigonal, small, white. Cardinal teeth of right valve small, the posterior short, left valve with a distinctly forked cardinal and with a short tooth at the angle of the cartilage; lateral teeth distinctly grooved.

TYPE SPECIES *Gnathodon parva* Petit, 1853.

Little Triangle

Notospisula parva (Petit, 1853)

(Figure 421)

Shell cream, smooth; triangularly ovate, somewhat ventricose, equivalve, inequilateral; anterior extremity rounded, posterior broadly angled. Lunule small. Interior white. Hinge teeth elongated, distinctly cross-grooved. Size: 1 inch.

Localities: Port Melbourne; South Melbourne; St Kilda; Western Port.

An exceedingly common species, generally located on the sands and sandy muds of river mouths. No outstanding feature is evident in this species excepting, perhaps, the characteristic cross-grooved hinge teeth. Beyond Victoria it is recorded from Queensland, New South Wales, South Australia and Tasmania.

cretacea (Angas, 1867); common in eastern Australia, this and the above species often occur together. It is easily distinguished by its more elongate form.

Electromactra Iredale, 1930

Shell elongately trigonal. Hinge plate deep, teeth short, forked cardinal of left valve small; resilium pit large carrying the hinge plate into the shell cavity.

TYPE SPECIES *Mactra parkesiana* Hedley, 1902.

Oval-shaped Trough Shell
Electromactra antecedens Iredale, 1930
(Figure 422)

Shell white, shining, invested with a fawn epidermis; ovate, flatly convex, equivalve, inequilateral, slightly gaping at the extremities; very finely, irregularly, concentrically striated; umbos not prominent, an indistinct ridge descending posteriorly from the umbos to the ventral margin. Lunule elongated. Ligament external. Interior white. Hinge normal, with the cartilage contained in a deep triangular pit near the centre. Size: 2 inches.

Localities: South Melbourne; St Kilda; Mordialloc; Dromana; Queenscliff; Western Port.

This shell is colourless and lacking in ornament; with its oval and flatly convex shape it should be easily distinguished from any Victorian members of the family. Inhabiting sand in shallow water these shells are generally distributed throughout Port Phillip, but more abundant in some localities than others. Besides Victoria, the species occurs in New South Wales.

Mactra Linné, 1767

Shell ovate, trigonal, equivalve. Sculptured with concentric striae which may be obsolete. Hinge of left valve with anterior and posterior lateral laminae and two cardinal teeth, the anterior deeply bifid, right valve with two longitudinally parallel laterals and two lamellate cardinal teeth.

TYPE SPECIES *Mactra stultorum* Linné, 1767.

Iredale, 1930, erected *Austromactra* with *Mactra contraria* Reeve, 1854, as type species. *contraria* and the allied species *rufescens* are more coarsely concentrically sculptured than *Mactra s.s.* but otherwise do not differ markedly. A series of the species illustrates graduations in sculpture strength within the genus. We are therefore of the opinion that it is of specific rather than generic significance. Also shape and disposition of the teeth of the Australian species do not differ from *stultorum* to any greater extent than they differ from each other.

Reddish Trough Shell

Mactra rufescens Lamarck, 1818
(Figure 423)

Shell very pale brown, sometimes with two or three radiating rays of reddish purple; solid, equivalve, equilateral; umbos prominent, invariably tinted with purple. Sculpture around the umbonal region with fine concentric striae becoming much coarser towards the ventral margin, the whole surface covered by a brownish periostracum. The anterior and posterior slopes well developed, flexuously wrinkled, and often stained with orange red. Interior white with occasional stains of purple or orange towards the umbo; muscle scars deep. Size: 2½ inches.

Localities: Anderson's Inlet; Cape Bridgewater; Portland; Apollo Bay; Lorne; Lakes Entrance; Port Albert; Snake Island; Mallacoota.

This species may be recognized by its peculiar shape and wrinkled sculpture. It is a very common shell living in sand and sandy mud. The presence of valves in kitchen middens indicates that the aborigines made use of the animal as an article of food, and it is also a good bait for fishing. The species occurs in Western Australia, South Australia, Tasmania and southern New South Wales.

Southern Trough Shell

Mactra australis Lamarck, 1818
(Figure 424)

Shell violet, shining; triangularly oval, anterior slightly rounded, ventral margin broadly convex. Umbos small, incurved. The colouring is composed of irregular bands of light grey and violet; the violet is dense at the umbo, and on the anterior slope where two radiating rays run from the umbo to the ventral margin. Sculptured with closely set concentric striae, the whole surface covered with a pale brown epidermis. Interior purplish, lighter towards the umbo; muscle scars and pallial line very distinct and with a roundly angulate pallial sinus. Hinge normal. Size: 1 inch.

Localities: Lorne; Port Fairy; Portland; Cape Bridgewater.

This Trough Shell is recognized by its interior and external violet colour. In sand at low water this species appears in great quantities and is subject to much attack by univalve molluscs which pierce the shell mostly at the umbonal region. The term Polished Trough Shell is applied to this species which occurs in Tasmania, South Australia and Western Australia. Previously known under the name of *polita* Chemnitz, 1782.

White Trough Shell

Mactra pura Deshayes, 1853
(Figure 425)

Shell white, solid; tumid, triangularly oval, equivalve, subequilateral, obtusely angled posteriorly. Sculptured with extremely fine, obsolete concentric striae; the whole surface enclosed in a light brown, silky periostracum. Interior white; muscle scars plainly seen but not deep; pallial line

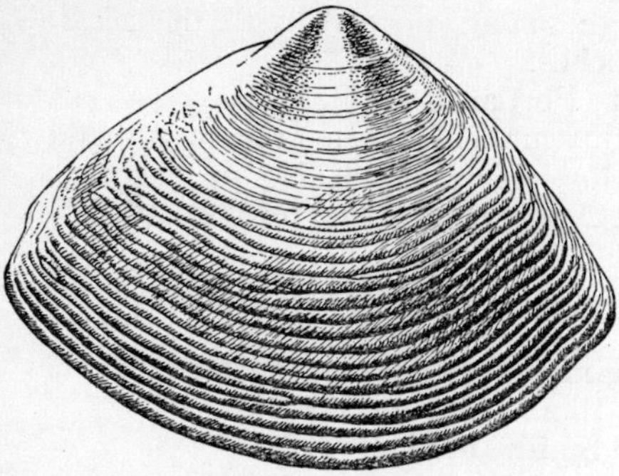

Figure 423

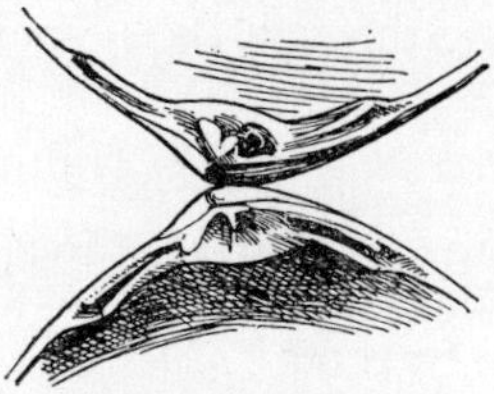

Figure 423A

Figure 424

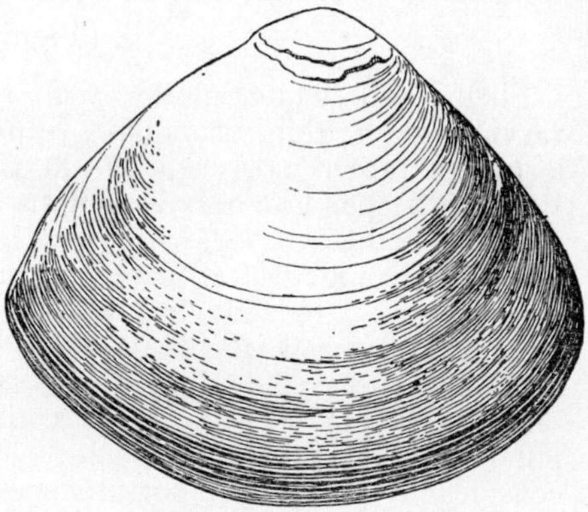

Figure 425

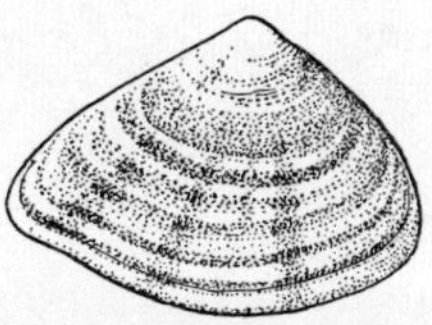

Figure 426

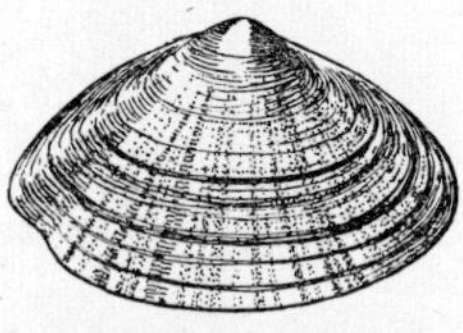

Figure 427

423. *Mactra rufescens* × 1
423A. Hinge
424. *Mactra australis* × 1
425. *Mactra pura* × 1
426. *Nannomactra jacksonensis* × 2
427. *Nannomactra pusilla* × 1

fairly evident; pallial sinus deep. Hinge strong, normal, with the central triangular pit well developed. Size: 2 inches.

Localities: San Remo, Western Port; Portland.

A pure white shell which by its tumidity and silky periostracum is readily identified. Occurs in sand patches at low tide. Found also in Tasmania, extending through Bass Strait to South Australia and Western Australia.

Nannomactra Iredale, 1930

Shell small, equivalve, inequilateral. Hinge plate narrow, delicate, cardinal teeth diverging at right angles, laterals thin.

TYPE SPECIES *Mactra jacksonensis* Smith, 1885.

Port Jackson Trough Shell

Nannomactra jacksonensis (Smith, 1885)

(Figure 426)

Shell small, white, glossy, with six to eight brown radiating rays of varying width; thin, triangular, equivalve, inequilateral, acuminate posteriorly and rather more rounded at the anterior end. Umbos small, white. Posterior slope more heavily sculptured than the anterior. Internally similar in colour to the exterior, with the radiating rays clearly visible; muscle scars very faint. Hinge white. Size: ½ inch.

Localities: Dredged (5-8 fathoms) off Point Cook; Portsea; between Phillip and French Islands, Western Port; Port Albert.

This is our smallest Trough Shell recognized by its glossy surface and brown radiating rays. It may be confused with young *pusilla* (A. Adams) but is distinguished by its pale coloured umbos and the absence of the conspicuous sulci on the dorsal slopes.

It also occurs in Tasmania and New South Wales.

Little Trough Shell

Nannomactra pusilla (A. Adams, 1855)

(Figure 427)

Shell light brown with numerous olive brown radiating rays of varying widths. Fine lines of growth are discernible under lens, the anterior and posterior slopes sculptured with rather strong sulci which become stronger as growth continues. The umbos are purple tipped, and a long white area about one-third the length of shell is a constant feature on the dorsal area. Interior purple, paler towards the ventral margin and whitish at the umbonal region; muscle scars plainly seen; pallial line and pallial sinus fairly distinct, the latter somewhat rounded at the end. Hinge plate with its well developed triangular pit, white. Size: 1⅛ inches.

Localities: Port Albert; Torquay.

It could not be confused with any Victorian Trough Shell, the consistent character of brown radiating rays providing a useful recognition mark.

It occurs in Tasmania, New South Wales and Queensland.

Lutraria Lamarck, 1799

Shell oblong, or oval, inequilateral, gaping at the two extremities, smooth or concentrically striated, epidermis grey, yellow, or white. Pallial sinus very deep, rounded. Hinge plate prominent with a large chondrophore; left valve with anterior bifid cardinal, and anterior and posterior laterals; right valve with two anterior diverging cardinals and small anterior and posterior laterals.

TYPE SPECIES *Mya lutraria* Linné, 1758.

Otter's Shell

Lutraria rhynchaena Jonas, 1844

(Figure 428)

Shell whitish, of solid texture, oblong, rather compressed, inequivalve, subequilateral, gaping at both ends, anterior end angularly rounded. Sculptured by fine lines of growth, the whole surface covered with a dull brown epidermis. Ligament short, subexternal, not separated from the cartilage. Interior white; muscle scars very deep and horizontal. Hinge well defined. Resilium resting in a rather large spoon-shaped chondrophore. Size: 4 inches.

Localities: San Remo, Western Port; off Lakes Entrance.

These shells in living condition are buried vertically in sand or sandy mud, especially of estuaries, at depths ranging from low water to several fathoms.

Cotton and Godfrey provide an interesting note: 'The short ligament admits of considerable movement of the ends of the shell without opening the shell throughout. With the valves in their ordinary positions the shell gapes equally at each end, but the arrangement of teeth and ligament is such that the front of the shell may be entirely closed by more widely expanding the rear, or the hinder gape may be nearly closed by separating the fronts of the valves. A closely allied form in the Channel Isles is dug out for food, and is locally known as "Clumps".'

Recorded also from Tasmania, New South Wales, South Australia and Western Australia.

Zenatiopsis Tate, 1879

Shell equivalve, inequilateral, oblong; umbo anterior, supported internally by a thick rib, extending, with a slightly backward inclination, half-way across valve; narrowly gaping at both ends. Chondrophore prominent; cardinal teeth distinct, no lateral teeth; pallial sinus deep, horizontal.

TYPE SPECIES *Zenatiopsis angustata* Tate, 1879.

Australian members of this genus were originally included in *Zenatia* Gray, 1853, but are immediately separated by the presence of the umbonal rib.

Victorian Trough Shell

Zenatiopsis victoriae (Pritchard & Gatliff, 1903)

(Figure 429)

Shell elongate-oval, thin and compressed, gaping at both ends, with a full and well-rounded anterior end, and slightly attenuate both ventrally and dorsally towards the posterior. Shell greyish white and shining, covered by a light greyish brown epidermis, which is very thin and glossy. Umbos small but well defined, situated at a distance of one-third the length of the shell from the anterior margin. Ventral margin almost straight, only slightly sinuated. Surface of the valves finely, concentrically striated, and under the lens an extremely fine radial striation will be noted. Ligament small and almost internal. Interior white; the internal thickening below the umbo forms at first a strong convex ridge, which gradually flattens and spreads out for a distance of two-thirds the height of the shell; muscle scars plainly seen. Cardinal teeth two, and one rudimentary posterior lateral in each valve. Size: 3 inches.

Locality: Port Albert.

Recorded also from New South Wales and south Queensland.

Family DONACIDAE

Pipis

Shell equivalve, triangular or elongate, more or less compressed, smooth or sculptured. Ligament short, arched, situated in a groove. Pallial line usually very distinct, interior of shell edge often toothed. Hinge as a rule with two cardinal teeth, and two laterals.

Deltachion Iredale, 1930

Shell small, triangular, posterior edge truncated and keeled, anterior produced. Strongly radially sculptured on the posterior, becoming less distinct toward the anterior. Lateral teeth well developed.

TYPE SPECIES *Deltachion virilis* Iredale, 1930 = *Donax brazieri* Smith, 1891.

Thiele, 1934, although accepting this genus, considers it very similar to *Chion* Scopoli, 1771; however comparison of the type species shows several well-marked and constant differences. In *Deltachion* the cardinal teeth are set more obliquely, the radial lines of sculpture are not punctate and continue on to the posterior edge, which may be latticed with concentric growth-lines, while in *Chion* this area has a distinctive wavy sculpture. Comparison of Smith's and Iredale's descriptions and figures and New South Wales material showed that *D. virilis* is a synonym of *D. brazieri.*

Chapman's Pipi

Deltachion chapmani (Gatliff & Gabriel, 1923)
(Figure 430)

Shell white, with discontinuous brown rays; solid, ovate-triangular, inequilateral; radiately striate, the striae becoming stronger at the posterior angle, but towards the anterior end are broader and flatter and not continued beyond the rounded angle. Ventral margin crenulate. Interior, in adults, usually shows a large, triangular, purple blotch descending from the umbos. Size: 1 inch.

Localities: Portsea; San Remo; Shoreham; Cowes, Phillip Island; Kilcunda; Anderson's Inlet.

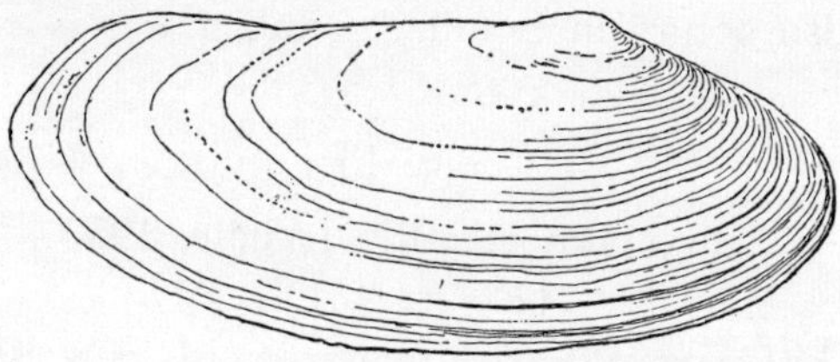

Figure 428

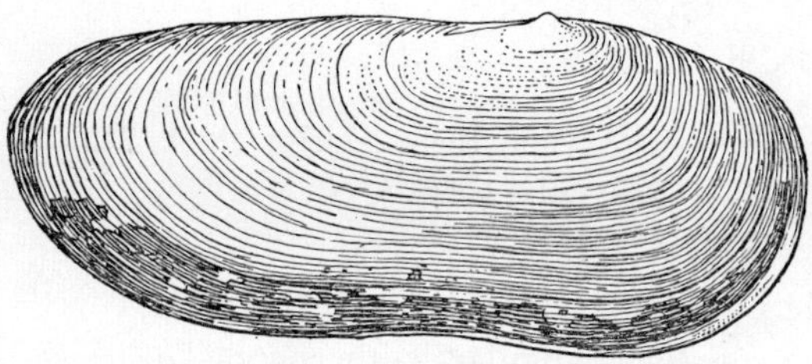

Figure 429

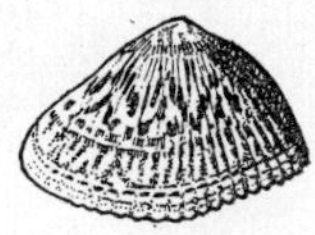

Figure 430

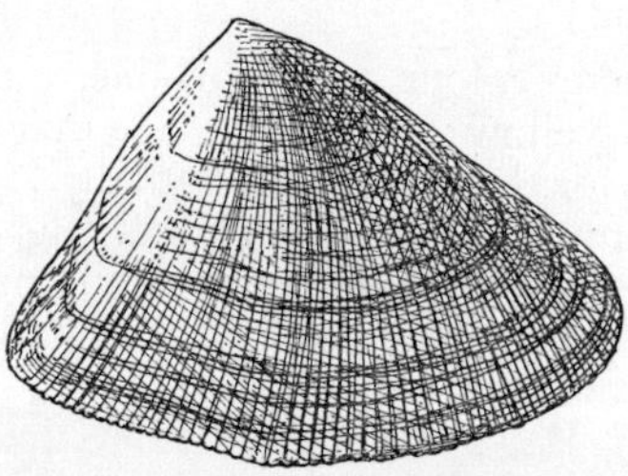

Figure 431

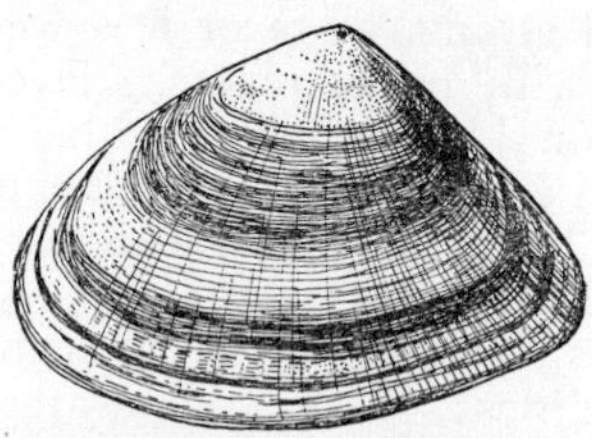

Figure 432

428. *Lutraria rhynchaena* × ⅔
429. *Zenatiopsis victoriae* × 1
430. *Deltachion chapmani* × ⅔
431. *Deltachion electilis* × 2
432. *Plebidonax deltoides* × ⅔

The species also occurs in South Australian waters where it is more commonly found.

Choice Pipi

Deltachion electilis Iredale, 1930

(Figure 431)

Shell bluish white, normally with two brownish rays; solid, triangularly oval, equivalve, inequilateral, anterior side the longer; anterior end rounded, posterior strongly keeled and with the end abruptly truncated. Surface of valves radiately striated, the striae becoming stronger towards the posterior angle but broader and flatter anteriorly; the posterior slope possesses much finer radial sculpture than the remainder of the shell; ventral margin flatly rounded, crenulate. Ligament external, prominent, short. Interior usually with a large purplish coloration descending from the umbo almost to the ventral margin. Hinge normal. Size: 1 inch.

Localities: Portland; Port Fairy.

This species is a close ally of *chapmani* (Gatliff & Gabriel) but easily distinguished by the finer radial sculpture at the posterior slope. *D. brazieri* Iredale, 1930 (non Smith) is an elongate form of this species. Recorded also from Queensland and New South Wales.

Plebidonax Iredale, 1930

Shell large and heavy, unequally triangular, with rounded edge. Sculptured with faint radial lines. Lateral teeth strong; right anterior dorsal edge furrowed, interior edge smooth.

TYPE SPECIES *Donax deltoides* Lamarck, 1818.

Pipi

Plebidonax deltoides (Lamarck, 1818)

(Figure 432)

Shell strong, more or less wedge-shaped; colour very variable, may be tinted with olive, pinkish, lavender or yellow; equivalve, inequilateral, posterior side much shorter than the anterior. Surface finely radially striate, covered with a horn-coloured epidermis, inner margin crenulated. Ligament external and prominent. Interior with purple or pinkish coloration; muscle scars clearly visible. Hinge composed of two cardinal teeth in the left valve, with a strong, bifid cardinal in the right valve. Size: 2½ inches.

Localities: Cape Bridgewater; Apollo Bay; Airey's Inlet; Kilcunda; Anderson's Inlet; Port Albert; Lakes Entrance; Mallacoota.

This species inhabits sandy shores and is particularly abundant at the last named locality. Usually found a few inches beneath the surface, near low-water mark. The name Goolwa Cockle is sometimes applied, and the animal is regarded as useful bait for fishing. Recently it has been canned for human consumption. Recorded also from South Australia, Tasmania and New South Wales where it is being fished commercially.

The vernacular name clam is sometimes applied to these shells but to save confusion with the tropical genus *Tridacna* we recommend preclusion of its use.

Family SANGUINOLARIIDAE
Sunset Shells

Shell equivalve, oval or elongate, smooth or sculptured, covered with a thick periostracum. Ligament always in a deep pit. Pallial sinus deep often with the pallial line connected, outer shell edge smooth. Hinge as a rule with two cardinal teeth in each valve, without laterals.

Soletellina Blainville, 1824

Shell fairly large, equivalve, more or less elongate, with a strong periostracum. Ligamental pit short and broad. Pallial sinus very deep, narrowing anteriorly. Hinge with two teeth in each valve.

TYPE SPECIES *Solen diphos* Gmelin, 1791.

Iredale, 1936, erected *Flavomala* with *Solen biradiata* Wood, 1815, as type species on the grounds that Australian species differed from *diphos* in shape. However, we do not consider this sufficient reason for generic separation of the two species.

Double-rayed Razor

Soletellina biradiata (Wood, 1815)
(Figure 433)

Shell fragile, flesh or cream colour, stained with rose or purple near the umbos, and with two light, whitish rays running obliquely posteriorly, from the umbos to the ventral margin; in life covered with a thin, deciduous, horny, shining epidermis; oval, equivalve, almost equilateral, flatly convex, gaping at both extremities, which are rounded. Sculpture consisting of rather obsolete, concentric growth-lines or striae and fine radial scratches. Ligament broad, resting on strong projections. Interior bearing a white callosity in the umbonal cavity. Size: 2½ inches.

Localities: Mornington; Rosebud; Sorrento; Western Port.

Washed up on the beach, this is a fairly common shell at the above localities, but is seldom seen at the northern end of Port Phillip Bay. In living condition it is found buried in sand. A considerable amount of colour variation is evident, more particularly internally, where some specimens are white, with traces of violet towards the outer edge; others possess a much darker colour with violet and yellow predominating. The glittering epidermis is very striking and quite easy to remove. The species occurs in South Australia, New South Wales and Tasmania.

Donax-like Razor

Soletellina donacioides Reeve, 1857
(Figure 434)

Shell fragile, smooth; colour purplish with two whitish rays at the posterior area passing from the umbos to the ventral margin; invested with a semi-transparent, horny epidermis; oblong, equivalve, almost equilateral; gaping at both extremities, which are rounded, the posterior slightly narrower. Surface of valves sculptured with obsolete, concentric striae. Liga-

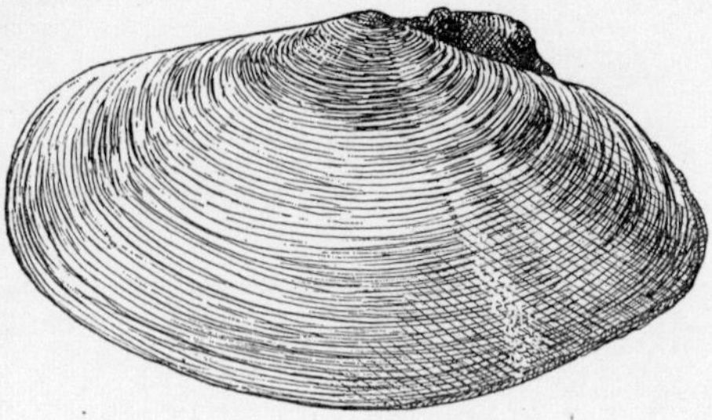

Figure 433

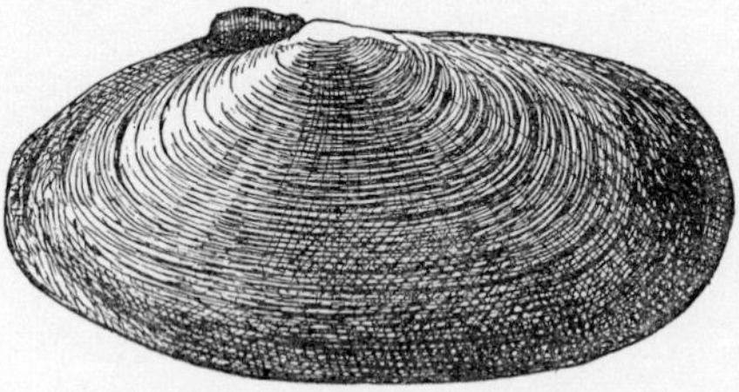

Figure 434

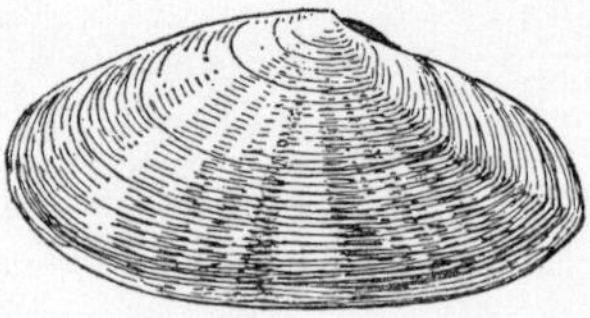

Figure 435

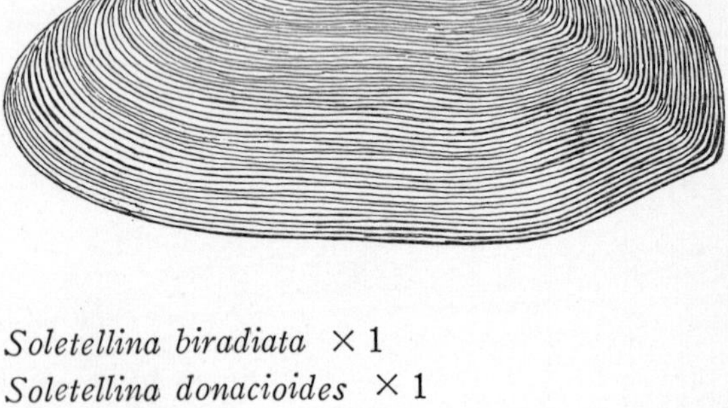

Figure 436

433. *Soletellina biradiata* × 1
434. *Soletellina donacioides* × 1
435. *Gari livida* × 1
436. *Gari kenyoniana* × 1

ment broad, inserted on strong projections. Interior purple, darkest in centre of valve. Size: 2 inches.

Localities: Port Melbourne; St Kilda; Sandringham; Frankston; Rosebud.

This is a fairly common species which may be separated from *biradiata* by its smaller size, darker colour and thinner texture. In fresh condition the shell is enveloped in an unusually strong epidermis, which, as is the case with *biradiata,* may be easily removed. In addition to Victoria it is recorded from South Australia, where the shell is somewhat more inflated.

Gari Schumacher, 1817

Shell more or less elongate, smooth or striated; valves gaping, anterior rounded, posterior usually truncated and angulate. Pallial sinus usually continuous with the pallial line. Hinge with two teeth in each valve, of which the left and the anterior right are bifid.

TYPE SPECIES *Gari vulgaris* Schumacher, 1817 = *Tellina gari* Linné, 1758.

Authors have differed as to whether *Gari* or *Psammobia* Lamarck, 1818, should be used for this genus, but *Gari* has priority and must take precedence. Shells of this genus are found living in sand or gravelly mud, and range from the littoral zone to depths of one hundred fathoms. Some of the species are of great delicacy and beauty.

Purple Sunset Shell

Gari livida Lamarck, 1818

(Figure 435)

Shell smooth, shining, compressed; colour creamy white zoned with livid brown or purple; equivalve, subequilateral. Surface finely striate. Ligament external. Interior tinted with heliotrope or pink; muscle scars well defined. Size: 1½ inches.

Localities: Hobson's Bay; Frankston; Portsea; dredged Western Port.

This is a smaller species than *kenyoniana* from which it may be distinguished by its much finer sculpture. Albinos are not infrequent. Occurs in South Australia, New South Wales and Tasmania.

menkeana Reeve, 1856, somewhat resembles *livida* but is readily distinguished by its comparative smoothness, straighter dorsal margin, and by the concentric sculpture which is less pronounced.

Kenyon's Sunset Shell

Gari kenyoniana (Pritchard & Gatliff, 1904)

(Figure 436)

Shell oblong, oval, fairly solid, inequivalve, almost equilateral, posterior side slightly longer and narrower than the anterior; both valves densely, concentrically striated, the striae occasionally anastomose, and at the ventral margin and each end, become erect, and sharp; under the lens, these striae are finely transversely striated on their lower side. The ventral margin is, under lens, very finely crenulated. Colour white, with a clouding of reddish yellow in the umbonal region, with about nine, narrow, somewhat indistinct rays, of the same colour, extending from thence to the ventral margin. Ligament large and prominent. Interior white, tinted with pink or pale purple; muscle scars plainly seen. Size: 2½ inches.

Localities: Airey's Inlet; Portsea; Rye; dredged off bank of Symond's channel, Port Phillip.

This is the largest Victorian species of the genus. It is also recorded from Tasmania and South Australia.

Family SEMELIDAE

Semeles

Shell usually rounded or oval. Ligament not in an interior pit but with an internal one situated behind the cardinal teeth; resilium more or less strong. Pallial sinus extended posteriorly. Hinge as a rule with one or two cardinal and lateral teeth.

Syndesmya Recluz, 1843

Shell small, oval or subtrigonal, equivalve; or slightly inequivalve, anterior usually more elongate, posterior terminating in a more or less oblique angle. Hinge with one weak cardinal in the left valve and two in the right valve; right valve with a small anterior and posterior lateral tooth.

TYPE SPECIES *Mactra alba* Wood, 1801.

Small Semele

Syndesmya exigua (A. Adams, 1861)

(Figure 437)

Shell thin, white, shining, somewhat compressed; triangularly oblong, inequivalve, inequilateral; posterior side shorter than the anterior and with a strong fold from the umbo to the ventral margin. Sculptured with fine concentric striae. Ligament external, short. Interior white; muscle scars distinctly visible. Size: ½ inch.

Locality: Cowes, Phillip Island.

Found alive in sand at low-tide level. A characteristic, shining shell which, with its peculiar shape, should be easily recognized. Recorded from Tasmania and South Australia.

Theora H. & A. Adams, 1866

Shell slender, transversely elongated oval, compressed, smooth and shining; posterior and also anterior gaping. Chondrophore oblique. Interior hinge with one small cardinal tooth in left; two in right valve, and a slender lateral tooth in the right valve; pallial sinus deep.

TYPE SPECIES *Theora nitida* Gould, 1861.

Shining Theora

Theora fragilis (A. Adams, 1855)

(Figure 438)

Shell thin, very fragile, small, hyaline, somewhat compressed; almost equilateral, attenuate, gaping at posterior end. Ventral margin rounded.

Interior white, shining; muscle scars faintly discernible, elongated; pallial line fairly distinct with a deep, somewhat angulated sinus. Hinge with chondrophore projecting posteriorly into interior of shell, cartilage internal. Size: ½ inch.

Locality: Dredged off Point Cook, Port Phillip.

This genus, numerically, is not very strong and the present species, one of our most highly polished bivalves, is the only representative in Victoria. It is an extremely frail shell easily recognized by its attenuate form and is, comparatively speaking, of rare occurrence. At the above locality it was dredged alive on a sea-bed of sandy mud at 8 fathoms. It occurs also in New South Wales and Queensland.

Semelangulus Iredale, 1924 (*Tellina tenuilirata* Sowerby, 1867).
dilutus (Smith, 1885) : a misidentification of the species below.
subdilutus (Tate, 1887) ; Western Port; 18 miles east of Lakes Entrance.
tenuiliratus (Sowerby, 1867) ; Port Phillip.

Family TELLINIDAE

Tellins

Shell usually oval, sometimes elongated and posteriorly beaked, very often unsymmetrical; smooth or variously sculptured. Ligament external. Pallial sinus large, often with the pallial line more or less continuous. Hinge narrow with two small cardinal teeth and anterior and posterior laterals in each valve.

Pseudarcopagia Bertin, 1878

Shell rounded, solid, posterior slightly asymmetrical. Sculptured with concentric and radial lines. Hinge with anterior lateral tooth shorter than the posterior and the left feebler than right.

TYPE SPECIES *Tellina decussata* Lamarck, 1818 (non Wood, 1815) = *Tellina victoriae* Gatliff & Gabriel, 1914.

Victorian Decussated Tellin

Pseudarcopagia victoriae (Gatliff & Gabriel, 1914)
(Figure 439)

Shell creamy white; solid, rounded, subequivalve, subequilateral. Sculpture consisting of fine concentric ridges and radiating striae, giving the shell a distinct decussated appearance; right valve angulated posteriorly, with a corresponding groove in the left valve; dorsal margin sloping. Anterior extremity much rounded. Ligament external. Interior white. Hinge with strong teeth. Size: 2 inches.

Localities: Mordialloc; Frankston; Point Lonsdale; Western Port.

A species easy to recognize. The characteristic decussated sculpture immediately separates it from any Victorian Tellen. These creatures live in sand or sandy mud, and after rough weather may be gathered in numbers. Common on the Victorian coastline and also occurs in Tasmania, South Australia and Western Australia.

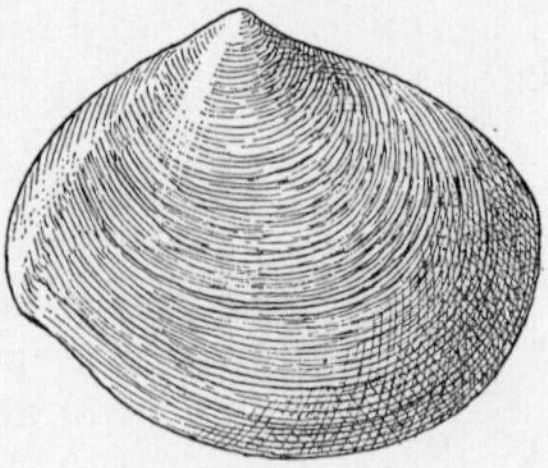

Figure 437

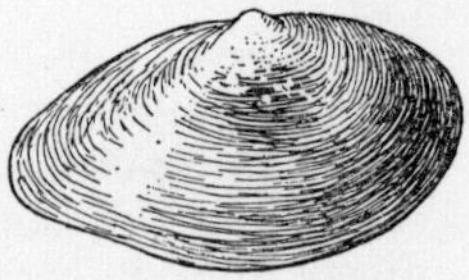

Figure 438

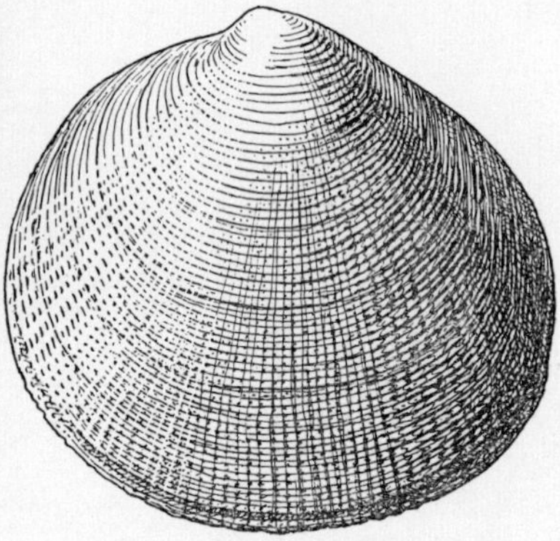

Figure 439

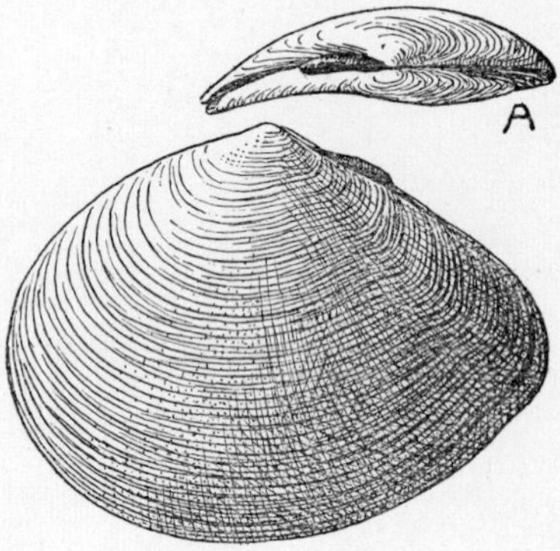

Figure 440

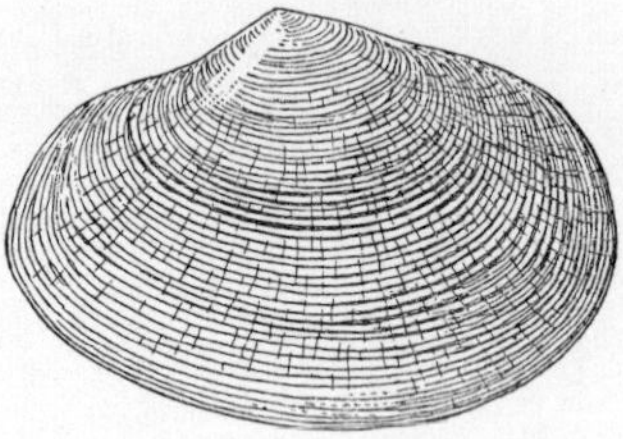

Figure 441

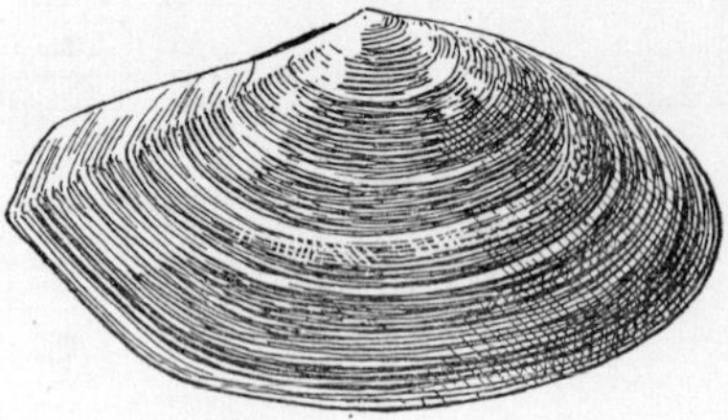

Figure 442

437. *Syndesmya exigua* × 2⅔
438. *Theora fragilis* × 2
439. *Pseudarcopagia victoriae* × 1
440. *Homalina deltoidalis* × 1
A Dorsal view × ⅔
441. *Homalina mariae* × 1⅓
442. *Tellina albinella* × 1

botanica Hedley, 1917; has finer sculpture and occurs at Point Leo; previously recorded from New South Wales and Tasmania.

Homalina Stoliczka, 1871

Shell oblong, compressed, inequivalve, inequilateral, anterior end short, rounded; posterior end flexuous, obsoletely carinate, somewhat produced, narrowed. Hinge with two cardinals, and right valve with a very short anterior and a weak posterior lateral at the end of the ligament. The anterior lateral may even be united with the anterior cardinal in some species.

TYPE SPECIES *Tellina triangularis* Chemnitz, 1782 = *Tellina trilatera* Gmelin, 1791.

The following species have been placed in *Macoma* Leach, 1819, and *Macomona* Finlay, 1927. As they possess lateral teeth in the right valve they cannot be retained in the former. The very short peg-like anterior lateral tooth close to the cardinal is typical of *Homalina* rather than *Macomona* which has an elongate anterior lateral, well separated from the anterior cardinal.

Triangular Tellin

Homalina deltoidalis (Lamarck, 1818)

(Figure 440)

Shell creamy white, moderately large, ovate, subinequilateral, right valve slightly more convex; anterior rounded, posterior slightly beaked. Sculpture consisting of fine concentric striae, which are hardly discernible on the upper half of shell; radial sculpture extremely fine, seen only with the aid of lens. Ligament external. Size: 1½ inches.

Localities: South Melbourne; Mordialloc; Frankston; Geelong.

Common on the whole Victorian coastline. Variation exists, particularly as regards the position of the umbos, which in some specimens are almost centrally situated. They live in sandy mud, and at the first-named locality may be dredged in considerable numbers. Occurs also in New South Wales, Tasmania and South Australia.

Maria's Tellin

Homalina mariae (Tenison Woods, 1875)

(Figure 441)

Shell milky white, shining; thin, fragile, ovate, somewhat compressed, inequivalve, subequilateral, gaping posteriorly; umbos acute; anterior end widely rounded; slightly attenuate and flexuous posteriorly. Surface of valves sculptured with fairly regular, concentric sulci and covered with a thin, yellowish epidermis. Ligament external, brown, prominent. Interior white; muscle scars very evident and rather elongated; pallial line well marked and with a very large sinus. Hinge with three small teeth in one valve. Size: 1 inch.

Localities: Portarlington; Geelong; dredged 8 fathoms off Point Cook; Western Port; Port Albert.

This is a very white, oval form without much character except for the concentric sculpture, which is discernible only with the aid of lens. Recorded also from Tasmania.

diemenensis Deshayes, 1854; Corio Bay, Port Phillip.

Tellina Linné, 1758

Shell usually ovate, compressed, rounded anteriorly, angular and slightly folded posteriorly, inequilateral. Ligament external, prominent. Hinge with two cardinals in each valve and two laterals, more distinct in the right valve.

Type Species *Tellina radiata* Linné, 1758.

Little White Tellin or Sunset Shell

Tellina albinella Lamarck, 1818

(Figure 442)

Shell white, sometimes pink or orange, oblong-oval, rather thin, flattish, inequivalve, almost equilateral, shining; umbos acute. Anterior end rounded, posterior subangular. Ligament external. Sculpture consisting of fine, dense, concentric lines and faint radial striae; a slight ridge descending from the umbos to the posterior extremity. Interior white. Hinge normal. Size: 2½ inches.

Localities: Sorrento; Point Nepean; Point Lonsdale.

For delicacy of colour few bivalves can surpass the Tellins, and on this account they are much sought after by collectors. Although described as white, numerous specimens appear in ever-variable tints of pink and orange, conspicuously rayed, hence the vernacular name for the coloured examples. The flat and elongate form is very characteristic, and no difficulty should be experienced in identifying the species. These molluscs live on sand, buried just beneath the surface, and by means of their muscular foot possess the power of leaping. The species is not uncommon in Western Australia, South Australia, New South Wales and Tasmania.

ensiformis Reeve, 1868; Port Albert; San Remo.

Family SOLENIDAE

Razor Shells

Shell elongate, cylindrical, gaping at both ends. Ligament external. Hinge variable, without lateral teeth.

Solen Linné, 1758

Shell very elongate, subcylindrical, straight, margins parallel, ends truncate, umbo very close to the anterior end; posterior muscle scar and end of pallial sinus extend to the posterior edge. Hinge always with one tooth.

Type Species *Solen vagina* Linné, 1758.

Razor Shells are among the most remarkable of all the bivalves, having a greater length proportionately, from the anterior to the posterior margins, than any other genus. They are found in a perpendicular position buried in sand. In other parts of the world, particularly Britain, Razor Fish have been used as food. They are regarded as excellent bait and are known by fishermen under the name of Finger Oyster.

Razor Shell

Solen vaginoides (Lamarck, 1818)

(Figure 443)

Shell fragile, transversely elongated, subcylindrical, equivalve, very inequilateral, gaping at both extremities; umbos close to the anterior end. Surface invested with an epidermis. Ligament long, external. Size: 2 inches.

Localities: Altona; Portarlington; Portsea; Western Port.

Common in all parts of Western Port where living examples are obtained by dredging in reasonably shallow water close to Cowes. This species is also recorded from Western Australia, South Australia, Tasmania and New South Wales.

Family HIATELLIDAE

Crypt Dwellers

Shell more or less elongate, often of irregular form and wrinkled, gaping at each end. Ligament external, strong, inserted in a deep furrow. Hinge with cardinal teeth weak or obsolete, laterals absent.

Hiatella Daudin, 1801

Shell usually small, elongate. Hinge with one or two weak or obsolete cardinal teeth.

TYPE SPECIES *Mya arctica* Linné, 1758.

The name *Saxicava* Bellevue, 1802, was originally used for members of this genus but the prior name *Hiatella* must take precedence.

Southern Crypt Dweller

Hiatella australis (Lamarck, 1818)

(Figure 444)

Shell white or cream; polymorphous, solid, oblong, inequilateral, valves gaping at each end, umbos prominent; valves very rugose covered by a light brown epidermis. Ligament external. Interior white; muscle scars well defined; pallial sinus not very deep. Hinge in the juvenile very narrow possessing a small tooth in each valve and a corresponding socket; when adult, edentulous. Size: 1 inch.

Localities: Lakes Entrance; Sorrento; Port Fairy; Portland.

An extremely common species on the whole Victorian coastline, occasionally attaining a length of two inches. No definite shape could be assigned to the shell as it appears to take a variety of forms according to the crevice in which it dwells. Usually found burrowing in limestone rocks, holes of old oysters and other shells, or amongst roots of seaweeds. The animals attach themselves by the byssus to the side of the crypt. It has a wide range of distribution, being found in all states of Australia.

Winged Crypt Dweller

Hiatella subalata (Gatliff & Gabriel, 1910)

(Figure 445)

Shell white; inflated, somewhat angular; subequal, inequilateral, valves slightly gaping anteriorly; umbos depressed, slightly incurved, produced anteriorly; dorsal margin straight, somewhat alate posteriorly. Post-dorsal border curved, ventral margin straight. Post-ventral margin subangulate, anterior end rounded. From the umbo to the posterior end runs an oblique ridge, steep on dorsal side and becoming more rounded as it approaches the ventral margin. The post-dorsal edge of the valves gaping. Sculptured at the umbonal region with very fine lines of growth, which develop into irregular, somewhat undulating rugae. Besides the concentric sculpture, the whole surface is studded with minute granular excrescences developing with age the form of tubercles, which may be plainly seen from the interior of the shell. These are arranged roughly parallel with the rugae. Ligament external, well developed, and faintly darker than the olivaceous epidermis. Interior somewhat iridescent; muscular impressions strong, the anterior oblong and the posterior subcircular; pallial line not sinuated. Hinge simple, the characteristic curved tooth of the juvenile form becoming obsolete in the adult. Size: ⅜ inch.

Localities: Dredged off Point Cook (8 fathoms); Portsea; Frankston; Dromana.

The prominent curved tooth in the right valve, with a corresponding cavity in the left, is entirely juvenile and presents a character differing in its entirety from the adult form; this is characteristic of the genus *Hiatella*. The angular form and granular surface serve as useful recognition marks, but with beach-worn specimens, the latter is difficult to discern. Taken alive at the first-named locality on a sea-bed of sandy mud; also occurs in South Australia and New South Wales.

angasi (Angas, 1865); Cotton & Godfrey, 1938, record this shell from all states.

Panopea Menard, 1807

Shell usually very large and thick, elongate, posterior truncated, very arched, ends gaping, particularly the posterior, irregularly wrinkled. Ligament attached to a very strong furrow. Hinge with one high conical tooth in each valve.

Type Species *Mya glycymeris* Born, 1780.

Southern Gaper

Panopea australis Sowerby, 1833

(Figure 446)

Shell whitish somewhat chalky looking; more or less solid, transversely oblong, equivalve, inequilateral, gaping widely at posterior and anterior ends, posterior end narrowly obliquely truncated, ventral margin concave; umbos acuminated. Sculpture composed of concentric furrows or wrinkles, the whole surface covered by a thick epidermis. Ligament external, short, prominent, attached to strong ridges. Muscle scars deep; pallial line well defined and with a deep posterior sinus. Hinge formed of a single tooth in each valve fitting into a cavity of the opposite valve. Size: 3-4 inches.

Localities: San Remo; Portland; Port Albert; dredged (8 fathoms) between Phillip and French Islands, Western Port; off Portsea, Port Phillip.

A strong, handsome, concentrically wrinkled shell, the only representative of the genus in Victoria. It resembles the genus *Hiatella* to a great extent

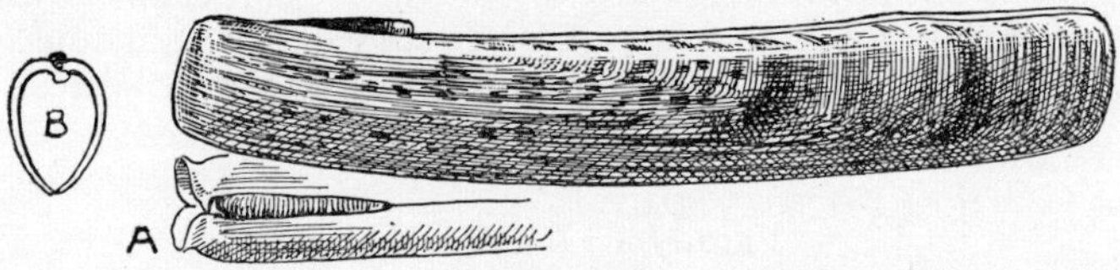

Figure 443

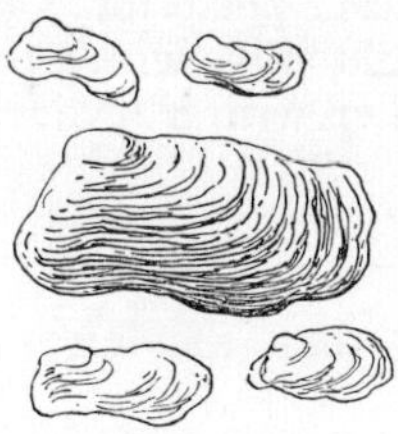

Figure 444

Figure 445

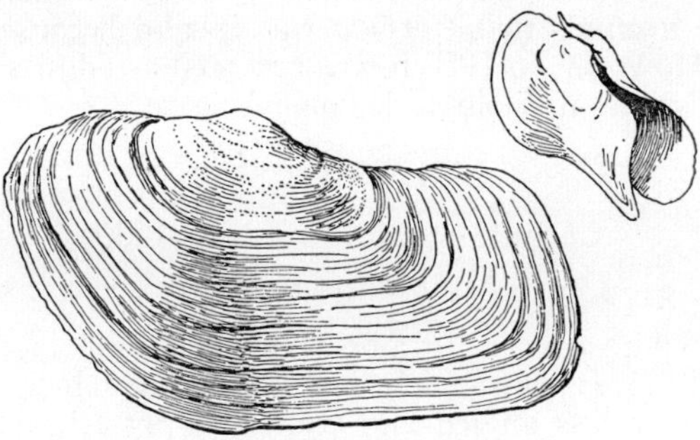

Figure 446

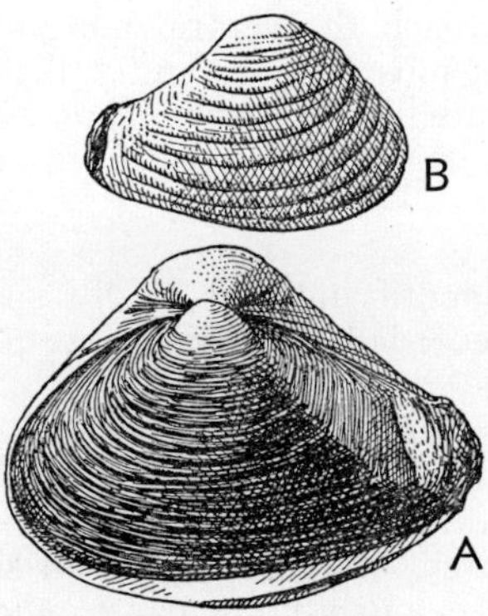

Figure 447

443. *Solen vaginoides* $\times 1$
    A Dorsal view
    B Anterior view

444. *Hiatella australis* $\times 1\frac{1}{3}$ and $\frac{2}{3}$

445. *Hiatella subalata* $\times 4$

446. *Panopea australis* $\times \frac{2}{3}$
    A Posterior lateral view

447. *Corbula stolata*
    A Shell viewed from left side $\times 2$
    B Right valve $\times 1\frac{1}{3}$

and is found from low-water mark to several fathoms, burrowing in sand. A narrower form is recorded from Queensland under the name *P. angusta* Hedley, 1915. The present species occurs in New South Wales, Tasmania and South Australia.

Family CORBULIDAE

Basket Shells

Shell usually small, more or less asymmetrical, posterior angular or beaked; often with strong concentric sculpture. Right valve with a prominent tooth in front of the cartilage pit; left valve smaller with a projecting cartilage process.

Corbula Bruguière, 1792

Shell inequivalve, rostrate, right valve larger than the left; sculpture concentric.

TYPE SPECIES *Corbula sulcata* Bruguière, 1792.

Aloidis Megerle, 1811, has been used for this genus but the prior name must take precedence. Iredale, 1930, erected a new genus *Notocorbula* for the southern Australian species (with *N. vicaria* Iredale, 1930, as type species) on the grounds that *Corbula* was invalid because its introduction was irregular. However, in 1950 it was accepted as admissible by the International Commission on Zoological Nomenclature.

Little Basket Shell

Corbula stolata (Iredale, 1930)

(Figure 447)

Shell creamy white, covered by a brownish epidermis more particularly on the left or upper valve; transversely oval, inequivalve, inequilateral, rounded anteriorly, more or less angled and truncated posteriorly, ventral margin convex; right valve overlapping the left at the ventral-posterior margin; umbos prominent; a ridge, running from the umbo to the ventral posterior margin, defines the posterior area. Sculpture consisting of flatly rounded concentric ridges, much coarser as they approach the ventral margin and separated by narrow grooves. Interior white; muscle scars and pallial line well defined; pallial sinus small, broad. Hinge with large prominent conical tooth just below the umbo of right valve, inserting its extremity into a pit in the opposite valve. Size: ¾ inch.

Locality: Port Albert.

No difficulty should be experienced in identifying this oddly shaped bivalve, found living gregariously in mud. Recorded also from New South Wales.

coxi Pilsbry, 1897; from Port Albert, of similar size but readily distinguished by its more elongate shape and finer concentric sculpture. It is very similar to *iredalei* Cotton, 1930, from South Australia.

flindersi Cotton, 1930; from Portland, larger, more triangular in shape and with very coarse concentric ribs. It was previously known as *scaphoides* Hind, 1843, a Philippines species.

Family GASTROCHAENIDAE

Gapers

Shell more or less elongate, sometimes small, gaping at the anterior end. Ligament external. Hinge toothless. Animal living partly free in sand and building a club-shaped tube in which the shell and siphons are enclosed, or boring in limestone or other molluscan shells.

Gastrochaena Spengler, 1783

Shell elongate, narrow, contained within a tube which is closed at the inflated lower end and tapers upwards; behind the valves the tube is furnished with a perforated diaphragm above which it is partially divided by a partition.

TYPE SPECIES *Gastrochaena cuneiformis* Spengler, 1783.

One-sided Gaper

Gastrochaena tasmanica Tenison Woods, 1876

(Figure 448)

Shell opaque, white; thin, fragile, inflated, elongated, irregular, inequivalve, inequilateral, widely gaping for the whole length of the ventral margin; shorter anteriorly, margin almost straight, wider and rounder posteriorly; concentrically and rugosely striate, striae irregular; valves somewhat unicostate obliquely in the middle with a groove by the side, ridge wide and ceasing in the umbos, which are inconspicuous. Ligament external, weak. Interior white. Hinge simple. The shell is contained in a flask-shaped tube, remaining there for all time. Size: ½ inch.

Localities: Off Point Cook (8 fathoms); off Lakes Entrance (20 fathoms).

The animal burrows in sandy mud and builds the protective 'flask' upon a base of shell or other material. Occurs in New South Wales, Tasmania and South Australia.

Family PHOLADIDAE

Angel Wing Borers

Shell white; usually large, elongate, thin, brittle, gaping at both ends, armed anteriorly with rasp-like imbrications. Adult shell usually has accessory valves, called protoplax when in front of the umbos, mesoplax when above the umbos and metaplax when situated posteriorly. Ligament absent. Hinge with single tooth.

Pholas Linné, 1758

Shell elongate, cylindrical, the overturned dorsal edge consists of two lamellae one of which is closely applied, the other free; a calcareous septum occupies the pedal gap. Sculpture of concentric ridges and weaker radials, often broken into diversely sculptured areas.

TYPE SPECIES *Pholas dactylus* Linné, 1758.

Of all the Victorian bivalves no genus is more remarkable and interesting than *Pholas*. They appear to form a natural link between the Shipworm or

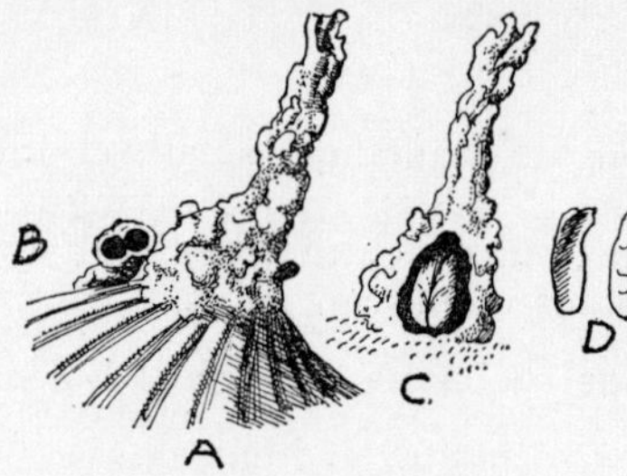

Figure 448

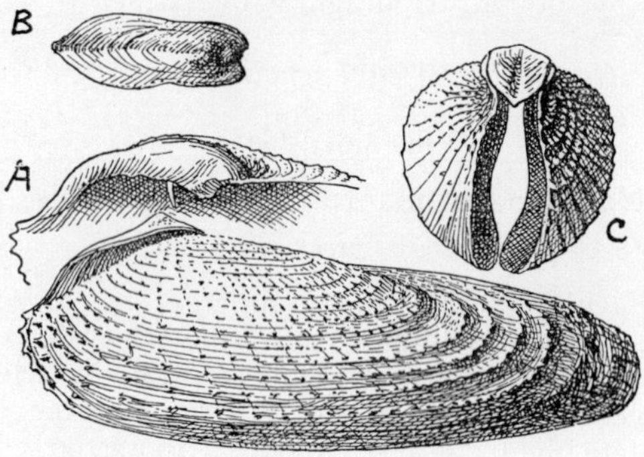

Figure 449

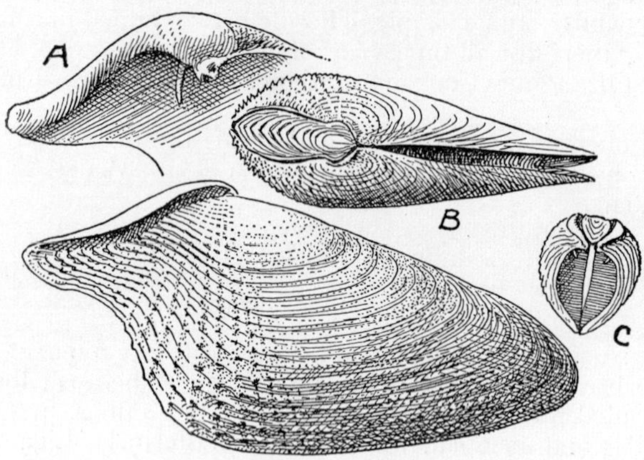

Figure 450

448. *Gastrochaena tasmanica* $\times 1$
A Flask-shaped tube B Open end of a flask
C Flask open to show shell
D Two valves of shell separated

449. *Pholas australasiae* $\times \frac{2}{3}$
A Hinge B Accessory plate C Anterior view

450. *Pholas obturamentum* $\times 1\frac{1}{3}$
A Hinge B Dorsal view C Anterior view

Teredo and the ordinary genera of bivalve mollusca. The majority of bivalves may be termed borers so far as the power of burying themselves in sand and mud is concerned, but the boring process adopted by the *Teredo* and the present genus appears to be effected by a very different means. That these molluscs enter rock for a home is undoubted, since the cavity is always smaller at the entrance, thus proving that they cannot leave the burrow.

In other countries these animals are used as food, being cooked with fine herbs and bread-crumbs. It is said that these Borers possess a phosphorescent property.

Australian Borer or Piercer or Angel Wing

Pholas australasiae Sowerby, 1849

(Figure 449)

Shell white; fragile, elongated, subcylindrical, equivalve, inequilateral, gaping at both ends, more so at the posterior. A large, shelly, accessory plate or shield rests upon the dorsal edges. Sculpture consisting of irregularly spaced lamellae, wider apart at the lower and posterior portion of the valves, and fine moderately distant, radiating lines; this crossing giving rise to short and rather broad spines, more evident at the anterior end. Interior white. Hinge margin reflected, a long, curved tooth protruding from beneath the umbo in each valve. Size: 3 inches.

Localities: St Kilda; Sandringham; Queenscliff; Western Port.

A few years ago, at the first-named localities, this species was fairly common, but is now less frequently met with, due possibly to drainage. It possesses a very wide distribution, being recorded from every state of Australia.

Tongue-shaped Borer or Angel Wing

Pholas obturamentum Hedley, 1893

(Figure 450)

Shell thin, fragile; dull white, covered with a straw-coloured epidermis; somewhat tongue-shaped, equivalve, inequilateral; umbos situated at about one quarter the length of shell from the anterior end; posterior end rounded. Ventral margin nearly straight except for a deep excavation at the anterior end which forms a heart-shaped space when the valves are closed. A shelly accessory plate or shield rests on the dorsal edges; this is flat and striated by divaricating lines of growth, with a shallow median furrow. The sculpture consists of irregularly spaced lamellae which are crossed by moderately distant radiating lines, giving rise to rather short, broad spines, more evident at the anterior region. Interior white. Hinge margin narrow, reflected, with a long curved process or tooth protruding from beneath the umbo in each valve. Size: 2 inches.

Localities: Lorne; Portland; Lakes Entrance.

The excavated ventral margin and the fewer radials and spines are associated features which distinguish the above species from *australasiae* Sowerby. It is found burrowing in clay or soft rock at low tide. Occurs also in Tasmania, New South Wales and South Australia.

Family TEREDINIDAE

Shipworms

Shell very small in comparison with the animal; equivalve, gaping. Hinge toothless but with a projecting interior spoon-shaped process. Animal vermiform, each siphon furnished at its extremity with a calcareous or horny pallet consisting of stalk and head of various shapes. This latter feature is the main basis of generic classification.

This family of bivalve molluscs is of considerable economic importance due to its destruction of piling and wooden ships in marine and brackish water, particularly in the tropics. These molluscs enter the timber through a pin hole as microscopic larvae and then proceed to grow and make long tunnels beneath the surface. In time almost the whole of the wood may be destroyed by the burrows which, although they may lie close together, never cross each other. Where the burrows lie very close together the walls may be strengthened by the secretion of lime on the side.

In the days of wooden ships these voracious molluscs had even more importance than today and they were given a number of common names including Cobra, Shipworm or Marine Borer.

Teredo Linné, 1758

The pallets elongate, compressed, cup-shaped.
TYPE SPECIES *Teredo navalis* Linné, 1758.

Austin's Shipworm

Teredo austini Iredale, 1932

(Figure 451)

Shell normally white, occasionally exhibiting a pink coloration in the young stage. Anterior auricle small, ornamented with fine ribs increasing to about thirty in the adult but becoming wider apart. Posterior auricle small, showing growth-lines. Interior usually whitish, sometimes light pink or greenish, and possessing a rather large knob; blade thin.

Pallets elongate, showing a hollow cone with a rather deep cut, flat on one side and convex on the other; stalk comparatively short, thick, and sometimes bent, about half the length of the cone. Size: Shell ⅕ inch; pallet ⅕ inch.

Locality: Lakes Entrance.

The characteristic cup-shaped pallets, flat on one side, and convex on the other, serve to identify the species; previously known as *navalis* Linné, 1758.

fragilis Tate, 1888, the Fragile Shipworm, from Brighton boatslip and Portsea pier, is the smallest marine borer in Victoria, but very destructive. It is difficult to separate from the juvenile valves of the larger Teredinids, but its peculiar small, shelly, clavate pallets, much attenuated stalk, and blade crowned with a periostracum which has a projecting horn at each end, will distinguish it from the above species. Occurs also in South Australia.

calmani Roch, 1931; Calman's Shipworm, from Lakes Entrance and Portsea pier, is similar to *fragilis* Tate but easily separated by the ovate

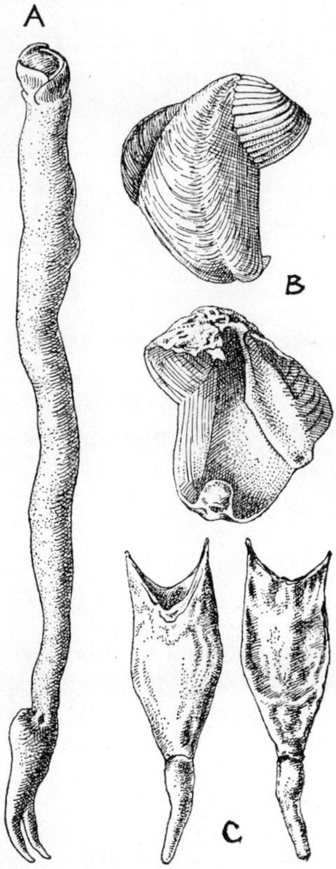

Figure 451

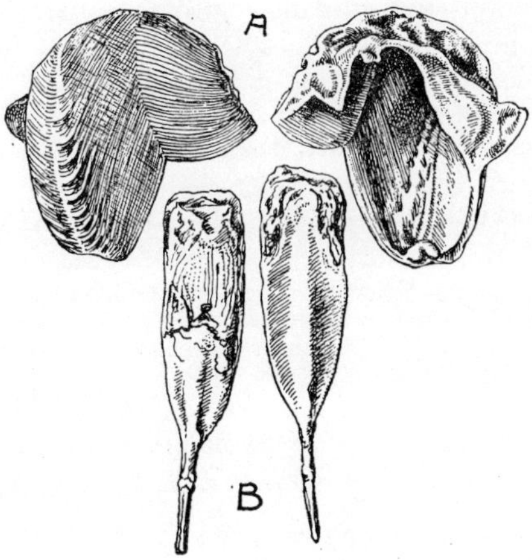

Figure 452

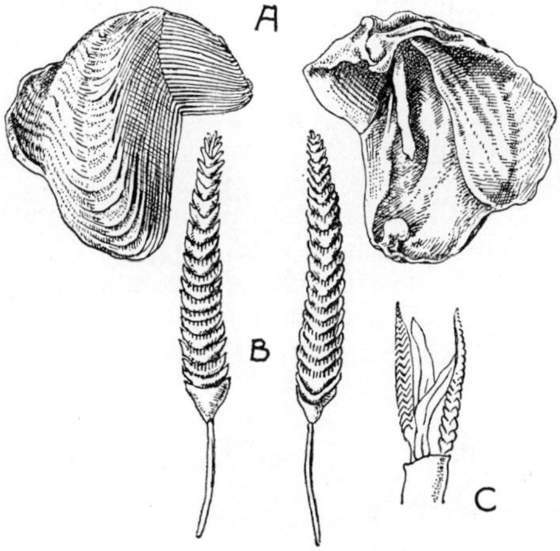

Figure 453

451. *Teredo austini*
   A Animal $\times 2$ B Valves $\times 6\frac{2}{3}$ C Pallets $\times 6\frac{2}{3}$

452. *Nototeredo edax*
   A Valves $\times 2$ B Pallets $\times 2$

453. *Bankia debenhami*
   A Valves $\times 2$ B Pallets $\times 2$
   C Posterior of animal, showing siphons and pallets $\times 1$

spoon-shaped pallet with the distal end covered by a dark periostracum ending in two points; the blade is much larger than that of *fragilis*. Previously known as *pedicillatus* Quatrefages, 1849. The species also appears in South Australia.

Nototeredo Bartsch, 1923

Shell with internal ridge of posterior auricle developed into a shelf; pallets elongate, flattened, not cupped, stalk short.

TYPE SPECIES *Teredo edax* Hedley, 1894.

Voracious Shipworm

Nototeredo edax (Hedley, 1894)

(Figure 452)

Shell greenish or light brown; strong, globose, trilobate, widely gaping, equivalve; anterior area sharply angled in front; umbonal area much eroded to the extent that the posterior auricle is practically absent; anterior auricle produced and concentrically sculptured by fifty to sixty fine closely set striae. Interior bluish white, edentulous; from the umbo, but not directly behind, hangs a characteristic narrow blade extending to about the centre of the valve; knob comparatively small.

Pallets elongate, concave on one side and convex on the other with a longitudinal ridge on each area extending about half or more of the length; the stalk varying in length and thickness; the whole showing some resemblance to a cricket bat. The calcareous tube exhibits a strong concamerated structure at the posterior extremity, vanishing anteriorly where the tube becomes fragile; these characteristics show better development in some cases than others.

Size: valves ¾ inch, pallets ¾ inch.

Localities: Portsea pier; San Remo; Lakes Entrance; Port Albert.

This species is readily separated from all Victorian borers by the concamerated structure of the tube lining the burrows, and its peculiar non-cupped cricket bat shaped pallet. Previously known as *T. bruguieri* Della Chiaje, 1828. Occurs also in South Australia and New South Wales.

Bankia Gray, 1840

Pallets very elongate, composed of cone-in-cone elements, edges separate, stalk long and slender.

TYPE SPECIES *Teredo bipalmulata* Lamarck, 1801.

Southern Shipworm or Cobra

Bankia debenhami Iredale, 1932

(Figure 453)

Shell whitish with a thin cream epidermis; globose, trilobate, auricles broad and fairly deep; widely gaping, equivalve, anterior area strongly angled in front, posterior auricle not large; umbos convex, not very prominent, incurved; anterior auricle with sculpture consisting of fine concentric striae; the anterior median area ornamented with fine, almost microscopic striae; posterior median area narrow, without sculpture except for faint growth-lines. Interior of shell white, edentulous, with a slender calcareous blade hanging from the umbo.

Pallets elongate, narrow towards the apex, rounder on the outer side and flat on the inner, composed of a series of imbricate, pectinate joints with a long cylindrical stalk about the same length. When disarticulated the pallet is seen to possess a long stalk over which are placed the little funnel-shaped cups one within another, the largest necessarily being at the base. Size: shell 2/5 inch, pallet 3/4 inch.

Localities: Lakes Entrance; Portsea pier.

Such variation exists that the species cannot be identified by the valves unless in living condition and then only with difficulty. But the characteristic, slender imbricating pallets readily distinguish it. Previously known mistakenly under the names *Xylotrya sauli* Wright, 1865, and *X. australis* Calman, 1920, the latter a New Zealand species.

gabrieli Cotton, 1934; occurs at Middle Brighton boatslip and Lakes Entrance. This was described by the author from Dennekin slip, Port Adelaide. He remarked: 'Pallets consisting of six funnel-in-funnel elements, well separated, edges smooth; sinuate on one side and not on the other; stalk long and thin, slightly longer than the funnel-in-funnel portion.' Pallet length about 2/5 inch and stalk about half that. The narrower pallet immediately distinguishes it from *debenhami*.

Nausitora Wright, 1864

Shell normal, posterior auricle generally very eroded; pallets quill-like, formed of flattened cones fused unequally, making them asymmetrical; outer surface convex, inner flat or slightly concave.

TYPE SPECIES *Nausitora dunlopei* Wright, 1864.

Messel's Brackish-Water Shipworm

Nausitora messeli Iredale, 1932
(Figure 454)

Shell of medium size, much eroded externally, all areas suffering; white. Anterior auricle large, basally eroded, but there remains thirty smooth ribs of sculpture. Posterior auricle entirely eroded away. Median area large. Anterior median very broad, though showing only thirty, strongly but closely denticulate ridges. Middle median very narrowed. Posterior median very narrow, smooth; faint growth-lines. Interior greenish. Knob small, rounded, almost separate. Shelf strongly indicated, almost a ridge. Hinge area much eroded and calloused, the umbos affected.

Pallets inequilateral, consisting of fused flattened cones, the outer surface convex, the inner flattened, the extremities broken; stalk rather long and thin. Size: shell 3/5 x 3/5 inch, pallets nearly 1 inch.

Localities: Brackish water Mitchell River, Bairnsdale; Mallacoota.

This brackish-water Borer alone represents the genus in Victoria, and the peculiar asymmetrical shape of the pallets provides a ready means of identification. It occurs in New South Wales and the author remarked: 'This species is referable to the group *Nausitora* in the broad sense, but the obliquity of the pallets and the fusion of the elements deserve that it be separated sub-generically as *Inequarista*.'

Family MYOCHAMIDAE

False Jingle Shells

Shell inequivalve, the dorsal margins of one valve overlapping the other valve. Ligament extending on both sides of the umbos; pallial sinus small. Hinge toothless.

Myochama Stutchbury, 1830

Shell triangular, right valve flattened and modified by the form of the surface to which it is attached, posterior side produced. Hinge with two tooth-like projections flanking the cartilage in each valve.

TYPE SPECIES *Myochama anomioides* Stutchbury, 1830.

False Jingle Shell

Myochama anomioides Stutchbury, 1830

(Figure 455)

Shell fawn; solid, adherent, irregular, somewhat triangular, inequivalve, almost equilateral; right or lower flat valve attached and conforming to the grooves or undulations of the shell or object to which it is attached; upper valve free, convex and sculptured with strong radiating ribs much wider than the intervening spaces. The whole surface crossed by irregular concentric striae, producing a somewhat nodulose character. Interior cream; muscle scars well excavated, anterior one curved, posterior rounded; pallial line with a small broad sinus. Hinge composed of two diverging tooth-like processes in each valve with a triangular pit between them to receive the internal cartilage. Size: ¾ inch.

Localities: Lakes Entrance; off East Moncoeur Island.

This is a remarkable genus, peculiar to the Australasian region and represented by very few species. Found attached to rocks or shells, the present form appears frequently on the shell *Eucrassatella kingicola* Lamarck, or species of *Glycymeris. keppelliana* A. Adams, 1852, from Lakes Entrance, may fall into synonymy. Occurs in New South Wales, Tasmania, through Bass Strait to South Australia.

Myadora Gray, 1840

Shell free, trigonal, rounded anteriorly, attenuated and truncated posteriorly, right valve vaulted, and overlapping the flattened left one. Cartilage narrow between two tooth-like ridges in the left valve.

TYPE SPECIES *Pandora brevis* Sowerby, 1829.

The White Myadora

Myadora albida Tenison Woods, 1875

(Figure 456)

Shell white; thin, transversely ovate, inequivalve, subequilateral; umbos acute. Right or lower valve shallow, left valve much flatter and a little smaller than the right; anterior-dorsal margin slightly convex, posterior-dorsal margin concave; posterior extremity almost vertically truncated. Both valves sculptured with very strong, distinct, regular, concentric ribs, and on

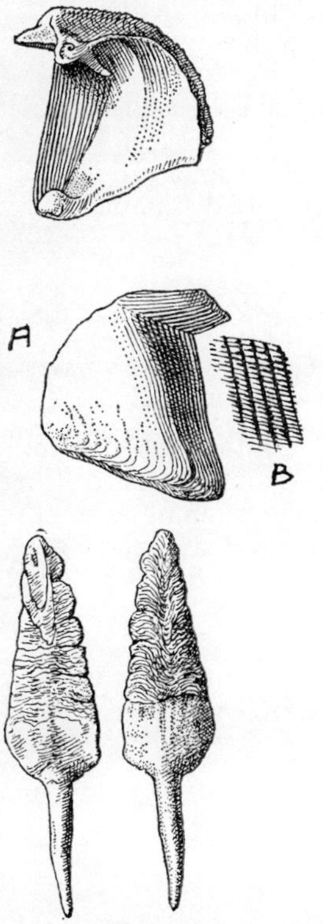

Figure 454

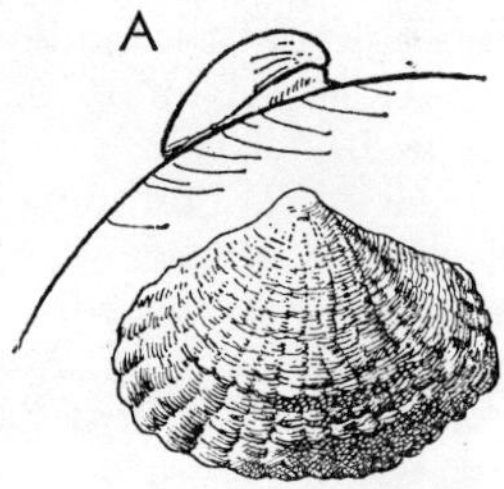

Figure 455

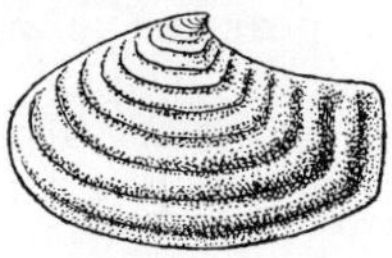

Figure 456

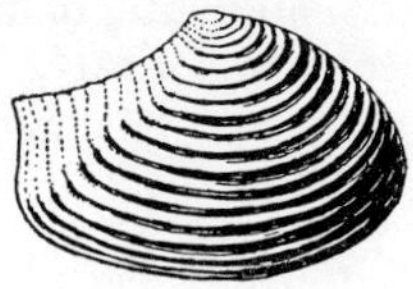

Figure 457

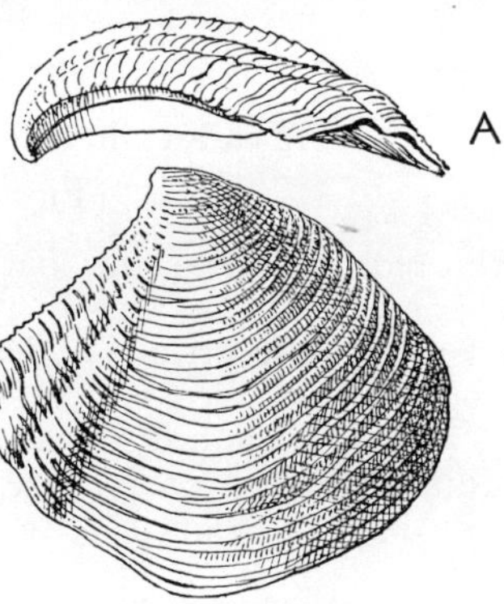

Figure 458

454. *Nausitora messeli*
A Valves × 2 B Sculpture (enlarged)
C Pallets × 2

455. *Myochama anomioides*
A Lateral elevation, showing attachment to substratum

456. *Myadora albida* × 2

457. *Myadora antipodum* × 2

458. *Myadora brevis* × 1
A Lateral elevation

each valve running from the umbo to the ventral margin is a definite ridge defining the posterior strongly sculptured slope. Interior white, shining, pearly. Size: ½ inch.

Locality: Dredged (8 fathoms) between Phillip and French Islands, Western Port.

This species was first recorded from Long Bay, Tasmania. It may be confused with juvenile *latilirata* Macpherson, but is readily distinguished by its more elongated form and less flat upper left valve.

Antipodes Myadora

Myadora antipodum A. E. Smith, 1880

(Figure 457)

Shell white, oblong-ovate, subequilateral; right valve rather shallow, concentrically lirate, lirae thickest near the anterior side and not so distinct posteriorly; posterior slope defined by a definite ridge running from the umbo to the ventral margin; left or flat valve slightly concave and finely, concentrically sulcate. Interior white, pearly; pallial sinus broad and not very deep. Size: ⅓ inch.

Locality: From Bass Strait cable.

This species can be confused only with *pandoriformis* (Stutchbury) which is more equilateral and its deep valve more convex. Recorded also from New Zealand.

Short Myadora

Myadora brevis (Sowerby, 1829)

(Figure 458)

Shell white or cream; somewhat triangular, inequivalve, almost equilateral; umbos prominent, acutely curved. Sculpture consisting of rather distant and wavy concentric ribs, less evident in the left valve; the right valve is moderately convex and has two posterior ridges descending from the umbos, left valve slightly concave; both valves minutely and distinctly granular. Interior white, pearly. Size: 1½ inches.

Localities: Port Melbourne; St Kilda; Sandringham; Altona; Point Cook.

This is a quaintly shaped bivalve, easily recognized and not likely to be confused with any Victorian member of the genus. Not infrequently the shell is washed up at the northern end of Port Phillip, and in living condition may be obtained by dredging in eight fathoms, on sandy mud, off Point Cook. The species is also recorded from South Australia, New South Wales and Tasmania.

Rounded Myadora

Myadora complexa Iredale, 1924

(Figure 459)

Shell white or cream; moderately strong, oblong-ovate, inequilateral; right valve fairly convex and entirely clasping the left valve which is flat; anterior end much rounded, posterior sloping and truncated. Both valves sculptured by strong concentric ridges with a fine microscopic radial sculpture crossing all the ribs; umbos sharp, the right projecting over the left; a posterior area is defined by a slightly raised rib, more evident in the right valve. Interior white; pearly muscle scars well marked. Size: 1 inch.

Localities: Portland; Western Port.

This species is obtained by dredging in shallow water on a sandy bottom. It is a common shell in Twofold Bay and Disaster Bay, New South Wales, and also in Tasmania. For many years it was known in Victoria under the name of *ovata* Reeve, 1844, a species of different shape and sculpture from the Philippines.

Flat-ribbed Myadora

Myadora latilirata Macpherson, 1958
(Figure 460)

Shell thin, white; oblong-ovate, abruptly truncated posteriorly, rounded anteriorly, concentrically ridged; ridges about sixteen in number, regularly spaced and inflated, the slightly flattened top being broader than the base. Umbos central, acute. Right valve convex, with an angle extending from the umbo to the posterior margin, defining the truncated area. A corresponding area in the flat left valve is slightly depressed. Interior of the shell white, shining and showing impressions of the ridges; muscle scars and pallial line indistinct. Size: ⅝ inch.

Localities: Dredged 8 fathoms between Phillip and French Islands, Western Port; off Wilson's Promontory.

This shell has some affinities to *complexa,* Iredale, but differs from it in the following characters—its more angular shape and elongate proportions, its thinner shell, its stronger, more inflated ridges which do not coalesce, its ridged interior and less distinct muscle scars and pallial sinus.

This species was originally described as *gabrieli* Macpherson, 1951, a name preoccupied by Chapman and Crespin, 1928, who used it for a fossil species from the Sorrento bore.

Pandora-shaped Myadora

Myadora pandoriformis (Stutchbury, 1830)
(Figure 461)

Shell white; thin, transversely elongate, inequivalve, almost equilateral; posterior-dorsal margin concave, anterior-dorsal margin slightly convex, posterior side truncated; umbos acute. Deep or right valve very convex, left or upper valve almost flat. Surface of deep valve sculptured by about thirty concentric ribs, a feature not so pronounced in the left valve. Defining the hollowed posterior slope is a definite ridge running from the umbo to posterior ventral margin, the ridge only faintly evident in the flat valve. The surface of both valves radially microscopically striated throughout. Interior white, shining, pearly; muscle scars not well impressed; pallial line ill-defined, with a somewhat rounded pallial sinus not very deep. Size: ¾ inch.

Locality: Port Phillip.

Easily separated from other Victorian representatives of the genus by its comparative smoothness and the extreme fineness of the microscopic radial

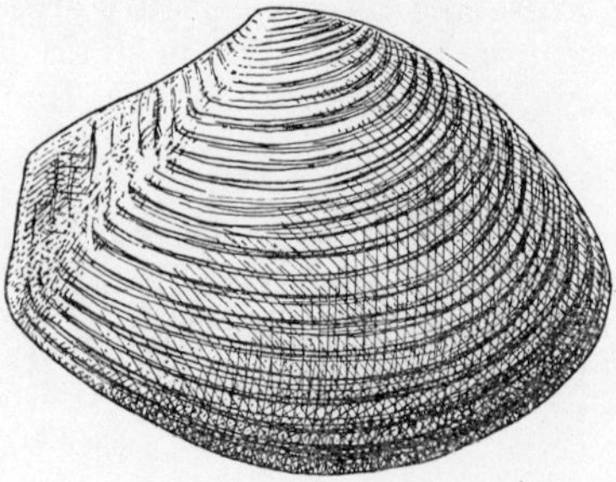

Figure 459

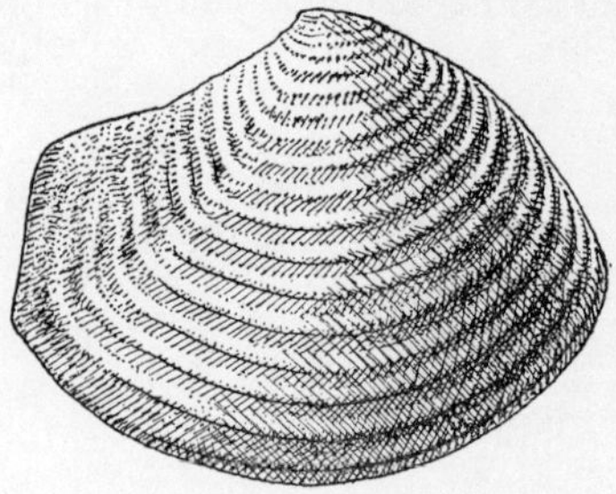

Figure 460

Figure 461

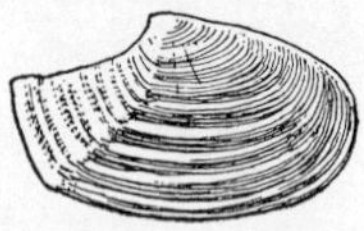

Figure 462

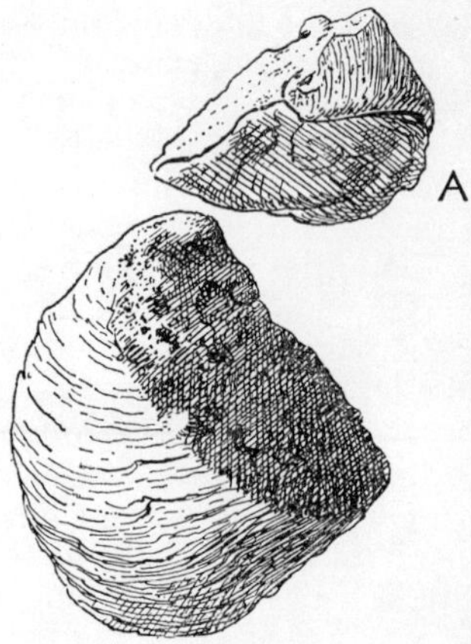

Figure 463

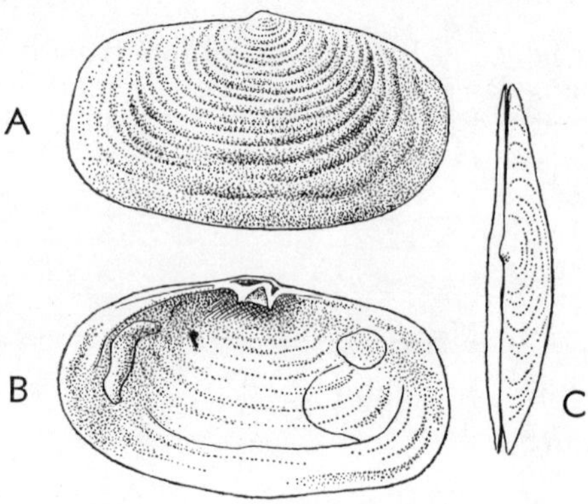

Figure 464

459. *Myadora complexa* × 2
460. *Myadora latilirata* × 2
461. *Myadora pandoriformis* × 1
462. *Myadora subalbida* × 3
463. *Cleidothaerus albidus* × ⅔
A Lateral elevation × ½
464. *Phragmorisma watsoni* × ⅔
A Exterior view B Interior view C Lateral elevation

sculpture. Recorded from Tasmania and very common in New South Wales and Queensland.

Creamy Myadora

Myadora subalbida Gatliff & Gabriel, 1914

(Figure 462)

Shell white; oblong, obliquely truncated posteriorly, rounded anteriorly, concentrically ridged, ridges somewhat rounded, regularly spaced, about fourteen in number; umbos central, acute. The whole shell is covered by very fine radial striae; under the microscope this sculpture will be seen to be divided into closely compacted, elongate, flattened, hexagonal areas, angularly defined posteriorly; right valve convex, with an angle extending from the umbos to the margin, and defining the truncated area; left valve similar to the right valve, but rather deeper. Interior pearly. Size: 2/5 inch.

Locality: Dredged (7 fathoms) between Phillip and French Islands, Western Port.

By no means a common shell, resembling *albida* Tenison Woods, but immediately separated by its more elongate form. It is recorded also from Tasmania.

elongata May, 1915; off Lakes Entrance.

royana Iredale, 1924; off Lakes Entrance.

Family CLEIDOTHAERIDAE

Rock Shells

Shell heavy, fixed by the right valve, irregular, inequivalve. Ligament internal; pallial sinus absent. Hinge with a small cardinal tooth in the left valve, fitting into a pit in the right valve.

Cleidothaerus Stutchbury, 1830

Description the same as for the family.

TYPE SPECIES *Chama chamoides* Stutchbury, 1830 = *Chama albidus* Lamarck, 1819.

White Cleidothaerus or Rock Shell

Cleidothaerus albidus (Lamarck, 1819)

(Figure 463)

Shell cream or brown; solid, triangularly globose, inequivalve, attached. Free valve concentrically, roughly striated; convex valve possessing similar sculpture excepting at the front slope, which is peculiarly wrinkled and which serves as the area of attachment. Interior pearly. Hinge with one conical tooth in the free valve, received into a corresponding pit in the other, and an oblong shelly appendage fixed in a groove under the umbos. Size: 3 inches.

Localities: Port Phillip; Western Port.

A shell somewhat resembling genus *Chama* in general outline, but readily distinguished by the internal clavicle or shelly appendage; it is fairly common adhering to small stones, and may be obtained by dredging from three to six fathoms. It is also found in South Australia, New South Wales, Tasmania and New Zealand.

Family THRACIIDAE

Thracias

Shell thin, inequivalve. Interior ligament more or less sunken; pallial sinus strong and deep. Hinge toothless.

Phragmorisma Tate, 1894

Shell large, thin; left valve flat; right valve compressed, convex, rounded anteriorly, posterior truncated. Interior with a broad plate pendant from the umbo, and attached by its anterior side to the interior of the valve.

TYPE SPECIES *Thracia watsoni* Smith, 1885.

Watson's Thracia

Phragmorisma watsoni (Smith, 1885)
(Figure 464)

Shell dirty white, elongate, nearly twice as long as high, subequilateral, inequivalve, anterior end rounded, posterior obliquely truncate; slightly gaping at both ends. Umbos not very prominent. Sculpture concentrically coarsely plicate-wrinkled, more evident on the central area of the valves. Surface rough to the touch, minutely granular. Ventral margin fairly straight. External ligament appears on both sides of the umbos. From the interior the sculpture is distinctly visible; muscle scars quite different in shape, the posterior subcircular and the anterior elongate with a bend in the middle; pallial line fairly straight, pallial sinus deep, broad and rounded at the extremity. Size: 2 inches.

Locality: Off East Moncoeur Island, Bass Strait.

The author in his description of this shell noted: 'This fine large species is remarkable on account of its flattened compressed character, and being almost equilateral. The ligament-pit is very strong, and the outer epidermal shell-layer is peculiar.' These characters should serve to identify the species. This rare shell was obtained by the *Challenger* Expedition in 38 fathoms on a bottom of sand and shells, over sixty years ago. It is unknown to the writers in any Victorian collection. It is recorded from New South Wales, three broken valves being dredged in fifteen to twenty fathoms at Twofold Bay, and also at St Francis Island, South Australia, where it was dredged in 1957 by Mr D. Howlett.

Eximiothracia Iredale, 1924

Shell elongate ovate, inequilateral; umbos posterior; anterior rounded, posterior truncate. Ligament external, short. Sculpture concentric, or granulated.

TYPE SPECIES *Thracia speciosa* Angas, 1869.

Iredale, 1924, erected this genus for Thracids with an external ligament as opposed to *Thraciopsis* Tate & May, 1900. Thiele, 1934, doubted the validity of the separation but we, on examination of the two groups, think it should be retained, as *Eximiothracia* also differs from *Thraciopsis* in the lack of the well-developed marginal ridges of the right valve.

Beautiful Thracia

Eximiothracia speciosa (Angas, 1869)
(Figure 465)

Shell small, dirty white, elongately ovate, moderately convex, thin, fragile, very inequilateral; posterior end the shorter, obliquely truncated, anterior end rounded, ventral margin flatly convex; a slightly curved, rounded angle runs from umbo to lower posterior end. Surface of valves sculptured with irregular growth-lines, and closely granulated beyond the umbonal ridge, features difficult to see without the aid of lens. Ligament external, yellowish brown. Interior slightly pearly; muscle scars not very distinct; pallial line deeply and widely sinuated, almost reaching centre of shell. Size: ¾ inch.

Localities: Dredged between Phillip and French Islands, Western Port (8 fathoms); Frankston.

Rather a frail shell and unknown to the writers from any other Victorian locality. Its finely granulated and truncated posterior provide useful recognition marks. Occurs also in New South Wales and South Australia.

Little Thracia

Eximiothracia modesta (Angas, 1867)
(Figure 466)

Shell white; fragile, triangularly oval; equilateral, inequivalve; anterior-dorsal border almost straight, posterior concave; anterior end somewhat rounded, posterior truncated. Sculptured with faint microscopic striae; surface of valves finely granulated. Interior white, shining; muscle scars faintly discernible; pallial sinus somewhat rounded. Size: ⅜ inch.

Locality: Dredged off Wilson's Promontory.

A fragile, white shell similar in size to *lincolnensis* (Verco) but a little more triangular in shape. Found also in New South Wales.

Myadora-shaped Thracia

Eximiothracia myodoroides (Smith, 1885)
(Figure 467)

Shell thin, dirty white; moderately convex, inequivalve, equilateral, obliquely truncated posteriorly, ventral margin convex and with a rounded angle running from the umbo to lower posterior end. Sculpture consisting of irregular concentric striae with the valves finely granulated. External ligament yellowish. Interior white, somewhat shiny; muscle scars faintly evident; pallial line plainly seen; pallial sinus broad but not very deep. Size: ¾ inch.

Localities: Dredged off East Moncoeur Island, Bass Strait; between Phillip and French Islands, Western Port (8 fathoms).

The shells are without much character but the present species should be identified by its shape and granulated surface. First recorded by the *Challenger* Expedition in 38 fathoms.

It occurs also in Tasmania and South Australia.

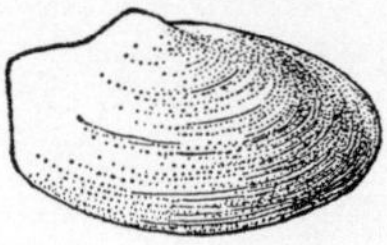

Figure 465

Figure 466

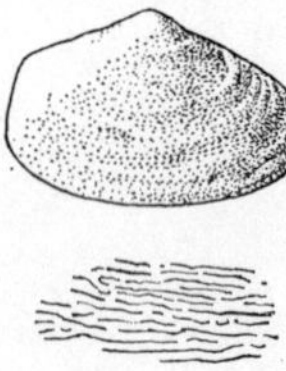

A

Figure 467

Figure 468

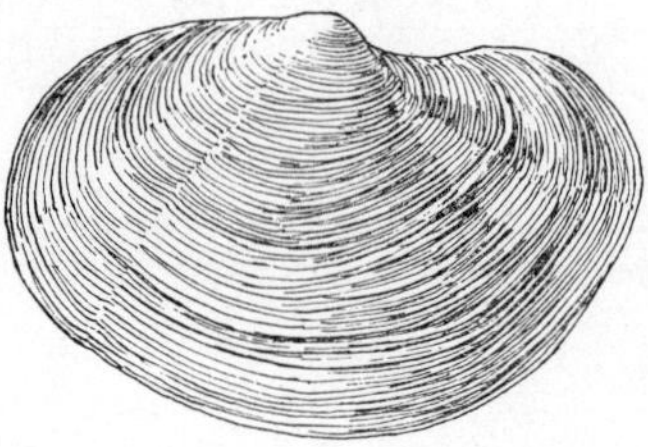

Figure 469

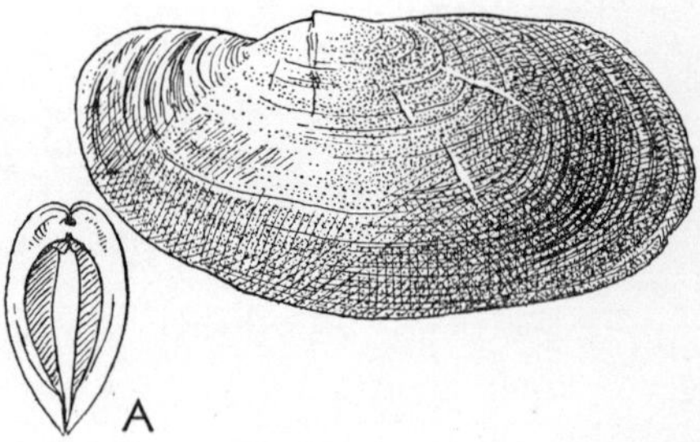

Figure 470

465. *Eximiothracia speciosa* × 2
466. *Eximiothracia modesta* × 3
467. *Eximiothracia myodoroides* × 1
    A Sculpture (enlarged)
468. *Eximiothracia lincolnensis* × 3
469. *Offadesma angasi* × ⅔
470. *Laternula creccina* × ⅔
    A Anterior view

Lincoln's Thracia

Eximiothracia lincolnensis (Verco, 1907)
(Figure 468)

Shell white; fragile, elongate-oval, inequilateral, anterior-dorsal border almost straight, scarcely convex; posterior concave; anterior end attenuately rounded, posterior end vertically, straightly truncate; right valve slightly more convex than the left. Sculpture consisting of quite obsolete, flat, irregular concentric ribs, with very faint microscopic striae. Interior white; pallial sinus wide and round, reaching nearly to centre of shell. Size: ¼ inch.

Localities: Frankston; Dromana.

The author of the species remarks that 'It somewhat resembles *myodoroides* (Smith) but is smaller, though its proportions are the same. It differs in having a concave instead of a straight post-dorsal line, in not being excavate in front of the umbo, and in having a deeper pallial sinus. It differs from *modesta* Angas, in being longer posteriorly, and so more equilateral, and in a vertical truncation instead of one sloping upwards and backwards. Recorded from South Australia and Tasmania.

Thraciopsis Tate & May, 1900 (*Thracia angustata* Angas, 1867).

angustata (Angas, 1867); from Port Albert, is a white species ¼ inch in length.

elongata (Stutchbury, 1835); Elongated Thracia, is a similar shell but a little flatter and narrower, from Western Port and Port Phillip dredgings. It occurs also in New South Wales and Tasmania. This species was previously recorded as *elegantula* (Angas, 1867). Iredale, 1924, drew attention to Stutchbury's earlier description and pointed out the slightly longer pallial sinus of his species. This is a feature of Victorian shells.

Family LATERNULIDAE

Lantern Shells

Shell ovate, posterior attenuated, gaping; umbo fissured. Pallial sinus wide and shallow. Hinge toothless, cartilage attached to a spoon-shaped plate which projects at an angle from under the umbos.

Offadesma Iredale, 1930

Shell thin, large, oval, concentrically striated, somewhat inequivalve, umbo of right valve swollen and projecting upon left valve. Ligament absent. Hinge edentulous but with a spoon-shaped process to reinforce the strong cartilage.

TYPE SPECIES *Periploma angasi* Crosse & Fischer, 1864.

Rounded Lantern Shell

Offadesma angasi (Crosse & Fischer, 1864)
(Figure 469)

Shell white, covered by a thin, light brown epidermis, translucent, very fragile, transversely ovate; right valve ventricose, left valve slightly convex, gaping at both ends, posteriorly obliquely truncated; anterior end rounded, ventral margin convex, inequivalve, inequilateral, anterior side the longer; umbos fissured. Sculpture consisting of irregular concentric lines of growth with well-marked rest periods, the whole surface crossed by very faint microscopic scratches. Interior white, shining; muscle scars faintly discernible, anterior elongate, posterior somewhat triangular connected by a deeply sinuated pallial line; running from the umbos are two diverging ridges, one to the base of the posterior margin and the other to the base of the anterior margin. Hinge edentulous but with a spoon-shaped process (lithodesma) in each valve. Size: 3 inches.

Localities: Port Albert; Frankston; Portarlington; off Point Cook.

Sometimes known as the Spoon-hinge Shell, it is extremely fragile and care should be taken in handling. Generally found in sheltered bays. At the last-named locality dredged in 8 fathoms on a sea-bed of sandy mud. Occurs in New South Wales, Tasmania and through Bass Strait to South Australia and New Zealand.

Some authors place this species in *Cochlodesma* Couthouy, 1839, but the lithodesma is lacking in that genus.

Laternula Röding, 1798

Shell elongate, thin, strongly or slightly inequivalve, inflated, posterior attenuated and gaping; umbos fissured with a supporting internal oblique plate. Pallial sinus large.

TYPE SPECIES *Mya anatina* Linné, 1758.

Elongated Lantern Shell

Laternula creccina (Reeve, 1860)

(Figure 470)

Shell white; thin, hyaline, semi-transparent, elongately oblong, subequivalve, very inequilateral; umbos prominent, fissured. Surface minutely granulated. Anterior end gaping, extremities reflexed; posterior end moderately gaping. Interior white. Hinge with a spoon-shaped process in each valve, containing the cartilage. Size: 3 inches.

Localities: Frankston; Portarlington; Point Cook (8 fathoms); Dromana; Western Port.

A pure white, fragile shell, not distinguished by any conspicuous ornaments. The minute granulated sculpture is characteristic of the genus.

It is also recorded from Western Australia, South Australia, Tasmania, New South Wales and Queensland.

tasmanica Reeve, 1818; from Port Phillip. It is readily distinguished by its smaller size, and in the position of the umbos which are almost centrally situated.

Recorded also from Tasmania, South Australia and New South Wales. Known as *recta* Reeve, 1860, in South Australia but the above name has page priority.

These molluscs live buried in mud or sand and, owing to their extreme fragility, should be handled carefully.

Family CLAVAGELLIDAE

Tube Shells

Shell inequivalve, gaping valves thin and free when young, but when adult one or both become cemented to an elongated calcareous tube. Ligament external. Hinge toothless.

Dacosta Gray, 1840

Shell oblong, right valve free, left embedded in the tube, which is cylindrical and more or less elongated, sometimes divided by a longitudinal partition, posterior and anterior ends of the tube simple. Valves flattened, often irregular and rudimentary.

TYPE SPECIES *Clavagella australis* Sowerby, 1829.

Southern Tube Shell

Dacosta australis (Sowerby, 1829)

(Figure 471)

Shell oblong, thin, flat, rough, irregular or rudimentary, inequivalve, left valve in the adult stage incorporated in the dilated portion of tube, right valve always free.

Tube comparatively smooth, calcareous, cylindrical, more or less elongated; posteriorly open, the margin of opening simple, not expanded; anterior or lower end of tube ovate, clavate, and forming a sort of disc. Size: disc 1 inch, tube 1 to 1½ inches, diameter of tube ¼ inch.

Locality: Victoria, exact locality not recorded.

A strikingly peculiar bivalve attached by the disc and to shell fragments and rock at low tide. At times very difficult to locate as only a small portion of the tube is visible, the remainder being surrounded by coralline, Bryozoa, seaweed, etc. Recorded from New South Wales, being common in Sydney Harbour. There is only one record from Victoria.

Many-angled Tube Shell

Dacosta multangularis Tate, 1886

(Figure 472)

Valves oblong, flat, inequilateral, translucent; umbos anterior, prominent. Surface with distant concentric growth-folds, and rayed with microscopic lines.

Disc roundly oblong, ventral and anterior sides prolonged into two rows of longish tubes; particles of shell or coarse sand are usually agglutinated to the disc, the left valve incorporated in the side of the burrow in the adult stage, and the right valve free inside the disc at the end of the tube. Tube, short, straight, or slightly curved, octagonal, obtuse, transversely irregularly striate. Size: disc ⅜ inch, tube ⅜ inch.

Localities: Dredged 8 fathoms between Phillip and French Islands, Western Port; off Lakes Entrance (20 fathoms).

A strange bivalve occasionally cast up on beaches, differing from the other Victorian member of the genus by its smaller size and in the possession of a 'many-angled' tube. There is variation in the tube which is often bent and at times twice the length of the disc. Recorded from South Australia.

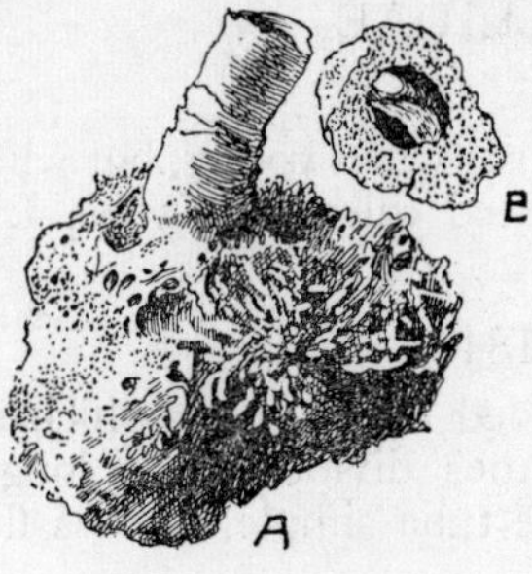

Figure 471

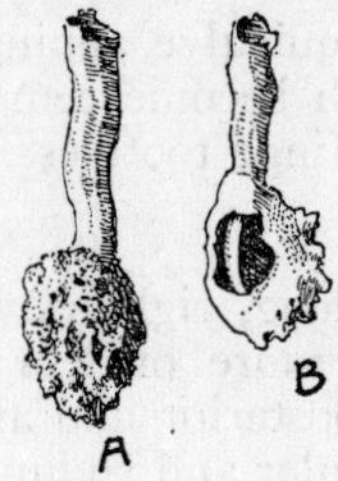

Figure 472

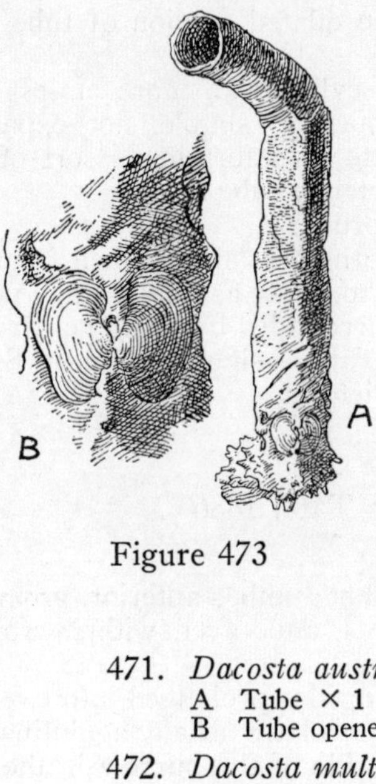

Figure 473

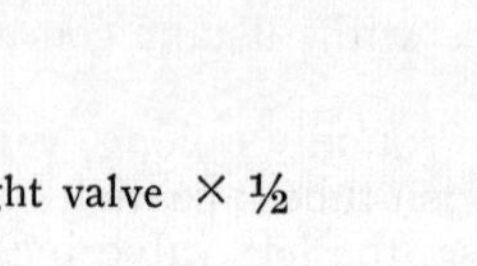

Figure 474

471. *Dacosta australis*
A Tube × 1
B Tube opened to show free right valve × ½

472. *Dacosta multangularis*
A Tube × 1
B Tube opened to show free right valve × 1

473. *Humphreyia strangei*
A Tube and valves × ⅜
B Enlargement of valve area × 2

474. *Ectorisma granulata* × 1⅓

Humphreyia Gray, 1858

Shell with both valves adherent; ovate broader posteriorly; embedded in the walls of a tube with the umbos visible externally.

Tube rounded or angular, base covered by a disc perforated by tubercles and frilled at the edge.

Type Species *Aspergillum strangei* A. Adams, 1852.

Strange's Watering-Pot or Tube Shell

Humphreyia strangei A. Adams, 1852

(Figure 473)

Shell small, valves flattened, squarely ovate, adherent at the base of the short stout tube.

Tube angular and obtusely keeled at the four angles, base distorted, perforations few, irregular and only slightly tubular. Size: 2-6 inches.

Localities: Balnarring and San Remo, Western Port; Port Phillip.

C.J.G., in 1905, at a depth of four fathoms on a stony bottom between Phillip and French Islands, obtained a juvenile form with the animal. A. E. Smith (1910) published a description of this specimen.

Occurs in South Australia, Tasmania and New South Wales.

Family VERTICORDIIDAE

Verticordias

Shell exterior ornamented with spinules, usually round, sometimes plainly ribbed. Interior nacreous. Hinge often with a swollen tooth for the cartilage in the right valve.

Thracidora Iredale, 1924 (*Thraciopsis arenosa* Hedley, 1904).
arenosa (Hedley, 1904) ; off Wilson's Promontory.

Verticordia Sowerby, 1844 (*Cryptodon verticordia* Wood, 1844).
tasmanica May, 1915; Bass Strait.

Family POROMYIDAE

Poromyas

Shell usually small and thin, round or oval. Interior nacreous. Often with a hinge tooth in the right valve; cartilage somewhat weak, situated just beneath the outer ligament.

Ectorisma Tate, 1892

Shell oval, thin, transparent, equivalve, inequilateral. Slightly nacreous within, valves gaping posteriorly; surface granulated. Ligament external, supplied anteriorly by a linear-lanceolate extension of the hinge-line. Right valve with a small cardinal tooth, fitting into a socket beneath the umbo of the left valve which has ridge-like lateral teeth over which the right valve fits.

Type Species *Ectorisma granulata* Tate, 1892.

Grained Ectorisma

Ectorisma granulata Tate, 1892

(Figure 474)

Shell white; transparent, thin, oval-oblong, inflated, equivalve, inequilateral; anterior side rounded, posterior side the longer with a narrow subtruncated margin; posterior margin slightly spreading to form a gape two millimetres at the widest and about six millimetres in length; umbos prominent. Surface of valves granulated, the granules arranged in definite radial rows except at the umbonal area which is more or less smooth and pearly. Ligament external. Interior dirty white; muscle scars and pallial line indistinct. Hinge small. Size: ⅜ inch.

Locality: Off Ninety Mile Beach (40 fathoms).

A fragile, interesting bivalve, rarely seen but easily recognized by its peculiar shape and characteristic granulated ornamentation. It is obtained in deep sea dredgings from 8 to 150 fathoms, and is recorded from New South Wales, Tasmania, through Bass Strait to South Australia.

Class CEPHALOPODA

Many-Armed Molluscs

THE CEPHALOPODS, like the Tusk Shells, are exclusively marine and inhabit all depths from intertidal rock-pools to the deepest ocean. They are separated immediately from all other molluscs by the circle of arms surrounding the well-developed head. There are either eight or ten arms which as a rule are furnished with suckers or hooks for clasping and to aid in crawling. The mouth is provided with horny beak-like jaws and a radula. The eyes are highly developed. The shape of the body varies in the different families; in Octopodidae it is round and sack-like while in some of the Loliginidae it is very elongated. Dorsally it is more or less continuous with the head, but on the ventral side there is a sack from which emerges the funnel or siphon. The sack is the mantle cavity and in it are situated the gills and ink sac as well as the opening of the excretory and reproductive organs. The siphon is the excretory opening of this sack and therefore of the whole body. Also it serves as a swimming organ as the expulsion of water causes the animal to move through the water, the speed being regulated by the force of the water (jet propulsion).

The ocean-living forms usually have fins or swimming membranes formed by expansion of the sides of the body. The shell is internal or absent in all living cephalopods except the Pearly or Chambered Nautilus which has an external shell coiled in one plane. The animal occupies a portion of the last convolution only, the remainder is divided up into a number of cavities by transverse septa connected by a thin tube or siphuncle. These chambers are believed to be filled with gas which gives buoyancy to the shell. The Pearly Nautilus, *Nautilus pompilius* Linné, 1758, has not been taken alive in Victoria but one specimen was found in South Australia in 1920, apparently a straggler from its home in the warmer waters of the Indian and Pacific Oceans.

When the shell is internal it may be coiled and well developed or reduced to a chitonous pen. The Cephalopods are carnivorous, catching or holding their prey by the arms and tearing and devouring it with the aid of the strong beak and radula.

All living Cephalopods, apart from Nautilus, are contained in the subclass Dibranchia, so named because all its members possess only one pair of gills. The Dibranchia are again divided on the number of arms. In the order Decapoda there are eight of varying lengths and two long tentacular arms which are more or less retractile. The body is variable in shape but always has well-developed fins and there is usually a vestigial internal shell (gladius). In the order Octopoda the body is short, rounded, usually without fins and the eight arms are alike (Figure 475).

In Dibranchia a single arm of the male is modified to form a copulatory organ; this is called the hectocotylus. At first it is enveloped in a sheath but on maturity it assumes its full length; usually it has few or no suckers and the tip is either spoon-like or very long and tapering. In *Loligo* and *Sepia*

Figure 475

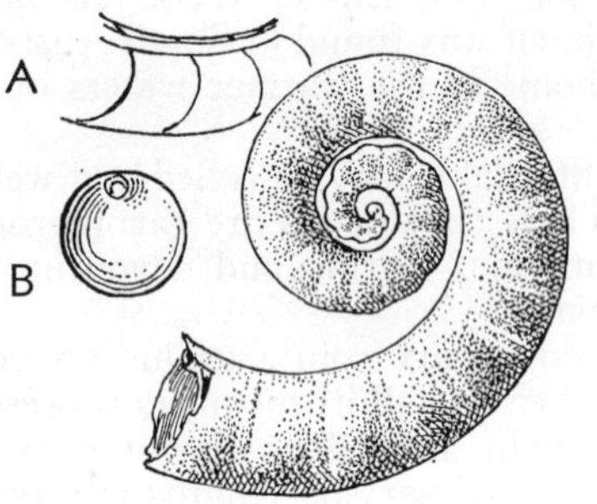

Figure 476

475. Octopus
476. *Spirula spirula* ×2
A Sculpture (enlarged)
B Cross section

the fourth left arm is hectocotylized, in *Octopus* the third right, and in *Spirula* both arms of the fourth pair.

Order DECAPODA

Family SPIRULIDAE

Spirulas

Shell a cone coiled in one plane, the coils not in contact. It is divided into chambers connected by a tube or siphuncle which passes through the septa towards their inner margin. The shell hangs free at the posterior end of the body and is held in position by two mantle flaps. A small portion of intestinal sack occupies the last chamber and a prolongation runs through the siphuncle. There are eight sessile arms and two long tentacles.

Spirula Lamarck, 1799

Generic characters the same as those of the family.
TYPE SPECIES *Nautilus spirula* Linné, 1758.

Ram's Horn

Spirula spirula (Linné, 1758)

(Figure 476)

Shell small, white, fragile, sculptured with fine reticulations which are usually lost on the convex side. The internal septa are indicated on the exterior by a band of greater opacity and the area may even have a slight constriction. The septa are convex with the siphuncle opening on the concave side of the shell. The apex of the shell is globose and slightly larger than the chambers immediately following. Size: 1½ inches.

Localities: Cape Otway; Bridgewater Bay.

Spirula shells are washed up on some beaches in great numbers from time to time, but there is very little known of the animal and its habitat. Dead animals have been taken at the surface of the sea and also brought up from great depths but living examples are unknown. The animal is sack-like with eight very short arms and two long tentacles. The shell is partially exposed by a dorsal and ventral window in the mantle through which portion of the last coil protrudes. Two small semicircular fins project from the posterior end of the animal.

Family SEPIIDAE

Cuttle-Fish

Shell or gladius broad, flat, thickened internally by numerous plates, terminating behind in a hollow, imperfectly chambered mucro or spine pointing downwards. The whole gladius is surrounded by a thin chitinous margin, forming a lateral expansion (Figure 477). The animal has an oval body with narrow, lateral fins as long as the body and generally united behind; eight short and two long tentacular arms.

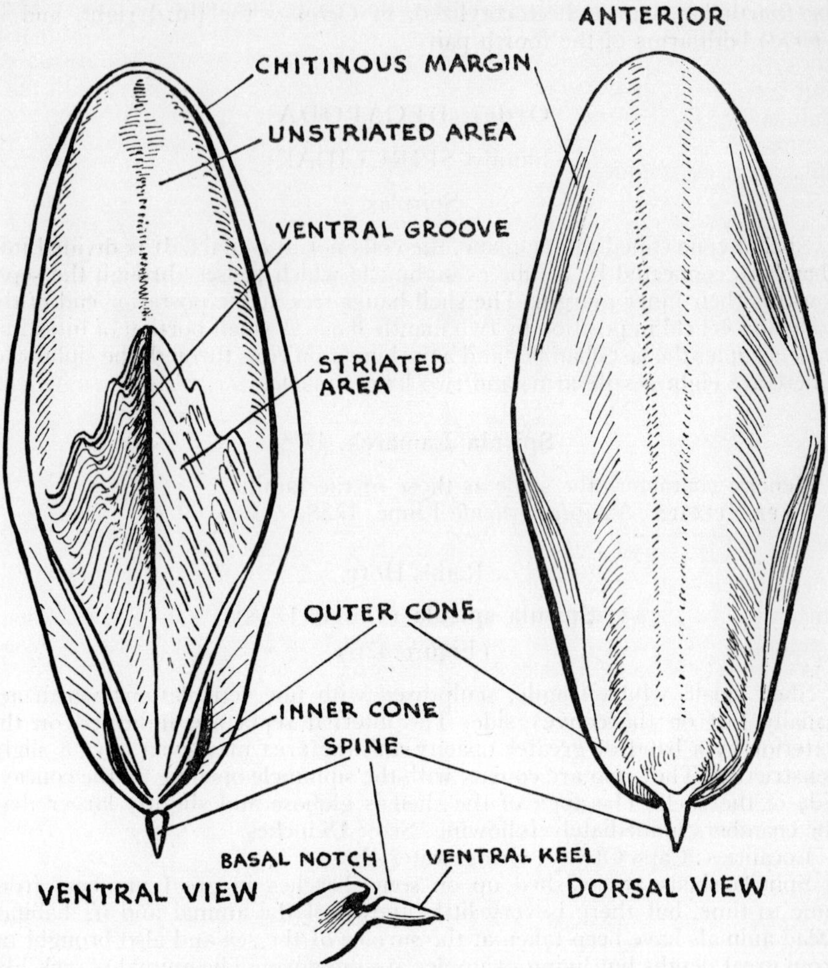

Figure 477. Parts of cuttle-fish gladius

Amplisepia Iredale, 1926

Gladius large, thick, elongate, outer cone well developed, inner cone elongate, with a coarse granulose glaze; spine small, delicate, but usually missing in adults.

TYPE SPECIES *Sepia apama* Gray, 1849.

Giant Cuttle-fish
Amplisepia apama (Gray, 1849)
(Figure 478)

Gladius oblong, elongate; posterior part much produced, sub-acute with a strong callosity at the posterior edge of the cavity; mucro blunt, rugose;

anterior extremity rounded, covered with a strong cartilaginous side; central portion rather convex. Inner and outer cone well developed. The animal is thick, fleshy and broadly ovate with a large head and short, thick arms with four rows of subequal suckers. The tentacular arms are also short with thick club ends. The lateral fins surround the posterior portion of the body except for a small rounded notch. Size: gladius 12 inches; animal 18 inches but may grow to 2 feet.

Locality: Coast generally.

This is the largest and commonest cuttle-fish in Victoria, the 'cuttle-bones' are very common on our beaches and the live animals are often taken by fishermen in their seine nets. It is recorded from the whole of southern Australia and New Zealand.

Arctosepia Iredale, 1926

Gladius small, narrow, swollen ventrally, posterior narrow; outer cone chitinous, fine, rounded, curved, and hood-like; inner cone absent.

TYPE SPECIES *Arctosepia limata* Iredale, 1926.

Elongate Cuttle-fish

Arctosepia braggi (Verco, 1907)

(Figure 479)

Gladius small, narrow, elongate; dorsum pink, smooth, central rib indistinct, striae close; outer cone narrow, chitinous; inner cone obsolete, mucro long, projecting at an angle of about 135 degrees. Size: 2½ inches.

Localities: Wilson's Promontory; Lorne.

This small cuttle-bone is the narrowest and most curved of any washed up on our coast, and is easily distinguished from all other species; the animal is unknown. It is common in southern Western Australia and South Australia as well as Victoria. There are a number of other Sepiidae recorded from Victorian waters but most of them are uncommon, and several, like the above species, are only known from cuttle-bones washed ashore.

limata Iredale, 1926; Norman Bay.

Mesembrisepia Iredale, 1926 (*Mesembrisepia macandrewi* Iredale, 1926).
novaehollandiae (Hoyle, 1909); Flinders; Torquay.

Decorisepia Iredale, 1926 (*Decorisepia rex* Iredale, 1926).
rex Iredale, 1926; Betka River Mouth; Norman Bay.

Glyptosepia Iredale, 1926 (*Glyptosepia opipara* Iredale, 1926).
hedleyi (Berry, 1918); western Victoria.
gemellus Iredale, 1926; Betka River Mouth.

Family SEPIOLIDAE

Dumpling Squids

Animal mostly small, mantle short, rounded with more or less large, rounded fins attached slightly towards the posterior. Shell rudimentary. Arms free, with spherical suckers, tentacles retractible into pockets.

Austrorossia Berry, 1918 (*Austrorossia australis* Berry, 1918).
australis Berry, 1918; Port Phillip; Gabo Island; Cape Everard.

Euprymna Steenstrup, 1887 (*Inioteuthis morsei* Verrill, 1881).
tasmanica (Pfeffer, 1884); Bass Strait.

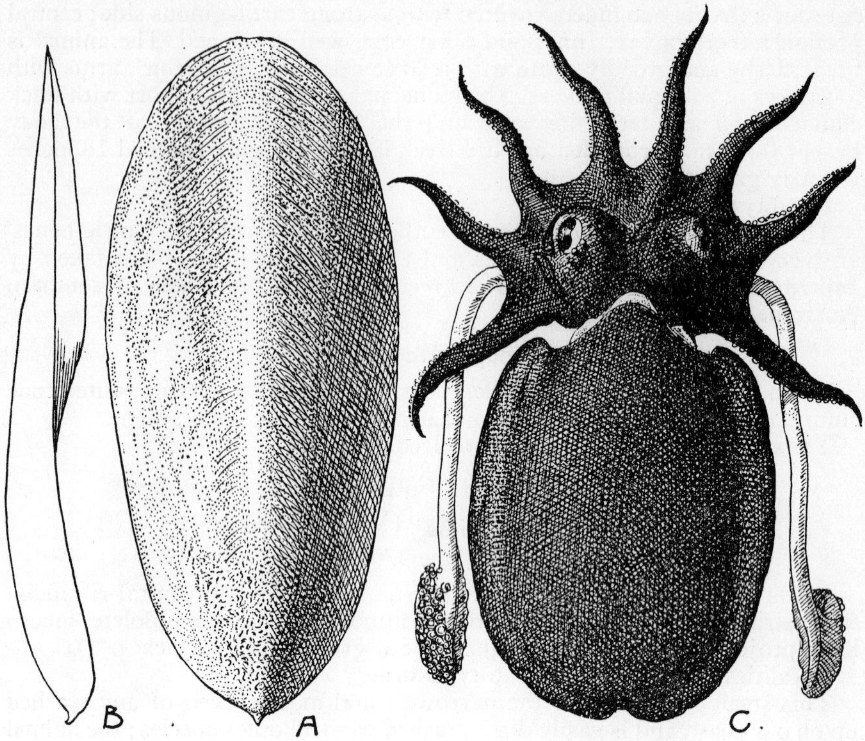

Figure 478

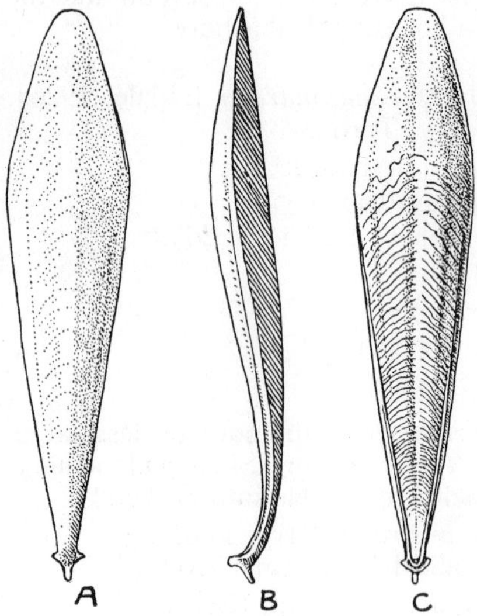

Figure 479

478. *Amplisepia apama*
A Dorsal view of gladius × 1/6
B Lateral view of gladius × 1/6
C Animal × 1/9

479. *Arctosepia braggi* × 1
A Dorsal view of gladius
B Lateral view of gladius
C Ventral view of gladius

Family LOLIGINIDAE

Calamaries

Body elongated, with lateral fins. Sessile arms with two rows of suckers, tentacular arms partly retractible. Gladius long, solid and chitinous, pointed in front, shaft keeled on the ventral side (Figure 480).

Calamaries are good swimmers living in the open ocean and some species grow to large size, the estimated body length being as much as 18 feet.

Septioteuthis Blainville, 1824

Body elongate, oval; fins very elongate, marginal, bordering nearly the whole mantle. Hectocotylized arm fourth left. Gladius feather-like, chitinous.

TYPE SPECIES *Sepioteuthis sepiacea* Blainville, 1824 = *Loligo sepioidea* Blainville, 1823.

Southern Calamary

Sepioteuthis australis Quoy & Gaimard, 1833
(Figures 481, 482C)

Body oblong, cylindrical, truncated in front, bluntly pointed behind; fins very broad and long, fleshy, subrhomboidal. Sessile arms elongate, unequal, tentacular arms very strong, compressed with large clubs with large suckers. Head small, narrower than body. Colour white, translucent, with irregular red, pink and black spots. Gladius or 'pen' chitinous, of the colour and consistency of a goose-quill. Size: 2 feet.

Localities: Western Port; Port Phillip; Bass Strait.

This calamary is found on most of the southern coast of Australia and is easily distinguished by its shape and colouring.

bilineata (Quoy & Gaimard, 1832); Western Port.

Loligo Schneider, 1784 (*Sepia loligo* Linné, 1758).

etheridgei Berry, 1918; Etheridge's Squid (Figure 482B), another common species distinguished by the torpedo-shaped body with triangular fins on the posterior half. Short, sessile arms and moderately long tentacular arms. It has a wide distribution in south-eastern Australia.

Family OMNASTREPHIDAE

Squids

Torpedo-shaped body usually of moderate size but some reach considerable dimensions; fins large, subterminal. Arms of moderate length with two rows of suckers. Gladius small, chitinous, narrowly lanceolate.

Nototodarus Pfeffer, 1912

Body large, subcylindrical; fins triangular on posterior third of body. Arms nearly equal ($2 > 3 > 4 > 1$); tentacles a third longer than arms, scarcely clubbed, largest suckers in the middle; large suckers with fifteen teeth, small ones and arm suckers with eight. Gladius slender, elongate, spoon-shaped.

TYPE SPECIES *Omnastrephes insignis* Gould, 1852.

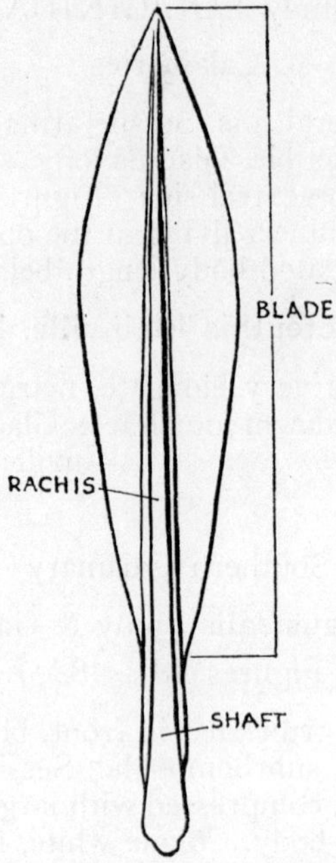

Figure 480

Figure 481

480. Parts of squid's pen or gladius

481. *Septioteuthis australis* × ⅙
A Animal B Gladius

Gould's Squid

Nototodarus gouldi (McCoy, 1888)

(Figures 482A, 483)

Animal with elongate, cylindrical body tapering to an acute point; fins large, triangular, a little prolonged posteriorly; head large, a little wider than the body. Sessile arms of moderate length about 60 per cent of body length, tentacular arms up to 90 per cent of body length. Gladius lanceolate, spoon-shaped. Size: 2 feet.

Locality: Bass Strait.

This squid is easily recognized by its flexible torpedo-shaped body and very narrow spoon-shaped pen.

Family ARCHITEUTHIDAE

Giant Squids

Animal gigantic, mantle slender, pointed posteriorly, with long ovate backward-pointed fins on the posterior. Arms long and strong with two rows of suckers with toothed rings; tentacles very long, club with four rows of cups with toothed rings, shaft with long rows of small sucker cups. Gladius, posterior end pointed, anterior portion of blade flat, posterior portion with a small terminal involuted cone; generally with two pronounced, and a number of smaller, ribs starting from the posterior end of the rachis diverging to the front.

Architeuthis Steenstrup, 1856 (*Architeuthis dux* Steenstrup, 1856).

kirkii Robson, 1887; a single specimen was washed up at Wingan Inlet in 1926.

Family ENOPLOTEUTHIDAE

Armed Squids

Animal comparatively small, mantle mostly pointed, with large fins almost or entirely attached at posterior end. Arms mostly with well-developed protective borders, suckers entirely or mostly transformed into hooks; basal part of the tentacle club with two rows of small cups and adhesive suckers. Gladius feather-shaped with clearly defined blade on either side of which near the edge is thickened rim, rachis grooved posteriorly with slightly raised median rib, apical portion flattened, spoon-shaped to elongate conical.

Enoploteuthis d'Orbigny, 1839 (*Loligo leptura* Leach, 1817).

galaxias Berry, 1918; Gabo Island to Cape Everard.

Family HISTIOTEUTHIDAE

Sail Squids

Mantle small in comparison with the head including arms, short cone-shaped, posteriorly more or less pointed; rounded medium-sized fins which are attached dorsally. Numerous light organs mainly on the ventral side of the mantle, head and arms. Two rows of suckers on sessile arms; tentacles long, club mostly with four to seven rows of suckers, shaft with one row of cups and suckers. Gladius feather-shaped without a terminal cone.

Calliteuthis Verrill, 1880 (*Calliteuthis reversa* Verrill, 1880).

miranda Berry, 1918; Gabo Island.

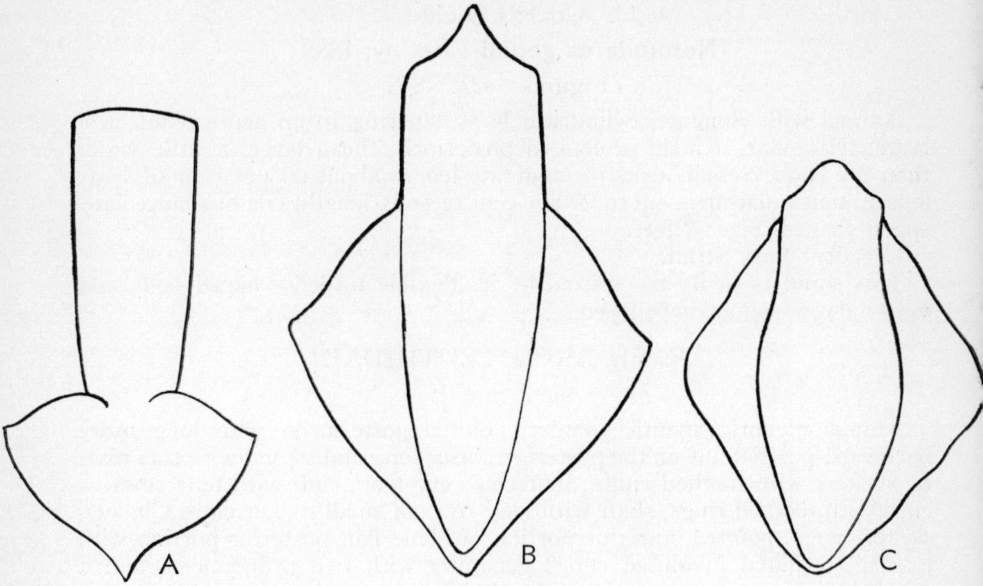

Figure 482

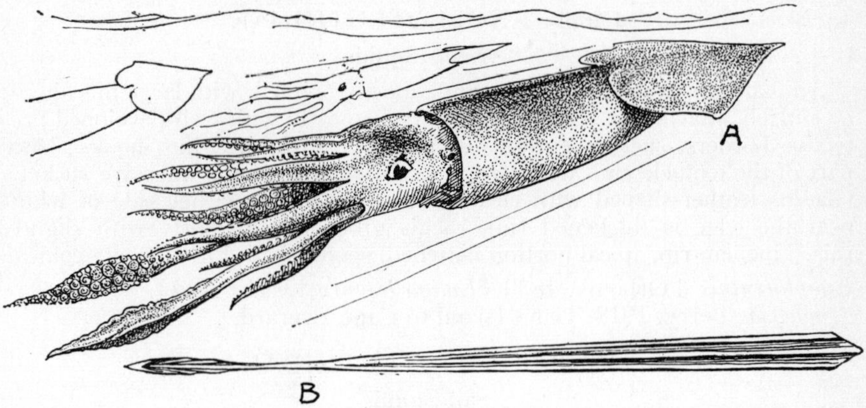

Figure 483

482. Outlines of body of common squids
A Gould's B Etheridge's
C Southern Calamary

483. *Nototodarus gouldi*
A Animal × 1/6 B Gladius × 2/3

Family CIRROTEUTHIDAE

Finned Octopus

Mantle medium to large, elongate; one pair of fins attached more or less medianly, longer than the diameter of the head. Arms long, web reaching well up the arms, secondary web present; suckers in a single row with two flanking rows of paired cirri mostly longer than the diameter of the cup. Shell-vestigial, saddle-shaped.

Teuthidiscus Berry, 1918 (*Teuthidiscus pluto* Berry, 1918).
persephone Berry, 1918; south-east of Genoa Peak (200 fathoms).

Order OCTOPODA

Family OCTOPODIDAE

Octopus or Devil-fish

Animals of small to medium size, short, round, body without fins. Eight arms of similar structure, one or two rows of suckers and a well-developed basal web or umbrella; without shell or gladius. Molluscs of this family are called Octopus or Devil-fish and the majority of the species live in shallow tropical and temperate seas, many of them inhabiting the pools and caverns of rock shores.

Octopus Lamarck, 1798

Body thick, short, rounded posteriorly; arms variable, usually long, web variable; with an ink sac; colour variable, fugitive.

TYPE SPECIES *Octopus vulgaris* Lamarck, 1798.

There has been some confusion over the name of this genus and Hoyle, 1901, considered *Polypus* Schneider, 1784, had priority but as the name is *nomen nudum* it cannot be used and therefore later authors have reverted to the better known *Octopus*.

Pale Octopus

Octopus pallidus Hoyle, 1885

(Figure 484)

Body oval but capable of considerable elongation in life, head narrower than the body. Arms lengths are 4 > 3 > 2 > 1 and about 75 per cent of body length, hectocotylized arm smaller than its pair; umbrella about one-quarter arm length. Sculpture with a pattern of rosette-shaped tubercles covering the body and web, rows of enlarged tubercles on the dorsum. Size: 18 inches total length.

Localities: Port Phillip; Western Port.

This species lives in deep water and has been taken in depth of up to 200 fathoms. It is often trapped in crayfish pots which it is fond of raiding. It is recorded also from New South Wales, Tasmania and South Australia.

australis Hoyle, 1885; a smaller species (12 inches), has a rounder body, deeper umbrella (⅓ arm length) and the granular body sculpture is finer.

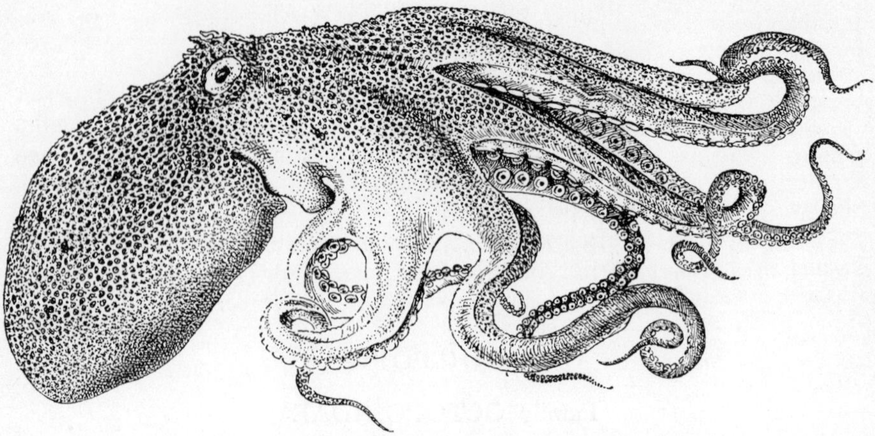

Figure 484

Figure 485

484. *Octopus pallidus* × ⅓
485. *Hapalochlaena maculosa* × 1⅓

It inhabits shallow water and often occurs in low-tide pools on rock platforms. Its range is southern Australia and New Zealand.

duplex Hoyle, 1885; Port Phillip.

superciliosus Quoy & Gaimard, 1832; Western Port.

variolatus Bainville, 1826; has been used for the species now known as *pallidus*. Robson, 1927, considers Blainville's species to be indeterminate.

Hapalochlaena Robson, 1929

Body thick, short, rounded; arms short, umbrella well marked; colour pattern of dark rings.

TYPE SPECIES *Octopus lunulatus* Quoy & Gaimard, 1832.

Ringed Octopus

Hapalochlaena maculosa (Hoyle, 1883)

(Figure 485)

Body small, round, head little narrower than body. Arms 75 per cent of total length in order 4 > 3 > 2 > 1; umbrella 20-30 per cent of arm length, uneven, longest between lateral arms. Sculptured with low warts which vary in number in each individual; colour yellow ochre splashed with patches of dark pigment in the form of rings, flecks or stripes, arms with transverse dark patches patterned with bluish rings. Ink sac small.

Localities: Port Phillip; Western Port; Lorne.

This small Octopus is readily distinguished by the prominent blue rings on the arms. It is quite common on rock platforms where there are ledges and loose stones to afford it some shelter, and in deeper water it will hide in dead shells. It has a wide distribution throughout southern Australia and into the Indian Ocean.

Family ARGONAUTIDAE

Paper Nautilus

Octopus of small to moderate size and pelagic habit, body rounded and without fins or shell. Males much smaller than the female. Eight arms with two rows of suckers, the dorsal arms of the female with broadly expanded glandular membranes at their extremities, which secrete and hold a delicate calcareous shell in which the female deposits her eggs and into which she herself partially retreats.

Argonauta Linné, 1758

Characters the same as the family.

TYPE SPECIES *Argonauta argo* Linné, 1758.

Paper Nautilus

Argonauta nodosa Solander, 1786

(Figure 486)

Shell fragile, compressed, thin, keeled; sides with transverse ribs broken up into tubercles; tubercles of the keel sharp, elevated.

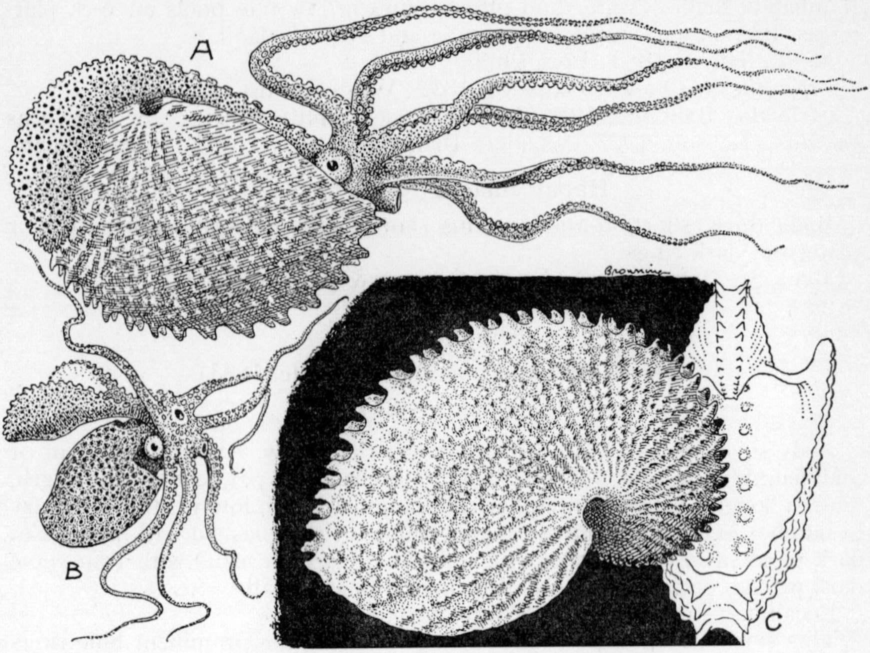

Figure 486. *Argonauta nodosa*
A Female and shell B Female out of shell
C Lateral and anterior view of shell × ½

Animal has a slender body with a narrow head and arms of unequal length. Colour white, flecked with closely set small, red brown dots. Size: the largest shell known is 14½ inches.

Locality: Coast generally.

The Paper Nautilus is an inhabitant of the open ocean, feeding on small floating animals. This species is recorded from the whole of the Indo-Pacific.

argo Linné, 1758; a smaller species with a narrow, keeled, compressed shell, sculptured with numerous bifurcating ribs. It is rarely found in eastern Victoria.

BIBLIOGRAPHY

THE FOLLOWING short list of books and papers records the most important recent reviews of temperate Australian Mollusca. The reference to the original description of most species can be found in Macpherson and Chapple (1951).

Abbott, R. T. (1955). *Introducing Sea Shells.* Van Nostrand, Princeton.

Allan, Joyce (1959). *Australian Shells* (rev. ed.). Georgian House, Melbourne.

Bennett, Isobel, and Pope, E. C. (1953). Intertidal Zonation of the Exposed Rocky Shores of Victoria, together with a Rearrangement of the Biogeographical Provinces of Temperate Australian Shores. *Aust. J. Mar. Freshw. Res.,* 4: 105-59.

Burn, Robert (1957). On Some Opisthobranchia from Victoria. *J. Mal. Soc. Aust.,* no. 1, 11-29.

——— (1958). Further Victorian Opisthobranchs. *J. Mal. Soc. Aust.,* no. 2, 20-36.

——— (1959). Comments on the Australian Umbraculacean Mollusca. *J. Mal. Soc. Aust.,* no. 3, 28-9.

——— (1960A). A Bivalve Gastropod. *Nature,* 186 (4719) : 179.

——— (1960B). Australian Bivalve Gastropods. *Nature,* 187 (4731) : 44-6.

——— (1960C). On *Tylodina corticalis* (Tate). A rare opisthobranch from south-eastern Australia. *J. Mal. Soc. Aust.,* no. 41, 64-9.

——— (1960D). New Names for two Victorian Opisthobranch Molluscs. *J. Mal. Soc. Aust.,* no. 4, 70.

——— (1961). On the Pleurobranch Subfamily Berthellinae (Mollusca: Gastropoda) ; A Revision and New Classification of the Species of New South Wales and Victoria. *Mem. nat. Mus. Vict.,* no. 25.

Cotton, B. C. (1943). Australian Shells of the Family Haliotidae. *Trans. roy. Soc. S. Aust.* 67 : 175-80.

——— (1944). Recent Australian Species of the Family Rissoidae (Mollusca). *Trans. roy. Soc. S. Aust.* 68: 286-314.

——— (1947A). Australian Recent and Tertiary Turridae. *Field Naturalists' Section of the Royal Society of South Australian Conchology Club,* no. 4.

——— (1947B). Some Southern Australian Turridae. *S. Aust. Nat.* 24 (3) : 13-16.

——— (1951). Australian Recent and Tertiary Mollusca Family Cerithiopsidae. *Rec. S. Aust. Mus.* 9: 383-96.

——— (1959). *South Australian Mollusca Archaeogastropoda.* Government Printer, Adelaide.

———and Godfrey, F. K. (1931). South Australian Shells, Part I. *S. Aust. Nat.* 12 (4) : 51-63.

——— ——— (1938). *The Molluscs of South Australia, Part I. Pelecypoda.* Government Printer, Adelaide.

——— ——— (1940). *The Molluscs of South Australia, Part II. Scaphopoda, Cephalopoda, Aplacophora and Crepipoda.* Government Printer, Adelaide.

——— and Ludbrook, N. H. (1938). Recent and Fossil Species of the Scaphopod Genus Dentalium in Southern Australia. *Trans. roy. Soc. S. Aust.* 62: 217-28.

Freame, M. E. (1940). *Vic. Nat.,* 56 (11) : 178.

Fretter, V., and Graham, A. (1949). The Structure and Mode of Life of the Pyramidellidae, Parasitic Opisthobranchs. *J. Mar. biol. Ass. U.K.,* 28: 493-532.

Gabriel, C. J. (1961). Additions to the Marine Molluscan Fauna of South-Eastern Australia. *Mem. nat. Mus. Vict.,* no. 25.

Gatliff, J. H. (1907). Additions to the Catalogue of the Marine Shells of Victoria. *Proc. roy. Soc. Vict.,* 1907, 20 (1) : 31-7.

——— and Gabriel, C. J. (1908). Additions to and Revision of the Catalogue of Victorian Marine Mollusca. *Proc. roy. Soc. Vict.,* 1908, 21 (1) : 368-91.

——— ——— (1909). Additions to the Catalogue of the Marine Shells of Victoria. *Proc. roy. Soc. Vict.,* 1909, 22 (1) : 37-46.

——— ——— (1911). Ibid. Ibid., 1910, 23 (1) : 87-98.

——— ——— (1912A). Additions to and Alterations in the Catalogue of Victorian Marine Mollusca. *Proc. roy. Soc. Vict.,* 1911, 24 (1) : 193-200.

——— ——— (1912B). Ibid. Ibid., 1912, 25 (1) : 169-75.

——— ——— (1913). Additions to the Catalogue of the Marine Mollusca of Victoria. *Proc. roy. Soc. Vict.,* 1913, 26 (1) : 71-87.

——— ——— (1914). Ibid. Ibid., 1914, 27 (1) : 99-103.

——— ——— (1916). Additions to and Alterations in the Catalogue of the Marine Shells of Victoria. *Proc. roy. Soc. Vict.,* 1916, 29 (1) : 106-13.

——— ——— (1917). Ibid. Ibid., 1917, 30 (1) : 21-31.

——— ——— (1922). Ibid. Ibid., 1921, 34 (2) : 128-61.

——— ——— (1926). Additions to the Catalogue of Victorian Marine Mollusca. *Proc. roy. Soc. Vict.,* 1925, 38: 88-94.

——— ——— (1931). Additions to and Alterations in the Catalogue of Victorian Marine Mollusca. *Proc. roy. Soc. Vict.,* 1930, 43 (2) : 202-32.

Grau, Gilbert (1959). Pectinidae of the Eastern Pacific. *Allan Hancock Pacific Expeditions,* 23: 1-308.

Griffiths, R. J. (1961). A Review of the Cypraeidae Genus *Notocypraea. Mem. nat. Mus. Vict.,* no. 25.

Hedley, C. (1902). Studies on Australian Mollusca, Part VII. *Proc. Linn. Soc. N.S.W.,* 27: 606-19.

——— (1919). A Review of the Australian Tun Shells. *Rec. Aust. Mus.,* 12: 329-36.

——— (1922). A Revision of the Australian Turridae. *Rec. Aust. Mus.,* 13: 213-359.

Hubendick, B. (1943). Systematic Monograph of the Patelliformia. *K. svenska Vetensk Akad. Handl.,* 23 (5) : 1-93.

Iredale, T. (1937). Middleton and Elizabeth Reefs South Pacific Ocean Mollusca. *Aust. Zool.,* 8 (4) : 232-61.

——— and Hull, A. F. B. (1923). A Monograph of the Australian Loricates (Phylum Mollusca—Order Loricata). *Aust. Zool.,* 3 (5): 186-94.

——— ——— (1924). Ibid. Ibid., 3 (6): 227-38.

——— ——— (1924). Ibid. Ibid., 3 (7): 277-97.

——— ——— (1925). Ibid. Ibid., 3 (8): 339-62.

——— ——— (1925). Ibid. Ibid., 4 (2): 75-111.

——— ——— (1926). Ibid. Ibid., 4 (3): 164-85.

——— ——— (1926). Ibid. Ibid., 4 (4): 256-76.

——— ——— (1926). Ibid. Ibid., 4 (5): 324-8.

——— ——— (1927). Ibid. Ibid., 4 (6): 339-59.

Kawaguti, S., and Baba, K. (1959). A Preliminary Note on a Two-valved Sacoglossan Gastropod, *Tamanovalva limax* n. sp. from Tamono, Japan. *Biol. J. Okayama.*

Laseron, C. F. (1948). New South Wales Marginellidae. *Rec. Aust. Mus.,* 22: 35-48.

——— (1950). Review of the Rissoidae of New South Wales. *Rec. Aust. Mus.,* 22: 257-87.

——— (1951A). Revision of the New South Wales Cerithiopsidae. *Aust. Zool.,* 11: 351-67.

——— (1951B). The New South Wales Pyramidellidae and the Genus Mathilda. *Rec. Aust. Mus.,* 22: 298-334.

——— (1953). Minute Bivalves from New South Wales. *Rec. Aust. Mus.,* 23: 33-53.

——— (1954A). Revision of the Liotidae of New South Wales. *Aust. Zool.,* 12: 1-24.

——— (1954B). Revision of the New South Wales Turridae (Mollusca). *Aust. Zool. Handb.* (Royal Zoological Society of New South Wales).

——— (1954C). Revision of the New South Wales Triphoridae. *Rec. Aust. Mus.,* 23: 139-58.

——— (1955A). Revision of the New South Wales Eulimoid Shells. *Aust. Zool.,* 12: 83-107.

——— (1955B). The Genus Tolema and Its Allies. *Proc. R. Zool. Soc. N.S.W.,* 1: 70-4.

——— (1955C). The New South Wales Cancellariidae. *Rec. Aust. Mus.,* 23: 267-72.

——— (1956A). The Family Cerithiopsidae (Mollusca) from the Solanderian and Dampierian Zoogeographical Provinces. *Aust. J. Mar. Freshw. Res.,* 7: 51-82.

——— (1956B). The Families Rissoinidae and Rissoidae (Mollusca) from the Solanderian and Dampierian Zoogeographical Provinces. *Aust. J. Mar. Freshw. Res.,* 7: 384-484.

——— (1956C). New South Wales Mussels. *Aust. Zool.,* 12: 263-83.

——— (1956D). A Revision of the New South Wales Leptonidae. *Rec. Aust. Mus.,* 24: 7-22.

——— (1957). A New Classification of the Australian Marginellidae (Mollusca), with a Review of Solanderian and Dampierian Zoogeographical Provinces. *Aust. J. Mar. Freshw. Res.,* 8: 274-311.

——— (1958). The Family Triphoridae (Mollusca) from Northern Australia; also Triphoridae from Christmas Island (Indian Ocean). *Aust. J. Mar. Freshw. Res.*, 9: 569-657.

Laursen, D. (1953). The Genus Ianthina. A Monograph. *Carlsberg Foundation's Oceanographical Expedition Round the World 1928–30 and Previous 'Dana'-Expeditions.* Dana-Report 38.

Lemche, H. (1957). A New Living Deep Sea Mollusc of the Cambro-Devonian Class Monoplacophora. *Nature,* 179: 413.

Ludbrook, N. H. (1953). Systematic Revision of the Volutid Genus *Amoria. Proc. Mal. Soc. Lond.*, 30 (4, 5) : 131-52.

——— (1954). The Molluscan Fauna of the Pliocene Strata Underlying the Adelaide Plains, Part I. *Trans. roy. Soc. S. Aust.*, 77 : 42-64.

——— (1955). Ibid., Part II. Pelecypoda. *Trans. roy. Soc. S. Aust.*, 78: 18-87.

——— (1956). Ibid., Part III. Scaphopoda, Polyplacophora, Gastropoda. *Trans. roy. Soc. S. Aust.*, 79: 1-36.

——— (1957). Ibid., Part IV. Gastropoda. *Trans. roy. Soc. S. Aust.*, 80: 17-58.

——— (1958). Ibid., Part V. Gastropoda. *Trans. roy. Soc. S. Aust.*, 81: 43-111.

MacGillivray, J. (1852). *Narrative of the Voyage of H.M.S. Rattlesnake.* Boone, London.

Macpherson, J. H. (1955). Preliminary Revision of the Families Patellidae and Acmaeidae in Australia. *Proc. roy. Soc. Vict.*, 1955, 67 (2) : 229-56.

——— (1961). Trochus obtusa Confusion. *Mem. nat. Mus. Vict.*, no. 25.

——— and Chapple (1951). A Systematic List of the Marine and Estuarine Mollusca of Victoria. *Mem. nat. Mus. Vict.*, 17 : 107-85.

May, W. L. (1923). *An Illustrated Index of Tasmanian Shells.* Government Printer, Hobart.

——— (1958). Ibid. (Revised by J. H. Macpherson.)

Powell, A. W. B. (1942). The New Zealand Recent and Fossil Mollusca of the Family Turridae. *Bull. Auckland Inst.*, no. 2.

——— (1944). The Australian Tertiary Mollusca of the Family Turridae. *Rec. Auckland* [*N.Z.*] *Inst.*, 3: 3-68.

Pritchard, G. B., and Gatliff, J. H. (1898). Catalogue of the Marine Shells of Victoria, Part I. *Proc. roy. Soc. Vict.*, 1897, 10 (2) : 236-84.

——— ——— (1899). Ibid., Part II. *Proc. roy. Soc. Vict.*, 1898, 11 (2) : 185-208.

——— ——— (1900A). Ibid., Part III. *Proc. roy. Soc. Vict.*, 1899, 12 (2) : 170-205.

——— ——— (1900B). Ibid., Part IV. *Proc. roy. Soc. Vict.*, 1900, 13 (1) : 139-56.

——— ——— (1902). Ibid., Part V. *Proc. roy. Soc. Vict.*, 1901, 14 (2) : 85-138.

——— ——— (1903). Ibid., Part VI. *Proc. roy. Soc. Vict.*, 1902, 15 (2) : 176-223.

——— ——— (1903A). Ibid., Part VII. *Proc. roy. Soc. Vict.*, 1903, 16 (1) : 96-139.

——— ——— (1904). Ibid., Part VIII. *Proc. roy. Soc. Vict.*, 1904, 17 (1) : 220-65.

——— ——— (1906). Ibid., Part IX; with Complete Index to the whole Catalogue. *Proc. roy. Soc. Vict.,* 1905, 18 (2) : 39-92.

Schenk, E. T., and McMasters, J. H. (1956). *Procedure in Taxonomy* (third ed.). Stanford University Press, Stanford, California.

Shaw, H. O. N. (1909). On the Dates of Issue of Sowerby's 'Conchological Illustrations' from the Copy Preserved in the Radcliffe Library, Oxford. *Proc. Mal. Soc. Lond.,* 8 (6) : 333-40.

Smith, A. E. (1910). Note on the Very Young Stage of the Genus Humphreyia. *Proc. Mal. Soc. Lond.,* 9 (1) : 23-5.

Soot-Ryen, T. (1955). A Report on the Family Mytilidae Pelecypoda. *Allan Hancock Pacific Expeditions,* 20 (1) : 1-175.

Thiele, J. (1929). *Handbuch der Systematischen Weichtierkunde,* 1 (1) : 1-376. Fischer, Jena.

——— (1931) : Ibid., 1 (2) : 377-778.

——— (1934). Ibid., 2 (3) : 779-1022.

——— (1935). Ibid., 2 (4) : 1023-1154.

Thompson, J. M. (1954). The Genera of Oysters and the Australian Species. *Aust. J. Mar. Freshw. Res.,* 5: 132-68.

Verco, J. C. (1909). Notes on South Australian Marine Mollusca, with Descriptions of New Species, Part XI. *Trans. roy. Soc. S. Aust.,* 33: 277-92.

Wenz., W. (1938). Gastropoda (Introduction and Prosobranchia 1). *Handb. Paläozool.,* 6 (1) : 1-240.

——— (1938). Ibid. (Prosobranchia 2). *Handb. Paläozool.,* 6 (1): 241-480.

——— (1939). Ibid. (Prosobranchia 3). *Handb. Paläozool.,* 6 (1): 481-720.

——— (1940). Ibid. (Prosobranchia 4). *Handb. Paläozool.,* 6 (1): 721-960.

——— (1941). Ibid. (Prosobranchia 5). *Handb. Paläozool.,* 6 (1): 961-1200.

——— (1943 . Ibid. (Prosobranchia 6). *Handb. Paläozool.,* 6 (1); 1201-1506.

——— (1944). Ibid. (Prosobranchia 7). *Handb. Paläozool.,* 6 (1): 1507-1639.

——— (1959). Ibid. (Euthyneura 1). *Handb. Paläozool.* (Ed. Adolph Zilch), 6 (2) : 1-701.

——— (1959). Ibid. (Euthyneura 2). *Handb. Paläozool.* (Ed. Adolph Zilch), 6 (2) : 702-1434.

——— (1960). Ibid. (Euthyneura 3). *Handb. Paläozool.* (Ed. Adolph Zilch), 6 (2) : 1435-2111.

——— (1960). Ibid. (Euthyneura 4). *Handb. Paläozool.* (Ed. Adolph Zilch), 6 (2) : 2112-2515.

GLOSSARY

Accessory plates *See* Accessory valves.

Accessory valves The smaller or less important testaceous plates found on the hinge or dorsal margins of the true valves of some shells.

Accremental striae Lines of growth.

Aculeated Furnished with prickles.

Acuminated Terminating in a point.

Adductor muscles Applied to bivalve shells; the anterior and posterior muscles which draw the valves of the shell together leaving marks on the inner surface of each called the muscular impressions or muscle scars or adductor scars.

Adductor scar *See* Adductor muscles.

Adherent A shell which attaches itself is adherent, e.g. oysters to rocks.

Agglomerated Gathered into a heap.

Agglutinated United or joined together.

Alate Winged.

Albinism Absence of colouring pigments.

Albino A white form, without pigment; in a true albino the shell and animal are both without colour.

Anastomose A running together.

Anterior In bivalves, it is the side on which the head, or part analogous to the head of the animal, lies; it is known in the shell by the umbos, which, if turned at all, are turned towards that part. The anterior of a spiral univalve is that part of the outer lip which is at the greatest distance from the apex. Of a conical univalve such as a Limpet, it is that part where the head of the animal lies.

Aperture The entrance to the spiral cavity of univalve shells; mouth.

Apex The tip, or small end of a shell.

Apical whorls Those near the apex.

Arcuate Arched.

Auriculate Having ears, or ear-like appendages.

Auriform Ear-shaped.

Axial A term applied to ribs or colour bands, which are longitudinal and not spiral.

Axis In a univalve shell the centre or pillar upon which the spire turns.

Beak The apices, or points of the valves of a bivalve shell, generally termed 'umbos'; also any part which is drawn out like a beak.

Benthic Bottom dwelling.

Biangulate Two-keeled or with two ledges.

Bicarinate Having two keels or ledges.

Bidentate Two-toothed.

Body-whorl The largest whorl of a spiral shell (anterior).

Brackish Saltish; or salt in a moderate degree.

Bryozoa A phylum of the animal kingdom often called 'moss animals'.

Byssiferous Producing a byssus.

Byssus The fibres by which some bivalve shells are anchored or moored to submarine substances.

Calcareous Limy or shelly matter.
Callosity Callus; the state of being hardened with a limy deposit.
Callus *See* Callosity.
Canal A groove which characterizes some spiral univalve shells, where the inner and outer lips unite at the front part of the aperture.
Canaliculate Channelled or grooved.
Cancellated Marked by ridges crossing each other.
Cardinal teeth Hinge teeth directly beneath the umbos of a bivalve, as distinguished from the lateral teeth which are placed at a distance on each side.
Carina Keel.
Carinated Having a raised, thin ledge; keeled.
Cartilage An internal, elastic substance found in bivalve shells which joins the valves.
Cartilage pit A pit formed in the centre of the hinge of a bivalve shell for the reception of the internal cartilage.
Cerulean Sky-coloured.
Chondrophore A projection of the hinge plate of some bivalves on which the spoon-shaped pit (resilifer) containing the internal ligamental cartilage is set.
Cicatrice A scar.
Cicatrix A scar.
Ciliate Having small hairs,
Cinereous Having the colour of ashes.
Clathrate Latticed; crossing to form a network.
Clavate Club-shaped.
Clypeiform Shield-shaped.
Columella The column formed by the inner sides of the volutions of a spiral univalve. It is sometimes described as the inner lip of the aperture, of which it forms a part.
Columella lip The inner edge of the aperture, including that part of it which covers the body-whorl.
Commensal Living as a tenant, but not a parasite upon another.
Concentric A term applied to sculpture which is spiral.
Conchology The science of shells.
Congener Belonging to same genus.
Conic Shaped like a cone.
Conoid Cone-like.
Conspecific Of the same species.
Contractile Having the power of shortening itself.
Cordate Heart-shaped.
Cordiform Heart-shaped.
Coriaceous Resembling leather; leathery.
Coronate Having the whorls of a spiral shell, surrounded by a row of spines or tubercles.
Costae Ribs.
Costated Ribbed.

Crenate Having the edge cut or notched.
Crenulated Finely crenated or notched.
Cuneate Wedge-shaped.

Decollated Apex broken off.
Decussated Intersected by striae crossing each other.
Dentate Toothed.
Denticle A small tooth or projecting point.
Denticulated Having little teeth or raised points.
Dextral Coiled in a right-hand spiral.
Disarticulated Joints separated.
Dorsal The margin of a bivalve shell on which the hinge is placed.
Dorsum Hump or upper surface (applied to Cowry Shells).

Edentate Toothless.
Edentulous Toothless.
Emarginate Indented at the edge.
Entire Uninterrupted, not emarginated. The pallial line in bivalves is entire, when continued without interruption or without a sinus, and the peritreme of a univalve is said to be entire when not interrupted by canals or by the body-whorl.
Entire margin Simple, not indented.
Epidermis External coating of shells. Also known as periostracum.
Equilateral Applied to bivalve shells when a line drawn perpendicularly from the apex would divide the shell into two equal parts.
Equivalve When the two valves are perfectly alike.
Escutcheon An elongated depression behind the umbos in a bivalve shell.
Estuary That portion of a river or inlet affected by tidal influence, producing varying degrees of salinity.
Evanescent Vanishing, imperceptible.
Exsert Protrude; projecting.

Fasciated Banded or striped.
Fissure A slit or cut; a narrow perforation.
Flagelliform Like a whip lash or whip-shaped.
Flamed Broadly splashed.
Flange A projecting edge.
Flexuous Winding, crooked.
Foliaceous Consisting of leaves or thin laminae.
Foot Portion of the animal's body used for locomotion. In most bivalves a hatchet-shaped muscular organ capable of protruding beyond valve margins.
Foramen Perforation.
Free shell An unattached shell, not attached by a byssus or growth of shelly matter.
Free valve In attached bivalve shells, one only is fixed; the other is then free, as far as to the action of opening and shutting.
Frondose Leafy.
Fulvous Tawny brown.

Fusiform Shaped like a spindle, swelling in the centre and tapering at the extremities.
Gape The opening between valves when in the closed position.
Gaping Bivalves which do not completely close.
Genus An assemblage of species, possessing certain characters in common.
Gibbous Humpbacked.
Glabrous Bald, smooth, possessing a surface without hairs or any unevenness.
Gladius Vestigial internal shell of Cephalopod.
Globose Rounded like a globe or ball.
Granose Granulated.
Granule A little grain or small particle.
Granulose Granulated or covered with minute grains.
Gregarious Associating or going together in herds.
Growth-lines The lines formed by the edges of the successive layers of shelly matter deposited by the animal as it increases the shell.

Hepatic Liver-coloured.
Heterostrophe Reverse or left-hand coiled, applied to protoconch of some univalves.
Hinge The edge of the bivalve shells near the umbones, including the teeth and ligament.
Holdfast Globe-shaped body at the base of tube shell.
Holotype A single specimen upon which a species is based.
Hyaline Glassy, transparent.

Imbricated When the superficial laminae are placed over each other like tiles or shingles.
Imperforate Without umbilicus.
Impressed Indented.
Inequivalve Not equivalve.
Inflated Swollen; applied to rotund shells of thin texture.
Infundibuliform Funnel-shaped.
Inner lip Edge of the aperture of a univalve shell which is near to the imaginary axis, as distinguished from the outer lip or that which is on the opposite side (columella lip).
Internal ligament Signifies that the ligament of a bivalve shell is placed within the closed part of the hinge, so as not to be seen when the valves are shut.
Internal shell One which is enclosed by the mantle of the animal.
Interstices Narrow spaces between things, closely set.

Keel A flattened ridge, resembling the keel of a ship.
Kitchen midden Aboriginal food refuse heap.

Labiate With a thickened margin.
Lamella Thin scale or plate.
Lamellated Covered with thin scales or plates.
Laminar tooth A plate-like structure.
Lanceolate Lengthened like a lance.

Lateral teeth The teeth in a bivalve, taking their rise near the umbos, proceed to some distance towards the sides of the shell, as distinguished from the cardinal teeth.
Left valve Sinistral valve of a bivalve shell may be known by placing the shell with its ligamentary or posterior part towards the observer; the sides of the shell will then correspond with his right and left side.
Ligament The true ligament is external, serving the purpose of binding the two valves of a shell together by the posterior dorsal margin.
Ligulate Strap-shaped.
Lineated Marked with lines.
Lines of growth *See* Growth-lines.
Lingual ribbon *See* Radula.
Linguiform Tongue-shaped.
Lip Edge of the aperture in a univalve shell.
Lirae Striae.
Lirate Striated.
Lithothamnion Limy or calcareous seaweed.
Littoral Pertaining to the shore; situated on a shore; the tidal zone; belonging to the interval on a sea coast between high- and low-water mark.
Livid Of a lead colour.
Lunule An impression on the anterior dorsal margin of some bivalve shells.

Maculation Spotting.
Malacology The study of mollusca.
Mamillated Applied to a univalve shell when the apex is rounded like a teat; bluntly rounded.
Mantle External tissue which secretes the shell.
Midden *See* Kitchen midden.
Mollusc A member of the phylum Mollusca.
Mouth Aperture or opening of a shell.
Murex Shaped like a Murex Shell.
Muscle The fleshy, contractile organ by which the animal is attached to the shell.
Muscle scar *See* Adductor muscle.
Muscular impression *See* Adductor muscle.

Nacre Mother-of-pearl. Inner layer of shell.
Nacreous Pearly; like mother-of-pearl.
Nebulous Cloudy, hazy.
Nidus A nest or hatching place.
Niveous Snow-like.
Nodose Having tubercles or knobs.
Nodulose Nodose.
Nucleus The first-formed part (of a shell).

Obtuse Blunt or rounded.
Octoradiate Eight-rayed.
Odontophore *See* Radula.
Oliviform Shaped like an Olive Shell.
Operculum The plate with which many molluscs close the aperture of their shell when retired within them. Cat's-eye of Turbo.

Ossicle An accessory shelly plate or protuberance.
Outer lip The edge of the aperture at the greatest distance from the axis.
Ovipositing Egg laying.
Ovoviviparous Producing young by eggs hatched within the body.

Pallets Two shelly projections at the posterior of *Teredo*; used to seal the burrow.
Pallial line In bivalve shells, the line of attachment of mantle to shell.
Pallial sinus A notch in the pallial line occasioned by siphons.
Papillary Shaped like a teat.
Papillose Papillary.
Parasite An organism that lives upon or within another and derives its nourishment from its host.
Paratype One of the original series from which a species was described.
Parietal whorl Broader upper part of inner lip in a univalve shell.
Patelliform Shaped like a dish; limpet-like.
Paucispiral Few whorls in the spiral, generally applied to the operculum.
Pectinate Like the teeth of a comb.
Pectiniform Shaped like a Pecten.
Pelagic Pertaining to, or inhabiting the open sea; free swimming.
Pelecypoda Alternative name for *Lamellibranchiata.*
Penultimate whorl The last whorl but one (anterior) of a spiral shell.
Periphery Greatest circumference of a whorl.
Periostracum *See* Epidermis.
Peristome Edge of aperture.
Peritreme A term used to express the whole circumference of the aperture of a spiral shell. In descriptions it may be said to be entire, notched, simple, reflected, round, oval.
Pilose Hairy, covered with hairs.
Plait A fold. A term applied to the prominences or folds on the pillar or columella lip of some univalve shells.
Planorbid Coiled in a flat spiral.
Polymorphous Having many forms.
Posterior The side known by the direction of the curve in the umbos, which is from the posterior towards the anterior. Posterior of univalve opposite end to the anterior (mouth).
Predatory Living by preying upon others.
Preoccupied Applied when a name has previously been used for another shell.
Prodissoconch Embryonic shell; the umbos in a bivalve.
Produced Elongate, applied to the spire of univalve shells.
Protoconch Embryonic shell, the apex in a univalve.
Pulmoniferous Breathing by means of lungs.
Punctate Dotted with small spots or depressions.
Pupaeform Shaped like a chrysalis.
Pyriform Pear-shaped.

Quadrate Square, applied when the outline of shells is formed by nearly straight lines meeting at right angles.
Quadruplicate With four folds or plaits—applied to the columella of some shells.

Radiating Applied to the ribs, striae, or bands of colours when they meet in a point at the umbos of a bivalve shell, and spread out towards the ventral margin.
Radula Lingual ribbon. Usually bearing on its upper surface numerous transverse rows of teeth. Used for the mastication of food. It is absent in bivalves.
Resilifer A spoon-shaped pit on the hinge of a bivalve to contain the resilium.
Resilium The internal ligamental cartilage in a bivalve hinge, allowing shell to open on the relaxing of the adductor muscles.
Reticulated Resembling network.
Retiolum A little net.
Right valve *See* left valve.
Rostrate Beaked or pointed at the end, usually the anterior end of bivalves.
Rostrum A beak-like process.
Rubiginous Rust coloured or brownish red.
Rugose Rough, rugged.

Sagittate Shaped like an arrow-head.
Salinity Degree of saltiness.
Semigranose Half-grained.
Semi-lunar Shaped like a half-moon.
Sensu lato (*s.l.*) In the broad sense.
Sensu stricto (*s.s.*) In the strict sense.
Septum A thin plate.
Shagreen Grained to resemble the skin of sharks.
Shell A calcareous or horny covering secreted by the mantle of a mollusc.
Sinistral Reversed; on the left side; coiled in a left hand spiral.
Sinuate Having a wavy edge.
Sinus A deep indentation; cavity.
Siphon Tube leading to the respiratory organ of some univalves.
Siphonal groove A groove in the shell in which the siphon lies.
Spatula Spoon-shaped area surrounded by the muscle scar of limpets.
Species A subdivision of a genus, a group into which is placed all individuals of the same kind.
Spinose Armed with spines or thorns; prickly.
Spire Includes all the volutions above the aperture of a univalve shell.
Squamiferous Scaly.
Squamose Scaly.
Stelliform Star-like.
Striae Very fine lines, either flat or raised, which cross the surface of shells in different directions.
Striated Marked with fine grooves or lines.
Submedian Almost central.
Suborbicular Nearly circular.
Subsutural Just under the suture.
Subtrigonal With the angles somewhat rounded.
Subulate Slowly tapering to a sharp point; applied to shells which are long and pointed, as in genus *Terebra*.
Sulcated Grooved or furrowed.

Sulci Grooves or furrows.
Suture The line which marks the joining of the whorls of the spire.
Synonyms Different names proposed by authors for the same animal.
Synonymy A collection of synonyms.

Taxodont With hinge plate composed of serially alternating teeth and sockets more or less similar in size and form, as in genera *Anadara, Glycymeris* and *Nucula.*
Tentoriform Tent-like.
Terraced Stepped.
Tesselated Applied to the colouring of shells, when arranged in regular defined patches like a tessellated pavement.
Testaceous Composed of shelly matter.
Transverse Bands, or striae running across a shell, are called transverse.
Transverse aperture At right angles with the axis of the shell.
Tricarinate Having three keels or ledges.
Tridentate Three-toothed.
Trochiform Shaped like a Trochus or Top Shell.
Truncate Cut off at the end; blunt.
Tubercle A protuberance, or knob, raised on the surface of some shells.
Tubiform In the form of a tube.
Tumid Swollen.
Tumulus Hillock, heap.
Tunicate Sea Squirt. Member of the Class Urochordata.
Turbinate Top-shaped like a Turbo.
Turreted When the whorls of a spire are regulated so as to have the appearance of little turrets rising above each other.
Type The specimen on which the original description of the species is based.

Umbilical area Around the umbilicus.
Umbilicus The hollow formed in spiral shells when the inner sides of the volutions do not join each other so that the axis is hollow.
Umbo Umbone; the point of a bivalve shell above the hinge, constituting the apex of each valve. Embryonic shell.
Undulated Waved.
Unicostate One-ribbed.
Univalve A shell consisting of a single piece, as distinguished from bivalves and multivalves.

Varicose Bearing one varix or more.
Varix A projecting ridge formed by the thickened, reflected edge of a former aperture, after increased growth of the shell.
Ventral The margin of a bivalve shell opposite the hinge.
Ventricose Swelled, rounded out.
Vertex Highest point or top.
Vitreous Resembling glass.
Volutions With spiral whorls.

Whorl A complete turn or revolution round the imaginary axis of a spiral shell.

INDEX

An index to families, genera, species and common names. Names that are considered synonyms or homonyms have been entered in italics. Bold type indicates main entry.